Generalized Linear Mixed Models

Modern Concepts, Methods and Applications

CHAPMAN & HALL/CRC
Texts in Statistical Science Series

Analysis of Failure and Survival Data
P. J. Smith

The Analysis of Time Series — An Introduction, Sixth Edition
C. Chatfield

Applied Bayesian Forecasting and Time Series Analysis
A. Pole, M. West, and J. Harrison

Applied Categorical and Count Data Analysis
W. Tang, H. He, and X.M. Tu

Applied Nonparametric Statistical Methods, Fourth Edition
P. Sprent and N.C. Smeeton

Applied Statistics — Handbook of GENSTAT Analysis
E.J. Snell and H. Simpson

Applied Statistics — Principles and Examples
D.R. Cox and E.J. Snell

Applied Stochastic Modelling, Second Edition
B.J.T. Morgan

Bayesian Data Analysis, Second Edition
A. Gelman, J.B. Carlin, H.S. Stern, and D.B. Rubin

Bayesian Ideas and Data Analysis: An Introduction for Scientists and Statisticians
R. Christensen, W. Johnson, A. Branscum, and T.E. Hanson

Bayesian Methods for Data Analysis, Third Edition
B.P. Carlin and T.A. Louis

Beyond ANOVA — Basics of Applied Statistics
R.G. Miller, Jr.

A Course in Categorical Data Analysis
T. Leonard

A Course in Large Sample Theory
T.S. Ferguson

Data Driven Statistical Methods
P. Sprent

Decision Analysis — A Bayesian Approach
J.Q. Smith

Design and Analysis of Experiments with SAS
J. Lawson

Elementary Applications of Probability Theory, Second Edition
H.C. Tuckwell

Elements of Simulation
B.J.T. Morgan

Epidemiology — Study Design and Data Analysis, Second Edition
M. Woodward

Essential Statistics, Fourth Edition
D.A.G. Rees

Exercises and Solutions in Biostatistical Theory
L.L. Kupper, B.H. Neelon, and S.M. O'Brien

Extending the Linear Model with R — Generalized Linear, Mixed Effects and Nonparametric Regression Models
J.J. Faraway

A First Course in Linear Model Theory
N. Ravishanker and D.K. Dey

Generalized Additive Models: An Introduction with R
S. Wood

Generalized Linear Mixed Models: Modern Concepts, Methods and Applications
W. W. Stroup

Graphics for Statistics and Data Analysis with R
K.J. Keen

Interpreting Data — A First Course in Statistics
A.J.B. Anderson

Introduction to General and Generalized Linear Models
H. Madsen and P. Thyregod

An Introduction to Generalized Linear Models, Third Edition
A.J. Dobson and A.G. Barnett

Introduction to Multivariate Analysis
C. Chatfield and A.J. Collins

Introduction to Optimization Methods and Their Applications in Statistics
B.S. Everitt

Introduction to Probability with R
K. Baclawski

Introduction to Randomized Controlled Clinical Trials, Second Edition
J.N.S. Matthews

Introduction to Statistical Inference and Its Applications with R
M.W. Trosset

Introduction to Statistical Limit Theory
A.M. Polansky

Introduction to Statistical Methods for Clinical Trials
T.D. Cook and D.L. DeMets

Introduction to the Theory of Statistical Inference
H. Liero and S. Zwanzig

Large Sample Methods in Statistics
P.K. Sen and J. da Motta Singer

Linear Models with R
J.J. Faraway

Logistic Regression Models
J.M. Hilbe

Markov Chain Monte Carlo — Stochastic Simulation for Bayesian Inference, Second Edition
D. Gamerman and H.F. Lopes

Mathematical Statistics
K. Knight

Modeling and Analysis of Stochastic Systems, Second Edition
V.G. Kulkarni

Modelling Binary Data, Second Edition
D. Collett

Modelling Survival Data in Medical Research, Second Edition
D. Collett

Multivariate Analysis of Variance and Repeated Measures — A Practical Approach for Behavioural Scientists
D.J. Hand and C.C. Taylor

Multivariate Statistics — A Practical Approach
B. Flury and H. Riedwyl

Multivariate Survival Analysis and Competing Risks
M. Crowder

Pólya Urn Models
H. Mahmoud

Practical Data Analysis for Designed Experiments
B.S. Yandell

Practical Longitudinal Data Analysis
D.J. Hand and M. Crowder

Practical Multivariate Analysis, Fifth Edition
A. Afifi, S. May, and V.A. Clark

Practical Statistics for Medical Research
D.G. Altman

A Primer on Linear Models
J.F. Monahan

Principles of Uncertainty
J.B. Kadane

Probability — Methods and Measurement
A. O'Hagan

Problem Solving — A Statistician's Guide, Second Edition
C. Chatfield

Randomization, Bootstrap and Monte Carlo Methods in Biology, Third Edition
B.F.J. Manly

Readings in Decision Analysis
S. French

Sampling Methodologies with Applications
P.S.R.S. Rao

Statistical Analysis of Reliability Data
M.J. Crowder, A.C. Kimber, T.J. Sweeting, and R.L. Smith

Statistical Methods for Spatial Data Analysis
O. Schabenberger and C.A. Gotway

Statistical Methods for SPC and TQM
D. Bissell

Statistical Methods in Agriculture and Experimental Biology, Second Edition
R. Mead, R.N. Curnow, and A.M. Hasted

Statistical Process Control — Theory and Practice, Third Edition
G.B. Wetherill and D.W. Brown

Statistical Theory, Fourth Edition
B.W. Lindgren

Statistics for Accountants
S. Letchford

Statistics for Epidemiology
N.P. Jewell

Statistics for Technology — A Course in Applied Statistics, Third Edition
C. Chatfield

Statistics in Engineering — A Practical Approach
A.V. Metcalfe

Statistics in Research and Development, Second Edition
R. Caulcutt

Stochastic Processes: An Introduction, Second Edition
P.W. Jones and P. Smith

Survival Analysis Using S — Analysis of Time-to-Event Data
M. Tableman and J.S. Kim

The Theory of Linear Models
B. Jørgensen

Time Series Analysis
H. Madsen

Time Series: Modeling, Computation, and Inference
R. Prado and M. West

Texts in Statistical Science

Generalized Linear Mixed Models

Modern Concepts, Methods and Applications

Walter W. Stroup

CRC Press is an imprint of the
Taylor & Francis Group an **informa** business
A CHAPMAN & HALL BOOK

CRC Press
Taylor & Francis Group
6000 Broken Sound Parkway NW, Suite 300
Boca Raton, FL 33487-2742

Printed in the United States of America on acid-free paper
Version Date: 20120813

International Standard Book Number: 978-1-4398-1512-0 (Hardback)

Visit the Taylor & Francis Web site at
http://www.taylorandfrancis.com

and the CRC Press Web site at
http://www.crcpress.com

Contents

Part II Estimation and Inference Essentials

Preface

Once upon a time "linear model" meant $\mathbf{y} = \mathbf{X}\boldsymbol{\beta} + \mathbf{e}$. In fact, $\mathbf{y} = \mathbf{X}\boldsymbol{\beta} + \mathbf{e}$, where $\mathbf{e}$ was assumed to have a Gaussian distribution, was routinely referred to as the "general" linear model.

Once upon a time is no more.

By contemporary standards, $\mathbf{y} = \mathbf{X}\boldsymbol{\beta} + \mathbf{e}$ is only a special case. Calling it "general" seems quaint. It is certainly misleading. "Linear model" now entails the following:

- A linear predictor, $\boldsymbol{\eta} = \mathbf{X}\boldsymbol{\beta} + \mathbf{Z}\mathbf{b}$, where $\mathbf{b}$ is a random vector.
- Observations whose distribution—or at least quasi-likelihood—is conditional on $\mathbf{b}$ and need not be Gaussian.
- A link function $\boldsymbol{\eta} = g(\boldsymbol{\mu}|\mathbf{b})$, where $\boldsymbol{\mu}|\mathbf{b}$ is the conditional expectation of the observations given $\mathbf{b}$. We fit the linear model to the link function, not directly to the data.
- Except for Gaussian data, the residual term $\mathbf{e}$ serves no purpose—in fact, for many distributions, it has no valid meaning nor interpretation.

In other words, as of 2012, "linear model" means "generalized linear mixed model" (GLMM), and all other linear models—linear mixed models (LMM, mixed models for Gaussian data), generalized linear models (GLM, linear models for non-Gaussian data with no $\mathbf{Zb}$ in the linear predictor), and linear models (LM, formerly "general" linear models before their demotion—no $\mathbf{Zb}$ in the linear predictor, Gaussian data only)—are subsumed as special cases of the GLMM.

Over the past two decades or more, GLMM—or LMM or GLM—courses have been taught as *second* courses in linear models, the first course devoted almost exclusively to the "general" linear model. The longer I am in this business, the more firmly convinced I become that we need to revisit this approach. What we are doing, I believe, is firmly embedding a "$\mathbf{y} = \mathbf{X}\boldsymbol{\beta} + \mathbf{e}$ mindset" in beginning statistics graduate students that is both an antiquated way to think about contemporary linear models and, for many students, appears to actively interfere with their ability to relearn linear models that work with random effects and non-Gaussian data while still working for Gaussian fixed effect models.

I have a colleague whose mantra is "Never knowingly teach something that you will have to unteach later." Much of a GLMM course consists of unlearning dysfunctional habits of mind accumulated while learning the "$\mathbf{y} = \mathbf{X}\boldsymbol{\beta} + \mathbf{e}$ mindset." The premise of this textbook is that we can—and should—do better. Accordingly, I have tried to write this not as a GLMM text to be used as a second course in linear models or for a GLMM special topics course—although it could be used for either—but as an introduction to linear models. I think that the first thing students new to linear models need to see is the big picture. What are we trying to accomplish, how does it fit with the rest of the core statistics curriculum, and what are the major issues that statistical modelers need to consider? These, with random model effects and non-Gaussian data firmly part of the picture, not withheld till later, should drive the conversation from day one.

Why does this matter? Consider the following example.

Imagine a study to compare two treatments—treatment "0" (the "control" treatment) and treatment "1" (the "test" treatment). At issue is the incidence of favorable outcomes. The researchers use a paired comparison design in which they observe 100 individuals on each treatment in each pair, and each observation is a binary response—either the outcome for a given individual is "favorable" or it is "unfavorable." Assuming mutually independent observations, the number of favorable outcomes for each pair–treatment combination has a binomial distribution, with $N = 100$ and the probability of a favorable outcome denoted by π_{ij} for the ith treatment and jth pair. The data are provided at the end of this preface. The object of any statistical analysis would be to estimate π_{ij} and use the estimates to compare treatments, an exercise that involves statistical modeling.

One approach would start with the normal approximation. With $N = 100$, this seems reasonable. Introductory statistics classes commonly use the binomial distribution with increasing N to demonstrate the central limit theorem. Histograms of simulated data from repeated sampling of a binomial distribution with $N = 100$ and π between 0.1 and 0.9 are virtually indistinguishable from the normal p.d.f. Even for π equal to 0.05 or 0.95, the normal approximation is visually convincing in these simulations. Using the normal approximation, the resulting linear model can be described as follows:

- Response variable: Let $p_{ij} = y_{ij}/100$ denote the sample proportion for the ith treatment and jth pair, where y_{ij} denotes the number of favorable outcomes out of the 100 observations.
- The model is $p_{ij} = \mu + \tau_i + \rho_j + e_{ij}$, where μ denotes an overall mean or intercept, τ_i denotes the ith treatment effect, ρ_j denotes the jth pair effect, and e_{ij} denotes residual error, assumed i.i.d. $N(0, \sigma^2)$.

We know that this model yields the same results as the analysis of variance accounting for pair and treatment effects, which in turn yields the same results as the paired t-test taught in introductory statistics courses. Here, the treatment means for the sample proportion are 0.738 and 0.895 for treatments "0" and "1," respectively, with a standard error of 0.062. The p-value for the test of equal treatment proportions is 0.1132—insufficient by most standards to conclude a statistically significant treatment effect on the expected incidence of a favorable outcome.

One obvious problem with the above analysis is as follows: normal approximation notwithstanding, the response variable remains binomial, meaning the variance must depend on π. Assuming π changes from treatment to treatment (or among pairs), the assumption of equal variance by definition does not hold. The "traditional" linear model "fix" calls for a variance stabilizing transformation. For a binomial response, the standard transformation is the arc sine square root. Using this approach, we define a transformed variable, $p^*_{ij} = \sin^{-1}\left(\sqrt{p_{ij}}\right)$ and fit it to the same model as above, that is, $p^*_{ij} = \mu + \tau_i + \rho_j + e_{ij}$. With p^*_{ij}, the p-value for the test of equal treatment proportions is 0.0605. Back transforming the mean p^*_{ij} for each treatment yields 0.760 ± 0.067 for treatment "0" and 0.916 ± 0.040 for treatment "1."

At this point in this example, we have reached the limits of the "general" linear model. In the 1970s, this was the state of the statistical art. More than three decades later, two mutually reinforcing developments have dramatically altered the landscape: the development of linear mixed model and generalized linear model theory and methods, at first parallel then increasingly interwoven, accompanied by rapid and sustained increase in computing capability. So in 2012, what would a generalized linear mixed model look like in this example? What results would it yield? How much does all this development matter?

Unlike the traditional linear model, an equation whose form is *observation* = (sum of explanitory variables) + *residual* and a probability assumption about the residual, the generalized linear mixed model involves three parts:

- The distribution of the observations: here $y_{ij} \mid \rho_j \sim Binomial(100, \pi_{ij})$.
- The linear predictor: here $\eta_{ij} = \eta + \tau_i + \rho_j$, where η denotes the intercept and the other terms are defined as before. Note the absence of a residual term—not an oversight but a necessity. More on that beginning in Chapter 1. If the pairs represent a larger population—and typically they do—then the model must include their assumed distribution. Here, assume the pair effects, ρ_j, are i.i.d. $N\left(0, \sigma_P^2\right)$.
- The link function: with non-normal data, the canonical parameter of the log likelihood is typically a better candidate for fitting a linear model than the mean itself. For the binomial, the canonical parameter is $\log[\pi/(1-\pi)]$. The link function is therefore $\eta_{ij} = \log[\pi_{ij}/(1-\pi_{ij})]$. When you estimate the parameters of the linear model, you estimate the link function—in this case the log-odds. The estimate of π_{ij} is therefore $\hat{\pi}_{ij} = 1/(1+e^{-\hat{\eta}_{ij}})$.

For this example, the generalized linear mixed model yields a *p*-value of 0.0355 for the test of equal incidence of a favorable outcome for the two treatments. The estimates of favorable outcome incidence are 0.781 ± 0.072 and 0.928 ± 0.030 for treatments "0" and "1," respectively.

Now we can frame the question, "why GLMMs?" more tangibly.

The three analyses in this example produce different estimates of the incidence of a favorable outcome for each treatment. Of these, only the estimate yielded by the GLMM is an estimate of a true binomial probability. The others are not—even though they are commonly interpreted as if they were. Understanding why requires an understanding of GLMM theory. Recalling boxer Joe Louis' famous quote, there is something of a "you can run but you cannot hide" aspect to GLMMs. Using ANOVA or "general" linear model methods does not exempt you from GLMM-related issues—it simply makes you oblivious to their existence.

The three analyses also yield different *p*-values. All three approaches control type I error roughly equally. However, GLMMs have greater power to identify model effects as statistically significant, accounting for the difference among the *p*-values in this example. Understanding the linear model theory at work here is essential for statistical science professionals in training.

"Why GLMMs?" They estimate explicitly what you think they estimate, not something else. If there is a nonzero treatment effect, the GLMM has at least as much power to find it as the traditional "general" linear model—more when the data are not normally distributed. The purpose of this text is to help you understand why.

The following table, my elaboration of a presentation by Locke and Devoe at the 2007 USCOTS meetings, provides additional perspective.

		Explanatory Variables and Error Structure			
Response Variable Type	**Typical Distributions**	**Fixed Effects**		**Random Effects (Split-Plot, Clustered, Multilevel Designs, Genetic Effects, etc.)**	**Correlated Errors (Repeated Measures, Longitudinal, Spatial)**
		Categorical	**Continuous**		
Categorical	Binomial Multinomial				
Counts	Poisson negative binomial				
Continuous symmetric	Gaussian (normal)	ANOVA $\mathbf{y} = \mathbf{X}\boldsymbol{\beta} + \mathbf{e}$	regression		
Continuous proportion	Beta				
Time to event	Exponential Gamma				

When we describe a linear model as "response variable = (explanatory variable list) + random," we need to think carefully what we are actually saying. We know that in contemporary statistical practice, response variables come in many forms. The table gives us an idea of the variety of response types, but it is hardly exhaustive. We also know that the right-hand, explanatory + random, side of the model also involves more than just treatment plus error. The "general" linear model effectively confines attention to two cells on this matrix and gives no inkling of the issues (such as the binomial paired comparison example) raised by attempting to use this limited view of linear modeling when the response and model reside in one of the other 18 cells of this matrix.

Organization of This Textbook

I have divided this textbook into three parts:

- Part I The Big picture
  Chapter 1 addresses the questions, "What is a model? What should a model do? What are the essential elements of a model? It concludes with how models are set up in matrix language.
  Chapter 2 is based on my observation over the years that modeling difficulties are disproportionately rooted in design issues. Even sample surveys and retrospective studies involve design concepts, especially when the goal is to assess the impact of predictor variables on a response. Absent a working understanding of the interplay between design principles and modeling, good modeling does not happen. Chapter 2 focuses on techniques to help clarify this interplay.

It features two techniques to construct models from data set structure. I call the first method "what's the experimental unit"—it is borrowed from Milliken and Johnson's *The Analysis of Messy Data,* Vol. 1, 2nd edn. (2008). Dallas Johnson used to emcee a game called "Stump the Statistical Consultant" during breaks at the Conference on Applied Statistics in Agriculture—the "what's the experimental unit" template is a loose adaptation of this approach. I call the second method *What Would Fisher Do?* (*WWFD*). I was inspired by Terence Speed's column, "And ANOVA Thing," that appeared in the May 2010 *IMS Bulletin.* Speed described the uneasy coexistence of analysis of variance and statistical modeling and Fisher's negative reception ("enraged" and "confused" to use Speed's words) of early statistical modeling work. Speed recommended reading Fisher's comments following Yates (1935) Royal Stalistical Society discussion paper, "Complex Experiments." Reading Fisher produced an "ah-ha" realization that the ANOVA *thought process,* properly understood, could be an effective tool for constructing an appropriate GLMM. On the other hand, ANOVA improperly understood—as an arithmetic exercise tied to ordinary least squares and quadratic forms—is a perfect example of the "$\mathbf{y} = \mathbf{X}\boldsymbol{\beta} + \mathbf{e}$ mindset," obstructing understanding of contemporary modeling. I have no idea what Fisher would actually do with modern theory and computing technology, but it is easy to believe that his distress with modeling was that it was headed toward what ultimately became the "general" linear model, and Fisher's intuition told him that this approach would prove to be too restrictive.

Chapter 3 is a late addition but may be this textbook's most important chapter. It covers issues in inference that simply do not arise in the "general" linear model. The first two—data versus model scale for generalized linear models and broad versus narrow inference for mixed models—have been discussed in other texts, although there is still not anything like widespread understanding of these issues. The third—conditional and marginal inference—is an issue the statistical science community is just beginning to appreciate. Let me put that more strongly. Truly useable GLMM software has only been around for a few years. SAS® PROC GLIMMIX, for example, did not appear until 2005, and the statistical science community is *just beginning* to come to grips with the implications of the conditional–marginal issue, which is at the root of the binomial example presented earlier in this preface. The more we explore this issue, the more compelling the case becomes to frontload GLMM concepts when we introduce linear models to statistics professionals in training.

- Part II Estimation and Inference Essentials
  These are the "theory" chapters.

  Chapter 4 develops the estimating equations, first for the generalized linear model and then for the linear mixed model. It then "marries" them for generalized linear mixed models. The overall approach is likelihood based. Needed matrix operations, such as generalized inverse and matrix derivatives, are introduced on a just-in-time basis.

  Chapters 5 and 6 develop tools for inference. Chapter 5 focuses on the linear predictor effects $\boldsymbol{\beta}$ and $\mathbf{b}$—standard errors, test statistics, degree-of-freedom approximations, bias correction such as the Kenward–Roger adjustment, sandwich estimators, etc. Chapter 6 focuses on variance and covariance components—formal tests and fit statistics. As in Chapter 4, needed distribution theory, including quadratic forms as they arise in GLMMs, are introduced on a "just-in-time" basis.

An appendix that follows the last chapter in this textbook reviews matrix essential matrix concepts and results.

As these chapters develop, it becomes clear that material in traditional first courses in linear models, the "general" linear model, consists largely of special cases of a much more overarching and inclusive linear model theory.

My experience has been that this material fits better *after* Part I. In the past, when the estimating equations came early in the course, students would say things like, "this is all well and good, but I don't have a clue why we're doing this. These estimating equations are so *complicated* compared to what we did last semester [their "general" linear model first course]." This sequence has gone better.

- Part III Working with GLMMs
  The remaining ten chapters form Part III. These chapters focus on common applications of GLMMs. This includes multifactor split-plot, clustered, and multilevel models for Gaussian and non-Gaussian data, correlated error models for repeated measures and spatial data, count data and overdispersion, special issues that arise with binary data, models for multinomial categorical data, and other applications of the GLMM.

  The third Part is necessarily a collection of surveys. There are entire courses taught on particular aspects of the GLMM—survival analysis, categorical data, time series, spatial statistics, etc. This textbook is not intended to replace or compete with in-depth textbooks on any of these topics. It is intended to be an introduction to give someone new to GLMMs sufficient literacy to work with common GLMMs and learn or self-teach specialized topics in more detail should the need arise.

  The concluding chapter in this part concerns power and sample size. Conventional wisdom regarding the design of experiments is largely rooted in a Gaussian data tradition. Commercial power and sample size software is largely untouched by GLMM concepts. As a result, power, sample size, and design decisions based on conventional design wisdom or commercial power and sample size software can be inappropriate—sometimes catastrophically so—especially when the study being planned involves non-Gaussian response variables and random model effects. GLMM theory is not difficult to use for design planning and power and sample size assessment—this chapter shows how.

Notation Used in This Textbook

Writing a textbook about GLMMs involves walking a notational tightrope. On one hand, there are conventions inherited from "general" linear model literature that are more or less standard. They are familiar to anyone who has worked with statistical models. On the other hand, some of the inherited conventions do not translate well to GLMMs or, worse, they tend to reinforce the "$\mathbf{y} = \mathbf{X}\boldsymbol{\beta} + \mathbf{e}$ mindset"—habits of mind that get in the way of learning and truly understanding GLMMs.

There are four notation decisions I feel the need to mention here to alert the reader and explain why, after several years of trial and error, I use them in this textbook.

1. Following Searle's lead, wherever possible I denote fixed model effects with characters from the Greek alphabet and random model effects with characters from the Latin alphabet. For example, the expression $\eta+\alpha_i+b_j+(ab)_{ij}$ means factor A is a fixed effect (α_i), factor B is a random effect (b_j), and hence the $A \times B$ interaction—$(ab)_{ij}$—is also random. Also, I prefer using the same symbols used by the factors comprising the interaction (switching as needed from Greek to Latin, as in α to a) instead of introducing a new character. I think it is easier for the reader to follow.
2. Characters in bold are vectors (if they are lower case) or matrices (upper case). For example, y is a scalar random variable; $\mathbf{y}$ is a random vector. X is a scalar; $\mathbf{X}$ is a matrix. β is a fixed effect parameter; $\boldsymbol{\beta}$ is a vector of parameters.
3. Linear model texts almost invariably present the model equation for analysis of variance type models as $\mu+\alpha_i+b_j+(ab)_{ij}$ (for example), where μ denotes the intercept—or the overall mean if requisite constraints are placed on the other parameters. For generalized models with non-Gaussian data, this becomes confusing and sometimes nonsensical. For example, with binomial data the canonical link is $\eta_{ij}=\log[\pi_{ij}/(1-\pi_{ij})]$ for the jth $A \times B$ combination. The standard notation for the link function is η_{ij}. The linear predictor (the GLMM analog of the model equation) is $\eta_{ij}=\text{intercept}+\alpha_i+b_j+(ab)_{ij}$. For years, I used μ to denote the intercept regardless of the distribution and link function. I finally despaired of hearing "η_{ij} is the logit link…where μ is the overall mean." The intercept on the link scale is *not* the overall mean, but it became clear that the notation worked against the message. When I started using η to denote the intercept, as in $\eta_{ij}=\eta+\alpha_i+b_j+(ab)_{ij}$, students reported far less confusion once they got used to it. This is the notation I use throughout this textbook.
4. Throughout this textbook, I use the generic notation $\boldsymbol{\eta}=\mathbf{X}\boldsymbol{\beta}+\mathbf{Z}\mathbf{b}$ for the matrix representation of the GLMM linear predictor. Some texts use $\boldsymbol{\gamma}$ for the random effect vector, but this violates the Greek/Latin rule. Some texts use $\mathbf{u}$ for the random effect vector, but this creates awkward moments in lecture when introducing the marginal distribution for a GLMM. Also, students tell me that my handwritten $\boldsymbol{\mu}$ and $\mathbf{u}$ are hard to distinguish when I write on the board, but my $\boldsymbol{\beta}$ and $\mathbf{b}$ are okay. Harville uses $\boldsymbol{\beta}$ and $\mathbf{b}$, and it seems to work pretty well.

Why GLIMMIX?

I have been asked why anchor the examples in this textbook to SAS® PROC GLIMMIX when the statistics world seems to be moving toward R. There are three reasons.

First, this textbook has part of its roots in a two-day short course Oliver Schabenberger and I team-taught at the 2008 Joint Statistical Meetings. Oliver was the developer of the GLIMMIX procedure, and we have been colleagues in thinking about what GLMM curriculum should look like for years. The thought processes embedded in GLIMMIX reflect the way we think about GLMMs.

Second, I have been a SAS user for more than 30 years and was department chair from 2001 to 2010. When I agreed to write this textbook, my response to R was, "I can only handle so many learning curves at any given time—if you are going to see this textbook anytime soon, GLIMMIX will be the anchor, not R."

Third, and most importantly, GLIMMIX has one syntax that, once learned, applies in a consistent (and—in my opinion—straightforward) manner to all of the GLMMs covered in this textbook. On the other hand, R has a package for this, another package for that, not clear if it even has a package for some things—all without dependably consistent mental architecture for turning the model into software instructions. The goal here is to introduce the main ideas and methodology associated with linear models that accommodate random model effects and non-Gaussian data to an audience that includes graduate students in statistics, statistics professionals seeking an update, and researchers new to the generalized linear model thought process. I want the *ideas* to be central and the software to *assist*, not for the software to get in the way or become an issue in itself. SAS remains the *lingua franca* of the statistical computing world. GLIMMIX, in particular, is the best way to minimize attention diverted to computing minutiae and maximize the focus on GLMM concepts and methods.

As an aside, I cannot resist a personal observation. I have come to regard GLIMMIX as part software and part work of art. It was obviously written to aid, abet, and reinforce the GLMM learning process.

Finally, Target Audience and How to Use This Textbook

As far as the target audience is concerned, I can only repeat what I wrote two paragraphs ago: my target audience is graduate students in statistics, statistics professionals seeking to get up to speed, and researchers new to the generalized linear model thought process. The reader should have a basic understanding of linear algebra, some previous exposure to probability distribution theory and likelihood-based estimation and inference, and a working knowledge of applied statistics, especially analysis of variance and regression.

In particular, I have written this textbook specifically to serve as a first course in linear models with the understanding that we need to rethink our current paradigm of indoctrinating students in the "$\mathbf{y}=\mathbf{X}\boldsymbol{\beta}+\mathbf{e}$ mindset" first and then—maybe—reteach a second linear models course with "generalized" and "mixed" added.

If your graduate program—like mine as of this writing—is locked into the "$\mathbf{y}=\mathbf{X}\boldsymbol{\beta}+\mathbf{e}$ first" approach, this textbook can also be used for a second course in linear models focusing on GLMMs.

I have also written this textbook to be a reference and a continuing education text for researchers, practitioners, and quantitatively literate nonstatisticians who want an introduction or want to know more about generalized, mixed, and generalized + mixed linear models.

If you use this textbook for a first course in linear models, I regard the first two parts as essential. If you cut corners, cut them in the third part.

If you use this textbook as a GLMM reference, I regard the first and third parts as essential. The second, focused on estimation and inference theory, can be browsed as needed.

The one part I regard as *not optional* under any circumstances is the first. As I mentioned earlier, we are just beginning to realize that there are fundamental linear model issues that never occurred to us when $\mathbf{y}=\mathbf{X}\boldsymbol{\beta}+\mathbf{e}$ was the "general" linear model. We are just beginning to realize that we need to come to terms with these issues. I think it is safe to say that we do not have anything close to general awareness of these issues even within the statistics community, much less in the wider research community. These first three chapters, especially Chapter 3, are, in my opinion, the most important chapters in this textbook.

Have fun!

Data Set 0.1 *Paired comparison, binomial data*

```
/* Binomial response variable                                  */
/* Treatment_0 is # favorable outcomes out of 100 for Trt=0  */
/* Treatment_1 is # favorable outcomes out of 100 for Trt=1  */
```

Pair	Treatment_0	Treatment_1
1	98	94
2	95	36
3	93	85
4	94	88
5	99	91
6	61	82
7	84	43
8	92	71

Acknowledgments

Above all, I thank my wife Ellen and son Frank for their love and inspiration. Without Ellen's support, encouragement, perspective, understanding, and *patience*, this would never have been possible.

Next, thanks to Oliver Schabenberger. I was deeply honored when Oliver invited me to co-teach the GLMM short course at the 2008 Joint Statistical Meetings. It was that course that prompted writing this book or, to put it another way, this book has its roots firmly in that course. It was a privilege to work with Oliver—on the course and on all of the other mixed model projects that preceded it.

Special thanks to my colleagues Mark West, Ed Gbur, Kevin McCarter, Mary Christman, Matt Kramer, Susan Durham, and Linda Young for innumerable challenging and rewarding conversations. They certainly provided the impetus for gaining a much richer understanding of GLMMs. Our North Central project turned into a learning community of the first order. Thanks also to George Milliken, Ramon Littell, Russ Wolfinger, Bill Sanders, and Dallas Johnson—colleagues and friends over many years—whose ideas are by now so much a part of me that I am sure they show up in various forms throughout this book even when I'm not fully aware of it.

I am very appreciative of the Kansas State University Department of Statistics, and in particular Jim Neill, who was chair at the time, for providing me with a sabbatical home during the time I wrote this book. It was close enough to home in Lincoln for me to see Ellen and Frank, at least on key weekends and holidays, but far enough away to "hide in my writing cave" (as my graduate students called it) for the extended periods of seclusion and concentration required to create this book. The K-Staters were great hosts. I could always count on them to provide a helpful sounding board when needed. Thanks especially to Nora Bello for her extensive review and immensely helpful feedback and suggestions.

Thanks to the many Nebraska and Kansas State graduate students who endured draft versions of this textbook in my linear model courses and who proofread, commented, suggested, and generally provided essential feedback at critical points. I am especially indebted to Elizabeth Claassen, Pamela Fellers, Martin Frenzel, and Trevor Hefley for their crucial help during the final stages of getting this book ready for publication.

Finally, thanks to Keith Richards, James McMurtry, Steve Earle, Alison Krause, Wynton Marsalis, Ali Farka Touré, Pierre Boulez, Jack White, and the thousands of other musicians dating back at least to the fifteenth century who provided music to keep me writing and, even more importantly, continuing inspiration about the power and majesty of the creative process.

Part I

The Big Picture

1

Modeling Basics

1.1 What Is a Model?

A *model* is a mathematical description of the processes we think give rise to the observations in a set of data. A *statistical model* includes an equation describing the presumed impact of explanatory variables and a description of the probability distributions associated with aspects of the process we assume to be characterized by random variation. It follows that statistical models include a *systematic* part and a *random* part.

To see how this works, consider two models familiar to all beginning statistics students: the "comparison of two means" model and the linear regression model. The former can be written as

$$y_{ij} = \mu_i + e_{ij} \tag{1.1}$$

where

- the subscript $i = 1,2$ denotes treatment
- $j = 1,2,\ldots n_i$ denotes the jth observation
- n_i denotes the number of observations on the ith treatment
- y_{ij} denotes the jth observation on the ith treatment
- μ_i denotes the mean of the ith treatment
- e_{ij} denotes random "error" (although "random variation" is arguably a better choice of words) associated with the ijth observation

We designate μ_i—that is, an observation's mean—the systematic part of model (1.1) because it is *determined* by the treatment it receives. For this reason, the systematic part of the model can also be called the *deterministic* part of the model. Whether we call it deterministic or systematic, this part of the model is a mathematical law, not subject to random variability. The *random* part of model (1.1) is e_{ij}. It tells us that observations are assumed to vary at random about their mean. The e_{ij} term is a blanket characterization of the uniqueness of the ijth individual. In theory, at least according to some schools of philosophy, if we knew everything there was to know about what made these individuals unique, we could write a completely deterministic model. However, we do not, so we characterize the distribution of these observations using probability. Typically, we assume e_{ij} to be i.i.d. $N(0, \sigma^2)$.

Now consider the linear regression model

$$y_{ij} = \beta_0 + \beta_1 X_i + e_{ij} \tag{1.2}$$

where
β_0 is the intercept
β_1 is the slope
X_i is the value of the predictor (a.k.a. independent) variable

The linear regression model is similar to (1.1). The only difference is that $\beta_0+\beta_1 X_i$ replaces μ_i as the systematic part of the model. The other terms, y_{ij} and e_{ij}, retain the same definitions and assumptions given previously.

As a final comment in this section, keep in mind that statistical models are necessarily approximations. George Box famously said, "All models are wrong, but some are useful." Even in these simple examples, there is much we do not know about the individuals that we choose not to pursue; instead, we settle for approximating variation among these individuals using the Gaussian probability distribution. Is this wrong? Technically, yes. Is it useful? My father has a favorite expression, "Good enough for what it's for." If the goal is to compare two means—as it is for model (1.1)—or to estimate the slope and intercept—as it is for model (1.2)—and the y_{ij}'s are continuous, independent, and more or less symmetrically distributed about their mean and the variance does not depend on the mean, then the i.i.d. $N(0, \sigma^2)$ approximation for e_{ij} is "good enough for what it's for."

At a minimum, a statistical model must include the three elements given in these two examples:

1. The observation
2. The systematic, or deterministic, part of the process giving rise to the observation
3. The random part, including a statement of the assumed probability distribution

Introductory texts typically stop with these three elements. As we will see in the following sections, these elements alone are inadequate to describe models commonly needed in contemporary statistics.

KEY IDEAS: *systematic/deterministic vs. random parts of the model*

1.2 Two Model Forms: Model Equation and Probability Distribution

The two examples in the previous section used what we refer to as the *model equation* form of a statistical model. That is, the basic form is "observation = systematic part + random part." Historically, the model equation form predominates in the literature of linear models. While it suffices for relatively simple models and may be acceptable for introducing models to students seeing them for the first time, the model equation form has severe limitations that render it inadequate for describing modeling situations common in contemporary statistics. To see this, we revisit models (1.1) and (1.2) from Section 1.1.

Start with the two-treatment model (1.1), given in Section 1.1 as $y_{ij}=\mu_i+e_{ij}$. We can rewrite (1.1) as $y_{ij}=\mu+\tau_i+e_{ij}$, where $\mu+\tau_i=\mu_i$. The former, expressing the systematic part of the model as a treatment mean, is called the *cell means* model; the latter, expressing the systematic component of the model as an intercept modified by a treatment or predictor effect, is called the *effects* model.

Consider what each model implies about the processes giving rise to the observed data. The effects model explicitly states all of the process generating the observations. Here is how. Individuals are sampled from a population whose probability distribution, with respect to the response variable we observe, we assume to be i.i.d. $N(\mu, \sigma^2)$. We then assign individuals (presumably at random) to a treatment, either 1 or 2. Assignment to the ith treatment "bumps" the mean response by τ_i. This results in observations with distribution $NI(\mu+\tau_i, \sigma^2)$ or, equivalently, $NI(\mu_i, \sigma^2)$, where we read NI as "normal and independent."

Expressing the model this way allows us to see that the systematic part of the model is actually a model of the expected value of the observations, that is, $E(y_{ij})=\mu_i=\mu+\tau_i$. Similarly, the linear regression model can be expressed as y_{ij} assumed to be distributed $NI(\beta_0+\beta_1 X_i, \sigma^2)$. The systematic component of the linear regression models the expected value of the observations as $E(y_{ij})=\mu_i=\beta_0+\beta_1 X_i$.

The important point here is that *as long as we assume a Gaussian* (normal) *distribution*—and this is an important stipulation—there are two ways to express statistical models:

- *Model equation form*, for example, $y_{ij}=\mu_i+e_{ij}$ or equivalently $y_{ij}=\mu+\tau_i+e_{ij}$
- *Probability distribution form*, for example, $y_{ij}\sim NI(\mu_i, \sigma^2)$ or equivalently $y_{ij}\sim NI(\mu+\tau_i, \sigma^2)$

The model equation form is more common in statistical literature, but it has unacceptable limitations, as you will see in the next example. For reasons that should be apparent by the end of this section, we will emphasize the probability distribution form for the rest of the text.

1.2.1 Twist Illustrating the Weakness of the Model Equation Form

Return to the linear regression example—model (1.2)—but suppose the observations on the ijth individual come from N_{ij} independent Bernoulli trials. That is, we observe N_{ij} "0/1" "either/or" "success/failure" outcomes on each individual, count the number of "successes," and denote them as y_{ij}. Thus, $y_{ij}\sim \text{Binomial}(N_{ij}, \pi_i)$, where π_i denotes the probability of a "success" for each Bernoulli trial on the ijth individual. If we assume that the probability of a success depends on the independent variable, X, then we would like to model the change in the probability of a success associated with changes in X.

The model equation approach gives us $y_{ij}=\beta_0+\beta_1 X_i+e_{ij}$. Immediately, we have a dilemma: What do we say about e_{ij}? We cannot assume that the e_{ij} are i.i.d. $N(0, \sigma^2)$, that is, we cannot assume the e_{ij} are Gaussian (why not?) nor can we assume they are i.i.d. (why not?—hint: focus on the "identically distributed" part).

We could redefine our response variable as $p_{ij}=y_{ij}/N_{ij}$, the sample proportion for the ijth individual. This gives us $p_{ij}=\beta_0+\beta_1 X_i+e_{ij}$ as a candidate model. Assuming sufficient N_{ij} for each individual, we could invoke the central limit theorem and claim p_{ij} have an approximate Gaussian distribution. However, unless $\beta_1=0$—that is, identical π_i for all X_i—we cannot assume constant variance σ^2 (why not?). This seems self-defeating given that our goal is estimating the effect of X on π_i. There must be a possibility that $\beta_1\neq 0$ or chances are that the data would never have been collected in the first place.

What if we proceed anyway? Consider the example data shown in Table 1.1.

TABLE 1.1

Example Data from Binomial Regression

Obs	x	N	y
1	0	11	0
2	1	7	0
3	2	9	2
4	3	11	2
5	4	12	2
6	5	15	5
7	6	11	7
8	7	15	12
9	8	11	10
10	9	16	16
11	10	10	9

Using the model $p_{ij} = \beta_0 + \beta_1 X_i + e_{ij}$, we obtain $\hat{\beta}_0 = -0.089$ and $\hat{\beta}_1 = 0.1115$. The p-value associated with the test $H_0{:}\beta_1 = 0$ is $p < 0.0001$ and the coefficient of determination, $R^2 = 0.917$. At first glance, this looks good. However, looking closer, the predicted value of p at $X=0$ is -0.089 and at $X=10$ is 1.026. What is one to make of predicted probabilities less than 0 or exceeding 1?

A traditional approach for handling such situations is to use a variance stabilizing transformation. For binomial data, the standard transformation is $\sin^{-1}(\sqrt{p_{ij}})$. However, this approach essentially amounts to a modeling version of the "when all you have is a hammer, try to make every problem look like a nail." In other words, the *model equation form* follows from a Gaussian mindset; if the data are not Gaussian, we have to make them "act Gaussian." If we know we have a binomial response variable, we would do better to deal with it as such and not try to force it to be normal when we know it is not. How? The *probability distribution form* offers a more viable approach.

There are a number of alternatives for modeling binomial data using the probability distribution approach. For now, we consider the two most common. They are as follows:

1. Inspect the probability distribution to identify a function of π_i suitable for fitting a linear model.
2. Define a process by which observations could plausibly arise in which π_i varies with X_i.

For this example, both approaches anticipate the *generalized linear model*, which we will fully develop in subsequent chapters. For now, we focus on the basic idea.

Alternative 1: Inspect the probability distribution

The p.d.f. of the binomial random variable y_{ij} is $\binom{N_{ij}}{y_{ij}}\pi_i^{y_{ij}}(1-\pi_i)^{N_{ij}-y_{ij}}$. The log likelihood is $\log\binom{N_{ij}}{y_{ij}} + y_{ij}\log(\pi_i) + (N_{ij}-y_{ij})\log(1-\pi_i)$ which we can reexpress as $y_{ij}\log[\pi_i/(1-\pi_i)] + N_{ij}\log(1-\pi_i) + \log\binom{N_{ij}}{y_{ij}}$. The key is the expression $y_{ij}\log[\pi_i/(1-\pi_i)]$—it

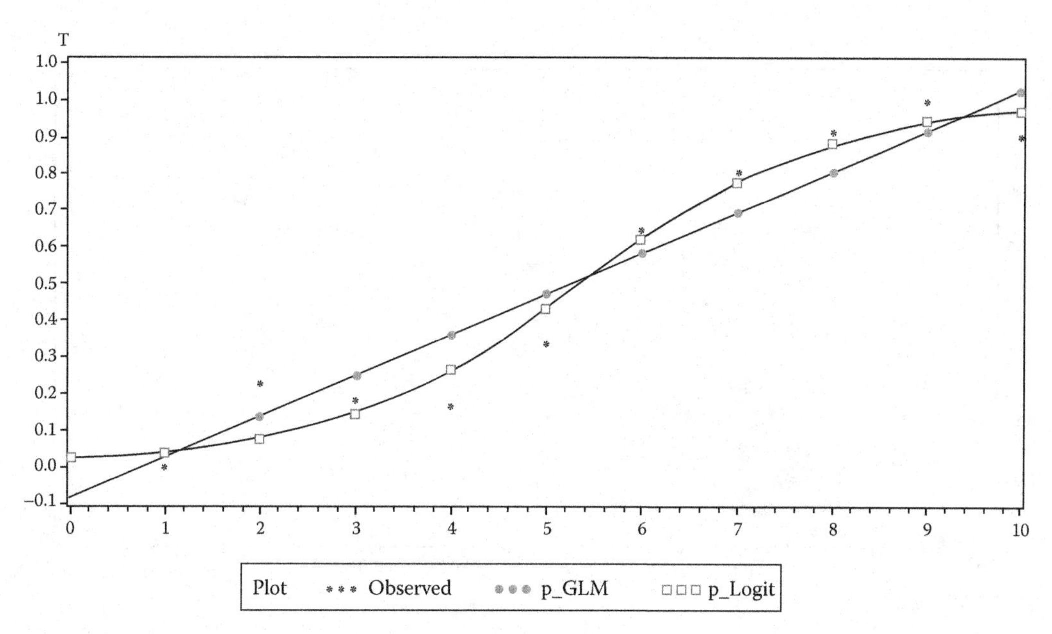

FIGURE 1.1
Linear model vs. logistic regression for binomial data.

reveals the random variable y_{ij} to be linear in $\log[\pi_i/(1 - \pi_i)]$. In statistical modeling, the expression $\log[\pi/(1 - \pi)]$ is known as the *logit* of π. A possible model is, thus, $\text{logit}(\pi_i) = \log[\pi_i/(1 - \pi_i)] = \beta_0 + \beta_1 X_i$.

This is the *logistic regression model,* familiar in many statistical applications. Using methods fully discussed in Chapter 4, the estimated intercept and slope are $\hat{\beta}_0 = -4.109$ and $\hat{\beta}_1 = 0.764$, respectively. We can determine the estimated π_i given X_i as $\hat{\pi}_i = 1/(1+e^{-(\hat{\beta}_0+\hat{\beta}_1 X_i)})$. For $X=0$, $\hat{\pi} = 0.016$ and for $X=10$, $\hat{\pi} = 0.972$. Figure 1.1 shows the comparative fit of the linear model fit to p_{ij} versus the fit of the logistic regression.

The correlation between to observed and predicted values in the linear model fit to p_{ij} is 0.957; for the logistic regression model, the correlation is 0.982. By any applicable criterion, the logistic model provides a better fit and, more importantly, it avoids the problem of trying to explain an impossible predicted value.

Alternative 2: Define a plausible process

Imagine that there is an unobservable process driven by X. We never see the process; we only see its consequences. If the process exceeds a certain value, we see a "failure." Otherwise, we see a "success." Now, denote the boundary between success and failure by η, and suppose that it varies linearly with X, that is, $\eta = \beta_0 + \beta_1 X$. This is a simple example of a *threshold model.* If we further imagine that the unobservable process has a standard normal distribution, then the probability of a "success" at X can be modeled as $\pi_i = \Phi(\beta_0 + \beta_1 X_i)$, where $\Phi(\bullet)$ is the inverse normal, or *probit*. Figure 1.2 illustrates the basic idea. As illustrated, for $X=3$ (right axis), $\eta \cong -1$ (horizontal axis). (You may find this disorienting at first—you are probably used to seeing X on the horizontal axis and the response variable on the vertical (Y) axis.) The shaded area under the curve $\Phi(-1)$ is the probability, π. As X increases, η increases, and, with it, the area under the normal curve, which corresponds

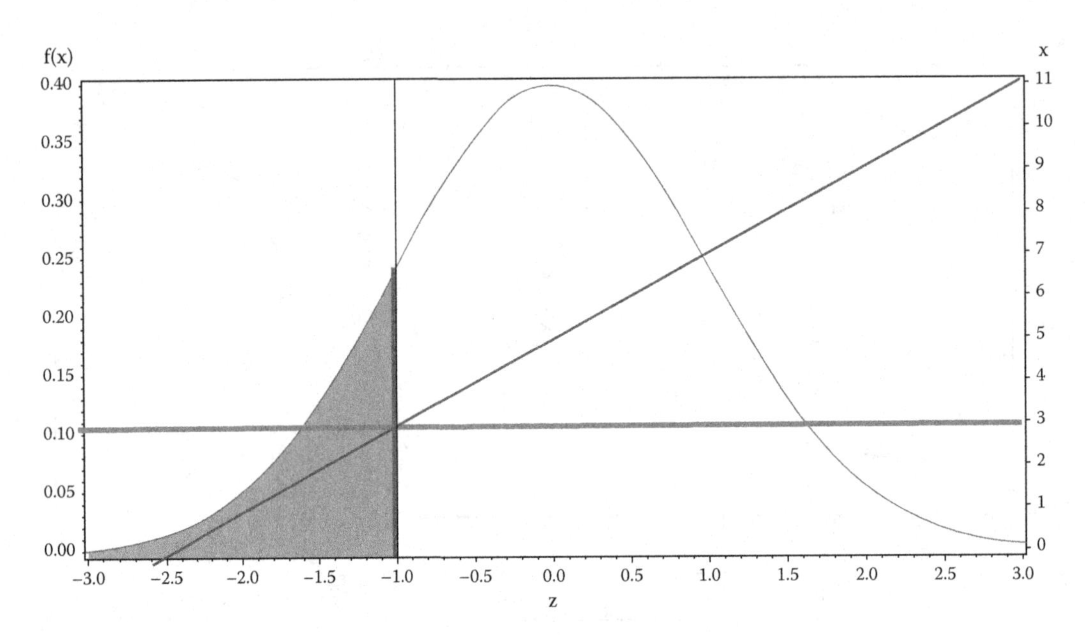

FIGURE 1.2
Probit model—basic idea illustrated.

to π, increases. Figure 1.2 illustrates a positive slope ($\beta > 0$). Note that negative slopes are possible: in such cases, as X increases, η would decrease and, hence, π would decrease as well. The location and slope of the of the regression line depend on β_0 and β_1.

This is the *probit* regression model. Like logistic regression, it is commonly used for binomial data. Both the logistic and probit models are examples of the *probability distribution form* of a linear model. Both share the following steps:

1. Identify the probability distribution of the observed data. In both the logistic and probit models, it is $y_{ij} \sim \text{Binomial}(N_{ij}, \pi_i)$.
2. Focus on modeling the expected value, that is, $E(y_j) = N_{ij}\pi_i$. Since N_{ij} is known, modeling effectively focuses on π_i.
3. State the *linear predictor*. In this example, the linear predictor is $\beta_0 + \beta_1 X$. *For now*, you can think of the linear predictor as the right hand side of the model equation form, but without e_{ij}. In this example, the linear predictor corresponds to the systematic part of the model.

> Note the emphasis on *"for now."* When we introduce random model effects—in Section 1.3—things get more complicated.

4. Identify the function that connects, or links, the expected value to the linear predictor. In the logistic model, this function is $\log[\pi_i/(1 - \pi_i)] = \beta_0 + \beta_1 X_i$. In linear model language, this is called the *link function*. You could also write the logistic model as $\pi_i = 1/\{1 + \exp[-(\beta_0 + \beta_1 X_i)]\}$. This is called the *inverse link* function. In the link, the linear predictor stands alone and the expected value is embedded in a

function; in the inverse link, the expected value stands alone and the linear predictor is embedded in a function. Both are valid and sufficient expressions. Note that we described the probit model using the inverse link, $\pi_i = \Phi(\beta_0 + \beta_1 X_i)$.

You can apply the same steps to the Gaussian model introduced in Section 1.1.

1. The distribution: $y_{ij} \sim NI(\mu_i, \sigma^2)$.
2. Model focus: $E(y_{ij}) = \mu_i$.
3. Linear predictor: $\beta_0 + \beta_1 X_i$.
4. Link: For Gaussian model, the link function is called the *identity link*, that is, the linear predictor itself predicts $E(y_{ij}) = \mu_i = \eta_i = \beta_0 + \beta_1 X_i$.

KEY IDEAS: *model equation form vs. probability distribution form, linear predictor, link and inverse link functions*

1.3 Types of Model Effects

In the previous sections, we considered the two-treatment mean comparison model and the linear regression model. The linear predictor for the former was $\mu_i = \mu + \tau_i$ and for the latter it was $\beta_0 + \beta_1 X_i$. In the two-treatment model, the predictor variable is *treatment* and the corresponding model effects are the τ_i. In the regression model, the predictor variable is X_i and the corresponding model effect is β_1. These illustrate two types of effects in statistical models. In the former model, treatment is an example of a predictor that is a *classification* variable; in the latter model, X_i is an example of a predictor that is a *direct* variable. Classification variables are defined by categories, for example, treatment 1 and treatment 2, whereas direct variables are defined on literal amounts.

This distinction can sometimes be confusing, because it is possible to define a predictor variable as either a direct or classification variable, depending on your objectives and the inference you want. For example, for the data in Table 1.1, you could define X as a classification variable and fit an effects model $\mu + \tau_i$; $i = 1,2,\ldots,10$, with the effects defined on the 10 levels of X instead of $\beta_0 + \beta_1 X_i$. You might do this if you did not expect the response to vary according to any defined function of X. If you have no preconceived idea of what relationship between y and X to expect, defining X as a classification variable allows you to explore.

Classification vs. direct is one way to distinguish between types of model effects. Another, and perhaps more fundamental, distinction has to do with how the levels of an effect are chosen and the scope of intended inference. To see how this works, consider the following.

1.3.1 Extension of the Linear Regression Example to Illustrate an Important Distinction between Types of Model Effects

In Table 1.1, there is only one observation per level of X. Suppose, instead, levels of X are observed at multiple locations or for multiple batches. Table 1.2 shows a hypothetical set of data. There are 11 levels of X in varying intervals from 0 to 48. For each of four batches, a continuous variable (Y) and a binomial variable (where N is the number of Bernoulli trials and *Fav* is the number of "successes") are observed at each level of X.

TABLE 1.2
Multi-Batch Data

	Batch 1			Batch 2			Batch 3			Batch 4		
X	*Y*	Fav	*N*	*Y*	Fav	*N*	*Y*	Fav	*N*	*Y*	Fav	*N*
0	95.6	15	21	96.8	18	21	96.6	16	19	96.6	18	21
3	96.9	13	17	96.7	14	16	96.5	13	19	96.8	18	21
6	98.5	19	23	96.3	14	17	97.7	17	22	96.4	14	20
9	99.0	14	17	96.3	17	20	98.3	23	27	96.6	14	19
12	100.2	18	23	96.5	15	20	99.1	16	21	96.8	14	19
18	101.9	19	27	96.4	14	22	100.5	11	16	96.9	11	20
24	104.1	15	20	95.7	12	22	101.2	13	18	97.1	11	16
36	107.8	14	21	95.2	11	25	103.3	10	17	97.4	10	21
48	111.5	13	18	94.6	5	26	105.9	6	16	97.4	5	19

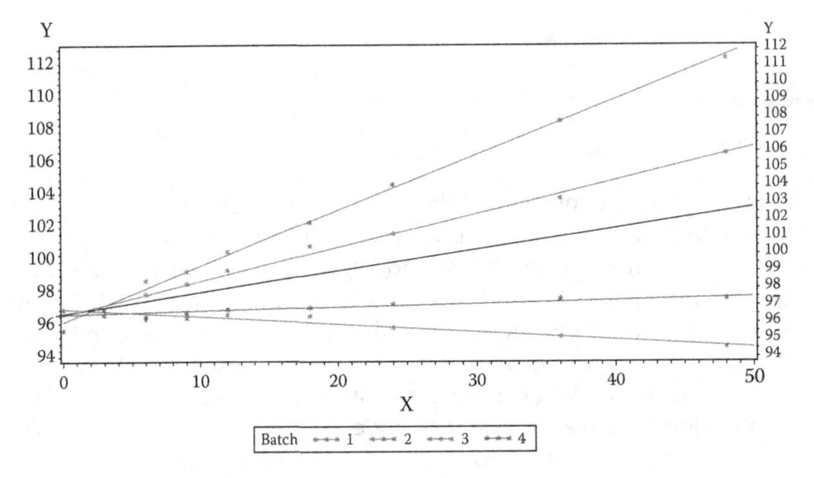

FIGURE 1.3
Plot of multi-batch regression data.

How should these data be modeled? Figure 1.3 shows a plot of the data with linear regression lines superimposed.

The solid line shows the average linear regression over all batches; the dashed lines show the regressions for each individual batch. Inspecting the graph suggests that assuming linear regression is reasonable but assuming a common regression for all batches may not be justified.

Assuming separate linear regressions by batch yields the linear predictor $\beta_{0i}+\beta_{1i}X_{ij}$, where β_{0i} and β_{1i} are, respectively, the intercept and slope for the ith batch, $i=1,2,3,4$, and X_{ij} is the jth value of X for the ith batch. Alternatively, this can be expressed as $\beta_0+b_{0i}+(\beta_1+b_{1i})X_{ij}$, where β_0 and β_1 are the overall intercept and slope (depicted by the solid black line) and b_{0i} and b_{1i} are the batch-specific deviations of the intercept and slope, respectively, from β_0 and β_1.

How we proceed with estimation and inference on b_{0i} and b_{1i} depends on the process giving rise to the batches and the inferences we wish to make. On one hand, the batches could represent a sample of a larger population of batches. In this scenario, we could have sampled any four batches—at least in theory. It follows that b_{0i} and b_{1i} are random variables and, thus, have probability distributions. On the other hand, the four batches in this data set could be the *entire* population. This makes sense if, for example, there are four suppliers of batches and batch i really means "supplier i." This is as opposed to four batches randomly sampled from a single supplier.

If the batches represent a larger population, we would naturally want to focus inference on the population, not just the four batches we observed. We would need to take into account the variance among batches. If the batches *are* the population, we would want to focus inference on batch effects and their impact on regression over X. In this case, we would regard b_{0i} and b_{1i} as model parameters, in the same sense as β_0 and β_1.

The distinction between batches as samples—and hence b_{0i} and b_{1i} as random variables—versus batches as a population—and hence b_{0i} and b_{1i} as model parameters—introduces the concept of *random* versus *fixed* model effects. If batches are samples, b_{0i} and b_{1i} are random effects; if batches are the population and, hence, b_{0i} and b_{1i} are parameters, then b_{0i} and b_{1i} are fixed effects.

Models that contain only fixed effects are called *fixed effects models* (as you would expect). Models that contain both fixed and random effects are called *mixed models*. In theory, the random effects in mixed models can have any plausible distribution. Lee and Nelder (1996) and Lee et al. (2006), for example, discuss doubly generalized linear models, in which random model effects as well as the response variable may have non-Gaussian distributions. In this text, we focus almost exclusively on random effects with Gaussian distributions. We do this for two reasons:

1. Computational methods for mixed models with Gaussian random effects are better developed, so focusing on the Gaussian case will allow us to concentrate on concepts and minimize the risk of computing issues becoming a distraction.
2. The vast majority of practical applications with mixed models assume Gaussian random effect—at least *at the time this textbook is being written.*

> The state of the art in statistical modeling is always advancing. Twenty-five years ago, generalized linear models and linear mixed models would have been considered beyond the state of the art. Now they are commonplace and any literate conversation about modeling must include them. In the future, non-Gaussian random effects will be similarly commonplace. For now, we are not at that point.

To summarize, the multi-batch regression example gives rise to the linear predictor $\beta_0 + b_{0i} + (\beta_1 + b_{1i})X_{ij}$. If you assume fixed batch effects, then β_0, β_1, b_{0i}, and b_{1i} are all model parameters. If you assume random batch effects, then only β_0 and β_1 are model parameters and you must state assumed probability distributions for the random variables b_{0i} and b_{1i}. Commonly (b_{0i}, b_{1i}) pairs from different batches are assumed independent and within each batch, the b_{0i} and b_{1i} pair are bivariate normal, that is,

$$\begin{bmatrix} b_{0i} \\ b_{1i} \end{bmatrix} \sim N\left(\begin{bmatrix} 0 \\ 0 \end{bmatrix}, \begin{bmatrix} \sigma_0^2 & \sigma_{01} \\ \sigma_{01} & \sigma_1^2 \end{bmatrix} \right)$$

where

σ_0^2 and σ_1^2 are the variances of b_{0i} and b_{1i}, respectively

σ_{01} is the covariance between b_{0i} and b_{1i}

Once you decide on the linear predictor, if any effects are random and, if so, their probability assumptions, then you complete the model much as you would in Section 1.2. That is, if the observed *data* have a Gaussian distribution, then the linear predictor is an estimate of the data's expected value, *conditional on the random effects* b_{0i} and b_{1i}. Formally, you state this as

1. *Observations:* $(y_{ij}|b_{0i},b_{1i}) \sim N(\mu_{ij}|b_{0i},b_{1i},\sigma^2)$
2. *Model focus:* $E(y_{ij}|b_{0i},b_{1i}) = \mu_{ij}|b_{0i},b_{1i}$
3. *Linear predictor:* $\beta_0 + b_{0i} + (\beta_1 + b_{1i})X_{ij}$
4. *Assumptions about* b_{0i} and b_{1i} (if random), for example, as shown previously
5. *Link: identity*, that is, $\mu_{ij}|b_{0i},b_{1i}$ modeled by $\beta_0 + b_{0i} + (\beta_1 + b_{1i})X_{ij}$

If you model the variable *Fav* in Table 1.2, then you adjust the distribution of the observations and the link, accordingly. That is,

6. *Observations:* $(Fav_{ij}|b_{0i},b_{1i}) \sim \text{Binomial}[N_{ij},(\pi_{ij}|b_{0i},b_{1i})]$
7. *Model focus:* $E(Fav_{ij}|b_{0i},b_{1i}) = \pi_{ij}|b_{0i},b_{1i}$
8. *Linear predictor:* $\beta_0 + b_{0i} + (\beta_1 + b_{1i})X_{ij}$
9. *Assumptions about* b_{0i} and b_{1i} (if random)
10. *Link: logit or probit*, for example, for logit, $\log[\pi_{ij}/(1-\pi_{ij})]$ modeled by $\beta_0 + b_{0i} + (\beta_1 + b_{1i})X_{ij}$, where π_{ij} is used here as a shorthand for $\pi_{ij}|b_{0i},b_{1i}$

The model assuming Gaussian data is our introduction to *linear models*. If the batch effects are fixed, we refer to it using the acronym LM (for "linear model"). If batch effects are random, it is a *linear mixed model*, whose acronym is LMM. The model assuming binomial data is our introduction to *generalized linear model* (acronym GLM). If batch effects are random, it is a *generalized linear mixed model* (acronym GLMM). Note that the GLMM is the general case; the LM, GLM, and LMM are all special cases. In Section 1.4, you will see how to write these models in matrix form, which is essential for developing the estimation and inference theory in Chapters 4 and 5. In Chapter 2, you will learn techniques to help translate a description of a data set and objectives into a plausible model. In Chapter 3, we will discuss basic concepts of inference and some subtleties essential for you to understand before we begin formal development of estimation and inference theory. At that point, we will be ready to fully develop the underlying theory and methodology needed to work with GLMMs (and, hence, with LM, GLM, and LMM).

KEY IDEAS: *classification vs. direct predictor variable, fixed vs. random model effect, linear model (LM), linear mixed model (LMM), generalized linear model (GLM), generalized linear mixed model (GLMM)*

1.4 Writing Models in Matrix Form

In the previous sections, we derived eight models. Table 1.3 summarizes them.

Note that the first two models on Table 1.3 are models (1.1) and (1.2) introduced in Section 1.1. For convenience, for the remainder of this chapter, we refer to the models in Table 1.3

TABLE 1.3

Complete Description of Models Discussed in This Chapter

Type of Model	Distribution	Linear Predictor	Link
LM	$y_{ij} \sim NI(\mu_i, \sigma^2)$	$\mu + \tau_i$ or equivalently μ_i	Identity: $\eta_i = \mu_i$
LM	$y_{ij} \sim NI(\mu_i, \sigma^2)$	$\beta_0 + \beta_1 X_i$	Identity: $\eta_i = \mu_i$
GLM	$y_{ij} \sim \text{Binomial}(N_{ij}, \pi_i)$	$\mu + \tau_i$ or equivalently μ_i	Logit: $\eta_i = \log[\pi_i/(1-\pi_i)]$ or probit: $\eta_i = \Phi^{-1}(\pi_i)$
GLM	$y_{ij} \sim \text{Binomial}(N_{ij}, \pi_i)$	$\beta_0 + \beta_1 X_i$	Logit or probit
LM	$y_{ij} \sim NI(\mu_{ij}, \sigma^2)$	$\beta_0 + b_{0i} + (\beta_1 + b_{1i})X_{ij}$	Identity: $\eta_{ij} = \mu_{ij}$
GLM	$y_{ij} \sim \text{Binomial}(N_{ij}, \pi_{ij})$	$\beta_0 + b_{0i} + (\beta_1 + b_{1i})X_{ij}$	Logit or probit
LMM	$y_{ij} \mid b_{0i}, b_{1i} \sim NI(\mu_{ij}, \sigma^2)$	$\beta_0 + b_{0i} + (\beta_1 + b_{1i})X_{ij}$ $\begin{bmatrix} b_{0i} \\ b_{1i} \end{bmatrix} \sim N\left(\begin{bmatrix} 0 \\ 0 \end{bmatrix}, \begin{bmatrix} \sigma_0^2 & \sigma_{01} \\ \sigma_{01} & \sigma_1^2 \end{bmatrix}\right)$	Identity: $\eta_{ij} = \mu_{ij}$
GLMM	$y_{ij} \mid b_{0i}, b_{1i} \sim \text{Binomial}(N_{ij}, \pi_{ij})$	$\beta_0 + b_{0i} + (\beta_1 + b_{1i})X_{ij}$ $\begin{bmatrix} b_{0i} \\ b_{1i} \end{bmatrix} \sim N\left(\begin{bmatrix} 0 \\ 0 \end{bmatrix}, \begin{bmatrix} \sigma_0^2 & \sigma_{01} \\ \sigma_{01} & \sigma_1^2 \end{bmatrix}\right)$	Logit or probit

as model 1 through model 8, where the model number reflects the order in which they appear above.

We can express each of these models in matrix form. In fact, we *must* do so for two important reasons:

1. The theoretical development of estimation and inference for these models—presented in Chapters 4 and 5—requires matrix algebra.
2. Statistical computing programs for these models—inside the "black box"—are essentially matrix algebra processors. When you specify a model in a software package (in our case, SAS® PROC GLIMMIX) what you are actually doing is specifying the matrices that define the response variable, its linear predictor, its link, and its probability distribution(s).

The purpose of this section is to show you how. We start with the fixed-effects-only models, and then consider the mixed models.

1.4.1 Fixed-Effects-Only Models

We begin with the first model, the two-treatment LM. Suppose there are three observations on each treatment. We can write the link function corresponding to each observation and their linear predictors as a system of equations:

$$\begin{aligned}
\mu_{11} &= \eta_{11} = \mu + \tau_1 \\
\mu_{12} &= \eta_{12} = \mu + \tau_1 \\
\mu_{13} &= \eta_{13} = \mu + \tau_1 \\
\mu_{21} &= \eta_{21} = \mu + \tau_2 \\
\mu_{22} &= \eta_{22} = \mu + \tau_2 \\
\mu_{23} &= \eta_{23} = \mu + \tau_2
\end{aligned}$$

This, in turn, can be expressed in matrix form as

$$\begin{bmatrix} \eta_{11} \\ \eta_{12} \\ \eta_{13} \\ \eta_{21} \\ \eta_{22} \\ \eta_{23} \end{bmatrix} = \begin{bmatrix} 1 & 1 & 0 \\ 1 & 1 & 0 \\ 1 & 1 & 0 \\ 1 & 0 & 1 \\ 1 & 0 & 1 \\ 1 & 0 & 1 \end{bmatrix} \begin{bmatrix} \mu \\ \tau_1 \\ \tau_2 \end{bmatrix}$$

This shows the general structure of a fixed-effects-only linear model—LM or GLM using the acronyms defined earlier. The general form if the fixed-effects-only linear predictor is

$$\boldsymbol{\eta} = \mathbf{X}\boldsymbol{\beta} \tag{1.3}$$

where

$\boldsymbol{\eta}$ is an $N \times 1$ vector of values of the linear predictors for each of the N observations in the data set

$\mathbf{X}$ is a $N \times p$ matrix, p being the number of parameters in the linear predictor

$\boldsymbol{\beta}$ is a $p \times 1$ vector of fixed effect model parameters

At this point, it is useful to partition $\mathbf{X}\boldsymbol{\beta}$ into components in order to establish a couple of general principles. Specifically,

$$\mathbf{X}\boldsymbol{\beta} = \begin{bmatrix} \mathbf{X}_\mu & \mathbf{X}_\tau \end{bmatrix} \begin{bmatrix} \mu \\ \boldsymbol{\tau} \end{bmatrix} = \mathbf{X}_\mu \mu + \mathbf{X}_\tau \boldsymbol{\tau}$$

where

$$\boldsymbol{\tau} = \begin{bmatrix} \tau_1 \\ \tau_2 \end{bmatrix} \quad \text{and} \quad \mathbf{X}_\tau = \begin{bmatrix} 1 & 0 \\ 1 & 0 \\ 1 & 0 \\ 0 & 1 \\ 0 & 1 \\ 0 & 1 \end{bmatrix}$$

Notice that $\mathbf{X}_\mu = \mathbf{1}_6$, that is, a 6×1 vector of ones. In other words, the $\mathbf{X}$ matrix is partitioned into a "μ" or "intercept part" and a "τ" or "effect part," as is the $\boldsymbol{\beta}$ vector. The intercept part of the $\mathbf{X}$ matrix is always a $N \times 1$ vector of ones and the effects part of the matrix always involves one column per level of the effect—in this case two since there are two treatments. The column for each level of the effect has a one in rows corresponding to that level and zero otherwise—in this example, the first column of $\mathbf{X}_\tau$ has ones in the first three rows (corresponding to observations receiving treatment 1), whereas the second column of $\mathbf{X}_\tau$ has ones in the next three rows (rows 4–6, corresponding to observations receiving treatment 2). Recall that treatment is a *classification* effect—$\mathbf{X}_\tau$ shows how you form the part of the $\mathbf{X}$ matrix for classification effects.

The $\mathbf{X}$ matrix is often called a *design matrix*, particularly when the model consists of classification effects. This is because you can tell exactly what treatments were assigned to what observations by looking at the $\mathbf{X}$ matrix—hence, the $\mathbf{X}$ matrix describes the design. The $\mathbf{X}$ matrix is also called the *derivative matrix*, since each element of $\mathbf{X}$ is the derivative with respect to the corresponding model parameter. That is, each row of $\mathbf{X}$ consists of

$$\begin{bmatrix} \frac{\partial \mathbf{X}_i\boldsymbol{\beta}}{\partial \mu} & \frac{\partial \mathbf{X}_i\boldsymbol{\beta}}{\partial \tau_1} & \frac{\partial \mathbf{X}_i\boldsymbol{\beta}}{\partial \tau_2} \end{bmatrix}$$

where $\mathbf{X}_i$ denotes the ith row of the $\mathbf{X}$ matrix. Thinking of $\mathbf{X}$ as a derivative matrix helps make the transition to *nonlinear* models. Nonlinear models are beyond the scope of this text, although in the last chapters of this book we briefly touch on certain nonlinear modeling tools. However, nonlinear models are an extension of statistical modeling beyond the GLMM.

Now consider the second model, the linear regression over levels of $X_{ij}=0,1,\ldots,10$. As with the first model, this sets up a set of equations

$$\begin{aligned}
\eta_1 &= \beta_0 \\
\eta_2 &= \beta_0 + \beta_1 \\
\eta_3 &= \beta_0 + 2\beta_1 \\
\eta_4 &= \beta_0 + 3\beta_1 \\
\eta_5 &= \beta_0 + 4\beta_1 \\
\eta_6 &= \beta_0 + 5\beta_1 \\
\eta_7 &= \beta_0 + 6\beta_1 \\
\eta_8 &= \beta_0 + 7\beta_1 \\
\eta_9 &= \beta_0 + 8\beta_1 \\
\eta_{10} &= \beta_0 + 9\beta_1 \\
\eta_{11} &= \beta_0 + 10\beta_1
\end{aligned}$$

which can be written in matrix form as

$$\begin{bmatrix} \eta_1 \\ \eta_2 \\ \eta_3 \\ \eta_4 \\ \eta_5 \\ \eta_6 \\ \eta_7 \\ \eta_8 \\ \eta_9 \\ \eta_{10} \\ \eta_{11} \end{bmatrix} = \begin{bmatrix} 1 & 0 \\ 1 & 1 \\ 1 & 2 \\ 1 & 3 \\ 1 & 4 \\ 1 & 5 \\ 1 & 6 \\ 1 & 7 \\ 1 & 8 \\ 1 & 9 \\ 1 & 10 \end{bmatrix} \begin{bmatrix} \beta_0 \\ \beta_1 \end{bmatrix}$$

As with the two treatment model, this has the general form given in Equation 1.3, $\boldsymbol{\eta} = \mathbf{X}\boldsymbol{\beta}$. For the regression model, the **X** matrix can be partitioned into the "intercept part" and the "effect part," that is, $\mathbf{X}\boldsymbol{\beta} = \mathbf{X}_{\beta_0}\boldsymbol{\beta}_0 + \mathbf{X}_{\beta_1}\boldsymbol{\beta}_1$. As before, the intercept part, $\mathbf{X}_{\beta_0}$, is just a vector of ones. In this model, the predictor is a direct variable, so the corresponding column in the **X** matrix, the effect part, $\mathbf{X}_{\beta_1}$, is literally the vector of values, 0 through 10, of the predictor variable.

You set the matrices up the same way for the two-treatment and regression models with binomial data. That is, the set up for model 3 is the same as model 1 and the set up for model 4 is the same as model 2. Only the definition of the link function, η_i, changes.

Now consider the fixed effects, multi-batch models (models 5 and 6 on Table 1.3). Equating the link to the linear predictor yields a set of equations of the form $\eta_{ij} = \beta_0 + b_{0i} + (\beta_1 + b_{1i})X_{ij}$. Recalling that there are four batches, this yields the matrix equation

$$
\begin{bmatrix} \eta_{11} \\ \eta_{12} \\ \eta_{13} \\ \eta_{14} \\ \eta_{15} \\ \eta_{16} \\ \eta_{17} \\ \eta_{18} \\ \eta_{19} \\ \eta_{21} \\ \cdots \\ \eta_{39} \\ \eta_{41} \\ \eta_{42} \\ \eta_{43} \\ \eta_{44} \\ \eta_{45} \\ \eta_{46} \\ \eta_{47} \\ \eta_{48} \\ \eta_{49} \end{bmatrix}
=
\begin{bmatrix} 1 \\ 1 \\ 1 \\ 1 \\ 1 \\ 1 \\ 1 \\ 1 \\ 1 \\ 1 \\ \cdots \\ 1 \\ 1 \\ 1 \\ 1 \\ 1 \\ 1 \\ 1 \\ 1 \\ 1 \\ 1 \end{bmatrix} \beta_0 +
\begin{bmatrix} 1 & 0 & 0 & 0 \\ 1 & 0 & 0 & 0 \\ 1 & 0 & 0 & 0 \\ 1 & 0 & 0 & 0 \\ 1 & 0 & 0 & 0 \\ 1 & 0 & 0 & 0 \\ 1 & 0 & 0 & 0 \\ 1 & 0 & 0 & 0 \\ 1 & 0 & 0 & 0 \\ 0 & 1 & 0 & 0 \\ \cdots & \cdots & \cdots & \cdots \\ 0 & 0 & 1 & 0 \\ 0 & 0 & 0 & 1 \\ 0 & 0 & 0 & 1 \\ 0 & 0 & 0 & 1 \\ 0 & 0 & 0 & 1 \\ 0 & 0 & 0 & 1 \\ 0 & 0 & 0 & 1 \\ 0 & 0 & 0 & 1 \\ 0 & 0 & 0 & 1 \\ 0 & 0 & 0 & 1 \end{bmatrix}
\begin{bmatrix} b_{01} \\ b_{02} \\ b_{03} \\ b_{04} \end{bmatrix} +
\begin{bmatrix} 0 \\ 3 \\ 6 \\ 9 \\ 12 \\ 18 \\ 24 \\ 36 \\ 48 \\ 0 \\ \cdots \\ 48 \\ 0 \\ 3 \\ 6 \\ 9 \\ 12 \\ 18 \\ 24 \\ 36 \\ 48 \end{bmatrix} \beta_1 +
\begin{bmatrix} 0 & 0 & 0 & 0 \\ 3 & 0 & 0 & 0 \\ 6 & 0 & 0 & 0 \\ 9 & 0 & 0 & 0 \\ 12 & 0 & 0 & 0 \\ 18 & 0 & 0 & 0 \\ 24 & 0 & 0 & 0 \\ 36 & 0 & 0 & 0 \\ 48 & 0 & 0 & 0 \\ 0 & 0 & 0 & 0 \\ \cdots & \cdots & \cdots & \cdots \\ 0 & 0 & 48 & 0 \\ 0 & 0 & 0 & 0 \\ 0 & 0 & 0 & 3 \\ 0 & 0 & 0 & 6 \\ 0 & 0 & 0 & 9 \\ 0 & 0 & 0 & 12 \\ 0 & 0 & 0 & 18 \\ 0 & 0 & 0 & 24 \\ 0 & 0 & 0 & 36 \\ 0 & 0 & 0 & 48 \end{bmatrix}
\begin{bmatrix} b_{11} \\ b_{12} \\ b_{13} \\ b_{14} \end{bmatrix}
$$

We can summarize the logic of these matrix equations as follows. The general form of a fixed effects LM or GLM is $\boldsymbol{\eta} = \mathbf{X}\boldsymbol{\beta}$. Partition $\mathbf{X}\boldsymbol{\beta}$ as

$$\begin{bmatrix} \mathbf{X}_{\beta_0} & \mathbf{X}_{b_0} & \mathbf{X}_{\beta_1} & \mathbf{X}_{b_1} \end{bmatrix} \begin{bmatrix} \beta_0 \\ \mathbf{b}_0 \\ \beta_1 \\ \mathbf{b}_1 \end{bmatrix} = \mathbf{X}_{\beta_0}\beta_0 + \mathbf{X}_{b_0}\mathbf{b}_0 + \mathbf{X}_{\beta_1}\beta_1 + \mathbf{X}_{b_1}\mathbf{b}_1$$

where $\mathbf{X}_{\beta 1}$ is an $N \times 1$ vector of ones, $\mathbf{X}_{b_0}$ is an $N \times 4$ matrix whose ijth row–column element is 1 if the corresponding observation is in the jth batch and 0 otherwise,

$$\mathbf{b}_0 = \begin{bmatrix} b_{01} \\ b_{02} \\ b_{03} \\ b_{04} \end{bmatrix}$$

$\mathbf{X}_{\beta 1}$ is an $N \times 1$ vector whose elements equal the value of the covariate X_{ij} for the corresponding observation, $\mathbf{X}_{\beta_1}$ is an $N \times 4$ vectors whose elements are X_{ij} if the corresponding observations is in the jth batch and zero otherwise, and $\mathbf{b}_1$ is a 4×1 vector whose elements are the b_{1i}; $i = 1,2,3,4$ for the four batches. Notice that the matrix $\mathbf{X}_{b_1}$ is the horizontal direct product of $\mathbf{X}_{b_0}$ and $\mathbf{X}_{\beta_1}$. Also notice that the predictor variable X_{ij} is a direct variable; hence, the columns in the $\mathbf{X}$ matrix are vectors consisting of literal values of X_{ij}, whereas batch is a classification variable, and hence, it has one column per batch and the columns have nonzero values only if the ijth observation is in the batch corresponding to that column.

This model contains essentially a microcosm of the major decisions involved setting up the matrix form. The intercept parameter is always a scalar and its corresponding element in the $\mathbf{X}$ matrix is a column vector of ones. Direct variables always have scalar parameters and their corresponding element in the $\mathbf{X}$ matrix is a column vector of the literal values of the direct variable. Classification variables have one parameter per effect level and hence one column per effect level in the $\mathbf{X}$ matrix, whose elements are one if the corresponding observation is in that level, zero otherwise. The $\mathbf{X}_{b_1}$ component of the $\mathbf{X}$ matrix and the $\mathbf{b}_1$ vector have the characteristics of classification and direct variables; hence, $\mathbf{X}_{b_1}$ is constructed as a direct product of the elements, in this case, $\mathbf{X}_{b0}$ and $\mathbf{X}_{\beta 1}$ from which it is composed.

To summarize, all fixed-effects-only models share the general form $\boldsymbol{\eta} = \mathbf{X}\boldsymbol{\beta}$. All probability statements reside in the distribution of the observation vector $\mathbf{y}$. A complete specification of a fixed-effects-only model is thus:

- $\mathbf{y} \sim$ some distribution with $E(\mathbf{y}) = \boldsymbol{\mu}$ and $\text{Var}(\mathbf{y}) = \mathbf{R}$
- The link function: $\boldsymbol{\eta} = g(\boldsymbol{\mu})$
- The linear predictor: $\boldsymbol{\eta} = \mathbf{X}\boldsymbol{\beta}$

We now turn to the multi-batch models with batches regarded as random effects.

1.4.2 Mixed Models: Models with Fixed and Random Effects

The last two models in Table 1.3 are mixed models. The linear predictor, $\beta_0+b_{0i}+(\beta+b_{1i})X_{ij}$, appears to be exactly the same as the linear predictor for models 5 and 6 discussed just above. The main difference is that in models 5 and 6, the effects associated with batches, b_{0i} and b_{1i}, are fixed parameters whereas in their model 7 and 8 forms, they have probability distributions. In matrix notation, the components of $\mathbf{X\beta}$ associated with random effects are removed from $\mathbf{X\beta}$ and placed in a new element, denoted $\mathbf{Zb}$, where the vector $\mathbf{b}$ consists of the random model effects.

Formally, the linear predictor is $\boldsymbol{\eta}=\mathbf{X\beta}+\mathbf{Zb}$, where $\mathbf{b}\sim N(\mathbf{0},\mathbf{G})$. For models 7 and 8, the fixed effects component of the linear predictor is $\beta_{0i}+\beta_{1i}X_{ij}$. Hence, its matrix form is

$$\begin{bmatrix}\eta_{11}\\ \eta_{12}\\ \eta_{13}\\ \eta_{14}\\ \eta_{15}\\ \eta_{16}\\ \eta_{17}\\ \eta_{18}\\ \eta_{19}\\ \eta_{21}\\ \cdots\\ \eta_{39}\\ \eta_{41}\\ \eta_{42}\\ \eta_{43}\\ \eta_{44}\\ \eta_{45}\\ \eta_{46}\\ \eta_{47}\\ \eta_{48}\\ \eta_{49}\end{bmatrix}=\begin{bmatrix}1\\1\\1\\1\\1\\1\\1\\1\\1\\1\\ \cdots\\1\\1\\1\\1\\1\\1\\1\\1\\1\\1\end{bmatrix}\beta_0+\begin{bmatrix}0\\3\\6\\9\\12\\18\\24\\36\\48\\0\\ \cdots\\48\\0\\3\\6\\9\\12\\18\\24\\36\\48\end{bmatrix}\beta_1$$

or

$$\begin{bmatrix}\mathbf{X}_{\beta_0} & \mathbf{X}_{\beta_1}\end{bmatrix}\begin{bmatrix}\beta_0\\ \beta_1\end{bmatrix}=\mathbf{X}_{\beta_0}\beta_0+\mathbf{X}_{\beta_1}\beta_1$$

where all terms are as defined previously. The remaining elements of $\mathbf{X\beta}$ from models 5 and 6 become $\mathbf{Zb}$ for models 7 and 8, that is,

$$\mathbf{Zb} = \begin{bmatrix} 1 & 0 & 0 & 0 \\ 1 & 0 & 0 & 0 \\ 1 & 0 & 0 & 0 \\ 1 & 0 & 0 & 0 \\ 1 & 0 & 0 & 0 \\ 1 & 0 & 0 & 0 \\ 1 & 0 & 0 & 0 \\ 1 & 0 & 0 & 0 \\ 1 & 0 & 0 & 0 \\ 0 & 1 & 0 & 0 \\ \cdots & \cdots & \cdots & \cdots \\ 0 & 0 & 1 & 0 \\ 0 & 0 & 0 & 1 \\ 0 & 0 & 0 & 1 \\ 0 & 0 & 0 & 1 \\ 0 & 0 & 0 & 1 \\ 0 & 0 & 0 & 1 \\ 0 & 0 & 0 & 1 \\ 0 & 0 & 0 & 1 \\ 0 & 0 & 0 & 1 \\ 0 & 0 & 0 & 1 \end{bmatrix} \begin{bmatrix} b_{01} \\ b_{02} \\ b_{03} \\ b_{04} \end{bmatrix} + \begin{bmatrix} 0 & 0 & 0 & 0 \\ 3 & 0 & 0 & 0 \\ 6 & 0 & 0 & 0 \\ 9 & 0 & 0 & 0 \\ 12 & 0 & 0 & 0 \\ 18 & 0 & 0 & 0 \\ 24 & 0 & 0 & 0 \\ 36 & 0 & 0 & 0 \\ 48 & 0 & 0 & 0 \\ 0 & 0 & 0 & 0 \\ \cdots & \cdots & \cdots & \cdots \\ 0 & 0 & 48 & 0 \\ 0 & 0 & 0 & 0 \\ 0 & 0 & 0 & 3 \\ 0 & 0 & 0 & 6 \\ 0 & 0 & 0 & 9 \\ 0 & 0 & 0 & 12 \\ 0 & 0 & 0 & 18 \\ 0 & 0 & 0 & 24 \\ 0 & 0 & 0 & 36 \\ 0 & 0 & 0 & 48 \end{bmatrix} \begin{bmatrix} b_{11} \\ b_{12} \\ b_{13} \\ b_{14} \end{bmatrix} = \begin{bmatrix} \mathbf{Z}_{b_0} & \mathbf{Z}_{b_1} \end{bmatrix} \begin{bmatrix} \mathbf{b}_0 \\ \mathbf{b}_1 \end{bmatrix}$$

where

$$\mathbf{b} = \begin{bmatrix} \mathbf{b}_0 \\ \mathbf{b}_1 \end{bmatrix} \sim N\left(\begin{bmatrix} 0 \\ 0 \end{bmatrix}, \begin{bmatrix} \mathbf{I}\sigma_0^2 & \mathbf{I}\sigma_{01} \\ \mathbf{I}\sigma_{01} & \mathbf{I}\sigma_1^2 \end{bmatrix} \right)$$

Notice that the elements of the **Z** matrix, $\mathbf{Z}_{b_0}$ and $\mathbf{Z}_{b_1}$, are constructed exactly as $\mathbf{X}_{b_0}$ and $\mathbf{X}_{b_1}$ were in models 5 and 6. All that changes is that, since they are random effects in models 7 and 8, they are placed in the **Z** matrix, their corresponding model effects are placed in **b**, the vector of random effects, and you must state a probability distribution for **b**.

To summarize, all mixed models share the general form $\boldsymbol{\eta} = \mathbf{X}\boldsymbol{\beta} + \mathbf{Zb}$. Probability statements apply both to the distribution of the observation vector, $\mathbf{y}|\mathbf{b}$, and to the random effects vector, **b**. Notice that in a mixed model, the observations are conditional on the random model effects. A look at several examples in Chapter 2 will make this clearer.

A complete specification of a mixed model is thus:

- Observation, conditional on the random effects: $\mathbf{y}|\mathbf{b} \sim$ some distribution with $E(\mathbf{y}|\mathbf{b}) = \boldsymbol{\mu}|\mathbf{b}$ and $\text{Var}(\mathbf{y}|\mathbf{b}) = \mathbf{R}$. Often, for convenience, we will use the shorthand $E(\mathbf{y}|\mathbf{b}) = \boldsymbol{\mu}$ to refer to the conditional mean of the observations.
- The link function: $\boldsymbol{\eta} = g(\boldsymbol{\mu}|\mathbf{b})$.
- The linear predictor: $\boldsymbol{\eta} = \mathbf{X}\boldsymbol{\beta} + \mathbf{Zb}$, where $\mathbf{b} \sim N(\mathbf{0},\mathbf{G})$.

KEY IDEAS: *matrix form of the model, how* **X** *matrix and* **β** *vector formed for fixed effects model, impact of classification vs. direct variable on matrix formation, what goes in* **Xβ** *vs.* **Zb** *in mixed model*

1.5 Summary: Essential Elements for a Complete Statement of the Model

Table 1.4 summarizes the types of models we have considered in this chapter.

To repeat a point made earlier, notice that the GLM, LM, and LMM are all special cases of the GLMM. This will be important to keep in mind in Chapters 4 and 5. These chapters develop linear model estimation and inference theory and methodology. They focus on the GLMM because the theory and methods for the GLMM apply to GLM, LM, and LMM as well.

As a historical note, our development in Chapters 4 and 5 is opposite the development of linear models theory and methods found in "traditional" texts, which typically start with the LM and then add layers of complication. To understand the rationale, we need to briefly review the history of ideas in linear modeling.

Linear models have undergone considerable development over the past three decades. The decade of the 1970s is a good point of comparison between the approach of this text and what we refer to as "more traditional" approaches. In the 1970s, the first widely accessible and genuinely comprehensive statistical software for linear models became available. Also during the decade, the classic textbooks were written. These texts used matrix algebra as a basis for linear model theory and methods; they focused heavily on quadratic forms and their associated distribution theory. During this period, what we call the LM in this text was presented in *model equation* rather than *probability distribution* form and it was referred to as "the General Linear Model." That is, the "General Linear Model" was defined as $\mathbf{y}=\mathbf{X}\boldsymbol{\beta}+\mathbf{e}$, where $\mathbf{e}\sim N(\mathbf{0}, \mathbf{I}\sigma^2)$. SAS developed a procedure, PROC GLM, that implemented the "General Linear Model." This SAS procedure now causes confusion, because it actually applies only to "LM" as we use the term, not *true* "GLMs"—*generalized* linear models as we refer to them in this text.

The "General Linear Model" has long been known not to be the last word on linear modeling. LMM and GLM tools have actually been around a long time.

Recovery of inter-block information—which is LMM analysis assuming random block effects—was introduced along with incomplete block designs in the 1930s. Variance

TABLE 1.4

Typology of Linear Models

Type of Model	Observations	Link	Linear Predictor	Mean Modeled by[a]
GLMM	$\mathbf{y}\mid\mathbf{b}\sim G(\boldsymbol{\mu},\mathbf{R})$ G denotes general distribution	$\boldsymbol{\eta}=g(\boldsymbol{\mu}\mid\mathbf{b})$	$\mathbf{X}\boldsymbol{\beta}+\mathbf{Z}\mathbf{b}$ $\mathbf{b}\sim N(\mathbf{0},\mathbf{G})$	$\hat{\boldsymbol{\mu}}=h(\hat{\boldsymbol{\eta}})=h(\mathbf{X}\hat{\boldsymbol{\beta}}+\mathbf{Z}\hat{\mathbf{b}})$
LMM	$\mathbf{y}\mid\mathbf{b}\sim N(\boldsymbol{\mu},\mathbf{R})$	Identity: $\boldsymbol{\eta}=\boldsymbol{\mu}$	$\mathbf{X}\boldsymbol{\beta}+\mathbf{Z}\mathbf{b}$ $\mathbf{b}\sim N(\mathbf{0},\mathbf{G})$	$\hat{\boldsymbol{\mu}}=\mathbf{X}\hat{\boldsymbol{\beta}}+\mathbf{Z}\hat{\mathbf{b}}$
GLM	$\mathbf{y}\sim G(\boldsymbol{\mu},\mathbf{R})$	$\boldsymbol{\eta}=g(\boldsymbol{\mu})$	$\mathbf{X}\boldsymbol{\beta}$	$\hat{\boldsymbol{\mu}}=h(\hat{\boldsymbol{\eta}})=h(\mathbf{X}\hat{\boldsymbol{\beta}})$
LM	$\mathbf{y}\sim N(\boldsymbol{\mu},\mathbf{R})$	Identity: $\boldsymbol{\eta}=\boldsymbol{\mu}$	$\mathbf{X}\boldsymbol{\beta}$	$\hat{\boldsymbol{\mu}}=\mathbf{X}\hat{\boldsymbol{\beta}}$

[a] $h(\boldsymbol{\eta})$ for GLMM and GLM denotes *inverse link* function, for example, $h(\eta)=e^{\eta}/(1+e^{\eta})$ for logit link.

component estimation has been a part of linear modeling since the 1940s. Use of multiple error terms (e.g., for split-plot experiments), an LMM procedure, was an extension of analysis of variance (ANOVA) that appeared shortly after ANOVA's introduction in the 1920s and has been a standard statistical tool ever since. Econometricians and geostatistics developed correlated error LMMs for specialized applications long before PROC GLM appeared. Harville (1976) published a general, unifying theory of the LMM.

On the GLM side, logistic and probit regression models, similar to our binomial regression example in Section 1.2, were in common use in the 1950s. Log-linear models for contingency tables were introduced in the 1960s. Survival analysis methods appeared about this time. Nelder and Wedderburn (1972) published a unifying theory of GLMs.

In the 1980s and 1990s two major developmental threads occurred simultaneously—in fact, drove and reinforced one another—to change our perspective on linear models. First, computers became smaller, faster, more capable, and more available—dramatically so. LMM and GLM computations, which were prohibitive by 1970s standards, became "easy" by 1990 (and certainly by 2010) standards. Paralleling the development of computers, GLM and LMM theory intermingled, so that by the mid-1990s a comprehensive theory of GLMMs had appeared in the literature, for example, Breslow and Clayton (1993). By the mid-2000s, comprehensive software to implement GLMMs, using syntax familiar to anyone with knowledge of LMs, had become widely available.

From a 2010 perspective, it seems quaint to refer to LM—especially LM where Var($\mathbf{e}$) is restricted to be $\mathbf{I}\sigma^2$—as the "General" linear model. By contemporary standards, LM is not at all "general" and it is misleading to refer to it as such. In addition, the *model equation* forms—$\mathbf{y}=\mathbf{X}\boldsymbol{\beta}+\mathbf{e}$ for fixed-effects-only models and $\mathbf{y}=\mathbf{X}\boldsymbol{\beta}+\mathbf{Z}\mathbf{b}+\mathbf{e}$ for mixed models—as we saw in Section 1.2 are useless and misleading for GLMs and GLMMs.

Hence, our approach in this text: Rather than developing an elaborate theory and methodology for working with $\mathbf{y}=\mathbf{X}\boldsymbol{\beta}+\mathbf{e}$—models in their LM model equation form—only to start over when we get to GLMs, LMMs, and GLMMs, we start with the general case from the outset.

In summary, the essential elements for a model are those given in Table 1.4. Any statement of a linear model must be given in such a way that the elements in each column of Table 1.4 are spelled out explicitly. Any statement of the model lacking any one of these elements is considered incomplete.

In Chapter 2, we will learn how to translate descriptions of data sets into plausible model statements. We actually began this process in Sections 1.2 and 1.3 of this chapter, but we will take it to a much higher level in Chapter 2. Then, in Chapter 3, we will begin to develop the concepts and tools needed for the theory and methodology to work with these models.

Exercises

1.1 Consider the two-treatment *paired* comparison. Consider two possible scenarios:

A. The response variable is continuous and can be assumed to have a Gaussian distribution.

B. For the *i*th treatment ($i=1,2$) on the *j*th pair ($j=1,2,\ldots,p$), N_{ij} observations are taken. Each observation has either a *favorable* or *unfavorable* outcome. Denote y_{ij} as the number of favorable outcomes observed on the *ij*th pair.

For both A and B do the following:

i. Write a complete specification of the model, including all required elements.
ii. Write your model from (i) in matrix form. Include matrix specification of the covariance assumptions for any random model effects.
iii. Given that the objective of a paired comparison is to find out if there is a difference between treatments, give a *formal* statement—*in terms of the model parameters you defined in (i)*—of the null hypothesis you would test to address the objective.

Suggested in class implementation of this exercise:

1. Divide into pairs. Select one person in your pair to be "A" and the other to be "B."
2. *Partner A*: Describe your work on Scenario A, as mentioned previously, to your partner and have your partner peer review your work.
3. *Partner B*: Present your solution to Scenario B and have your partner peer review.
4. *Both*: After peer review, rewrite your proposed answers to (i), (ii), and (iii) mentioned earlier, incorporating revisions suggested by your peer review that you think are warranted.
5. Once you are both happy with your work on the two scenarios, turn in your written work.
6. Take *at most* 15 min on all of the aforementioned. Presentation and peer review for each scenario should take 5 min *max*. Revision should take 5 min *max*.
7. For peer review, focus on
   a. Is the model clearly stated?
   b. Are all terms defined?
   c. Are all required elements given clearly?
   d. Are assumptions stated?
   e. Are the answers *concise* (no wasted words, no memory dump ...)

1.2 Similar to multi-batch scenario shown in Figure 1.2, except there are two treatments. Three batches are assigned (randomly, of course!) to each treatment. As with the Figure 1.2 scenario, assume that changes over time are linear. Consider the following response variable cases:

A. Response variable Gaussian
B. Response variable is y_{ijk} favorable outcomes out of N_{ijk} observations on the *ijk*th treatment-batch-time combination
C. Response variable is count, assumed to have a Poisson distribution

For each scenario:

i. Write a complete specification of the model, including all required elements.
ii. Suppose the linear response over time corresponds to a degradation of the material (e.g., food, drug, pesticide) over time (e.g., it loses effectiveness, there is a buildup of toxins, etc.). The researchers want to know "which treatment is better?" In terms of the model parameters from (i), write a formal null hypothesis that gives this question a statistically tractable operating definition and will answer the researchers' question.

You only need to do part (ii) once. Unlike the model specification, your answer here should be equally applicable to each response variable scenario (if you did part (i) correctly ...)

Hint: your "hypothesis" may actually be a sequence of hypotheses and an associated protocol for working through them to answer the research question. Do this as parsimoniously as possible and explain it as clearly and *concisely* as possible. If your answer takes more than half a page, you are not being parsimonious, clear, or concise.

2

Design Matters

2.1 Introductory Ideas for Translating Design and Objectives into Models

In Chapter 1, we saw that completely specifying a statistical model requires

- A linear predictor, $\mathbf{X}\boldsymbol{\beta} + \mathbf{Zb}$
- The distribution of the random model effects, $\mathbf{b} \sim N(\mathbf{0}, \mathbf{G})$
- The distribution of the data, conditional on the random effects, $\mathbf{y}|\mathbf{b}$
- The link function, $g(\boldsymbol{\mu}|\mathbf{b}) = \mathbf{X}\boldsymbol{\beta} + \mathbf{Zb}$

Additionally, in Chapter 1 we saw that the model serves two purposes. First, it describes a plausible process giving rise to the observations. Second, it does so in a way that enables the hypothesis testing or interval estimation needed to address the objectives that motivated collecting the data being modeled.

In this chapter, we consider the all important bridge between the way a given study is organized and data are collected and the resulting model. *Design*, broadly understood, means the organization and data collection protocol of a study. Design, as we use the term in this book, may apply to formally designed experiments, but we also intend the term to include survey design, quasi-experimental design, etc. We call this chapter "Design Matters" because translating the study design into a statistical model is perhaps *the* most important facet in the practice of statistical modeling. It may also be the most undervalued and overlooked. The purpose of this chapter is to define design terminology and concepts crucial to statistical modeling and to present techniques to help you accurately translate study design to statistical model.

Most modeling issues are in reality "poor understanding of design" issues. Among the major causes of modeling problems are the following:

- (Unfortunate) decisions in constructing the linear predictor.
- (Unfortunate) decisions about which model effects are fixed and which are random.
- (Unfortunate) decisions about the probability distribution of the observations and—to a lesser extent—the random model effects.
- Mismatch between the model specification and software instructions.
- Not knowing what to do with parameter estimates once obtained. Rarely is the objective merely to estimate the model parameters. Usually, objectives stated in "discipline-speak" (the language of the subject matter motivating the study) translate to "stat-speak": linear combinations of model parameters for interval estimation or hypothesis testing.

In this chapter, we focus on constructing the linear predictor, distinguishing between fixed and random effects, making decisions about the distribution of the observations, and writing software instructions that accurately describe the model you intent to fit. The last point, "what to do with parameter estimates once obtained," we leave to Chapter 3, which covers inference issues in detail.

2.1.1 Chapter Organization

Section 2.2 covers essential preliminaries for model construction—picturing the study design and essential concepts and terminology. Section 2.3 presents techniques for translating the study design into a linear predictor. This involves returning to modeling's roots in the analysis of variance, but reenvisioning the analysis of variance (ANOVA) thought process in a GLMM-appropriate way. Section 2.4 focuses on distribution issues, including assigning components of the linear predictor to the fixed ($\mathbf{X\beta}$) or random ($\mathbf{Zb}$) part and guidelines for the distribution of the observations—technically, the distribution of $\mathbf{y}|\mathbf{b}$. Section 2.5 extends the concepts developed in previous sections to more complex designs. Appendix 2.B, subtitled "How GLIMMIX thinks," concerns writing software instructions. While the appendix specifically addresses SAS, the larger lessons—that software commands are fundamentally matrix operators, and for any software you must know how your commands are translated into model components—are generally applicable.

Two comments before we proceed. First, in many modeling situations, there is no "one and only one" correct $\mathbf{X\beta}$. Often we have a choice of reasonable alternatives. However, among our choices, we generally find that one alternative better enables us to address the objectives of the study, whereas other choices, while providing technically accurate description of the data, fail in some sense to clarify adequately. For example, recall the example shown in Table 1.1 with observations at times 0, 3, 6, 9, 12, 18, 24, 36, and 48. The objective was to estimate the linear change in response over months. Two possible linear predictors are $\beta_0 + \beta_1 X_i$, where X_i is the time (i.e., $X_1 = 0$, $X_2 = 3, \ldots, X_9 = 48$) and $\mu + \tau_i$, $i = 1, 2, \ldots, 9$. The former defines time as a direct linear regression effect. The latter defines time as a class effect. As descriptions of the data, both are technically correct. However, the former allows a straightforward answer to the research question; the latter obscures things and makes interpretation needlessly difficult.

Second, this chapter may at times look like it belongs in a design textbook and you may say to yourself, "I thought this was supposed to be a textbook about *modeling*." Strictly speaking, you *can* study statistical models with limited knowledge of the thought process underlying design of experiments (and, for that matter, probability distribution theory). However, just because you *can* does not mean you *should*. If you do, you will be working by rote and by cookbook. Working knowledge of both design and probability theory are essential for the work we do in this book. This chapter focuses on design matters—even for models that do not involve designed experiments. Experience shows that the most serious errors in statistical modeling—models that make no sense given the data architecture or in view of the objectives—almost invariably result from a deficient understanding of design. This is true even for data not from designed experiments. Even surveys and "quasi-experiments" must exploit design principles in order to obtain believable answers for the questions asked of the data. As a result, this chapter will necessarily draw on design concepts. Fair warning. Let us now proceed.

2.2 Describing "Data Architecture" to Facilitate Model Specification

To see how this process works, let us consider a couple of examples.

First, return to the two-treatment comparison discussed in Chapter 1. Most introductory statistics classes discuss two ways to collect data for this comparison: two independent samples or paired observations with each treatment observed on one member of the pair. In Chapter 1, we considered the independent sample case and saw that the linear predictor is $\eta + \tau_i = \eta_i$. For the paired comparison, the linear predictor is $\eta + \tau_i + p_j = \eta_i + p_j$, where p_j is the effect of the jth pair.

For the second example, let us consider something more complex. Suppose that a school district is implementing a professional development program, but before requiring every school in the district to participate, it wants evidence of the program's effectiveness. To find out, it conducts a study, set up as follows. Ten schools are selected at random from the schools in the district (so we know this is a fairly large district). In each school, four teachers are selected. Assume for simplicity that they teach the same grade and same subjects. Two of these teachers participate in the professional development program, two do not. At the end of the study, student gains in achievement are measured (if you know something about designing studies to assess professional development programs and their impact on student achievement, undoubtedly you see major flaws in the plan for this study. For this example, suspend judgment—our purpose here is to work through the design architecture to construct a linear predictor).

How do we translate this into a linear predictor? In order to approach this question, we need to put some basic concepts and language in place.

2.2.1 Every Data Set Has a "Plot Plan"

We start by visualizing the study design. A picture is worth a thousand words. When mountaineers describe their route to the summit, there is always one part of the route they call "the crux"—the most difficult part of the climb. If you can negotiate the crux, you can make it to the summit. In translating from data and objectives to model, the crux of the process is making sure linear predictor accurately describes the study design's impact on the data you observe. To negotiate the crux, pictures can be your most valuable tool.

In visualizing the study design, we borrow from the design of agricultural experiments. When agricultural researchers do experiments in the field, they sketch the layout of their experiment in what they call a "plot plan." The plot plan is simply a map showing where all the observations are located, where the treatment factor levels are to be applied, if there is any pairing or blocking, etc. There is nothing specific to agriculture in a plot plan. You can draw a "plot plan" for any data set, even if it has nothing whatsoever to do with agriculture.

For the school example we have 10 schools, so we draw a rectangle for each school. Within each school, we selected four teachers. We represent each teacher with a box—drawn within each school rectangle, one per teacher. Two of these teachers were selected to participate in the profession development program. Label two of the boxes "participant." Label the other two "nonparticipant." You now have a plot plan, shown in Figure 2.1.

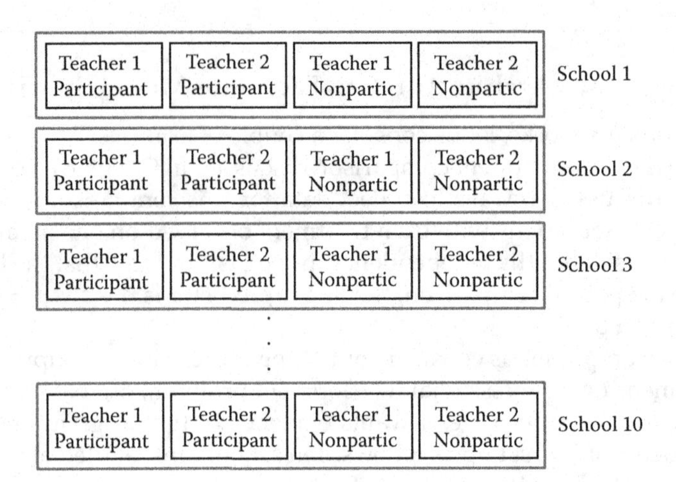

FIGURE 2.1
Plot plan for school professional development study.

2.2.2 Terminology for Treatment and Design Structure

Now that we have a plot plan, we need language to describe the features relevant to constructing the linear predictor. These are as follows:

- *Factor*: A factor is an independent variable whose impact on the response we want to estimate or test. In the two-treatment case, independent sampling or paired, the factor is treatment. In the linear regression examples in Chapter 1, the factor was the X_i variable used in the regression equation. In the school example, the factor is the professional development program. Factors can be *classification* or *direct* variables.
- *Level*: The levels are the specific categories (for classification variables) or amounts (for direct variables) of the factor. In the two-treatment case, the levels are "treatment 1" and "treatment 2." In the multi-batch linear regression example, the levels are $X = 0, 3, 6, 9, 12, 18, 24, 36, 48$. In the school example, the levels are "participant" or "nonparticipant" in the professional development program.
- *Unit of replication*: Alternatively, the *unit of randomization*. In design of experiments, this is the "*experimental unit*." In the social sciences, linear mixed models are called "hierarchical linear models" or "multilevel" models. In "hierarchical-linear-model" terminology, "level" does not mean "level" as we used it earlier, but rather it corresponds to "unit of replication." The unit of replication is the smallest entity that is independently assigned a factor level, the key phrase being "independently assigned." In a formally designed experiment, we achieve independent assignment of units to factor levels through randomization. In an observational study, we cannot assign units to factor levels; nature does the assigning. From a modeling perspective, the resulting linear predictor is similar. In the two-treatment case, if an independent sample is used, the individuals assigned to a given treatment are the units. For the paired comparison, the unit of replication is a single member of the pair. In the school example, the unit of replication is the teacher—it is the teacher that is randomly assigned to participate or not.

Studies can have more than one factor. In multifactor experiments, different factors can have different sized units of replication associated with them. We will discuss multifactor models in Sections 2.3 and 2.5. Also note that observations are not necessarily replications. In the school example, we test individual students, but individual students are not the unit of replication with respect to the professional development factor. This distinction generates additional terminology:

- *Unit of observation*: In design of experiments, this is called the "sampling unit." In "hierarchical-linear-model" language, the unit of observation is the "level 1" observation. In some cases, such as the two-treatment paired design, the units of replication and observation are identical. In others, like the school example, they are not: the teacher (or equivalently, the teacher's classroom) is the unit of replication; the student is the unit of observation. Notice that units of observation are always contained within units of replication.
- *Blocking or clustering or stratification criterion*: In some studies, units of replication are grouped together—ideally because they are similar in some relevant way. In design of experiments, these groups are called *blocks* (pairing is a special case—pairing is just blocking when there are exactly two factor levels). In survey sampling, these groups may be *strata* or they may be *clusters*. The terminology is different, but from a modeling perspective, the idea is the same. Unlike factors, levels, units of replication and units of observation, which are always present and must be identified in any modeling exercise, blocking is optional. If it is used, we must account for it in the model, but only if blocking is actually part of the study design. In the two-treatment case, the independent sample involves no blocking. Hence no term for block appears in the linear predictor. The paired case does involve blocking, so a term for block must appear in the linear predictor. In the multi-batch linear regression example in Chapter 1, batch is the blocking criterion, and it appears in the model in two places: b_{0i} and b_{1i}, the effects of the ith batch on intercept and slope, respectively. In the school example, teachers are grouped by school, so school is the blocking criterion.

2.2.3 Nested and Cross-Classification: Alternative Ways of Organizing Treatment and Design Structure

Before we move on to constructing a linear predictor from the plot plan and its features, we need to say something about two important ways of organizing treatment factors, blocks, and units of replication and observations: *nesting* and *cross-classification*.

To illustrate, suppose we organized the school example using a slightly different design. Ten schools were randomly sampled. Five were assigned at random to participate in the professional development program. The other five served as nonparticipating controls. Two teachers were selected at random from each school and their students measured for "gain in achievement." The resulting plot plan is shown in Figure 2.2.

Notice two things about Figure 2.2. First, there is no blocking. Schools are assigned to the professional development treatment in the same manner as a two-treatment independent sample. Second, the unit of replication is now the school.

The difference in study organization shown in Figures 2.1 and 2.2 illustrates the distinction between a cross-classification and a nested classification. In Figure 2.1, school and participation are combined so that both levels of the professional development are observed at all ten schools. In Figure 2.2, either a school participates in the professional development

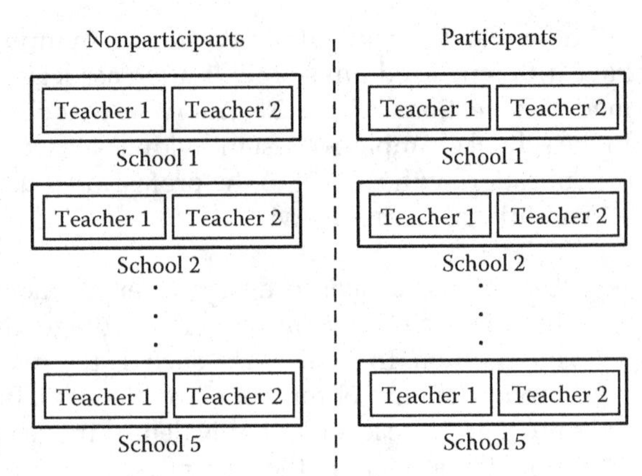

FIGURE 2.2
Alternative plot plan for school professional development study.

or it does not. A school cannot simultaneously be a participating and a nonparticipating school: the five schools in each group are completely separate. School 1 in the participating group is not the same school (indeed, *cannot* be the same school) as school 1 in the nonparticipating group.

If we sketch a two-way table to show schools observed in conjunction with each professional development factor level, Figure 2.1 yields

Professional Development Factor Level	School									
	1	2	3	4	5	6	7	8	9	10
Participating	✓	✓	✓	✓	✓	✓	✓	✓	✓	✓
Nonparticipating	✓	✓	✓	✓	✓	✓	✓	✓	✓	✓

The check mark (✓) shows where an observation occurs.

By contrast, Figure 2.2 yields

Professional Development Factor Level	School									
	1(P) (1)	2(P) (2)	3(P) (3)	4(P) (4)	5(P) (5)	1(N) (6)	2(N) (7)	3(N) (8)	4(N) (9)	5(N) (10)
Participating	✓	✓	✓	✓	✓					
Nonparticipating						✓	✓	✓	✓	✓

School 1(P) refers to school 1 in the participating group in Figure 2.2. School 1(N) refers to school 1 in the nonparticipating group. The numbers in parentheses (e.g., (1) for school 1(P) and (6) for school 1(N)) indicate the equivalent school in Figure 2.1.

Figure 2.1 illustrates a cross-classification pattern. Cross-classification is defined as a pattern such that all possible combinations of levels of two or more factors may occur in the design and hence in the data set. Figure 2.2 illustrates a nesting pattern. Nesting is defined as a pattern such that if a level of one factor is observed in combination with a level of another factor, it cannot be observed in conjunction with any other level of that factor.

KEY IDEAS: *plot plan, cross-classification, nested classification*

2.3 From Plot Plan to Linear Predictor

With the plot plan and terminology from Section 2.2 in mind, we are now in a position to assemble a linear predictor for the school example. In this section, we present two techniques. The first lists the blocking criteria and treatment effects with their associated units of replication and observations, similar to an approach used in *SAS for Mixed Models*, 2nd edn. (Littell et al., 2006). We call this process *What's the Unit of Replication*? The second approach returns to modeling's ANOVA roots (Fisher and Mackenzie, 1923) to reframe the ANOVA thought process adapted to the generalized linear mixed setting. We call this process *What Would Fisher Do*? or *WWFD*?

2.3.1 Unit of Replication Approach

Table 2.1 shows the template of a table. Once you complete this table, the linear predictor virtually writes itself.

Table 2.2 shows how you fill in Table 2.1 for the school example.

The linear predictor needs to address each item in the blocking and factor rows—that is, the blocking criterion (if any), the factor, *and* its unit of replication. The *unit of replication is* handled when the probability distribution is specified—*this is a subsequent step, not part of the linear predictor.*

Following the items in Table 2.2, the resulting linear predictor is $\eta_{ijk} = \eta + s_i + \rho_j + t(s\rho)_{ijk}$, where η_{ijk} is the value of the linear predictor for the ith school ($i = 1,2,\ldots,10$), jth professional development factor level ($j = 1,2$) (i.e., participant or nonparticipant), and kth teacher for the ijth school-participant combination ($k = 1,2$); η denotes the intercept; and s_i denotes the effect of the ith school, ρ_j denotes the effect of participation in the professional development program, and $t(s\rho)_{ijk}$ denotes the effect of the kth teacher at the ijth school-participant combination. Table 2.3 shows how each term in the linear predictor relates to terms in the table. Notice that the elements come from the unshaded cells: the blocking criterion, the factor, and the unit of replication with respect to the factor. Also, notice that the linear predictor *does not* include a subscript for the unit of replication: its purpose is to model expected response at the unit of replication and higher level—teacher and school—not at the unit of observation level.

TABLE 2.1

Template for Unit of Replication Approach

Blocking Criteria	Yes or No?	If Yes, Identify:
Factor	Identify each factor (for good measure, identify the levels for each factor as well)	Identify the unit of replication with respect to each corresponding factor
Unit of observation		

TABLE 2.2

Completed Template for School Professional Development Example

Blocking Criteria	Yes	If Yes, Identify: *School*
Factor	Professional development (two levels: participant or not)	*Unit of replication:* teacher (technically, teachers nested within schools)
Unit of observation	Students (technically, students nested within teachers)	

TABLE 2.3

Template for School Professional Development Example with Model Terms Identified

Blocking Criteria	Yes	School: s_i
Factor	Professional development ρ_j	*Unit of replication:* teacher [technically, teachers nested within schools]: $t(s\rho)_{ijk}$
Unit of observation	Students (technically, student nested within teachers)	

2.3.2 "What Would Fisher Do?"

A column in *IMS Bulletin* by Speed (2010) indirectly inspires the title of this section. Fisher's remarks following Yates (1935) *Complex Experiments* along with Fisher's reaction to early modeling work (Speed uses the terms "confused" and "enraged") provide the direct impetus. Fisher said that complex study designs could be broken into their "topographical" and "treatment" components and then combined to yield a sensible approach to analysis. Fisher used the term "topographical" because he was talking about agricultural experiments. We can understand "topographical" more broadly to include all nontreatment aspects of the study design. In the same spirit, later authors, for example, Federer (1955), Milliken and Johnson (2008), refer to the "experiment" and "treatment" designs.

ANOVA, properly understood as a thought process in thinking through how data have arisen, can be a powerful tool for constructing linear predictors for GLMMs. On the other hand, ANOVA, *improperly* understood as the arithmetic of sums of squares, mean squares and their associated quadratic forms, can be a virtually insurmountable obstacle to understanding contemporary modeling. Obviously, we can never know what Fisher would actually do, but this method takes his remarks following Yates' paper and attempts to follow his thinking in view of contemporary developments in statistical theory. See also Gelman (2005) and Nelder (1977, 1994, 1998) for further discussion of the ANOVA-modeling connection.

How does it work? For each aspect of the design, "topographical" and "treatment," we write the sources of variation and associated degrees of freedom only. We call these *skeleton ANOVA* tables. For the school example, these two ANOVAs appear in Table 2.4.

Note that the total degrees of freedom must be 39 because there are 40 units of replication altogether: 10 schools × 4 classrooms per school. Also, we do not include students

TABLE 2.4

"Topographical" and "Treatment" Skeleton ANOVA Tables

Source of Variation	d.f.
Topographical	
School	9
Classroom(school)	30
Total	39
Treatment	
Professional development	1
"Parallels"	38
Total	39

TABLE 2.5

"Topographical" and "Treatment" Skeleton ANOVA Tables Arranged Side by Side

Topographical		Treatment	
Source	**d.f.**	**Source**	**d.f.**
School	9		
		Professional development	1
Classroom(school)	30		
		Parallels	38
Total	39	Total	39

within classrooms in this exercise—the object of this exercise is the construction of a linear predictor to model expected response at the unit of replication level. No degree of freedom formulae appear here—none are needed and, in fact, such formulae tend to cause more confusion than clarity. The sources of variation completely specify what is happening on the topographical side and the degrees of freedom are (or should be) obvious. On the treatment side, the term "parallels" is Fisher's, used to include all variation on the treatment side not specifically attributable to the treatment design.

Now we assemble the combined skeleton ANOVA. We do this by placing the topographical and treatment ANOVA tables side by side, being careful to place the rows for treatment sources of variation next to their respective units of replication on the topographical side. Table 2.5 shows the side-by-side arrangement.

The arrangement of the rows is important. The blocking criterion, school, appears in the first row. Professional development appears second and classroom(school) appears in the third row to indicate that classroom(school) serves as the unit of replication for professional development, and hence degrees of freedom for professional development in the combined ANOVA will be removed from classroom(school) to reflect this. We are now ready to construct the combined ANOVA, shown in Figure 2.6.

Note the modification of classroom(school) to classroom(school × professional development). This reflects the assignment of treatments to units of replication. From the structure of the combined ANOVA, we see that terms from school, professional development, and classroom(school × professional development) need to be included in the linear predictor.

Following the Fisher thought process more rigorously, our modeling objective is to estimate μ_{ijk} the expected response of the *ijk*th school–professional development–classroom combination conditional on the design structure. We do this through the link function $\eta_{ijk} = g(\mu_{ijk})$, where the link function depends on the distribution of the responses. The combined ANOVA tells how to decompose η_{ijk}. Specifically, $\eta_{ijk} = \eta + s_i + \rho_j + \tau(s\rho)_{ijk}$, where all terms are defined as before.

2.3.2.1 *Complication*

Up to this point, we have developed the linear predictor exclusively from the physical units—school and classroom—and the treatment factor levels. In some cases—and the school example mentioned earlier is arguably such a case—there may be other processes that affect the responses we observe. From a design perspective, one could argue that the

TABLE 2.6

Combined Skeleton ANOVA Tables Arranged Side by Side

Topographical		Treatment		Combined	
Source	d.f.			Source	d.f.
School	9	------------------------	→	School	9
		Professional development	1--→	Professional development	1
Classroom(school)	30	------------------------	→	Classroom (school × prof. development)	30 − 1 = 29
		Parallels	38		
Total	39	Total	39	Total	39

unit of replication in Figure 2.1 is actually the *pair* of classrooms that receive each professional development treatment level thus requiring us to amend Table 2.6 to the following:

Topographical		Treatment		Combined	
Source	d.f.			Source	d.f.
School	9			School	9
		Professional development	1--→	Professional development	1
Pair(school)	10			Pair, a.k.a. school × professional development	10 − 1 = 9
Classroom(pair)	20	------------------------	→	Classroom(school × prof. development)	20
		Parallels	38		
Total	39	Total	39	Total	39

We would amend the linear predictor accordingly to $\eta_{ijk} = \eta + s_i + \rho_j + (s\rho)_{ij} + \tau(s\rho)_{ijk}$, where $(s\rho)_{ij}$ denotes the school × professional development effect. This would be a particularly important term to include in the model if we think there may be school-specific variation in the treatment effect that is distinct from variation among individual classrooms. Failing to include this term would deprive us of important insight about the impact of the professional development program and the extent to which we might be able to generalize conclusions about the program's benefits. We will return to this issue in Chapter 3 when we discuss inference space.

2.3.2.2 Linear Predictor for Nested Schools

Figure 2.2 showed an alternative design for the school study, with schools nested within the professional development treatments rather than cross-classified. To develop the linear predictor, we use the same strategies—either complete the template provided by Table 2.1 or use the WWFD skeleton ANOVA approach.

Table 2.7 shows the template for this design.

For this design, we also need to recognize that classrooms within schools will contribute additional variation that should be accounted for in the linear predictor.

Table 2.8 shows the skeleton ANOVA approach.

Note the reorganization relative to Table 2.7 where schools were crossed. In the cross-classification design, professional development treatment was assigned to teachers'

TABLE 2.7

Template for School Professional Development Example with Nested Schools

Blocking Criteria	No	
Factor	Professional development (two levels: participant or not)	*Unit of replication:* school (technically, schools nested within treatment)
Unit of observation	Students (technically, students nested within classrooms)	

TABLE 2.8

Skeleton ANOVA of Professional Development Example with Nested Schools

Topographical				Combined	
Source	d.f.	Treatment		Source	d.f.
		Professional development	1---→	Professional development	1
School	9	-----------------------→		School(professional development)	9 − 1 = 8
Classroom(school)	30	-----------------------→		Classroom(school × professional development)	30
		Parallels	38		
Total	39	Total	39	Total	39

classrooms, whereas in the nested design, treatments are assigned at the school level. Hence, the placement of professional development above school rather than above classroom, signifying where the assignment is and how the degrees of freedom are affected.

The resulting linear predictor is thus $\eta_{ijk} = \eta + \rho_i + s(\rho)_{ij} + t(s\rho)_{ijk}$, where $s(\rho)_{ij}$ denotes the effect of the jth school assigned to the ith professional development treatment.

2.3.3 Matching the Objective

In many studies, we have a choice of defensible models, but some are better suited to address our objectives. Designs with quantitative levels illustrate this point.

Recall the multi-batch regression example from Chapter 1. The study involved selecting three batches. For each batch, observations are taken on a quantitative factor, X, with nine levels, 0, 3, 6, 9, 12, 18, 24, 36, and 48. Following this description, Figure 2.3 shows the implied "plot plan."

We use the plot plan in Figure 2.3 along with either the Table 2.1 template or the skeleton ANOVA to identify the elements of the linear predictor. Table 2.9 shows a possible way to complete the table.

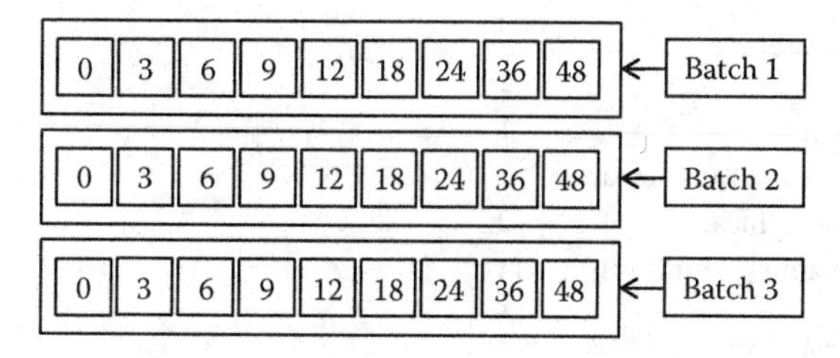

FIGURE 2.3
Plot plan for multi-batch study with quantitative levels.

TABLE 2.9

Template for Multi-Batch Quantitative Level Study

Blocking criteria?	Yes	If Yes, Identify: Batch
Factor	X (i.e., whatever level of X represents)	*Unit of replication:* units represented by square in each batch
Unit of observation	Not distinct from units of replication	

TABLE 2.10

Skeleton ANOVA for Multi-Batch Study with Quantitative Levels

Topographical		Treatment		Combined	
Source	**d.f.**	**Source**	**d.f.**	**Source**	**d.f.**
Batch	2			Batch	2
		X = treatment	8	X = treatment	8
Units(batch)	24			Units(batch)	16
		Parallels	18		
Total	26	Total	26	Total	26

Alternatively, we can complete the topographical, treatment, and skeleton ANOVAs. These are shown in Table 2.10.

The implied linear predictor from this procedure appears to be $\eta_{ij} = \eta + b_i + \tau_j(+u(b)_{ij}?)$, where b_i denotes the effect of the ith batch and τ_j denotes the effect of the jth treatment. The last term, $u(b)_{ij}$, the unit(batch) term, appears in parentheses because it is not at all clear whether it should be part of the linear predictor. It turns out that the decision whether to include $u(b)_{ij}$ in the linear predictor depends on the distribution of the observations. We leave this question for now but will return to it again in Section 2.4.

Notice that the linear predictor does not agree with what we did in Chapter 1. This is an example of a linear predictor that is technically correct as a description of the study design, but fails to allow us to address the study's objective in a straightforward manner. Factor X is a direct variable and linear regression was the stated objective when this example was introduced in Chapter 1. It follows that the linear predictor should include $\beta_0 + \beta_1 X$ to account for factor X. Batch can potentially affect both intercept and slope, so there must be b_{0i} and b_{1i} terms to account for batch effects on intercept and slope, respectively. We can use the skeleton ANOVA to help visualize the needed changes—shown in Table 2.11.

TABLE 2.11

Skeleton ANOVA for Multi-Batch Study Modified to Accommodate Linear Regression

Combined		
Source		**d.f.**
Batch		2
X = treatment	Linear X	1
	Departure from linear	7
Units(batch)	Batch × linear X	2
	Remaining unit(batch)	14
	Total	26

The resulting linear predictor is $\eta_{ij} = \beta_0 + b_{0i} + (\beta_1 + b_{1i})X_j + \tau_j$, where X_j is the direct variable reflecting the levels of the quantitative factor given earlier, b_{0i} is the effect of the ith batch on the intercept of the linear regression, b_{1i} denotes the effect of the ith batch on the slope of the linear regression, and τ_j denotes the remaining effect of treatment, after accounting for the linear regression. Note that we modify the batch effect from a generic b_i to specific b_{0i} and b_{1i} terms to distinguish batch effect on intercept and slope, respectively.

In principle, we could add additional regression terms, for example, quadratic and cubic, removing degrees of freedom from the remaining treatment effect (which we would call "departure from regression," "lack of fit," or words to that effect) and from remaining unit(batch). Each new regression term would have two slope terms, one regression coefficient and one batch effect on the regression coefficient. For example, a linear predictor for quadratic regression would be $\eta_{ij} = \beta_0 + b_{0i} + (\beta_1 + b_{1i})X_j + (\beta_2 + b_{2i})X_j^2 + \tau_j$. Note that all three versions of the linear predictor are technically accurate descriptions of the study design, but depending on the nature of the treatment and the objectives, one of the three forms will be better suited to the task at hand.

As with the school example, a subtle change in the architecture of the study profoundly affects the linear predictor. Suppose instead of using the same three batches throughout the study, three batches are used for each level of X, but these batches are not reused for different levels of X. The modified plot plan is shown in Figure 2.4.

Note that batches 1, 2, and 3 for $X = 0$ are not the same batches as the ones labeled 1, 2, and 3 for any other level of X. The Table 2.1 template and skeleton ANOVAs are modified accordingly, as shown in Tables 2.12 and 2.13.

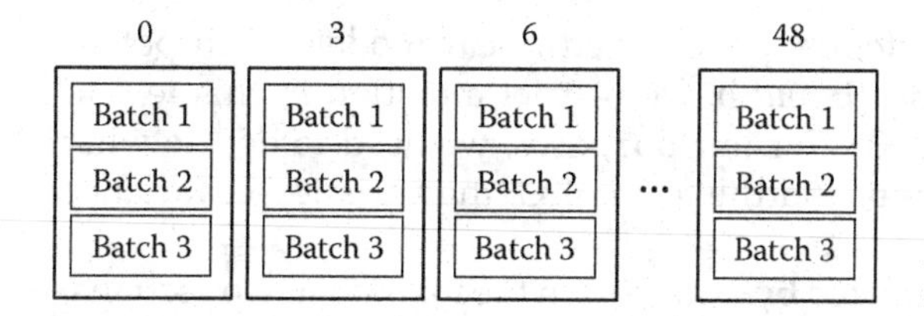

FIGURE 2.4
Alternative plot plan for multi-batch linear regression study.

TABLE 2.12

Template for Alternative Multi-Batch Regression Design

Blocking Criteria?	**No**	
Factor	X	*Unit of replication:* units represented by a "batch" within each level of X
Unit of observation	Not distinct from units of replication	

TABLE 2.13

Skeleton ANOVA for Alternative Multi-Batch Regression Design

Topographical		**Treatment**		**Combined**		
Source	**d.f.**	**Source**	**d.f.**		**Source**	**d.f.**
Lot	8	X = Treatment	8	X = treatment	Linear	1
					Departure from linear	7
Batch (lot)	18	Parallels	18		Batch (lot)	18
Total	26	Total	26		Total	26

No blocking criterion means that the b_{0i} and b_{1i} terms are removed from the linear predictor. Not blocking on batch changes the structure from batches and levels cross-classified (Figure 2.3) to batches nested within levels of X (Figure 2.4). Effects of nested terms can never be estimated independently of the factor within which they are nested. In this case, this precludes estimating the batch effects on intercept and slope. The resulting linear predictor—assuming linear regression—is $\eta_{ij} = \beta_0 + \beta_1 X_j + \tau_j$. Technically, the linear predictor for this design could drop the subscript i.

2.4 Distribution Matters

In Section 2.3, we focused on constructing the linear predictor exclusively as an equation partitioning the link into its component parts. In this section, we turn our attention to the final issues required in model construction. These are

- Distinguishing between fixed and random model effects, that is, what goes in $\mathbf{X}\boldsymbol{\beta}$ and what goes in $\mathbf{Zb}$
- Specifying the probability distributions of the random effects
- Specifying the probability distribution of the observations, that is, the distribution of $\mathbf{y}|\mathbf{b}$

Although Lee and Nelder (1996) discuss doubly generalized linear models, with potentially non-Gaussian distributions of both $\mathbf{y}|\mathbf{b}$ *and* $\mathbf{b}$, we restrict attention in this text to Gaussian random effects, that is, $\mathbf{b} \sim N(\mathbf{0}, \mathbf{G})$. For our purposes, we need to decide what effects comprise the random effect vector, $\mathbf{b}$, and the form of $\mathbf{G}$, the variance–covariance matrix of $\mathbf{b}$.

The required decision processes are illustrated by examples in the following two sections.

2.4.1 Model Effects: Fixed or Random

As an illustration of the decisions involved and how they are made, consider the school and professional development intervention example. The linear predictor for the design shown in Figure 2.1 was $\eta_{ijk} = \eta + s_i + \rho_j + t(s\rho)_{ijk}$. The effect η is, by definition, the fixed intercept, so it is automatically part of $\mathbf{X}\boldsymbol{\beta}$. For the remaining terms, the key questions are as follows:

- Do each effect's levels *represent* a larger population of interest or are the levels observed in the study of explicit interest and hence the entire population?
- Is there a probability distribution associated with the effect in question?

For the professional development effect, ρ_j, the two levels (participant or nonparticipant) are clearly the only two levels of interest—they are the explicit motivation for the study and do not represent a larger population. Thus, the ρ_j are fixed and belong in $\mathbf{X}\boldsymbol{\beta}$. On the other hand, the description of the study says, "ten schools were selected at random." This also seems clear. The schools observed in the study represent all the schools in the district (and possibly, by extrapolation, all schools from similar districts); other schools could just as well have been selected; and the 10 observed here represent one possible random draw. Similar wording seems to apply to the teachers. It follows that the school and teacher effects, s_i and $t(s\rho)_{ijk}$, are random and belong in **Zb**.

Assuming school and teacher effects are random, the next step is to specify their probability distribution. Typically, in linear mixed models, they are assumed to be normal and mutually independent, that is, s_i are assumed i.i.d. $N\left(0,\sigma_S^2\right)$, $t(s\rho)_{ijk}$ are assumed i.i.d. $N\left(0,\sigma_T^2\right)$, and s_i and $t(s\rho)_{ijk}$ are assumed mutually independent.

Putting it all together in matrix terms, the linear predictor is thus

$$\eta = \mathbf{X\beta} + \mathbf{Zb} = \begin{bmatrix} \mathbf{1} & \mathbf{X}_\rho \end{bmatrix} \begin{bmatrix} \eta \\ \boldsymbol{\rho} \end{bmatrix} + \begin{bmatrix} \mathbf{Z}_S & \mathbf{Z}_T \end{bmatrix} \begin{bmatrix} \mathbf{s} \\ \mathbf{t} \end{bmatrix}$$

where

$$\begin{bmatrix} \mathbf{s} \\ \mathbf{t} \end{bmatrix} \sim N\left(\begin{bmatrix} 0 \\ 0 \end{bmatrix}, \begin{bmatrix} \mathbf{I}\sigma_S^2 & 0 \\ 0 & \mathbf{I}\sigma_T^2 \end{bmatrix} \right)$$

1 is an $N \times 1$ vector of ones

$\mathbf{X}_\rho$ is an $N \times 2$ design matrix for program participation

$\boldsymbol{\rho} = \begin{bmatrix} \rho_1 \\ \rho_2 \end{bmatrix}$ is the vector of participation effects

s and **t** are the vectors of school and teacher effects

$\mathbf{Z}_S$ and $\mathbf{Z}_T$ are the design matrices for school and teacher effects (question for reader: what is the dimension of $\mathbf{Z}_S$? and $\mathbf{Z}_T$?).

For the multi-batch regression example, using the design shown in Figure 2.3 and the linear predictor $\eta_{ij} = \beta_0 + b_{0i} + (\beta_1 + b_{1i})X_j + \tau_j$, β_0, β_1, and τ_j would be considered fixed effects because the objective of the study involves understanding how the response varies with changing levels of X. Levels of X must be chosen to facilitate estimation of this effect. Random selection would never be used if it could be avoided. On the other hand, the batches presumably represent a population, and any three representative, randomly selected batches would do. Therefore, the batch effects would be considered random. The resulting linear predictor, in matrix form, is

$$\eta = \mathbf{X\beta} + \mathbf{Zb} = \begin{bmatrix} \mathbf{1} & \mathbf{x} & \mathbf{X}_\tau \end{bmatrix} \begin{bmatrix} \beta_0 \\ \beta_1 \\ \boldsymbol{\tau} \end{bmatrix} + \begin{bmatrix} \mathbf{Z}_0 & \mathbf{Z}_1 \end{bmatrix} \begin{bmatrix} \mathbf{b}_0 \\ \mathbf{b}_1 \end{bmatrix}$$

where

x is an $N \times 1$ vector of the numeric values of X_j for each observation in the data set

$\mathbf{X}_\tau$ is an $N \times 9$ matrix, one column per level of X, where the ijth element is 1 if the corresponding observation receives the jth level of X and 0 otherwise

$\boldsymbol{\tau}$ is a 9×1 vector of treatment effects

$\mathbf{b}_0$ is a 3×1 vector of batch effects on the intercept

$\mathbf{b}_1$ is a 3×1 vector of batch effects on the slope

$\mathbf{Z}_0$ is an $N \times 3$ design matrix for batches

$\mathbf{Z}_1 = \mathbf{I}_3 \otimes \mathbf{x}$ is the $N \times 3$ matrix of constants for the random effect of batch on slope

Typically, the random batch effects on intercept and slope are assumed to be at least potentially correlated, so

$$\begin{bmatrix} \mathbf{b}_0 \\ \mathbf{b}_1 \end{bmatrix} \sim N\left(\begin{bmatrix} \mathbf{0} \\ \mathbf{0} \end{bmatrix}, \begin{bmatrix} \sigma_0^2 \mathbf{I}_3 & \sigma_{01}\mathbf{I}_3 \\ & \sigma_1^2 \mathbf{I}_3 \end{bmatrix}\right)$$

where σ_{01} denotes $\mathrm{Cov}(b_{0i}, b_{1i})$. Often, it is better to rearrange the random effects vector and the **G** matrix so that the random intercept and slope terms are together by batch. We hold that discussion for later, when we discuss translating models into software statements and when we discuss estimation procedures.

2.4.2 Response Variable Distribution

The remaining decisions about the model involve the distribution of the observed data. Because the linear predictor has random effects, this is a mixed model. The observation vector is thus formally denoted $\mathbf{y}|\mathbf{b}$, e.g. for the school example $y_{ijkl}|s_i, t(s\rho)_{ijk}$, the observation on the lth student in the classroom of the ijkth teacher. Notice first that we do not begin any discussion of the observed data until we have fully specified the linear predictor. (Beginning students tend to want to do this in the opposite order, which tends to get them into trouble.) At this point, we would want to know more precisely what observations have been taken on students. How exactly is "gain in achievement" measured? Is it the difference between a pretest and a posttest? Is it categorical, for example, "satisfactory/unsatisfactory."

Note that the linear predictor remains valid regardless of the distribution of $\mathbf{y}|\mathbf{b}$. What is affected is the distribution. This in turn affects the link function. Following our discussion in Chapter 1, "gain" is defined in terms of pre–post score differences that could be regarded as normally distributed, and then we would use an identify link and write the model $[y_{ijkl}|s_i, t(s\rho)_{ijk}] \sim N(\eta_{ijk}, \sigma^2)$ or, in more detail, $N(\eta + s_i + \rho_j + t(s\rho)_{ijk}, \sigma^2)$. On the other hand, if the data are binomial ("satisfactory/unsatisfactory") and we use a logit link, then the model is $[y_{ijkl}|s_i, t(s\rho)_{ijk}] \sim \mathrm{Binomial}(N_{ijk}, \pi_{ijk})$, where N_{ijk} is the number of students in the ijkth teacher's class, π_{ijk} is the probability of a student having satisfactory gain in the ijkth teacher's class, and the link is $\eta_{ijk} = \log[\pi_{ijk}/(1 - \pi_{ijk})] = \eta + s_i + \rho_j + t(s\rho)_{ijk}$. If you want to write the entire model in one statement, you use the inverse link:

$$\left[y_{ijkl}|s_i, t(s\rho)_{ijk}\right] \sim \mathrm{Binomial}\left(N_{ijk}, \frac{1}{\left[1+e^{-(\eta+s_i+\rho_j+t(s\rho)_{ijk})}\right]}\right)$$

To summarize, the three key steps in writing a model are

1. Identifying the linear predictor
2. Identifying the distribution of the observed data
3. Identifying the link

Of these, identifying the linear predictor is the crucial first step. It is also where attempts to specify a plausible and useful model run afoul. In the next section, we discuss some "visual aids" you can use to help keep the linear predictor identification process on track.

2.4.3 Fixed or Random?: Tough Calls

Deciding whether an effect is fixed or random is not always clear cut. Many "fixed or random" controversies have gone on for decades. Examples include locations in multi-center clinical trials, years in long-term agronomic trials and blocks in randomized block designs. Most of the controversies are unlikely to be resolved definitively because a single "right answer" does not exist. The only all-encompassing right answer is "it depends" on the specifics of each study.

What is relevant to our discussion in this chapter is how we think about the issues bearing on our "fixed or random" decision in the context of a given experiment. To illustrate, we use a seemingly extreme incomplete block design and pose the question, "does it make more sense for block effects to be fixed or random—or does it matter?"

Consider the following hypothetical (but realistic) scenario. We want to compare six treatments—denote them 0, 1, 2, 3, 4, and 5. There is a natural reason to block, but blocks are limited in size. Only three treatments can be assigned to any given block. Suppose that there are 10 blocks available. There are many ways to design a study within these constraints. Figure 2.5 shows three alternatives that would likely receive serious consideration. The first is a true balanced incomplete block (BIB) design. Its main advantage is that all treatment comparisons have equal precision. The second design might be considered if one of the treatments—say treatment 0—is a control or reference standard and the objective is to maximize precision for pairwise comparisons between each treatment and the control. The third design is called a *disconnected* design because treatments 0, 1, and 2 always appear together in the same blocks and never appear in the same block with treatment 3, 4 or 5. Surprisingly, the disconnected design is widely used (possibly the most widely used of the three design shown here) in practice. From a modeling perspective, it is also the most interesting and certainly the most relevant to this discussion.

The obvious linear predictor for each of the three designs is $\eta_{ij} = \eta + \tau_i + b_j$, where τ_i denotes the ith treatment effect and b_j denotes the jth block effect. For the first two designs the consequences of assuming fixed or random block effects is relatively minor as far as inference on treatment effects is concerned. However, for the disconnected design, the impact is dramatic. If we assume fixed block effects, only comparisons among treatments 0, 1, and 2 or among treatments 3, 4, and 5 are possible. None of the treatment means are estimable, a concept we will begin to discuss explicitly in Chapter 3. This much is consistent with traditional, ordinary least squares based linear model theory. The example also sheds light

	Designs		
Block	**BIB**	**Control vs Others**	**Disconnected**
1	0 1 2	0 1 2	0 1 2
2	0 1 3	0 1 3	0 1 2
3	0 2 4	0 1 4	0 1 2
4	0 3 5	0 1 5	0 1 2
5	0 4 5	0 2 3	0 1 2
6	1 2 5	0 2 4	3 4 5
7	1 3 4	0 2 5	3 4 5
8	1 4 5	0 3 4	3 4 5
9	2 3 4	0 3 5	3 4 5
10	2 3 5	0 4 5	3 4 5

FIGURE 2.5
Three alternative designs for 6 treatment, 10 block, 3 treatment/block example.

on why Fisher's reaction to early linear model work was "confused" and "enraged" to use Speed's characterization.

If we regard block effects as random, *all* of the treatment means are estimable, using the formal criteria of estimability. Moreover, *all* of the mean differences can be estimated, although with varying precision depending on whether we compare means of treatments observed together in the same blocks or means in different disconnected blocks.

For the disconnected design, the fixed or random block decision has extreme consequences. Why the discrepancy? Our WWFD skeleton ANOVA procedure, implemented carefully, helps explain. Tables 2.14 and 2.15 shows the topographical and treatment skeleton ANOVAs.

Notice that we have been more precise in specifying the treatment skeleton ANOVA than we apparently were when we constructed the linear predictor $\eta_{ij} = \eta + \tau_i + b_j$ (Table 2.14). Treatments are divided into two sets, {0,1,2} and {3,4,5}, and the integrity of the set is maintained throughout the design. Therefore, we must account for it as a source of variation in the treatment design. Table 2.15 shows the combined skeleton ANOVA.

The resulting linear predictor is $\eta_{ij} = \eta + \alpha_i + \tau(\alpha)_{ij} + b(\alpha)_{ik}$, where α_i denotes the effect of the ith set, $\tau(\alpha)_{ij}$ denotes the effect of the jth treatment within the ith set, and $b(\alpha)_{ik}$ denotes the effect of the kth block within the ith set.

Before we discuss the question of fixed or random block effects, two comments are in order.

First comment: Viewed through the skeleton ANOVA lens, we see that the disconnected design can also be understood as a nested factorial. In fact, if we add additional structure

TABLE 2.14

Topographical and Treatment ANOVA for Disconnected Design

Topographical		Treatment	
Source	**d.f.**	**Source**	**d.f.**
		Set	1
Block	9		
		Treatments(set)	4
Units(block)	20		
		"Parallels"	24
Total	29		29

TABLE 2.15

Combined ANOVA for Disconnected Design

Combined	
Source	**d.f.**
Set	1
Block(set)	9 − 1 = 8
Treatment(set)	4
Units(block)	20 − 4 = 16
Total	29

to the treatment design by defining the two "sets" as two levels of treatment factor A, and treatments 0, 1, and 2 as three levels of factor B crossed with the first level of A and treatments 3, 4, and 5 as the same three levels of factor B crossed with the second level of A, then we have a 2 × 3 factorial treatment design and our "disconnected" design goes by a much more familiar name: it is a split-plot.

Second comment: It is instructive to review how Fisher and company approached the problem of assessing "set" and treatment within set effects. As a general principle, we know that we assess treatment or treatment group effects by comparing their differences to naturally occurring differences among units to which those treatments (or treatment groups) were applied. In this design, sets are applied to blocks. Therefore, we compare the difference between sets with variation among blocks that occurs naturally even without the application of treatments. Formally, we estimate the mean for set 1 by averaging the five blocks to which set 1 was applied and the mean for set 2 by averaging the five blocks to which set 2 was applied. Then we evaluate the mean difference between set 1 and set 2 relative to the standard error of the difference, which is based on the variance of the difference between the average of the five blocks for set 1 and the five blocks for set 2, Var(mean of set 1 blocks) + Var(mean of set 2 blocks)/5.

For comparisons between any two treatments within the same set, for example, treatments 0 and 1, we average the five units within a block to which treatment 0 was applied and the five units within that block to which treatment 1 was applied. The resulting standard error depends on Var(mean of trt 0 units) + Var(mean of trt 1 units)/5.

For the mean difference between two treatments not in the same block, for example, treatments 0 and 4, we average over the five units within the set 1 blocks to which treatment 0 was applied and the five units within the set 2 blocks to which treatment 4 was applied. The resulting variance involves variance among the blocks *and* variance among the units.

The split-plot structure, illustrated by this example, historically, was a primary impetus for the development of mixed model theory. Speed (2010) describes the split-plot as the "litmus test" of one's understanding of the thought processes involved in this discussion. The failure of fixed-effect-only linear model theory to adequately address this type of design—which in practice is arguably more the rule than the exception—provides insight into why Fisher's reaction to early modeling work, which was largely what we now call the LM (fixed effects only, Gaussian observations), was so negative.

Returning to the task at hand, as we have worked through this example, we have in essence mandated that block effects *must* be random. We did this when we said "we assess the difference between sets by comparing it to naturally occurring variation among the blocks to which sets were applied." Ergo, blocks must have a probability distribution. They must be random effects.

One criterion, then, for deciding if a model effect is fixed or random is whether it has naturally occurring variation that can be used as a defensible criterion for assessing the statistical significance of another effect in the model—or for constructing an interval estimate in lieu of a hypothesis test. Notice that this does *not* resolve all ambiguous "is it fixed or random?" situations. Here, however, the interplay between the topographical and treatment aspects of the design makes the decision *un*ambiguous for this particular case. Parenthetically, it also raises an interesting question: If the physical nature of the blocks is identical no matter which of the three design options shown in Figure 2.5 is used, how would a "block effects are fixed" advocate reconcile treating block effects as random—clearly appropriate for the third design and universally the case with split-plot experiments—and not regarding them as random for the first two designs?

2.5 More Complex Example: Multiple Factors with Different Units of Replication

We now consider a more complex data set, which we will return to throughout the text as a running example. This particular example is from an experiment conducted to evaluate different seed mixes grown under various methods of cultivation to identify effective ways of restoring damaged grassland. What is important for students of statistical modeling to pay attention to is structure of this data set. Although this example is agricultural, its modeling structure occurs in many contexts—medical, engineering, educational, etc. It is a good example for you to pay close attention to, because it illustrates many important features modeling students must master. It is also a structure that is very commonly analyzed using misspecified models—as we said earlier in this chapter, the most common cause of modeling gone awry is poor understanding of design principles.

The data set consists four blocks. Each block is divided into seven rectangular areas. There are seven cultivations methods. One of the seven areas in each field is randomly assigned to each of the seven cultivation methods. Each area is divided into four subareas. There are four seed mixes. Each subarea is randomly assigned to receive one of the seed mixes. Figure 2.6 shows the resulting plot plan for a sample block.

For readability, the methods of cultivation are shown in order, from Trt 1 through Trt 7; in the actual study, the order of these methods was randomized for each of the four blocks. As with the school and multi-batch example, the plot plan is the first step in constructing the linear predictor. The next step is to complete Table 2.1. As we do this, we introduce additional detail required by the fact that this data set has two factors. We will need to deal with each factor and correctly identify its unit of replication.

Here is a start:

Blocking Criteria?	**Yes**	**Unit of Replication**
Factors	Trt (method of cultivation)	
	Mix	
	Trt × Mix	
Unit of observation		

At this point, we note that there is a blocking criterion, so there will have to be a corresponding term in the model. There are also two factors, "Trt" and Mix. Each is a classification variable, not a direct quantitative variable. Also, they are cross-classified, meaning that interaction terms are also possible. So far, then, the linear predictor must include $\eta + b_i + \tau_j + \gamma_k + (\tau\gamma)_{jk}$,

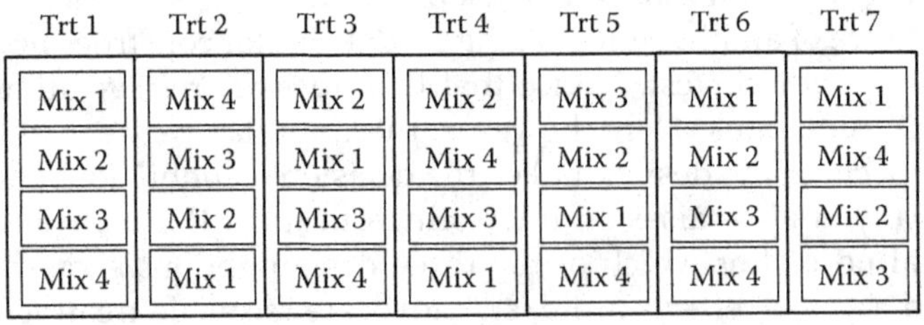

Trt 1	Trt 2	Trt 3	Trt 4	Trt 5	Trt 6	Trt 7
Mix 1	Mix 4	Mix 2	Mix 2	Mix 3	Mix 1	Mix 1
Mix 2	Mix 3	Mix 1	Mix 4	Mix 2	Mix 2	Mix 4
Mix 3	Mix 2	Mix 3	Mix 3	Mix 1	Mix 3	Mix 2
Mix 4	Mix 1	Mix 4	Mix 1	Mix 4	Mix 4	Mix 3

FIGURE 2.6
Plot plan for seed-mix × method of cultivation study (Trt denotes method of cultivation).

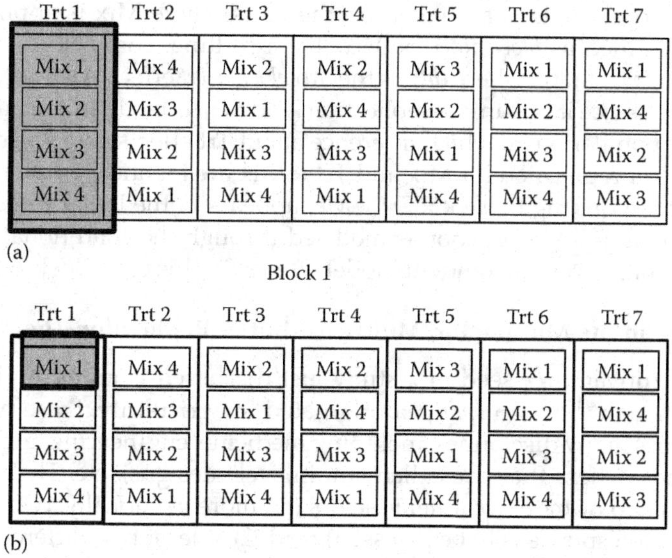

Trt 1	Trt 2	Trt 3	Trt 4	Trt 5	Trt 6	Trt 7
Mix 1	Mix 4	Mix 2	Mix 2	Mix 3	Mix 1	Mix 1
Mix 2	Mix 3	Mix 1	Mix 4	Mix 2	Mix 2	Mix 4
Mix 3	Mix 2	Mix 3	Mix 3	Mix 1	Mix 3	Mix 2
Mix 4	Mix 1	Mix 4	Mix 1	Mix 4	Mix 4	Mix 3

(a)

Block 1

Trt 1	Trt 2	Trt 3	Trt 4	Trt 5	Trt 6	Trt 7
Mix 1	Mix 4	Mix 2	Mix 2	Mix 3	Mix 1	Mix 1
Mix 2	Mix 3	Mix 1	Mix 4	Mix 2	Mix 2	Mix 4
Mix 3	Mix 2	Mix 3	Mix 3	Mix 1	Mix 3	Mix 2
Mix 4	Mix 1	Mix 4	Mix 1	Mix 4	Mix 4	Mix 3

(b)

FIGURE 2.7
(a) Plot plan showing unit of replication for method of cultivation study (Trt). (b) Plot plan showing unit of replication for mix.

where η is the intercept, b_i is the effect of the ith block, τ_j is the effect of the jth cultivation method treatment (Trt), γ_k is the effect of the kth mix, and $(\tau\gamma)_{jk}$ is the Trt × Mix interaction effect. Note that our choice of Greek letters to denote Trt and Mix effects is somewhat arbitrary. The one convention we will attempt to adhere to is that Greek letters will denote fixed model effects and Latin letters will denote blocking criteria and random model effects. Note that a factor can *never* be a blocking factor *and* a treatment factor—it must be one or the other.

At this point, our linear predictor is not complete, because we have not identified units of replication. It is at this point that we must be careful, because the units of replication for Trt and Mix are not the same. First, consider Trt. Recall how the randomization occurred. Levels of Trt were randomly assigned to the seven areas created by dividing the block. The resulting unit of replication with respect to Trt is shown on Figure 2.7.

How can we identify the unit of replication with respect to Trt? If you know which block and which method of cultivation, then that would uniquely identify the shaded rectangle representing the unit of replication. Thus, we can identify block × Trt as the unit of replication with respect to Trt.

What about Mix?

Using the same thought process, we see that the unit of replication with respect to Mix is *block × Trt × Mix*. Note also that the unit of replication for the *Trt × Mix* effect is also *block × Trt × Mix*. Now we can complete Table 2.1.

Blocking Criteria?	**Yes**	**Unit of Replication**
Factors	Trt (method of cultivation)	*block × Trt*
	Mix	*block × Trt × Mix*
	Trt × Mix	*block × Trt × Mix*
Unit of observation	*block × Trt × Mix*	

Notice that the observations are taken on the plots where Mix is applied, that is, the unit we have identified as *block* × *Trt* × *Mix*. The resulting, complete linear predictor is $\eta_{ijk}=\eta+b_i+\tau_j+\gamma_k+(\tau\gamma)_{jk}+(bt)_{ij}$, where $(bt)_{ij}$ is the *ij*th *block* × *Trt* effect. Also note that $(bt)_{ij}$ must be a random effect, since it is a unit of replication (and hence, by definition, a representative of a larger target population, to which inference about Trt effects are to be drawn). Finally, because the units of replication for Mix and Trt × Mix are identical to the unit of observation, no term corresponding to *block* × *Trt* × *Mix* appears in the linear predictor. Variation attributable to the unit of observation is modeled through the conditional distribution of $\mathbf{y}|\mathbf{b}$, the observations given the random model effects (in this case block and block × Trt).

2.5.1 Variations on the Multifactor, Multisize Unit of Replication Theme

The method of cultivation × seed mix study just discussed is an example of a *split-plot* experiment. Although the term *split-plot* originated in agriculture, these types of designs are common across many disciplines, including medicine, engineering, molecular biology, the social sciences (where they are called "multilevel" designs), etc. Their defining characteristics are (1) two or more treatment factors (or, more generically, two or more factors whose impact on a response is to be assessed), and (2) at least two different sized units of replication for the various factors or factor combinations.

There is no one "split-plot design." Federer and King (2007) published an entire textbook describing variations on the split-plot. Littell et al. (2006) describe seven ways to conduct a 2 × 2 factorial (two factors, each at two levels), four of which are different ways of setting up a split-plot. This merely scratches the surface. In other words, memorizing a one-size-fits-all "cookbook" for these designs is futile. Instead, sticking to the discipline of drawing a plot plan, using it to complete Table 2.1, and completing the linear predictor from there is the best approach.

In this section, we look at three of the variations discussed by Littell et al. (2006) to see how this works. In each scenario, there are two treatment factors, each with two levels. We denote these A1 and A2 for factor A and B1 and B1 for factor B.

Scenario 1: Eight units are randomly sampled from a target population. Four of these units are assigned, at random, to receive A1 and the other four receive A2. Each unit is subdivided into two subunits. One subunit is assigned at random to receive B1 and the other B2. The plot plan is as follows.

A1		A2	
B1	B2	B2	B1
B2	B1	B2	B1
B1	B2	B1	B2
B1	B2	B1	B2

Completing Table 2.1 yields

Blocking Criteria?	No	Unit of Replication
Factors	A	*unit*(*A*)
	B	*subunit*
	A × B	*subunit*
Unit of observation	*subunit*	

The following figures add shaded areas to the plot plan to depict the unit of replication with respect to A, that is, "large unit nested within A" or *unit(A)* and the unit of replication with respect to B and A × B, that is, the *subunit*. The light gray shading surrounded by the solid black border shows *unit(A)*. The darker gray shading surrounded by the dashed gray border is the subunit.

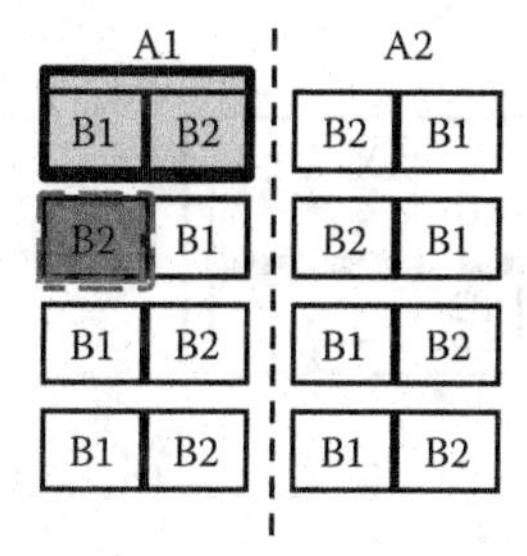

The resulting linear predictor is $\eta_{ijk} = \eta + \alpha_i + \beta_j + (\alpha\beta)_{ij} + u(\alpha)_{ik}$, where α and β denote A and B effects, and $u(\alpha)_{ik}$ denotes the unit(A) effect of the kth unit assigned to the ith level of factor A. Being units of replication, the $u(\alpha)_{ik}$ are assumed to be random effects, typically i.i.d. $N\left(0, \sigma_U^2\right)$. The intercept, A and B effects, and A × B interaction effects, $\eta + \alpha_i + \beta_j + (\alpha\beta)_{ij}$ can be combined and referred to simply as η_{ij}. You can also model the η_{ij} in ways other than main effects and interaction. We will show examples in subsequent chapters, but not here.

Scenario 2: Similar to Scenario 1, except the eight units sampled from the original population are paired together into blocks. Typically, pairing would be based on some relevant criterion. In each block, one unit gets A1 and the other gets A2, assignment being done at random. Each unit assigned to a level of A is subdivided and each subunit is assigned to a level of B, much as in Scenario 1. The following shows the plot plan for a typical block.

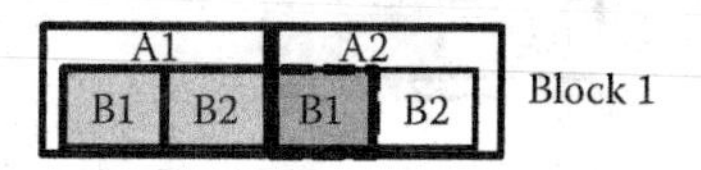

Table 2.1 is thus

Blocking Criteria?	Yes	Unit of Replication
Factors	A	*unit* with respect to *A*(=*block* × *A*)
	B	*subunit*(=*block* × *A* × *B*)
	A × B	*subunit*
Unit of observation	*subunit*	

The light shaded area under A1 illustrates the size of the "unit with respect to A." Notice that it is uniquely identified by a *block* × *A* intersection. The darker shaded area, illustrated by the unit for B1 under A2, is the subunit, or unit of replication with respect to both A and A × B. Notice that it is uniquely identified by a *block* × *A* × *B* intersection. Also note that the subunit is the unit of observation.

The linear predictor is thus $\eta_{ijk} = \eta + \alpha_i + \beta_j + (\alpha\beta)_{ij} + r_k + r\alpha_{ik}$ or, alternatively, $\eta_{ijk} = \eta_{ij} + r_k + r\alpha_{ik}$, where r_k is the effect of the kth block, $r\alpha_{ik}$ is the ikth unit of replication with respect to factor A, and the rest of the terms are as defined previously. The main difference here is that there is blocking, so a block effect is added, and the unit with respect to A, which is physically the same unit as in Scenario 1, can be defined in terms of *block* × *A* intersections.

Scenario 3: This scenario starts similarly to Scenario 2. There are four blocks, each consisting of two units of replication with respect to A. One such unit in each block is assigned, randomly, to A1; the other unit receives A2. At this point, the block is divided in half, with this division perpendicular to the first division. This created two units whose orientation is perpendicular to the units with respect to A. One of these units receives B1, the other B2. The plot plan is shown as follows.

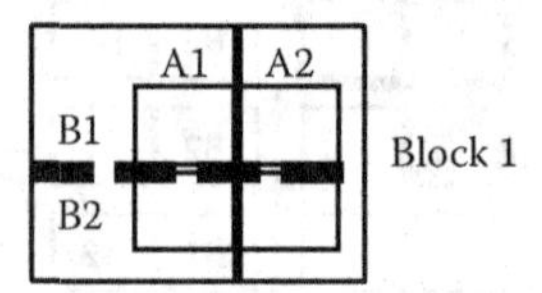

Now, notice that for A, the unit of replication is the left/right half of the block, as shown in the following.

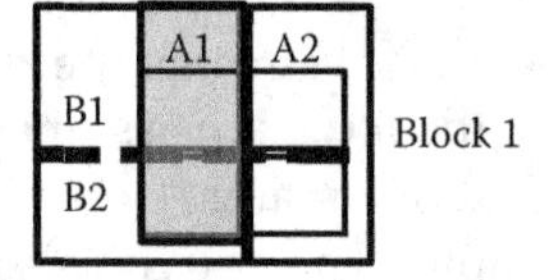

For A, the unit of replication is uniquely identified by a block × A intersection. The unit of replication with respect to B is the up/down half of the block, uniquely identified by a block × B intersection, as shown in the following.

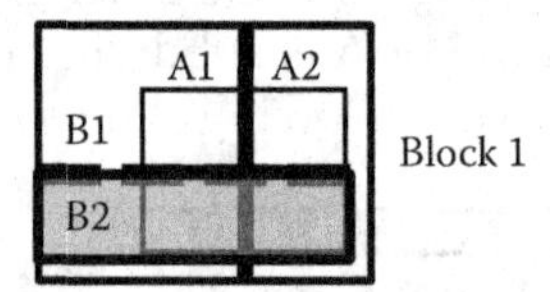

In this scenario, for A × B, the unit of replication is the intersection of the units of replication for A and B, as shown in the following. This unit is uniquely identified by a block × A × B intersection.

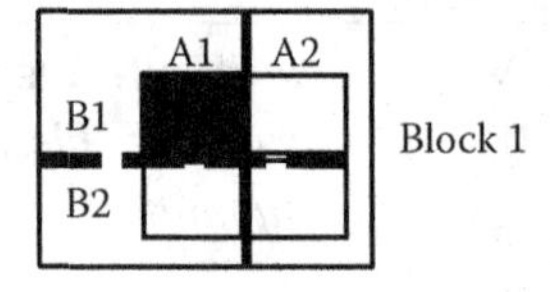

Now we can complete Table 2.1.

Blocking Criteria?	Yes	Unit of Replication
Factors	A	*block* × *A*
	B	*block* × *B*
	A × B	*block* × *A* × *B*
Unit of observation	*block* × *A* × *B*	

This illustrates a design for which the size of the unit of replication is different for each main effect (both factor A and B) and the size of the unit of replication for the interaction is different for both of the main effects units. The resulting linear predictor is $\eta_{ijk} = \eta_{ij} + r_k + r\alpha_{ik} + r\beta_{jk}$ or, alternatively, $\eta_{ijk} = \eta + \alpha_i + \beta_j + (\alpha\beta)_{ij} + r_k + r\alpha_{ik} + r\beta_{jk}$, where $r\beta_{jk}$ is the jkth block × B, unit with respect to factor B, and all other terms are as defined in Scenario 2.

Proceeding in this manner, you can determine a linear predictor that validly describes any data architecture, no matter how complex. The main lessons to take from these examples are as follows:

- Do not try to memorize cookbooks; your data set very likely has quirks that defy any memorizable cookbook.
- Do not worry about the *name* of the design at this point. As we have just seen, the name "split-plot" can mean any of an essentially endless variety of layouts. Even seemingly well-defined design formats like randomized complete block or completely randomized designs are understood—and misunderstood—in so many ways that a picture in the form of a plot plan is essential for accurate understanding of exactly how the data were collected.
- Draw an accurate plot plan.
- Identify the unit of replication for *all* factors *and* their interactions. They all may well be different. Be careful—attention to detail is vital at this point.
- Complete Table 2.1. This may be tedious, but the self-discipline involved in doing it greatly improves the accuracy of your linear predictor.

Once you have the linear predictor, it is applicable to any distribution and any link function amenable to generalized linear model methodology. In the next four chapters, we will develop this methodology and its underlying theory in detail.

Exercises

2.1 Consider the following six scenarios. Each describes a study—how it was conducted, the response variable of interest, and the objective of the study (in some cases, the objective is implicit and not stated in so many words). For each scenario, do Exercises 2.1 and 2.2.

Scenarios:

A. Ten clinics are randomly sampled from a population. At each clinic, volunteers are randomly assigned into two groups. One group receives treatment 1; the other group receives treatment 2. Let N_{ij} be the number of patients receiving the ith treatment at the jth clinic. The response of interest is Y_{ij}, the number of patients out of N_{ij} showing a favorable outcome to the treatment.

B. Seven locations are sampled. Each location is divided into four parts. Each parts is randomly assigned a "level of exposure" to an environmental hazard (the levels are 0, 0.5X, X, and 2X, where "X" is the "nominal maximum safe exposure level"). The response variable is the number of living organisms of a species affected by the hazard (the theory is that the number of organisms decreases linearly as the level of exposure increases).

C. The country of Statsylvania has three distinct soil types. One location of each soil type is selected. Each location is split into 10 parts, called "plots." Five are randomly assigned to treatment 1, the other five to treatment 2. Corn is grown on each plot; the response of interest is the corn yield suitable for ethanol production, and a total yield measurement is obtained for each plot. The objective is to estimate the affect of soil type and treatment on yield.
D. A court wants to determine if there is evidence of racial discrimination in the state's sentencing practices for murder convictions. Records for a period the court deems relevant are assembled into a three-way, 2^3, contingency table. The first category is race of victim (white or nonwhite). The second category is race of the convict (white or nonwhite). The third category is type of sentence (death penalty or no death penalty).
E. Twelve counties are randomly selected from a population of "urban, middle-American" counties. Two types of vaccines are to be compared. Six counties are assigned to vaccine type A; the other six are assigned to vaccine type B. In each county, the patients are randomly assigned to two groups. The first group receives the assigned vaccine at a low dose; the second group receives the assigned vaccine at a high dose. Let N_{ijk} be the number of patients at the k^{h} county receiving the j^{h} dose level (low or high) of the ith vaccine type (A or B). Let Y_{ijk} be the number of patients in the ijk^{h} group who show symptoms of the disease the vaccine is intended to protect against. The objective is to assess the impact of vaccine type and dose level on the effectiveness of protection (as measured by the likelihood of showing symptoms).
F. Twelve schools are sampled, one per county, from the same population defined in part (D). Six schools participate in a professional development program for math teachers; six do not (the schools are randomly assigned to the development or no-development group). Students in the schools' math courses are given math proficiency tests at four times during the school year (at the beginning, after the first term, after the second term, and at the end of the school year). The objective is to see if the professional development program improved student learning in math.

For each scenario, determine the required elements of the model:

- Response variable and its assumed distribution
- The link function
- The linear predictor

Note: $\mathbf{X\beta}$ or $\mathbf{X\beta} + \mathbf{Zb}$ is not an adequate statement of the linear predictor—you need to spell it out, so your reader knows what you mean be each symbol and what each subscript signifies.

Also note that there is no *one right answer* here—give what you consider to be a plausible model and be prepared to explain/defend your reasoning.

- The assumed distribution of any effects in the linear predictor that you assume to be random
- *Extra*: what parameters in your model would inform you relative to the objectives? How? (you do not need to go into a lot of detail—just think about how this would work.)

2.2 For each scenario in Exercise 2.1, complete the PROC GLIMMIX statements given in the following that would be minimally necessary to start on analysis.

```
CLASS ...; (if needed)
MODEL ... = .../(essential options, if any);
RANDOM ...; (if needed)
```

For this problem, focus on these three statements, only, and do not include any others.

2.3 For the data sets used in Tables 1.1 and 1.2, write PROC GLIMMIX statements to implement the analyses described in Chapter 1.

a. For Table 1.1, the two analyses both used the linear predictor $\eta_i = \beta_0 + \beta_1 X_i$. One assumed the sample proportion to be approximately normal; the other used a generalized linear model assuming a binomial response.

b. For Table 1.2, there were four analyses. All assumed the linear predictor $\eta_{ij} = (\beta_0 + b_{0i}) + (\beta_1 + b_{1i})X_j$ but made different assumptions as follows:

i. Response variable Y, assumed Gaussian, batch effects fixed

ii. Response variable Y, assumed Gaussian, batch effects random

iii. Response variable assumed binomial, specified by Fav and N, batch effects fixed

iv. Response variable assumed binomial, specified by Fav and N, batch effects random

2.A Appendix A: Common Response Variable (y|b) Distributions

Distribution	Type of Variable	Range of Variable	Mean	Commonly Used Link(s)	Variance
Gaussian/ normal	Continuous, symmetric	$-\infty < y < \infty$	μ	Identity: $\eta = \mu$	σ^2
Log normal	Continuous	$-\infty < y < \infty$	μ	$\eta = \log(\mu)$	σ^2
Gamma	Continuous	$y > 0$	μ	$\eta = \log(\mu), \frac{1}{\mu}$	$\phi\mu^2$
Exponential	Continuous	$y > 0$	μ	$\eta = \log(\mu), \frac{1}{\mu}$	μ^2
Binomial	*Discrete* proportion	$\frac{0}{N}, \frac{1}{N}, \ldots, \frac{N}{N}$	$\pi = \frac{\mu}{N}$	Logit: $\eta = \log\left(\frac{\pi}{1-\pi}\right)$, Probit: $\eta = \Phi^{-1}(\pi)$	$N\pi(1-\pi)$
Beta	*Continuous* Proportion	$0 < y < 1$	μ	$\eta = \log\left(\frac{\mu}{1-\mu}\right)$	$\frac{\mu(1-\mu)}{1+\phi}$
Poisson	Discrete count	$y = 0, 1, 2, \ldots$	λ	$\eta = \log(\lambda)$	λ
Geometric	Discrete count	$y = 0, 1, 2, \ldots$	λ	$\eta = \log(\lambda)$	$\lambda(1+\lambda)$
Negative binomial	Discrete count	$y = 0, 1, 2, \ldots$	λ	$\eta = \log(\lambda)$	$\lambda(1+\phi\lambda)$

2.B Appendix B: Communicating Your Model to Software or "How SAS® PROC GLIMMIX 'Thinks'"

The statements you write in a SAS® procedure statement (or the analogous statements you write in any other statistical software package) are essentially translations from the descriptions of the model in statistical notation (what will henceforth be referred to as "model-speak") to language understood by the computer software (referred to here as "computer-speak" or, when appropriate, "SAS-speak").

In this section, we focus on SAS's GLIMMIX procedure. Why? Recall from Chapter 1 that the GLMM is the most general of the linear models. The GLM is less general in that it can handle non-Gaussian response variables but not random model effects. The LMM is also a special case, handling random model effects but not non-Gaussian data. The LM is the most specific, handling Gaussian data only and no random model effects. The linear model procedures in SAS developed historically in the opposite order. PROC GLM was written for the LM (referred to as the "general" linear model at the time it was written, although "general" is now a dated adjective). Strictly speaking, PROC GLM cannot perform any LMM-specific computations; GLM does have several "band-aids" that address *some* LMM requirements, but at best PROC GLM's handling of LMMs is partial and inadequate. PROC MIXED was written for LMM and can therefore do LM analyses as well. PROC GENMOD was written for GLM (the *generalized* linear model—the *true GLM* in contemporary acronyms) and, since the LM is a subset of the GLM as well as the LMM, GENMOD can therefore do LM analyses as well. Finally, PROC GLIMMIX was written for GLMMs and, since the GLM, LMM, and LM are all special cases, GLIMMIX can handle any model the other PROCs can do and more. All of these procedures share a common approach to syntax. When they were developed, MIXED and GENMOD made use of PROC GLM syntax wherever possible, creating new terms only for items such as random model effects or link functions not found in the LM. GLIMMIX, with minor exceptions, used LMM syntax from MIXED and GLM syntax from GENMOD.

For these reasons, if you are a beginner, start with PROC GLIMMIX. Once you master GLIMMIX, SAS's other linear model procedures are easy to learn, because the syntax is, with few exceptions, the same. The procedures GLM, GENMOD, and MIXED do have special-interest features unique to certain LM, GLM, or MIXED applications, respectively, so GLIMMIX does not *completely* replace the other procedures. However, for most situations, including almost all examples covered in this textbook, GLIMMIX provides everything you need.

2.B.1 General Principles

Recall that the GLMM consists of the following:

- A vector of observed responses, $\mathbf{y}|\mathbf{u}$
- An assumed distribution of $\mathbf{y}|\mathbf{u}$ such that $\mu|u = E(\mathbf{y}|\mathbf{u})$ and $\mathbf{R} = \mathrm{Var}(\mathbf{y}|\mathbf{u})$
- A link function, $\boldsymbol{\eta} = g(\boldsymbol{\mu}|\mathbf{u})$
- A linear predictor, $\mathbf{X}\boldsymbol{\beta} + \mathbf{Z}\mathbf{u}$ that directly models $\boldsymbol{\eta}$
- An assumed distribution of the random vector $\mathbf{u}$—specifically $\mathbf{u} \sim N(\mathbf{0}, \mathbf{G})$

Three basic statements in PROC GLIMMIX work together to specify each element in the aforementioned list: the CLASS, MODEL, and RANDOM statements. From this perspective, you can view these statements as matrix defining and organizing statements. Their functions are as follows:

- The CLASS statement lists all factors in the linear predictor to be considered *classification variables*. If any factor in the linear predictor, fixed or random, does not appear in the CLASS statement, it is considered a *direct* (regression) variable.
- The MODEL statement has the basic form

  `model` *`response variable = list of fixed model effects/`* `options`

  The observation variable y is identified by `response variable`. The `list of fixed model effects` specifies what terms comprise the fixed effects parameter vector $\boldsymbol{\beta}$ and the $\mathbf{X}$ matrix. The `options` include `dist=` and `link=`. The distribution is assumed Gaussian unless you specify otherwise with the `dist=` option. Distributions have a default link (e.g., the logit is the default link for the binomial). If you do not include a `link=` option, the default link is used. So, for example, if you want to fit a probit regression model, you would have to include `dist=probit` as an option in the MODEL statement.

- The RANDOM statement has the basic form

  `random list of random model effects/options`

  This specifies the terms that comprise the random vector $\mathbf{u}$, the $\mathbf{Z}$ matrix, and the form of the variance–covariance matrix $\mathbf{G}$. Several options allow you alter the shape and correlation structure of $\mathbf{G}$. The two most frequently used are `subject=`, which allows you to form a block-diagonal $\mathbf{G}$ matrix and `type=`, which allows you to specify correlation structures.

- A variation on the RANDOM statement,

  `random _residual_ /options`

  allows you to specify the shape and structure of the $\mathbf{R}$ matrix. If you have previously used PROC MIXED or PROC GENMOD, you need to know that REPEATED statement in those PROCs does not exist in PROC GLIMMIX. Use the `random_ residual_`statement where you would use REPEATED in MIXED or GENMOD.

Example 2.B.1

Simple linear regression, Gaussian data. The linear predictor is $\alpha + \beta X$. Observations are Gaussian, that is, $\mathbf{y} \sim N(\boldsymbol{\mu}, \mathbf{I}\sigma^2)$. The link is the identity, $\eta = \mu$. The PROC GLIMMIX statements are

```
proc glimmix;
model y=x;
```

No CLASS statement appears because the only variable in the linear predictor is X, a direct variable. Y identifies the response variable. The intercept (α in this case) is assumed on default, so a column of ones is automatically put in the $\mathbf{X}$ matrix and the intercept parameter is automatically placed in the $\boldsymbol{\beta}$ vector unless you specifically turn the intercept default off (a subsequent example will demonstrate when you would want to do this and how to do it). The variable X is direct, so the elements of X appear literally as a column in the **X** matrix. There are no random models effects, so no need for a RANDOM statement.

You can see from this example that the only *mandatory* statements in PROC GLIMMIX are the PROC and MODEL statements.

To visualize the effect of these statements, consider the data in Table 1.1.

The statement `model y = x` does the following.

First, the elements in the column variable y are placed in a vector

$$\mathbf{y} = \begin{bmatrix} 0 \\ 0 \\ 2 \\ 2 \\ 2 \\ 5 \\ 7 \\ 12 \\ 10 \\ 16 \\ 9 \end{bmatrix}$$

Note that if the response variable had a different name, for example, `weight`, then the model statement would be `model weight = x`.

Second, the **X**-matrix is formed from a column of ones and the first parameter in the $\boldsymbol{\beta}$ vector is the intercept—these occur by default. Then the contents of the column variable x become the second column of the **X** matrix—the elements of x are inserted literally because x is a direct variable. This results in the following:

$$\mathbf{X}\boldsymbol{\beta} = \begin{bmatrix} 1 & 0 \\ 1 & 1 \\ 1 & 2 \\ 1 & 3 \\ 1 & 4 \\ 1 & 5 \\ 1 & 6 \\ 1 & 7 \\ 1 & 8 \\ 1 & 9 \\ 1 & 10 \end{bmatrix} \begin{bmatrix} \alpha \\ \beta \end{bmatrix}$$

Example 2.B.2

Simple logistic linear regression. Binomial data. As in Example 2.B.1, the linear predictor is $\alpha + \beta X$. The observations are $\mathbf{y} \sim \text{Binomial}(\mathbf{N}, \boldsymbol{\pi})$. The logit link is $\eta = \log[\pi/(1 - \pi)]$. The PROC GLIMMIX statements are

```
proc glimmix;
model y/n=x;
```

Note that the absent class variable and the presence only of the variable x on the right hand side of the equal sign in the MODEL statement have the same effect as in Example 2.B.1: they set up $\mathbf{X}\boldsymbol{\beta}$ consistent with the linear predictor. The only difference is the response variable *Y/N*. Giving the response variable as a ratio means that the distribution is binomial. The logit is the default link for the binomial, so no `link=` option is needed.

If you want to use a different link, you must specify it. For example, if you want to fit the model using the probit link, the PROC GLIMMIX statements are

```
proc glimmix;
model y/n=x/link=probit;
```

Example 2.B.3

One-way treatment design, Gaussian data, linear predictor $\mu + \tau_i$, where μ is the intercept and τ_i denotes the ith treatment effect. The observations are $\mathbf{y} \sim N(\boldsymbol{\mu}, \mathbf{I}\sigma^2)$, where $\boldsymbol{\mu}$ is the vector of treatment means $\mu_i = \mu + \tau_i$ (do not confuse the treatment mean *vector* $\boldsymbol{\mu}$ with the intercept *parameter* (a scalar) μ!). The link is the identity, $\boldsymbol{\eta} = \boldsymbol{\mu}$. The PROC GLIMMIX statements are

```
proc glimmix;
class trt;
model y = trt;
```

As with Example 2.B.1, the "`y =`" part of the `model` statement cause the response vector to be created from the contents of the column variable y. Also, as with both Examples 2.B.1 and 2.B.2, the first column in the $\mathbf{X}$ matrix is a column of ones. The variable *trt* is a CLASS variable, so when it occurs in the MODEL statement, a column of dummy variables is created for each distinct value of *trt*. Table 2.B.1 shows data from a one-way design to illustrate the impact of the CLASS and MODEL statement.

The MODEL statement reads the TRT column vector

$$\begin{bmatrix} 1 \\ 1 \\ 2 \\ 2 \\ 3 \\ 3 \end{bmatrix}$$

TABLE 2.B.1

One-Way Gaussian Data

Obs	trt	y
1	1	19.0
2	1	19.2
3	2	21.9
4	2	20.8
5	3	21.2
6	3	23.3

and the CLASS statement converts it to the "design" matrix

$$\begin{bmatrix} 1 & 0 & 0 \\ 1 & 0 & 0 \\ 0 & 1 & 0 \\ 0 & 1 & 0 \\ 0 & 0 & 1 \\ 0 & 0 & 1 \end{bmatrix}$$

Ones appear in column 1 in rows corresponding to the observations that received $trt = 1$, in column 2 in rows corresponding to the observations that received $trt = 2$, etc. Thus, the final **X** matrix and $\boldsymbol{\beta}$ vector are

$$\begin{bmatrix} 1 & 1 & 0 & 0 \\ 1 & 1 & 0 & 0 \\ 1 & 0 & 1 & 0 \\ 1 & 0 & 1 & 0 \\ 1 & 0 & 0 & 1 \\ 1 & 0 & 0 & 1 \end{bmatrix} \begin{bmatrix} \mu \\ \tau_1 \\ \tau_2 \\ \tau_3 \end{bmatrix}.$$

Example 2.B.4

Identical to Example 2.B.3 except observations are binomial, that is, y_{ij} = Binomial(N_{ij}, π_i). Same linear predictor as Example 2.B.3. The default link is the logit. Table 2.B.2 shows data to illustrate the required form for SAS.

The required PROC GLIMMIX statements are

```
proc glimmix;
class trt;
model Y_ij/N_ij = trt;
```

Note that in the data set, N_{ij} has the variable name `N _ ij` and y_{ij} has the variable name `Y _ ij` so in the model statement create the binomial response variable and the default logit link. The variable `trt` in the CLASS statement and to the right of the equal sign in the MODEL statement has the same effect as it did in Example 2.B.3.

TABLE 2.B.2

One-Way Binomial Data

Obs	trt	N_ij	Y_ij
1	1	13	2
2	1	7	2
3	2	13	4
4	2	13	9
5	3	12	8
6	3	15	7

Example 2.B.5

Multi-batch linear regression, as discussed in Section 1.3. Data shown in Table 1.2. Two forms of the linear predictor were considered: $\alpha_i + \beta_i X_{ij}$ and $\alpha + a_i + (\beta + b_i)X_{ij}$. First consider $\alpha_i + \beta_i X_{ij}$.

In Table 1.3, LM and GLM forms of this model were given, the LM with a response variable assumed to have a $N(\mu_i, \sigma^2)$ distribution and an identity link, the GLM assumed to have a Binomial(N_{ij}, π_i) distribution and a logit link. First, consider the LM. To get the computer to understand the model we want, we need to do two new things: we need to give each batch a unique intercept (the α_i) and we need to specify a separate slope for each batch. For the latter, we introduce the nested effect command, in this case `X(A)`. SAS understands this command to mean "X has different parameter values for each level of factor A." The matrix command it sets up is a direct product of the vector or matrix called for by the effect x and the vector or matrix called for by the effect A. In this case, x is a direct variable, so the **X** matrix is the 9×1 forms from the values of x, 0, 3, 6,…, 48 and $\boldsymbol{\beta}$ vector is simple with slope coefficient β. "Factor A" in this case is `Batch`, which is a classification variable with four levels. You can see the form of the **X** matrix and $\boldsymbol{\beta}$ vector for `Batch` in Section 1.4. Taking the direct product yields

$$\begin{bmatrix} 0 \\ 3 \\ 6 \\ 9 \\ 12 \\ 18 \\ 24 \\ 36 \\ 48 \end{bmatrix} \otimes \begin{bmatrix} 1 & 0 & 0 & 0 \\ 1 & 0 & 0 & 0 \\ 1 & 0 & 0 & 0 \\ 1 & 0 & 0 & 0 \\ 1 & 0 & 0 & 0 \\ 1 & 0 & 0 & 0 \\ 1 & 0 & 0 & 0 \\ 1 & 0 & 0 & 0 \\ 1 & 0 & 0 & 0 \\ 0 & 0 & 0 & 0 \\ \cdots & \cdots & \cdots & \cdots \\ 0 & 0 & 0 & 0 \\ 0 & 0 & 0 & 1 \\ 0 & 0 & 0 & 1 \\ 0 & 0 & 0 & 1 \\ 0 & 0 & 0 & 1 \\ 0 & 0 & 0 & 1 \\ 0 & 0 & 0 & 1 \\ 0 & 0 & 0 & 1 \\ 0 & 0 & 0 & 1 \\ 0 & 0 & 0 & 1 \end{bmatrix} = \begin{bmatrix} 0 & 0 & 0 & 0 \\ 3 & 0 & 0 & 0 \\ 6 & 0 & 0 & 0 \\ 9 & 0 & 0 & 0 \\ 12 & 0 & 0 & 0 \\ 18 & 0 & 0 & 0 \\ 24 & 0 & 0 & 0 \\ 36 & 0 & 0 & 0 \\ 48 & 0 & 0 & 0 \\ 0 & 0 & 0 & 0 \\ \cdots & \cdots & \cdots & \cdots \\ 0 & 0 & 48 & 0 \\ 0 & 0 & 0 & 0 \\ 0 & 0 & 0 & 3 \\ 0 & 0 & 0 & 6 \\ 0 & 0 & 0 & 9 \\ 0 & 0 & 0 & 12 \\ 0 & 0 & 0 & 18 \\ 0 & 0 & 0 & 24 \\ 0 & 0 & 0 & 36 \\ 0 & 0 & 0 & 48 \end{bmatrix} \begin{bmatrix} \beta_1 \\ \beta_2 \\ \beta_3 \\ \beta_4 \end{bmatrix}$$

With this is mind, the SAS statements for this model are

```
proc glimmix;
  class batch;
  model y=batch x(batch)/noint;
```

Note the new option in the MODEL statement: `noint`. Recall that the MODEL statement creates an intercept by default. With the default (i.e., without the `noint` option), the aforementioned statements would produce the linear predictor $\mu + \alpha_i + \beta_i X_{ij}$—that is, a linear predictor with an unwanted and superfluous intercept term (μ). The noint option suppresses the creation of the intercept, yielding the desired model $\alpha_i + \beta_i X_{ij}$.

For binomial data with the logit link function, replace y on the left hand side of the equal sign in the MODEL statement by Fav/N_ij. If you want to use the probit link, add link=probit as an option in the MODEL statement, just as you did in Example 2.B.2.

Now consider the alternative form of the model $\alpha + a_i + (\beta + b_i)X_{ij}$. Here, you *want* the intercept default to be active, since the intercept parameter is α. You also want the slopes to be partitioned into the overall (β) and batch-specific (b_i) components. You do this using the statements

```
proc glimmix;
  class batch;
  model y=batch x x*batch;
```

Here, the absence of the noint option allows the intercept (α) to be created. The variable batch is defined by the statement class batch to be a classification variable, so a 36×4 design matrix (as shown in Section 1.4) is created for the **X** matrix corresponding to the four batch parameters in the $\boldsymbol{\beta}$ vector. The variable x is a direct variable, so a 36×1 vector containing the value of x for each observation is created corresponding to the overall slope parameter β. The term x*batch creates a 36 × 4 matrix equal to the direct product of the design matrix for batch and the vector for x. The full matrix form of this model is shown in Section 1.4.

If the response variable is *Binomial*, you make the same changes shown earlier.

Question: What model would the following statements create?

```
proc glimmix;
  class batch;
  model y=batch x*batch/noint;
```

Answer: The model $\alpha_i + \beta_i X_{ij}$. Note that the terms x(batch) and x*batch both create matrix direct products, so they effectively create the same model.

Question: With this in mind, what model does

```
proc glimmix;
  class batch;
  model y=batch x*batch;
```

create?

Example 2.B.6

Same linear predictor equation as Example 2.B.5, but batch effects a_i and b_i are random. So far, we have only seen fixed effects models, and hence only the CLASS and MODEL statements. Recall that the MODEL statement defines the $\mathbf{X}\beta$ component of the linear predictor. The RANDOM statement defines the **Zu** component. The basic syntax of the RANDOM statement is essentially the same as the syntax for the right-hand side of the equation in the MODEL statement.

The simplest way to write the model $\alpha + a_i + (\beta + b_i)X_{ij}$ with the batch effects random is

```
proc glimmix;
  class batch;
  model y = x;
random batch x*batch;
```

This creates the same matrices as the fixed effects version of the model shown in Example 2.B.5. Note, however, that the batch and x* batch effects move to the RANDOM statement. This simultaneously creates the appropriate components of the **Z** matrix and **u** vector—noting that

$$\mathbf{u}=\begin{bmatrix}\mathbf{a}\\\mathbf{b}\end{bmatrix}=\begin{bmatrix}a_1\\a_2\\a_3\\a_4\\b_1\\b_2\\b_3\\b_4\end{bmatrix}$$

and it creates the implied **G** matrix, recalling

$$\mathbf{G}=\mathrm{Var}(\mathbf{u})=\mathrm{Var}\left(\begin{bmatrix}\mathbf{a}\\\mathbf{b}\end{bmatrix}\right)=\begin{bmatrix}\mathbf{I}\sigma_A^2 & \mathbf{I}\sigma_{AB}\\\mathbf{I}\sigma_{AB} & \mathbf{I}\sigma_B^2\end{bmatrix}$$

An alternative—and preferable—way to define the random part, **Zu**, of the linear predictor uses the SUBJECT= option. This option defines a block-diagonal structure for the **Z** and **G** matrices. So—how does it work, and why is it preferable? First, how does it work?

Note that each batch has an intercept and a slope. In matrix terms, **Zu** for the ith batch is

$$\begin{bmatrix}1 & 0\\1 & 3\\1 & 6\\1 & 9\\1 & 12\\1 & 18\\1 & 24\\1 & 36\\1 & 48\end{bmatrix}\begin{bmatrix}a_i\\b_i\end{bmatrix}$$

The **G** matrix for the ith batch is thus $\mathrm{Var}\begin{bmatrix}a_i\\b_i\end{bmatrix}=\begin{bmatrix}\sigma_A^2 & \sigma_{AB}\\\sigma_{AB} & \sigma_B^2\end{bmatrix}$. Denote these $\mathbf{Z}_i$, $\mathbf{u}_i$, and $\mathbf{G}_i$, respectively. Now note that you can construct the full **Z** and **G** as block-diagonal matrices formed from the $\mathbf{Z}_i$ and $\mathbf{G}_i$, respectively. That is,

$$\mathbf{Z}=\begin{bmatrix}\mathbf{Z}_1 & \mathbf{0} & \mathbf{0} & \mathbf{0}\\\mathbf{0} & \mathbf{Z}_2 & \mathbf{0} & \mathbf{0}\\\mathbf{0} & \mathbf{0} & \mathbf{Z}_3 & \mathbf{0}\\\mathbf{0} & \mathbf{0} & \mathbf{0} & \mathbf{Z}_4\end{bmatrix}$$

and

$$G = \begin{bmatrix} G_1 & 0 & 0 & 0 \\ 0 & G_2 & 0 & 0 \\ 0 & 0 & G_3 & 0 \\ 0 & 0 & 0 & G_4 \end{bmatrix}$$

It follows that

$$\mathbf{u} = \begin{bmatrix} \mathbf{u}_1 \\ \mathbf{u}_2 \\ \mathbf{u}_3 \\ \mathbf{u}_4 \end{bmatrix}$$

You can quickly verify that these are simply rearranged versions of the **u** vector and **Z** and **G** matrices shown earlier. The SAS statements defining the linear predictor using this approach are

```
proc glimmix;
  class batch;
  model y = x;
  random intercept x/subject=batch;
```

The `subject=batch` option causes the **Z** and **G** matrices to be defined as block diagonal with `batch` used to define the blocks. Within each block of the matrices, there is an intercept and a slope coefficient for x, resulting in the matrices $\mathbf{Z}_i$, $\mathbf{u}_i$, and $\mathbf{G}_i$ shown earlier.

Now for the second question: Why go to all this trouble? Why is using the SUBJECT= option preferable? First, the block-diagonal matrix structure allows for more efficient computing. This is not only true of SAS PROC GLIMMIX, but it is generally true of matrix-based statistical computing algorithms. All GLMM software, SAS or otherwise, is necessarily matrix based. When we get to more complex GLMMs in the latter chapters of the text, there are models that simply will not run unless we use the SUBJECT= option to streamline the required computing. The second advantage will become clear when we begin to consider estimates of linear combinations of fixed and random effects (*"best linear unbiased predictors"* or *BLUP*). In this example, $\alpha + \beta X_{ij}$ gives the *population averaged* regression equation, that is, the expected value of the linear predictor over the entire population that the batches represent. However, in some applications, predicted values for specific batches, $\alpha + a_i + (\beta + b_i)X_{ij}$, called the *batch-specific BLUP*, are of interest. Note that this is a linear combination of fixed and random model effects. If you use the `subject=batch` option, you enable the procedure to compute these BLUPs relatively conveniently.

Example 2.B.7

A × B factorial, that is, two factors, factor A having $a \geq 2$ levels and factor B having $b \geq 2$ levels. All possible A × B combinations observed. In this example, we focus on defining the linear predictor. Sample SAS statements here are all shown with a normally distributed response variable, but as with the Examples 2.B.1 through 2.B.6, you can easily adapt these statements to binomial or other non-normally distributed response variables using the same adaptations previously shown.

The first thing to understand—and this is a point that introductory design of experiment classes frequently miss—is that there is not *one* model for factorial structures. Several models are possible. Each has its uses, depending on the situation and objectives. This section will be more terse than the previous examples. The basic principles of how SAS "understands" model formation commands have been present. In this section, SAS statements and their model consequences are given. You are encouraged to work through each example in sufficient detail so that you can visualize and describe exactly the contents of each models **X** matrix and $\boldsymbol{\beta}$ vector. In Example 2.B.8, we will revisit these models with additional random effects resulting from various blocking structures and split-plot features.

1. Classical main effects and interaction: $\mu + \alpha_i + \beta_j + (\alpha\beta)_{ij}$

   ```
   proc glimmix;
     class a b;
   model y=a b a*b;
   ```

   or equivalently `model y=a|b;`
2. Consider every $A_i \times B_j$ combination a "treatment" and reduce the model to a one way treatment effects model: $\mu + (\alpha\beta)_{ij}$. **Note!** The effect $(\alpha\beta)_{ij}$, in this model, is *not* an interaction effect! It is simply the effect—or departure from the intercept—of the *ij*th treatment combination.

   ```
   proc glimmix;
     class a b;
   model y = a*b;
   ```

3. *Cell means model*: Suppress the intercept in the previous model. This gives you $(\alpha\beta)_{ij}$. Note this is just another way of writing μ_{ij}, the *ij*th treatment means.

   ```
   proc glimmix;
     class a b;
   model y = a*b/noint;
   ```

4. *Nested model*: In this case, the main effects of A, then the effects of B nested within A. The model is $\mu + \alpha_i + \beta(\alpha)_{ij}$. You use this model when the main effects of B are not of interest, only the effects of B within specific levels of A. Sometimes, the objectives call for this model directly; often, this model is used once a non-negligible A × B interaction has been established and your focus shifts to simple effects, for example, of B within specific levels of A.

   ```
   proc glimmix;
     class a b;
   model y = a b(a);
   ```

   Recalling our discussion of Example 2.B.5, note that the following statements

   ```
   proc glimmix;
     class a b;
   model y = a a*b;
   ```

   yield exactly the same model. This is because both cause the same direct product matrices to be computed and the same resulting parameter vectors. *SAS does not assume a B main effect if none is given.*
5. *Regression model*: What if B is a quantitative factor? Then there is no reason not to treat it as a regression (direct) variable—and considerable justification *for* doing so. You can use essentially the same modeling strategy as Example 2.B.5:

```
proc glimmix;
  class a;
model y =a b(a)/noint;
```

or, alternatively,

```
proc glimmix;
  class a;
model y =a b a*b;
```

What models do these produce? Be careful! Be sure to state the linear predictor implied by each set of statements precisely!
How would you modify these statements if you suspected a quadratic effect of B?

Example 2.B.8

School professional development effectiveness study from Section 2.1. Recall that the linear predictor was $\eta_{ijk} = \mu + s_i + \rho_j + t(s\rho)_{ijk}$, with the school effects, s_i, and teacher effects, $t(s\rho)_{ijk}$, random. Thus, the MODEL statement defines the **X** matrix and $\boldsymbol{\beta}$ vector for the fixed effects, intercept, μ, and professional development ρ_j and the RANDOM statement defines the **Z** matrix, **u** vector, and covariance matrix **G** for the school and teacher effects. Professional development and school and teacher effects are all classification variables. Using the same logic as in the random batch regression example (Example 2.B.6), the following SAS statements define the linear predictor:

```
proc glimmix;
  class pd school teacher;
  model y=pd;
  random school teacher(school*pd);
```

The variable pd denotes the professional development effect. Notice that the component of the **Z** matrix corresponding to the teacher effect [teacher(school*pd)] is defined by the direct product of the design matrices for the classification variables teacher, school, and pd.

As with previous examples, if you write the random statement as

```
random school teacher*school*pd;
```

the resulting model is identical, because teacher*school*pd creates the same component of the **Z** matrix, and associated elements of the **u** vector and **G** matrix, as teacher(school*pd).

Alternatively, and preferably, you can use the SUBJECT= option in the RANDOM statement. As with Example 2.B.6, this creates a more computationally efficient form of the model and allows you to compute school- or teacher-specific BLUPs if called for by the objectives of the study. The SAS statements are

```
proc glimmix;
  class pd school teacher;
  model y=pd;
  random intercept teacher*pd/subject=school;
```

Recall that subject=school defines a block-diagonal structure for the **Z** matrix, one block per school and the columns within each block correspond to the intercept (a column of ones) and teacher*pd (the design matrix defined by the direct product of teacher and pd).

You are only allowed one MODEL statement, but you can have multiple random statements. To make teacher-specific BLUPs more convenient, you could rewrite the SAS statements as

```
proc glimmix;
  class pd school teacher;
  model y=pd;
  random intercept/subject=school;
  random pd/subject=teacher(school);
```

or equivalently `random pd/subject=teacher*school;`
When we begin to work with BLUPs in detail (starting in Chapter 6), you will see that the second RANDOM statement makes it convenient to compute an effect specific to each teacher.

Example 2.B.9

The three split-plot scenarios from Section 2.2 describing different ways to conduct the A × B treatment design.

Scenario 1: Split-plot, no blocking. Linear predictor: $\eta_{ijk} = \mu + \alpha_i + \beta_j + (\alpha\beta)_{ij} + u(\alpha)_{ik}$. SAS statements:

```
proc glimmix;
  class a b exp_unit;
  model y=a|b;
```

or equivalently `model y=a b a*b;`

```
random exp_unit(a);
```

or equivalently `random exp_unit*a;`
or equivalently `random intercept/subject=exp_unit(a);`
or equivalently `random intercept/subject=exp_unit*a;`
The variable `exp_unit` denotes the whole-plot experimental unit. Note that it *must* be a classification variable.

Recalling our discussion of the $A \times B$ treatment design earlier in this section (see Example 2.B.7), remember that the decomposition $\mu + \alpha_i + \beta_j + (\alpha\beta)_{ij}$ of the $A \times B$ treatment design is *not* mandatory. You could use the cell means approach, μ_{ij}, whose associated SAS statements are

```
proc glimmix;
  class exp_unit a b;
  model y = a*b/noint;
  random intercept/subject=exp_unit(a);
```

All the "or equivalently" forms of the RANDOM statement have been skipped, but you could use any of them here as well.

Alternatively, you could use decomposition, $\mu + \alpha_i + \beta(\alpha)_{ij}$, to focus on the simple effects of B within specific levels of factor A. The SAS statements are

```
proc glimmix;
  class exp_unit a b;
  model y = a b(a);
```

or equivalently `model y = a a*b;`

```
random intercept/subject=exp_unit(a);
```

Scenario 2: Split-plot. The whole-plot experimental units (units of replication with respect to levels of factor A) are arranged in blocks. Linear predictor: $\eta_{ijk} = \mu + \alpha_i + \beta_j + (\alpha\beta)_{ij} + r_k + r\alpha_{ik}$. SAS statements:

```
proc glimmix;
  class block a b;
  model y = a|b;
  random intercept a/subject=block;
```

note this is equivalent to `random block block*a;` **why?**

The variable `block` denotes the blocking factor. As with Scenario B.1, you can use all of the alternative decompositions of the A × B treatment design (cell means, nested, etc.) in the MODEL statement. Also, there are several equivalent ways to write the RANDOM statements. The MODEL statements are identical to those shown in Scenario B.1. The RANDOM statements are similar, differing only in that they need the block effect as well as the whole plot error. It is left as an exercise to the reader to determine their exact form.

Scenario 3: Strip-plot. Factor A, factor B, and A × B combinations each have different sized units of replication. Linear predictor: $\eta_{ijk} = \mu + \alpha_i + \beta_j + (\alpha\beta)_{ij} + r_k + r\alpha_{ik} + r\beta_{jk}$. SAS statements:

```
proc glimmix;
  class block a b;
  model y = a|b;
  random intercept a b/subject=block;
```

note this is equivalent to `random block block*a block*b;` **why?**

As with Scenarios 1 and 2, you can use all of the alternative decompositions of the A × B treatment design (cell means, nested, etc.) in the MODEL statement and the resulting MODEL statements are identical to those shown in Scenario B.1. Similarly, you can use either form of the RANDOM statement shown earlier. With standard analyses of observations with a Gaussian distribution, it does not make much difference which form you use. For non-Gaussian response variables, the form with the `subject=block` option is preferable for computational reasons.

KEY IDEAS: *Functions of CLASS, MODEL, and RANDOM statements. What does NOINT option do and when do you use it? What does SUBJECT= option do and why do you use it? Matrix operations are implicit in the various terms in the MODEL and RANDOM statements. What do B(A) and A*B syntax do?*

3

Setting the Stage

In Chapter 1, the essential elements of a linear statistical model were introduced. At a minimum, a model must specify the distribution of the observations, the linear predictor, and the link function. If the linear predictor is entirely deterministic, we have a fixed-effects-only linear predictor, $\mathbf{X\beta}$. If the model also includes random effects, the linear predictor is $\mathbf{X\beta} + \mathbf{Zb}$, and the distribution of the random effects, $\mathbf{b}$, must be specified.

Chapter 2 explored the process leading from data set structure to model construction. The essential feature of the resulting model is that it must be a plausible description of the process that gave rise to the observations. The wording here is important: "...*a* plausible description..." not "...*the* one and only true description...." In science, we can—and often do—have two (or more) plausible descriptions of a process. The object of statistical inference is, at least in part, to determine which of the competing explanations is more consistent with the data.

The purpose of this chapter is to set the stage for statistical inference with linear models. This involves two general themes.

First, once we specify a model and estimate its parameters, what do we *do* with these estimates? How do we use them to answer the questions that motivated us to collect the data in the first place? This in itself is an important—and underappreciated—modeling topic. Too often, surprising as it sounds, statistical analysis and scientific inquiry are compartmentalized in ways that impoverish both. One goal of this chapter is to help reduce such compartmentalization. Framing parameter estimates in terms of scientific questions—and vice versa—is one essential step. It is important to be able to move easily between "science-speak" and "stat-speak."

The second theme of this chapter concerns issues that would not occur to us in a fixed-effects-only, Gaussian-data-only world. These are model scale/data scale, inference space, and conditional/marginal issues. All are to some extent underappreciated but, as you will see, vitally important. They are introduced in Section 3.2 and dealt with in detail in Sections 3.3 through 3.5.

3.1 Goals for Inference with Models: Overview

A helpful mnemonic for the types of linear models is (G)LM(M). Common to all is the LM, linear model, part: the fixed effects component of the linear predictor $\mathbf{X\beta}$ present in every linear model. The (G) prefix means the distribution of the observations may be non-normal; the (M) suffix means the linear predictor includes random effects, a $\mathbf{Zb}$ term. The fixed linear predictor, $\mathbf{X\beta}$, is important because, as we saw in Chapter 2, the fixed effects usually describe the treatment design, which in turn is determined by the objectives—the fundamental questions we want to answer—of the study. For this reason, if we are doing a good job modeling, we should be able to express every objective as a question about a model parameter or a linear combination of model parameters.

Consider some examples.

1. *Motivating question*: How does a stability limiting characteristic of a drug change over time?

   Assuming that the characteristic, Y, changes linearly with time, $\mathbf{X\beta}$ follows from the linear regression equation $Y = \beta_0 + \beta_1 T$, where T denotes time. The questions "is there change over time" or "how does Y change as a function of T" would both be answered by inference on the slope parameter, β_1. What if linear regression does not adequately characterize change over time? We could replace $\beta_0 + \beta_1 T$ with a more suitable linear predictor (perhaps a quadratic or higher-order polynomial or a nonlinear function). Inference would still focus on the relevant change-over-time parameters.

2. *Motivating question*: How do three or more treatments of interest affect response?

   Here $\mathbf{X\beta}$ would follow a one-way model $\eta + \tau_i$, where τ_i is the effect of the ith treatment. The scientific question may translate into one of more statements about the treatment effect. Interest may focus on assessing the overall equality of the treatments' expected response, that is, $\tau_1 = \tau_2 = \tau_3$. Alternatively, we may want to characterize the expected mean performance of individual treatments. In model terms, $\eta + \tau_i$ provides a starting point for expected treatment performance. If the data are Gaussian, $\eta + \tau_i$ also gives a direct estimate of the treatment mean. If the data are non-Gaussian, $\eta + \tau_i$ gives an estimate in terms of the link function, but does not directly estimate the treatment mean. This anticipates Section 3.3: the model scale vs. data scale issue.

   Interest in treatment differences typically goes beyond merely assessing $\tau_1 = \tau_2 = \tau_3$ to more specific questions. Often we want to assess the difference between pairs of means, $\tau_i - \tau_{i'}$ in model terms, or sets of means, for example, $\tau_1 - 1/2(\tau_2 + \tau_3)$. As with expected treatment performance, these directly estimate mean differences for Gaussian data. For non-Gaussian data, they are estimates in terms of the link function and require translation if one desires mean differences.

3. *Motivating question*: In a two-treatment growth curve study, how do treatments affect changes over time?

   To keep this discussion simple, as with the first motivating question, assume linear change over time. The generic form of $\mathbf{X\beta}$ is $\beta_{0i} + \beta_{1i}T$, where β_{0i} denotes the intercept for the ith treatment and β_{1i} is the slope for the ith treatment. Inference might focus on the difference in slopes between the two treatments, $\beta_{1,1} - \beta_{1,2}$, or the difference in the expected response to the two treatments at a particular time, $\beta_{0,1} - \beta_{0,2} + (\beta_{1,1} - \beta_{1,2})T$, and so forth.

4. *Motivating question*: How do *two* factors affect response, either individually or as a system?

   Note that addressing this objective requires a factorial treatment structure. For simplicity, consider a 2 × 2 factorial—two levels of each factor, all possible combinations. Here $\mathbf{X\beta}$ follows a two-way model with interaction, $\eta_{ij} = \eta + \alpha_i + \beta_j + (\alpha\beta)_{ij}$. As with all models, the linear predictor is expressed in terms of the link function—η_{ij} may directly estimate the treatment combination mean, μ_{ij} (as with Gaussian data), or not (as is usually the case for non-Gaussian data).

TABLE 3.1

Estimable Functions of Interest in Factorial Treatment Structures

Estimation Objective	Estimate in Terms of Linear Predictor Parameter	Objective Stated in Terms of Expected Value[a]
A × B combination	$\eta + \alpha_i + \beta_j + (\alpha\beta)_{ij}$	μ_{ij}
Main effect mean, e.g., factor A	$\bar{\eta}_{i\bullet} = \eta + \alpha_i + \frac{1}{2}\sum_j \beta_j + \frac{1}{2}\sum_j (\alpha\beta)_{ij}$	$\bar{\mu}_{i\bullet} = \frac{1}{2}\sum_j \mu_{ij}$
Main effect difference, e.g., A	$\bar{\eta}_{1\bullet} - \bar{\eta}_{2\bullet} = \alpha_1 - \alpha_2 + \frac{1}{2}\left[\sum_j (\alpha\beta)_{1j} - \sum_j (\alpha\beta)_{2j}\right]$	$\bar{\mu}_{1\bullet} - \bar{\mu}_{2\bullet}$
Simple effect $A \mid B_j$	$\eta_{1j} - \eta_{2j} = \alpha_1 - \alpha_2 + (\alpha\beta)_{1j} - (\alpha\beta)_{2j}$	$\mu_{1j} - \mu_{2j}$
Simple effect $B \mid A_i$	$\eta_{i1} - \eta_{i2} = \beta_1 - \beta_2 + (\alpha\beta)_{i1} - (\alpha\beta)_{i2}$	$\mu_{i1} - \mu_{i2}$
Interaction $= A \mid B_1 - A \mid B_2$ $= B \mid A_1 - B \mid A_2$	$(\eta_{11} - \eta_{21}) - (\eta_{12} - \eta_{22})$ $= (\alpha\beta)_{11} - (\alpha\beta)_{21} - (\alpha\beta)_{12} + (\alpha\beta)_{22}$ $= (\eta_{11} - \eta_{12}) - (\eta_{21} - \eta_{22})$	$\mu_{11} - \mu_{12} - \mu_{21} + \mu_{22}$

[a] Expectations estimated directly by expressions in column two for identity link *only*.

Inference might focus on one or more of the following: a treatment combination mean; a main effect mean, for example, the mean of a level of factor A averaged over all levels of B or vice versa; a main effect difference; and a simple effect, that is, the difference between two levels of A at a given level of B, the difference between two levels of B at a given level of A, and so forth. Each of these objectives can be expressed in terms of the parameters of the linear predictor, as shown in Table 3.1.

If the data are Gaussian, the expressions in column two do in fact estimate the expectations in column three. For non-Gaussian data, column two directly estimates the expectations in column three only if the model uses the identity link (which is generally not the case). For nonidentity link functions, the estimates in column two require translation.

In general, you should start by assessing interaction between the two factors. By definition, interaction occurs when the simple effects are not equal. If interaction is negligible, the main effects provide useful information; otherwise, main effects tend to be misleading, and it is best to focus on simple effects.

> There is no statistical rule that says *any* of the aforementioned tests is *required*. For any of the aforementioned to be the valid target of statistical inference, you *must* be able to paraphrase it strictly in terms of the question motivating the study and in the language of the discipline conducting the study.
>
> *Suggested exercise*: Paraphrase each linear combination of parameters defined earlier in "discipline-speak" rather than as the "stat-speak" expression given here.

KEY IDEAS: *expressing objectives or motivating questions as linear combinations of parameters; objectives associated with regression, one-way treatment design, and factorial treatment structure; simple effects, main effects, and interaction*

3.2 Basic Tools of Inference

All of the examples in Section 3.1 illustrate variations on a common theme. Every statement involves either estimating or testing a linear combination of model parameters. In this section, we present the formal structure and terminology associated with these linear combinations.

3.2.1 Estimable Functions

The generic form of every linear combination of parameters shown in Section 3.1 is $\mathbf{K}'\boldsymbol{\beta}$, where $\mathbf{K}$ is a $p \times k$ matrix and $\boldsymbol{\beta}$ is the $p \times 1$ vector of fixed effects. Either we estimate $\mathbf{K}'\boldsymbol{\beta}$ or we test $H_0\colon \mathbf{K}'\boldsymbol{\beta} = \boldsymbol{\theta}$, where $\boldsymbol{\theta}$ is a $p \times 1$ vector of constants. Typically, we structure tests so that $\boldsymbol{\theta} = \mathbf{0}$. In most of the examples shown previously, $\mathbf{K}$ is a vector—we denote such vectors by $\mathbf{k}$.

In example 1, $\boldsymbol{\beta} = \begin{bmatrix} \beta_0 \\ \beta_1 \end{bmatrix}$; interest centered on testing $H_0\colon \beta_1 = 0$, therefore $\mathbf{k}' = \begin{bmatrix} 0 & 1 \end{bmatrix}$.

In example 2, $\boldsymbol{\beta} = \begin{bmatrix} \mu \\ \tau_1 \\ \tau_2 \\ \tau_3 \end{bmatrix}$. For estimating treatment means, for example, treatment 1, $\mathbf{k}' = \begin{bmatrix} 1 & 1 & 0 & 0 \end{bmatrix}$. For estimating or testing treatment differences, for example, treatment 1 vs. treatment 2, $\mathbf{k}' = \begin{bmatrix} 0 & 1 & -1 & 0 \end{bmatrix}$. For the contrast, treatment 1 vs. the average of treatments 2 and 3, $\mathbf{k}' = \begin{bmatrix} 0 & 1 & -1/2 & -1/2 \end{bmatrix}$. For the overall equality of treatment means, $\mathbf{K}$ must be composed of any two vectors that guarantee $\tau_1 = \tau_2 = \tau_3$. Why two? Because there are three treatment levels, hence two degrees of freedom for differences among treatments—one contrast vector per degree of freedom.

The matrix $\mathbf{K}' = \begin{bmatrix} 0 & 1 & 0 & -1 \\ 0 & 0 & 1 & -1 \end{bmatrix}$ satisfies this requirement (why?).

So does $\mathbf{K}' = \begin{bmatrix} 0 & 1 & -1/2 & -1/2 \\ 0 & 0 & 1 & -1 \end{bmatrix}$. (Why?)

> Suggested exercise: What other matrices would qualify to define the comparison $\tau_1 = \tau_2 = \tau_3$?

To summarize, when the objectives of a study focus on fixed factor effects, inference begins by translating these objectives into statements about linear combinations of the fixed parameters of the linear predictor. These linear combinations are called *estimable functions* and are denoted by $\mathbf{K}'\boldsymbol{\beta}$. As the name implies, $\mathbf{K}'\boldsymbol{\beta}$ must be *estimable*—Chapter 5 presents the formal theory associated with *estimability*. For now, it suffices to note that when the $\mathbf{X}$ matrix in the linear predictor is full rank, the estimability requirement is always satisfied. When $\mathbf{X}$ is not of full rank, the condition $\mathbf{K}' = (\mathbf{X}'\mathbf{X})^{-}(\mathbf{X}'\mathbf{X})\mathbf{K}'$, where $(\mathbf{X}'\mathbf{X})^{-}$ denotes a generalized inverse, must hold.

3.2.2 Linear Combinations of Fixed and Random Effects: Predictable Functions

Refer to example 2, comparing three or more treatments, from Section 3.1. We now add a twist: suppose we apply each treatment at multiple sites. These could be different clinics, different schools, different laboratory facilities, and different farms—generically, think of them as different locations. A linear predictor is $\eta + \tau_i + L_j$, where L_j denotes the jth location effect. Suppose there are L locations, thus $j = 1,2,\ldots,L$.

At this point in the construction of the model, we need to decide whether location effects are fixed or random. This decision has important implications regarding the estimability criteria as well as how we interpret resulting estimates and tests. These implications matter, so we need to understand them. In order to do so, we consider both possibilities, starting with location effects fixed.

If location effects are fixed, the linear combination $\eta + \tau_i$ fails the estimability criterion given earlier. Estimating the expected response of a treatment, a treatment mean, requires an average over all locations—thus the estimable function $\eta + \tau_i + \bar{L}_{\bullet}$, where $\bar{L}_{\bullet} = 1/L \sum_j L_j$. For comparisons among treatments, for example, treatment 1 vs. treatment 2 or treatment 1 vs. the average of treatments 2 and 3, the average over locations subtracts out, yielding the same estimable functions shown earlier with this treatment design. The location effects appear in estimable functions for treatment means but not treatment differences.

If location effects are random, the L_j are no longer model parameters; they are random variables, assumed i.i.d. $N\left(0,\sigma_L^2\right)$. For this reason, the L_j no longer enter into the estimability picture: With random location effects, $\eta + \tau_i$ is estimable and is the starting point for assessing treatment means. However, we might ask what would happen in the location-effects-random model if we did try to estimate $\eta + \tau_i + \bar{L}_{\bullet}$? Is it possible? If so, how do we interpret it?

The answer to the first question is "yes, it is possible." The function $\eta + \tau_i + \bar{L}_{\bullet}$ is now a linear combination of fixed and random effects. The generic notation in linear model inference is $\mathbf{K}'\boldsymbol{\beta} + \mathbf{M}'\mathbf{b}$. This is called a *predictable function* provided $\mathbf{K}'\boldsymbol{\beta}$ satisfies the estimability requirement given earlier. We defer "how would we interpret it?" until Section 3.4 when we discuss *inference space*. Note that for mixed models, Gaussian linear mixed model (LMM) as well as non-Gaussian generalized linear mixed model (GLMM), the estimable function is a special case of the predictable function where $\mathbf{M} = \mathbf{0}$. Again, this has ramifications that we will discuss in detail in Section 3.4.

The multilocation, multi-treatment example becomes even more interesting when the data allow adding a location × treatment term to the linear predictor, that is, $\eta + \tau_i + L_j + (\tau L)_{ij}$, where $(\tau L)_{ij}$ denotes the location × treatment interaction effect. One could now frame objectives in terms of a treatment difference—for example, treatment 1 vs. treatment 2—averaged over *all* locations or a treatment difference at a *specific* location. The estimable function for the former is $\tau_1 - \tau_2 + \overline{(\tau L)}_{1\bullet} - \overline{(\tau L)}_{2\bullet}$ for the location-effects-fixed model and $\tau_1 - \tau_2$ for the location-effects-random model. For both the location-effects-fixed and -random models, $\tau_1 - \tau_2 + (\tau L)_{1j} - (\tau L)_{2j}$ estimates the treatment 1 vs. treatment 2 difference at location j, but for the fixed location model this is an *estimable* function whereas for the random location model it is a linear combination of fixed and random effects—a *predictable* function. This has implications we explore in Section 3.4.

3.2.3 Three Issues for Inference

There are three primary issues in working with the types of functions we have discussed in this section. They are as follows.

3.2.3.1 Model Scale vs. Data Scale

This is an issue unique to models with nonidentity link functions. With Gaussian data, or more generally, with the identity link, the estimable function literally estimates the expected value or the expected difference. This is not true of the nonidentity link functions typically used with non-Gaussian data. For example, in a binomial data logit model, the estimable function $\eta + \tau_i$ estimates a logit or, equivalently, a log odds. Expressing this as the estimate of the expected value for subjects receiving the *i*th treatment means expressing $\eta + \tau_i$ as a *probability*, not as a *logit*. This requires converting the estimate to a probability using the inverse link—that is, $\pi_i = 1/\left(1+e^{-(\eta+\tau_i)}\right)$. Thus, for nonidentity links, there are two ways to express estimates: in terms of the model parameters directly (the *model scale*) or in terms of the expected value of the response variable (the *data scale*). We consider this issue in more detail in Section 3.3.

3.2.3.2 Inference Space

This issue arises only when the linear predictor contains random effects. In these models, the estimates yielded by linear combinations of the fixed-effects-*only* $\mathbf{K}'\boldsymbol{\beta}$ represent inference applicable to the entire population represented by the random model effects. This is called *broad* inference—so named because it applies *broadly* to the entire population. This is also called "population-averaged" (PA) inference by some authors, although broad inference and PA inference are not the same. As we will see in Section 3.5, PA inference is a subset of broad inference, but not all broad inference is PA.

What about *predictable* functions? For $\mathbf{K}'\boldsymbol{\beta} + \mathbf{M}'\mathbf{b}$, with $\mathbf{M} \neq \mathbf{0}$, the nonzero coefficients in the $\mathbf{M}$ matrix specifically limit inference to just those levels of $\mathbf{b}$ defined by the $\mathbf{M}$ matrix. The three-treatment, multilocation example illustrates such a case. The *estimable* function $\tau_1 - \tau_2$ provides broad inference for the difference between treatments 1 and 2, whereas the *predictable* function $\tau_1 - \tau_2 + (\tau L)_{1j} - (\tau L)_{2j}$ limits the scope of inference about treatments 1 and 2 to location *j* only. This is called *narrow* inference, so named because the nonzero coefficients of $\mathbf{M}$ narrow the scope of inference from the entire population to just those levels identified by $\mathbf{M}$. In certain applications, narrow inference is referred to as *subject-specific* (*SS*) inference.

3.2.3.3 Inference Based on Conditional and Marginal Models

This issue is unique to mixed models and most importantly for mixed models with non-Gaussian data. This issue is also widely misunderstood (or not understood at all), even among experienced statistical practitioners.

We specify GLMMs in terms of two probability distributions: the conditional distribution of the observations given the random effects, $\mathbf{y}|\mathbf{b}$, and the distribution of the random effects, $\mathbf{b}$. GLMMs so specified are called *conditional models*. Neither of the distributions in the conditional model can be observed directly. We know from probability theory that the marginal distribution of the data, $\mathbf{y}$, can be obtained by taking the joint distribution of $\mathbf{y}$ and $\mathbf{b}$ and integrating over $\mathbf{b}$. The marginal distribution is the only distribution we can observe directly. In non-Gaussian mixed model settings, many models that *seem* reasonable do not distinguish between the distribution of $\mathbf{y}|\mathbf{b}$ and $\mathbf{b}$. Models that do not make this distinction are called *marginal models*. Estimates produced by marginal models yield estimates with different expected values than conditional models—marginal models do not estimate the same thing as conditional models.

Marginal vs. conditional model inference is very much a case of what you do not know can (and often does) hurt you. Whenever non-Gaussian data arise from a setting in which there are random effects, the observed data follow the marginal distribution. All models other than conditional GLMMs produce some variation of a marginal estimator. However, regardless of the model used, data analysts more than likely *conceptualize* the problem as it would be defined in a conditional GLMM. This is true even when the data analyst makes a point of *not* using a GLMM in an effort to "avoid GLMM issues." The frequent result is an estimate the data analyst *thinks* is estimating one thing when it is in fact estimating something else. If this is unclear, Section 3.5 presents a familiar example that should clarify the issue and the problem.

KEY IDEAS: *estimable function, estimability criterion, predictable function, three GLMM inference issues: data/model scale, inference space, and conditional/marginal*

3.3 Issue I: Data Scale vs. Model Scale

As we saw in the previous section, inference begins with the estimable function, $\mathbf{K}'\boldsymbol{\beta}$. Because all linear models are defined in terms of the link function, $\boldsymbol{\eta} = g(\boldsymbol{\mu}) = \mathbf{X}\boldsymbol{\beta}$ (+$\mathbf{Zb}$ if there are random effects), $\mathbf{K}'\boldsymbol{\beta}$ yields results in terms of the link function.

For Gaussian models, **LMs** and **LMMs**, the link function is effectively invisible. This is because these models use the identity link. Linear combinations of the model parameters *directly* estimate expected values, differences between expected values and contrasts among expected values. Inference for the LM is straightforward. Indeed, linear model theory for Gaussian case can proceed entirely without the concept of a link function.

For non-Gaussian linear models, generalized linear models (**GLMs**) and **GLMMs**, this is not the case. Here, the estimate of $\mathbf{K}'\boldsymbol{\beta}$ yields a linear combination of elements of $\boldsymbol{\eta}$, that is a linear combination of $g(\boldsymbol{\mu})$, typically a nonlinear function of $\boldsymbol{\mu}$, not of $\boldsymbol{\mu}$ itself. For example, with binomial data, $\mathbf{K}'\boldsymbol{\beta}$ is usually a function of logits or probits; for Poisson data it is a function of logs. However, in most cases, researchers and decision makers want to see binomial results expressed in terms of the probability of the outcome of interest and Poisson results expressed in terms of counts. This is the model scale/data scale—or link scale/mean scale—issue.

To illustrate the model scale/data scale issue, we consider five examples. The first two examples are from a two-treatment completely randomized design, one with Gaussian data, the second with binomial data. The next two examples illustrate a more complex data structure: four treatments and multiple locations, again one with Gaussian data, the other with binomial data. The issues we see with the binomial data are typical of non-Gaussian data so the lessons from these examples apply to other non-Gaussian data. The fifth example revisits the four-treatment scenario. In the third and fourth example, no structure is imposed on the four-treatment design; in the fifth, we assume the four treatments have a 2×2 factorial structure. This allows us to consider some useful strategies for modeling factorial treatment designs. Along the way, this section should provide additional tools for working with estimable functions as well as understanding the model and data scales.

Example 3.1 Gaussian data, two-treatment independent sample

We start with a familiar "intro stat" example, the independent sample to compare two means, more formally known as the two-treatment completely randomized design. We developed the model in previous chapters. To review, we describe it as follows:

- Linear predictor: $\eta_i = \eta + \tau_i$ (model 3.3.1)
- Distribution: $y_{ij} \sim N(\mu_i, \sigma^2)$
- Link: identity—that is, $\eta_i = \mu_i$

The data for this example appear in the SAS Data and Program Library as Data Set 3.1. Note that Data Set 3.1 shows a response variable Y assumed to be Gaussian and another response, defined by the variables N and F, to be used later in the binomial version of this example.

The parameter estimates, which you could obtain in SAS using any of the linear model software procedures, GLM, GLIMMIX, or MIXED, are

Parameter Estimates

	Effect	Trt	Estimate
$\hat{\eta}$	**Intercept**		12.1880
$\hat{\tau}_0$	**Trt**	0	−1.6640
$\hat{\tau}_1$	**Trt**	1	0
$\hat{\sigma}^2$	*Scale*		9.9444

Strictly speaking, except for $\hat{\sigma}^2$, we should refer to these as "solutions" not estimates, since this is an overparameterized model and a generalized inverse—in this case the SAS sweep operator that sets the last class effect to zero—must be used.

In this example, consider the estimable functions for treatment means and treatment mean difference. These are $\eta + \tau_0$ for "treatment 0" (a "control" treatment, we might surmise), $\eta + \tau_1$ for "treatment 1" (the "test" treatment?), and $\tau_0 - \tau_1$ for the mean difference. From the solutions, we can easily see that the "treatment 0" estimate is 12.188 − 1.664 = 10.524, the "treatment 1" estimate is 12.188, and the mean difference is −1.664. In SAS, we could obtain these using either LSMEANS or ESTIMATE statements. Figure 3.1 shows a GLIMMIX program.

The LSMEANS statement is sufficient to compute the treatment means and mean difference. The E option gives you a listing of the coefficients of the estimable functions SAS uses to compute the LSMEANS, so you can make sure it is doing what you think it is doing. The ESTIMATE statements specify the desired terms more explicitly: coefficients for INTERCEPT refer to η and for TRT refer to τ_i with the first and second coefficient corresponding to the order in which they appear in the parameter estimate listing shown earlier. Note that you can lead with a negative coefficient, as in "`reverse diff`," allowing you to compute the magnitude of the difference without the sign. You can also compute the overall mean, defined by the estimable function $\eta + 1/2(\tau_0 + \tau_1)$.

```
proc glimmix data=CRD_2Trt_Ex;
 class Trt;
 model Y=Trt / solution;
 lsmeans Trt / Diff e;
 estimate 'LSM Trt 0' intercept 1 Trt 1 0;
 estimate 'LSM Trt 1' intercept 1 Trt 0 1;
 estimate 'overall mean' intercept 1 Trt 0.5 0.5;
 estimate 'Overall Mean' intercept 2 Trt 1 1 / divisor=2;
 estimate 'Trt diff' Trt 1 -1;
 estimate 'reverse diff' Trt -1 1;
```

FIGURE 3.1
Example GLIMMIX program for two-treatment CRD with Gaussian data.

The output for these statements appears as follows:

Trt Least Squares Means

Trt	Estimate	Standard Error
0	10.5240	0.9972
1	12.1880	0.9972

Differences of Trt Least Squares Means

Trt	_Trt	Estimate	Standard Error	DF	*t* Value	Pr > |*t*|
0	1	−1.6640	1.4103	18	−1.18	0.2534

Estimates

Label	Estimate	Standard Error	DF	*t* Value	Pr > |*t*|
LSM Trt 0	10.5240	0.9972	18	10.55	<.0001
LSM Trt 1	12.1880	0.9972	18	12.22	<.0001
overall mean	11.3560	0.7051	18	16.10	<.0001
Overall Mean	11.3560	0.7051	18	16.10	<.0001
Trt diff	−1.6640	1.4103	18	−1.18	0.2534
Reverse diff	1.6640	1.4103	18	1.18	0.2534

The important point to stress here is that all of these estimates are linear combinations of the model parameters themselves, and they all estimate linear combinations of the observation variables' expected values—$\mathbf{K}'\hat{\boldsymbol{\beta}}$ yields linear combinations of $\hat{\mu}_i$, directly.

Example 3.2 Two-treatment CRD with binomial response

We now consider the binomial response in Data Set 3.1 defined by the variables *N* and *F*, where *N* denotes the number of independent Bernoulli trials (e.g., the number of subjects in a clinical trial or the number of students taking a test) observed on the *ij*th unit and *F* denotes the number of "successes" (e.g., subjects whose symptoms improve or students who pass an exam). The linear predictor is identical to model 3.3.1 we just considered, but the distribution and link are different:

- Distribution: $y_{ij} \sim Binomial(N_{ij}, \pi_i)$ (model 3.3.2)
- Link: $\eta_{ij} = \text{logit}(\pi_i) = \log[\pi_i/(1-\pi_i)]$

3.3.1 Model Scale Estimation

The effect solutions, which we can obtain in SAS with PROC GENMOD or GLIMMIX, are

Parameter Estimates

Effect	Trt	Estimate
Intercept		−0.8860
Trt	0	−1.2682
Trt	1	0

Similar to the Gaussian example, we can estimate the difference in treatment effects, $\tau_0 - \tau_1$ by $-1.2682 - 0 = -1.2682$, and the "treatment 0" and "treatment 1" linear predictors, $\hat{\eta} + \hat{\tau}_0 = -0.886 - 1.2682 = -2.1542$ and $\hat{\eta} + \hat{\tau}_1 = -0.886 - 0 = -0.886$, respectively. We could also estimate the "overall total" $\hat{\eta} + 1/2(\hat{\tau}_0 + \hat{\tau}_1) = -0.886 + 1/2(-1.2682 + 0) = -1.5201$. The question is, what do these actually estimate?

In the case of the "treatment 0" and "treatment 1," we see from the model's linear predictor and link that, by definition, they estimate $\log[\pi_0/(1 - \pi_0)]$ and $\log[\pi_1/(1 - \pi_1)]$, the log odds for treatments 0 and 1, respectively. We can obtain the estimates of π_0 and π_1, the probability of a favorable outcome for treatments 0 and 1, by applying the inverse link: $\hat{\pi}_0 = 1/\left(1 + e^{-(\hat{\eta}+\hat{\tau}_0)}\right) = 1/\left(1 + e^{2.1542}\right) = 0.104$ and $\hat{\pi}_1 = 1/\left(1 + e^{0.886}\right) = 0.292$.

In the case of the difference between treatment 0 and treatment 1, we can see that $\tau_0 - \tau_1$ estimates

$$\log\left(\frac{\pi_0}{1-\pi_0}\right) - \log\left(\frac{\pi_1}{1-\pi_1}\right) = \log\left(\frac{\pi_0}{1-\pi_0} \Big/ \frac{\pi_1}{1-\pi_1}\right)$$

the log odds ratio. However, if we apply the inverse link to the log odds ratio, we get $1/\left(1 + e^{-(\hat{\tau}_0 - \hat{\tau}_1)}\right) = 1/\left(1 + e^{1.2682}\right) = 0.220$. It is not clear what this is, but it clearly is *not* the mean difference $\hat{\pi}_0 - \hat{\pi}_1$. In most cases, applying the inverse link to an estimable function defining a difference yields nonsense and little more.

3.3.2 Data Scale

Example 3.2 makes it clear that for non-Gaussian data, estimation and inference occur on two different scales. The linear predictor, $\mathbf{X\beta}$, and estimable functions, $\mathbf{K'\beta}$, are expressed in terms of the link function—the logit in Example 3.2. These are said to be *model scale* (or *link scale*) estimates. In some models, these estimates have a recognizable interpretation. The logit model is one such case—the linear predictor for a given treatment estimates the log odds and the difference estimates the log odds ratio, both familiar tools in categorical data analysis. However, the logit link is more the exception than the rule. Typically, model scale estimates in GLMs are not easily interpreted without some translation. This is where the *data scale* enters into the picture.

In its most basic form, the data scale involves applying the inverse link to the estimable function, as we did to convert the log odds for each treatment to a probability. However, the inverse link cannot be applied indiscriminately, as we saw with the difference. In general, we use the inverse link to transform model scale estimates of *location*—"means" on the link scale—to data scale estimates, but we do *not* use the inverse link for estimates of *difference*. Link functions are typically nonlinear; difference is not preserved in a nonlinear function. Applying the inverse link to a difference usually produces a nonsense result. To estimate differences on the data scale—for example, differences between probabilities rather than differences between log odds—we have to be smarter.

In the logit model, consider two approaches to the difference problem. First, we could simply take the difference between $\hat{\pi}_0$ and $\hat{\pi}_1$ after applying the inverse link to each treatment's linear predictor. In formal terms, we estimate the difference between π_0 and π_1 by $\left[1/\left(1 + e^{-(\eta+\tau_0)}\right)\right] - \left[1/\left(1 + e^{-(\eta+\tau_1)}\right)\right]$, not $1/\left(1 + e^{-(\hat{\tau}_0 - \hat{\tau}_1)}\right)$. This is one approach.

TABLE 3.2

Essential Features of the Model and Data Scales

	Model Scale	Data Scale
Linear predictor	$\eta = g(\mu \mid b) = X\beta + Zb$	$\mu \mid b = h(X\beta + Zb)$
Estimable function	$k'\beta$	Location: $h(k'\beta)$ Difference: typically, first get location inverse links, then take difference between them; some links permit other operations—e.g., logit odds(diff) → odds ratio
Parameter of interest	Link: $\eta = g(\mu \mid b)$—in this case logit(π)	Inverse link: $h(\eta) = \mu \mid b$—in this case π
Standard error[a]	$s.e.(k'\hat{\beta}) = \sqrt{k'V(\hat{\beta})k}$	Delta rule: $s.e.(\hat{\pi}) = \left.\frac{\partial h(\eta)}{\partial \eta}\right\vert_{\eta=k'\hat{\beta}} \times s.e.(k'\hat{\beta})$

[a] Formal methods to compute standard error in GLMMs presented in Chapter 6.

Alternatively, we know that $\tau_0 - \tau_1$ estimates the log odds ratio. Taking $e^{(\tau_0 - \tau_1)}$ yields an estimate of the odds ratio, a familiar and often-used quantity in categorical data analysis. Either approach can be defended. Which one you use depends on the requirements of a particular study and, to some extent, the culture of the discipline in which data analysis is taking place.

We can summarize the model scale—data scale distinction with Table 3.2.

In SAS PROC GLIMMIX, you can implement the data scale using ILINK, EXP, and ODDSRATIO options. The GLIMMIX program in Figure 3.2 illustrates this, modifying the program shown for the CRD with Gaussian data (Example 3.1) for the binomial case.

The ILINK option appears in the LSMEANS statement and in the ESTIMATE statements where a *mean*, not a *difference*, is computed. In the LSMEANS statement, GLIMMIX computes ILINK for the least squares means *only*, not their differences. For the "overall mean," use of the ILINK option is questionable; it is shown here for demonstration purposes. The inverse link of the mean does not equal the mean of the inverse links—*caveat emptor.*

The ODDSRATIO options appear in the MODEL and LSMEANS statements and in the ESTIMATE statements where a difference is computed. Alternatively, you can use the EXP option—as the name implies it computes e^{estimate}. It is the same computation as the ODDSRATIO. The result is the same and the output listing is identical.

```
proc glimmix data=CRD_2Trt_Ex;
 class Trt;
 model F/N=Trt / solution oddsratio;
 lsmeans Trt / Diff ilink oddsratio e;
 estimate 'LSM Trt 0' intercept 1 Trt 1 0/ilink;
 estimate 'LSM Trt 1' intercept 1 Trt 0 1/ilink;
 estimate 'overall mean' intercept 1 Trt 0.5 0.5 /ilink;
 estimate 'Overall Mean' intercept 2 Trt 1 1 / divisor=2 ilink;
 estimate 'Trt diff' Trt 1 -1 / oddsratio;
 estimate 'reverse diff' Trt -1 1 / exp;
```

FIGURE 3.2
GLIMMX statements for two-treatment CRD with binomial response variable.

The following show selected GLIMMIX output:

Odds-Ratio Estimates

Trt	_Trt	Estimate	DF	95% Confidence Limits	
0	1	0.281	18	0.172	0.460

ODDSRATIO output from MODEL statement. Note 95% confidence interval for odds ratio. The confidence interval is default output.

Trt Least Squares Means

Trt	Estimate	Standard Error	Mean	Standard Error Mean
0	−2.1542	0.1962	0.1039	0.01827
1	−0.8860	0.1274	0.2919	0.02634

"Estimate" in least squares means is model scale $\hat{\eta}+\hat{\tau}_i$. MEAN is inverse link estimate $\hat{\pi}_i = 1/\left(1+e^{-(\hat{\eta}+\hat{\tau}_i)}\right)$.

Differences of Trt Least Squares Means

Trt	_Trt	Estimate	Standard Error	DF	t Value	Pr > |t|	Odds Ratio
0	1	−1.2682	0.2339	18	−5.42	<.0001	0.281

"Estimate" is model scale $\hat{\tau}_0-\hat{\tau}_1$. "OddsRatio" is $e^{(\hat{\tau}_0-\hat{\tau}_1)}$. You can add CL option to obtain confidence interval. No CI on default.

Estimates

Label	Estimate	Standard Error	DF	t Value	Pr> |t|	Mean	Standard Error Mean	Exponentiated Estimate
LSM Trt 0	−2.1542	0.1962	18	−10.98	<.0001	0.1039	0.01827	
LSM Trt 1	−0.8860	0.1274	18	−6.95	<.0001	0.2919	0.02634	
overall mean	−1.5201	0.1170	18	−13.00	<.0001	0.1795	0.01722	
Overall Mean	−1.5201	0.1170	18	−13.00	<.0001	0.1795	0.01722	
Trt diff	−1.2682	0.2339	18	−5.42	<.0001			0.2813
Reverse diff	1.2682	0.2339	18	5.42	<.0001			3.5545

In the "Estimates" listing, the odds ratio appears as "Exponentiated Estimate" regardless of whether you use the ODDSRATIO or EXP option in the ESTIMATE statement. Note that the inverse link applied to the "overall average" link function $\hat{\eta}+1/2(\hat{\tau}_0+\hat{\tau}_1)$, 0.1795, is clearly not the same as averaging the $\hat{\pi}_i$ for each treatment—(0.1039 + 0.2919)/2 = 0.1979.

Example 3.3

In this example, there are four treatments, labeled 0, 1, 2, and 3. The study is conducted at eight locations. All four treatments are observed at every one of the eight locations. In design of experiments terms, this is a randomized complete block design with locations as blocks. However, data sets with this structure can arise from studies that are not formally designed experiments.

The data appear in the SAS Data and Program Library as Data Set 3.2. There are several response variables in this data set. In this example, we consider the Gaussian response, labeled Y in the data set.

A candidate model for these data is as follows:

- Linear predictor: $\eta_{ij} = \eta + \tau_i + L_j$, where τ_i: $i = 0,1,2,3$ denotes the ith treatment effect, and L_j: $j = 1,2,\ldots,8$ denotes the jth location effect
- Distribution: $y_{ij} \sim NI(\mu_{ij}, \sigma^2)$ where "$\sim NI$" means "normally and independently distributed"
- Link: identity—$\eta_{ij} = \mu_{ij}$

Design of experiments students will notice that this is standard linear model for a randomized complete block, but presented in "GLMM-ready" form. In this example, location effects are fixed. In Section 3.4, we revisit this decision. In practice, we would need to ask how these locations were selected and what they are supposed to represent. These questions raise inference space issues that we will deal with in the next section, not here.

As with Examples 3.1 and 3.2, we begin with treatment means and treatment differences. The treatment mean immediately raises a question. In the previous examples, we defined it as $\eta + \tau_i$. In this example, however, $\eta + \tau_i$ is not estimable. Why? We know that we estimate the treatment mean by $\bar{y}_{i\bullet} = 1/8\sum_{j=1}^{8} y_{ij}$ and that, expressed in model terms, $\bar{y}_{i\bullet}$ estimates $1/8\sum_{j=1}^{8}(\eta + \tau_i + L_j) = \eta + \tau_i + 1/8\sum_j L_j$. For the last term $1/8\sum_j L_j$ we can use the shorthand notation $\bar{L}_\bullet$. This term must appear in the estimable function for treatment mean.

For treatment mean difference, we define its estimable function by taking the difference between least squares means, that is,

$$diff = \eta + \tau_i + \bar{L}_\bullet - (\eta + \tau_{i'} + \bar{L}_\bullet) = \tau_i - \tau_{i'}.$$

Depending on the objectives, we can also define averages of several means. For example, consider the average of treatments 0, 2, and 3. Start by using $\bar{\eta}_{i\bullet} = 1/8\sum_j \eta_{ij}$ to denote the ith treatment least squares mean. Take the average of these and express them in model terms:

$$avg = \frac{\bar{\eta}_{0\bullet} + \bar{\eta}_{2\bullet} + \bar{\eta}_{3\bullet}}{3} = \frac{(\eta + \tau_0 + \bar{L}_\bullet) + (\eta + \tau_2 + \bar{L}_\bullet) + (\eta + \tau_3 + \bar{L}_\bullet)}{3} = \eta + \frac{\tau_0 + \tau_2 + \tau_3}{3} + \bar{L}_\bullet$$

If an objective is to compare treatment 1 with the average of treatments 0, 2, and 3, we use the estimable function

$$diff\,(\text{Trt 1 vs. Trts } 0,2,3) = \bar{\eta}_{1\bullet} - \left(\frac{\bar{\eta}_{0\bullet} + \bar{\eta}_{2\bullet} + \bar{\eta}_{3\bullet}}{3}\right) = \tau_1 - \left(\frac{\tau_0 + \tau_2 + \tau_3}{3}\right)$$

The overall equality of treatment means is expressed, in estimable function terms, as

$$\bar{\eta}_{0\bullet} = \bar{\eta}_{1\bullet} = \bar{\eta}_{2\bullet} = \bar{\eta}_{3\bullet} \Rightarrow \tau_0 = \tau_1 = \tau_2 = \tau_3$$

As we saw early in this chapter, several **K** matrices can be constructed that, when $\mathbf{K'\beta} = \mathbf{0}$, imply $\tau_0 = \tau_1 = \tau_2 = \tau_3$. To illustrate, three possibilities are

$$\mathbf{K}' = \begin{bmatrix} 0 & 1 & 0 & 0 & -1 & 0 & 0 & 0 & 0 & 0 & 0 & 0 & 0 \\ 0 & 0 & 1 & 0 & -1 & 0 & 0 & 0 & 0 & 0 & 0 & 0 & 0 \\ 0 & 0 & 0 & 1 & -1 & 0 & 0 & 0 & 0 & 0 & 0 & 0 & 0 \end{bmatrix}$$

$$\mathbf{K}' = \begin{bmatrix} 0 & 1 & -1 & 0 & 0 & 0 & 0 & 0 & 0 & 0 & 0 & 0 & 0 \\ 0 & 0 & 1 & -1 & 0 & 0 & 0 & 0 & 0 & 0 & 0 & 0 & 0 \\ 0 & 0 & 0 & 1 & -1 & 0 & 0 & 0 & 0 & 0 & 0 & 0 & 0 \end{bmatrix}$$

$$\mathbf{K}' = \begin{bmatrix} 0 & -1 & 3 & -1 & -1 & 0 & 0 & 0 & 0 & 0 & 0 & 0 & 0 \\ 0 & 2 & 0 & -1 & -1 & 0 & 0 & 0 & 0 & 0 & 0 & 0 & 0 \\ 0 & 0 & 0 & 1 & -1 & 0 & 0 & 0 & 0 & 0 & 0 & 0 & 0 \end{bmatrix}$$

Note that the first column, corresponding to the intercept η, always has a coefficient of zero, as do the last eight columns, corresponding to $L_1, L_2, \ldots, L_8$. Only the τ_i have possibly nonzero coefficients. Note that the third **K** matrix is simply the completion of a set of orthogonal contrasts beginning with the *"treatment 1 vs. average of treatments 0, 2, and 3"* contrast defined earlier. The **K** matrix can be composed of orthogonal contrasts, but, as the first two **K** matrices illustrate, they do not have to be. The two requirements are as follows:

1. One row in $\mathbf{K}'$ per treatment degree of freedom.
2. The rows must be mutually independent, but need not be mutually orthogonal.

Mutually independent means no row's comparison can be determined by the outcome of the other two. For example, in the second matrix rows 1 and 2 define $\tau_0 - \tau_1$ and $\tau_1 - \tau_2$. If they both equal zero, then $\tau_0 = \tau_1 = \tau_2$ must be true, but it says nothing about τ_4. An additional comparison that involves τ_4 without contradicting $\tau_0 = \tau_1 = \tau_2$ must be added.

Why the attention to the use of the **K** matrix to define comparisons usually associated with analysis of variance (ANOVA) test of overall treatment effect? First, it is a common tactic used to compute ANOVA sums of squares in Gaussian, fixed-effect-only linear model software (e.g., SAS PROC GLM). More importantly, for other linear models—GLM, LMM, and GLMM—sums of squares lose their usual meaning (in fact, have *no* useful meaning in most of these models), so we need alternatives to sums of squares in order to construct a test for overall equality of treatment effects. The estimable function approach shown here is an important and widely used tactic.

Figure 3.3 shows the GLIMMIX statements to implement this example.

The program contains the usual CLASS, MODEL, and LSMEANS statements we have seen before. The ESTIMATE statements define the estimable functions explicitly. The first two, "LSM Trt 0" and "LSM Trt 0 alt," define the mean for treatment 0. The first is the correct estimable function; the second shows what happens when you try to set $\bar{L}_\bullet$ to zero. The statement "Trt 0 v Trt 1" speaks for itself. The statements "avg Trt 0+1" and "avg Trt 0_2_3" define estimable functions for the mean of treatments 0 and 1 and the mean of treatments 0, 2, and 3, respectively. Recall how we constructed "avg Trt 0_2_3" earlier by starting with $(\bar{\eta}_{0\bullet} + \bar{\eta}_{2\bullet} + \bar{\eta}_{3\bullet})/3$ and then expressing it in terms of the effects in the linear predictor. Use this tactic when you are unsure what coefficients to assign the effects in the ESTIMATE statement. The next ESTIMATE statement shows how to compute several estimates in a single statement. You would want to do this if you want to adjust for multiplicity (control for *experiment*-wise rather than *comparison*-wise type I error) using the ADJUST= option. Finally, there are three contrast statements. These implement the three **K** matrices for $\tau_0 = \tau_1 = \tau_2 = \tau_3$ we discussed earlier.

```
proc glimmix data=MultiLoc_4Trt;
 class loc trt;
 model Y=loc trt;
 lsmeans trt / diff e;
 estimate 'LSM Trt 0' intercept 8 Trt 8 0 0 0 Loc 1 1 1 1 1 1 1 1 / divisor=8;
 estimate 'LSM Trt 0 alt' intercept 1 Trt 1 0 0 Loc 0 0 0 0 0 0 0 0;
 estimate 'Trt 0 v Trt 1 diff' Trt 1 -1 0 0;
 estimate 'avg Trt 0+1' Intercept 8 Trt 4 4 0 0 Loc 1 1 1 1 1 1 1 1 / divisor=8;
 estimate 'avg Trt 0_2_3' Intercept 24 Trt 8 0 8 8 Loc 3 3 3 3 3 3 3 3 /
     divisor=24;
 estimate 'Trt 0 v Trt 1 diff' Trt 1 -1 0 0,
          'Trt 2 v Trt 3 diff' Trt 0 0 1 -1,
          'Trt 1 v Trt 2 diff' Trt 0 1 -1 0,
          'Trt 1 v Trt 3 diff' Trt 0 1 0 -1,
          'avg Trt 0+1 v avg Trt 2+3' Trt 1 1 -1 -1,
          'trt 1 v avg Trt 0+1+2' Trt -1 3 -1 -1 / divisor=1,1,1,1,2,3
     adjust=sidak;
 contrast 'Type 3 Trt SS' Trt 1 -1 0 0,
                          Trt 1 0 -1 0,
                          Trt 1 0 0 -1;
 contrast 'Type 3 Trt Test' Trt 1 -1 0 0,
                            Trt 0 1 -1 0,
                            Trt 0 0 1 -1;
```

FIGURE 3.3
GLIMMIX statements for four-treatment multilocation study with Gaussian data.

The selected output appears as follows:

Parameter Estimates

	Effect	Loc	Trt	Estimate
$\hat{\eta}$	Intercept			28.2500
$\hat{L}_j$	Loc	1		−0.9750
	Loc	2		−3.2500
	Loc	3		−1.6750
	Loc	4		−0.3500
	Loc	5		0.8250
	Loc	6		−0.3000
	Loc	7		−3.5750
	Loc	8		0
$\hat{\tau}_i$	Trt		0	−2.1375
	Trt		1	−2.9500
	Trt		2	0.2875
	Trt		3	0
$\hat{\sigma}^2$	Scale			2.8770

To be technically correct, the "parameter estimates" shown earlier are really *solutions* using the SAS sweep operator, which sets the last effect equal to zero, to compute a generalized inverse.

Coefficients for Trt Least Squares Means						
Effect	**Loc**	**Trt**	**Row1**	**Row2**	**Row3**	**Row4**
Intercept			1	1	1	1
Loc	1		0.125	0.125	0.125	0.125
Loc	2		0.125	0.125	0.125	0.125
Loc	3		0.125	0.125	0.125	0.125
Loc	4		0.125	0.125	0.125	0.125
Loc	5		0.125	0.125	0.125	0.125
Loc	6		0.125	0.125	0.125	0.125
Loc	7		0.125	0.125	0.125	0.125
Loc	8		0.125	0.125	0.125	0.125
Trt		0	1			
Trt		1		1		
Trt		2			1	
Trt		3				1

The "Coefficients for Trt Least Squares Means" show how the software uses the parameter solutions to compute treatment means. You obtain this listing by using the E option in the least squares means. This listing verifies that $\bar{L}_{\bullet}$ is used in the least squares mean estimable function.

Trt Least Squares Means		
Trt	**Estimate**	**Standard Error**
0	24.9500	0.5997
1	24.1375	0.5997
2	27.3750	0.5997
3	27.0875	0.5997

Differences of Trt Least Squares Means						
Trt	**_Trt**	**Estimate**	**Standard Error**	**DF**	***t* Value**	**Pr > |*t*|**
0	1	0.8125	0.8481	21	0.96	0.3489
0	2	−2.4250	0.8481	21	−2.86	0.0094
0	3	−2.1375	0.8481	21	−2.52	0.0199
1	2	−3.2375	0.8481	21	−3.82	0.0010
1	3	−2.9500	0.8481	21	−3.48	0.0022
2	3	0.2875	0.8481	21	0.34	0.7380

Estimates					
Label	**Estimate**	**Standard Error**	**DF**	***t* Value**	**Pr > |*t*|**
LSM Trt 0	24.9500	0.5997	21	41.60	<.0001
LSM Trt 0 alt	**Nonest**				
Trt 0 v Trt 1 diff	0.8125	0.8481	21	0.96	0.3489
Avg Trt 0+1	24.5438	0.4240	21	57.88	<.0001
Avg Trt 0_1_2	26.4708	0.3462	21	76.45	<.0001

Notice that the "LSM Trt 0 alt" contrast is labeled "Nonest"—SAS-speak for not estimable. Recall that this was the estimate we attempted to define as $\eta + \tau_i$ with $\bar{L}_\bullet = 0$. GLIMMIX evaluates the expression $\mathbf{K}' = (\mathbf{X}'\mathbf{X})^{-}(\mathbf{X}'\mathbf{X})\mathbf{K}'$. If the equality does not hold, you get "Nonest" in the listing.

Estimates Adjustment for Multiplicity: Sidak

Label	Estimate	Standard Error	DF	*t* Value	Pr> |*t*|	*Adj P*
Trt 0 v Trt 1 diff	0.8125	0.8481	21	0.96	0.3489	0.9238
Trt 2 v Trt 3 diff	0.2875	0.8481	21	0.34	0.7380	0.9997
Trt 1 v Trt 2 diff	−3.2375	0.8481	21	−3.82	0.0010	0.0060
Trt 1 v Trt 3 diff	−2.9500	0.8481	21	−3.48	0.0022	0.0134
Avg Trt 0+1 v avg Trt 2+3	−2.6875	0.5997	21	−4.48	0.0002	0.0012
Trt 1 v avg Trt 0+1+2	−2.3333	0.6925	21	−3.37	0.0029	0.0173

The estimates obtained from the ESTIMATE statement with multiple estimable functions and the multiplicity adjustment appears in a separate table. Two columns with *p*-values appear. The first, "Pr> |*t*|," is the unadjusted *p*-value based on the *t*-distribution with 21 degrees of freedom. The second, "*Adj P*," is adjusted using the multiplicity adjustment you specify in the ADJUST= option.

Contrasts

Label	Num DF	Den DF	*F* Value	Pr > *F*
Type 3 Trt SS	3	21	7.04	0.0019
Type 3 Trt Test	3	21	7.04	0.0019
Type 3 Trt H_0	3	21	7.04	0.0019

Type III Tests of Fixed Effects

Effect	Num DF	Den DF	*F* Value	Pr > *F*
Loc	7	21	3.41	0.0135
Trt	3	21	7.04	0.0019

Notice that the contrasts defined by the three **K** matrices and the *F*-value in the listing for "Type III tests of Fixed Effects" yield identical results.

Example 3.4 Multilocation four-treatment with binomial response

Suppose we want to model the same four-treatment, multilocation randomized block study as the previous example, but with a binomial response. The changes we make to the model relative to the Gaussian case are similar to the two-treatment completely randomized design. The distribution and link change; the linear predictor does not. The model is thus as follows:

- Linear predictor: $\eta_{ij} = \eta + \tau_i + L_j$, where τ_i: $i = 0,1,2,3$ denotes the *i*th treatment effect and L_j: $j = 1,2,\ldots,8$ denotes the *j*th location effect.
- Distribution: $y_{ij} \sim Binomial(N_{ij}, \pi_{ij})$. In this example, to keep things simple, $N_{ij} = 60$ for every treatment location.
- Link: logit—$\eta_{ij} = \log[\pi_{ij}/(1 - \pi_{ij})]$.

The same estimable functions defined for the Gaussian response in Example 3.3 can be used here. No further discussion is needed, except the reminder that all estimable functions yield *model scale* estimates and that we need decide what data scale conversions we want. Here is a list of the estimable functions included in the GLIMMIX program for the Gaussian data and the appropriate data scale conversion. Notice that for some estimable functions, there is no appropriate conversion applied directly to $\mathbf{K}'\hat{\boldsymbol{\beta}}$:

- Least squares means: inverse link.
- Differences between pairs of least squares means: odds ratio.
- Estimate "LSM trt 0": inverse link.
- Estimate "LSM trt 0 alt": none—it is not estimable so drop it.
- Estimate "Trt 0 v Trt 1": exponentiate $\left(\text{i.e., } e^{k'\hat{\beta}}\right)$ (or odds ratio).
- Multiple ESTIMATE statement: divide into two statements. The first four define treatment differences; exponentiation (or the odds-ratio conversion) is appropriate. The last two are averages over two or more treatments. No direct conversion is appropriate. You can average the estimates of specified $\hat{\pi}_i$ from the inverse link transformed least squares means.
- Contrasts: no conversion to data scale needed. These just compute *F*-statistics.

Figure 3.4 shows a GLIMMIX program to implement this model.

```
proc glimmix data=MultiLoc_4Trt;
 class loc trt;
 model S_1/N_Bin=loc trt / solution oddsratio;
 lsmeans trt / diff oddsratio e;
 estimate 'LSM Trt 0' intercept 8 Trt 8 0 0 0 Loc 1 1 1 1 1 1 1 1 /
      divisor=8 ilink;
 *estimate 'LSM Trt 0 alt' intercept 1 Trt 1 0 0 Loc 0 0 0 0 0 0 0 0;
 estimate 'Trt 0 v Trt 1 diff' Trt 1 -1 0 0 / exp;
 estimate 'avg Trt 0+1' Intercept 8 Trt 4 4 0 0 Loc 1 1 1 1 1 1 1 1 /
      divisor=8;
 estimate 'avg Trt 0_1_2' Intercept 24 Trt 8 0 8 8 Loc 3 3 3 3 3 3 3 3 /
      divisor=24;
 estimate 'Trt 0 v Trt 1 diff' Trt 1 -1 0 0,
          'Trt 2 v Trt 3 diff' Trt 0 0 1 -1,
          'Trt 1 v Trt 2 diff' Trt 0 1 -1 0,
          'Trt 1 v Trt 3 diff' Trt 0 1 0 -1 / exp adjust=sidak;
 estimate 'avg Trt 0+1 v avg Trt 2+3' Trt 1 1 -1 -1,
          'trt 1 v avg Trt 0+1+2'     Trt -1 3 -1 -1 / divisor=2,3;
 contrast 'Type 3 Trt SS' Trt 1 -1 0 0,
                          Trt 1 0 -1 0,
                          Trt 1 0 0 -1;
 contrast 'Type 3 Trt Test' Trt 1 -1 0 0,
                            Trt 0 1 -1 0,
                            Trt 0 0 1 -1;
 contrast 'Type 3 Trt H_0'  Trt -1 3 -1 -1,
                            Trt  2 0 -1 -1,
                            Trt  0 0  1 -1;
```

FIGURE 3.4
GLIMMIX statements for four-treatment multilocation study with binomial data.

This program uses the binomial response variable denoted S_1 in Data Set 3.2. Notice that you can use the ADJUST= option to adjust for multiplicity for non-Gaussian GLM just as you did with the Gaussian linear model. The other statements are unchanged from the program for Gaussian data, except for the addition of ILINK, ODDSRATIO, or EXP options where needed and appropriate.

The selected output is as follows:

Parameter Estimates

	Effect	Loc	Trt	Estimate
$\hat{\eta}$	Intercept			−0.3743
$\hat{L}_j$	Loc	1		−0.8496
	Loc	2		0.6890
	Loc	3		0.9368
	Loc	4		−0.8496
	Loc	5		0.1257
	Loc	6		0.7727
	Loc	7		−1.9837
	Loc	8		0
$\hat{\tau}_i$	Trt		0	−1.8631
	Trt		1	−1.7764
	Trt		2	−0.2811
	Trt		3	0

Note that the last-effect-set-to-zero convention for the generalized inverse applies to GLMs as well as Gaussian models. These are *model* scale *solutions* so any $\mathbf{K}'\hat{\boldsymbol{\beta}}$ applied to them will be a model scale estimate. Finally, no $\hat{\sigma}^2$ appears. The variance for the binomial distribution is not an independent parameter, it is $\pi_{ij}(1 - \pi_{ij})$.

Odds-Ratio Estimates

Loc	Trt	_Loc	_Trt	Estimate	DF	95% Confidence Limits	
1		8		0.428	21	0.250	0.731
2		8		1.992	21	1.269	3.126
3		8		2.552	21	1.630	3.994
4		8		0.428	21	0.250	0.731
5		8		1.134	21	0.711	1.808
6		8		2.166	21	1.381	3.395
7		8		0.138	21	0.066	0.286
	0		3	0.155	21	0.106	0.227
	1		3	0.169	21	0.117	0.245
	2		3	0.755	21	0.559	1.020

These are the odds-ratio estimates produced by the ODDSRATIO option in the MODEL statement. Two sets appear, one effect in the model, LOC and TRT. All are odds ratios of the effect level indicated in the first column compared to the last factor level—a consequence of the set-to-zero convention for obtaining a solution for $\boldsymbol{\beta}$. Odds ratios for location may be of little or no interest but those for treatment clearly are.

Trt Least Squares Means

Trt	Estimate	Standard Error	Mean	Standard Error Mean
0	−2.3822	0.1558	0.08454	0.01206
1	−2.2955	0.1515	0.09149	0.01259
2	−0.8002	0.1075	0.3100	0.02299
3	−0.5191	0.1042	0.3731	0.02438
	Model Scale		Data Scale	

The column labeled "Estimate" and its standard error give results for $\hat{\eta}+\hat{\tau}_i+\hat{\bar{L}}_{\bullet}$, the model scale estimates. The columns labeled "Mean" and "Standard Error Mean" apply the inverse link and the delta rule, respectively, to get $\hat{\pi}_i$ and its standard error.

Differences of Trt Least Squares Means

Trt	_Trt	Estimate	Standard Error	DF	*t* Value	Pr > |*t*|	Odds Ratio
0	1	−0.08665	0.2082	21	−0.42	0.6815	0.917
0	2	−1.5820	0.1827	21	−8.66	<.0001	0.206
0	3	−1.8631	0.1820	21	−10.23	<.0001	0.155
1	2	−1.4953	0.1790	21	−8.35	<.0001	0.224
1	3	−1.7764	0.1784	21	−9.96	<.0001	0.169
2	3	−0.2811	0.1446	21	−1.94	0.0654	0.755

The column labeled estimate gives model scale results for $\hat{\tau}_i - \hat{\tau}_{i'}$. Applying $e^{(\hat{\tau}_i - \hat{\tau}_{i'})}$ yields the odds ratios given in the last column.

Estimates

Label	Estimate	Standard Error	DF	*t* Value	Pr > |*t*|	Mean	Standard Error Mean	Exponentiated Estimate
LSM Trt 0	−2.3822	0.1558	21	−15.29	<.0001	0.08454	0.01206	.
Trt 0 v Trt 1 diff	−0.08665	0.2082	21	−0.42	0.6815			0.9170
Avg Trt 0+1	−2.3389	0.1130	21	−20.69	<.0001			.
Avg Trt 0_1_2	−1.2338	0.07653	21	−16.12	<.0001			.
Avg Trt 0+1 v avg Trt 2+3	−1.6792	0.1285	21	−13.06	<.0001			.
Trt 1 v avg Trt 0+1+2	−1.0617	0.1613	21	−6.58	<.0001			.

Estimates Adjustment for Multiplicity: Sidak							
Label	**Estimate**	**Standard Error**	**DF**	***t* Value**	**Pr > |*t*|**	***Adj P***	**Exponentiated Estimate**
Trt 0 v Trt 1 diff	−0.08665	0.2082	21	−0.42	0.6815	0.9897	0.9170
Trt 2 v Trt 3 diff	−0.2811	0.1446	21	−1.94	0.0654	0.2370	0.7550
Trt 1 v Trt 2 diff	−1.4953	0.1790	21	−8.35	<.0001	<.0001	0.2242
Trt 1 v Trt 3 diff	−1.7764	0.1784	21	−9.96	<.0001	<.0001	0.1692

All of the estimates without a multiplicity adjustment appear in a single table. The inverse link results only appear for estimable functions where they were requested, and the odds ratio ("Exponentiated estimate") only appears where it was requested. All of the other estimates appear in model scale form only.

Two *p*-values appear in the multiplicity-adjusted table: the comparison-wise "Pr > |*t*|" and the experiment-wise "*Adj P*." These *p*-values apply equally to the model scale estimate and the data scale odds ratio ("Exponentiated estimate").

Contrasts				
Label	**Num DF**	**Den DF**	***F* Value**	**Pr > *F***
Type 3 Trt SS	3	21	58.25	<.0001
Type 3 Trt Test	3	21	58.25	<.0001
Type 3 Trt H_0	3	21	58.25	<.0001

Type III Tests of Fixed Effects				
Effect	**Num DF**	**Den DF**	***F* Value**	**Pr > *F***
Loc	7	21	22.07	<.0001
Trt	3	21	58.25	<.0001

As with the Gaussian data, the contrasts for all three **K** matrices yield the same *F*-value as the "Type III Tests for Fixed Effects" treatment *F*-value.

Example 3.5 The four-treatment multisite revisited as a 2 × 2 factorial

Suppose there is more structure in the treatment design of the multisite Data Set 3.2 than simple treatment 0, 1, 2, and 3. Specifically suppose the treatments have the following factorial structure:

		Factor B	
	↓ Factorial Levels →	$\mathbf{B_0}$	$\mathbf{B_1}$
Factor A	$\mathbf{A_0}$	Treatment 0 = AB_{00}	Treatment 2 = AB_{01}
	$\mathbf{A_1}$	Treatment 1 = AB_{10}	Treatment 3 = AB_{11}

AB_{ij} denotes the ijth $A_i \times B_j$ treatment combination, with $i = 0,1$ and $j = 0,1$. To accommodate this structure, we replace $\eta + \tau_i$ in the linear predictor with η_{ij}, which can be decomposed into factorial components as $\eta + \alpha_i + \beta_j + (\alpha\beta)_{ij}$. The full linear predictor is thus

$$\eta_{ijk} = \eta_{ij} + L_k = \eta + \alpha_i + \beta_j + (\alpha\beta)_{ij} + L_k$$

Think of the $\eta_{ijk} = \eta_{ij} + L_k$ form of the linear predictor as the GLM version of a cell means model. In fact, for Gaussian data—or any GLM that uses the identity link—$\eta_{ijk} = \eta_{ij} + L_k = \mu_{ijk} = \mu_{ij} + L_k$ *is* the cell means model.

Recall that Table 3.2 showed the estimable functions of interest in the standard analysis of a factorial treatment design. Table 3.2 focused on estimable functions defined on η_{ij} only. For the multilocation data, with location effects fixed, we also have location effects. They enter into estimable functions exactly as they did in Examples 3.3 and 3.4. For example, the estimable function for the main effect "mean" of A_0 is $\bar{\eta}_{0\bullet} = \eta + \alpha_0 + 1/2\sum_{j=0}^{1}\beta_j + 1/2\sum_{j=0}^{1}(\alpha\beta)_{0j}$ from Table 3.2, the η_{ij} part, plus $\bar{L}_{\bullet}$, required to guarantee estimability. The complete estimable function is thus $\eta + \alpha_0 + 1/2\sum_{j=0}^{1}\beta_j + 1/2\sum_{j=0}^{1}(\alpha\beta)_{0j} + 1/8\sum_{k=1}^{8}L_k$. Using the "cell means" form of the linear predictor, this estimable function is $1/2\sum_{j=0}^{1}\eta_{0j} + 1/8\sum_{k=1}^{8}L_k = \bar{\eta}_{0\bullet} + \bar{L}_{\bullet}$. Notice that this estimable function is just $\eta + 1/2\sum_{i=0}^{1}\tau_i + \bar{L}_{\bullet}$ from Examples 3.3 and 3.4, with $\alpha_i + \beta_j + (\alpha\beta)_{ij}$ replacing τ_i. Other estimable functions for this analysis proceed along the same lines. As with Examples 3.3 and 3.4, $\bar{L}_{\bullet}$ appears in estimable functions for means and averages of means, but it generally subtracts out of estimable functions for differences.

As noted earlier, but a point worth stressing, there is a logical order to interpreting a factorial treatment structure. *You should always assess the interaction first.* To borrow a phrase from Nelder, main effects are "uninteresting" estimable functions in the presence of nonnegligible interaction. Note the phrase "*nonnegligible*" rather than "*nonsignificant*." The word choice matters. They are not synonyms. Statistically significant interactions of little practical consequence can and do happen. These do not preclude assessing main effects. *However*, in general, it *is* true that if you assess main effects, you should *not* assess simple effects and vice versa—at least not in the same report. Consider main effect and simple effect analyses to be mutually exclusive for the reported analysis of a given response variable in a given data set.

Figure 3.5 shows GLIMMIX statements to implement an analysis of the Gaussian response in Data Set 3.2 using the factorial treatment design model.

```
proc glimmix data=MultiLoc_4Trt;
 class loc a b;
 model y=a|b loc;
 lsmeans a b / diff e;
 lsmeans a*b / slicediff=(a b);
 lsmestimate a 'a0 marg mean' 1 0 / e;
 lsmestimate b 'b0 marg mean' 1 0 / e;
 lsmestimate a*b 'a|b0 simple effect' 1 0 -1 0,
                 'b|a0 simple effect' 1 -1 0 0 / e;
 estimate 'a0 marginal mean' intercept 2 a 2 0 b 1 1 a*b 1 1 0 0,
          'b0 marginal mean' intercept 2 a 1 1 b 2 0 a*b 1 0 1 0,
          'a|b0 simple effect' a 1 -1 a*b 1 0 -1 0,
          'b|a0 simple effect' b 1 -1 a*b 1 -1 0 0 / e divisor=2,2,1,1;
```

FIGURE 3.5
GLIMMIX statements for multilocation factorial with Gaussian response.

Compared to the Gaussian analysis of this same data in Example 3.3, A B replace TRT in the CLASS statement and A|B, SAS shorthand for A B A*B replaces TRT in the model statement. Two LSMEANS statements appear. One is for the main effects means A and B, the other is for the treatment combination mean, denoted by A*B. We could use one statement. In fact, we could write the statement LSMEANS A|B/ DIFF SLICEDIFF=(A B). The reason we split the two statements is that LSMEANS A*B /DIFF gives an indiscriminant listing of all possible pairs of means, which is generally not of interest in a factorial treatment structure. In a standard factorial analysis, comparisons of the A*B treatment combination means, when they are of interest, focus on simple effects only—comparisons of the A levels with B held constant or vice versa—rather than all possible pairs. SLICEDIFF limits the differences between A*B combinations to simple effects only and organizes the listing accordingly. LSMESTIMATE allows you to define estimates in terms of least squares means estimable functions rather than explicitly in terms of model effects. For example, defining $\bar{\eta}_{ij\bullet} = 1/8 \sum_k \eta_{ijk}$, the average estimable function for the ijth AB combination over the eight locations, the estimable function for the simple effect of A given B_0 is $\bar{\eta}_{00\bullet} - \bar{\eta}_{10\bullet}$ when defined in terms of least squares means. The LSMESTIMATE allows you to specify estimable functions the way you want. If you use the ESTIMATE statement instead, you have to spell out the coefficients for all of the effects in the model. The estimable function defined explicitly on all of the effects is

$$\eta + \alpha_0 + \bar{\beta}_\bullet + \left(\overline{\alpha\beta}\right)_{0\bullet} + \bar{L}_\bullet - \left(\eta + \alpha_1 + \bar{\beta}_\bullet + \left(\overline{\alpha\beta}\right)_{1\bullet} + \bar{L}_\bullet\right) = \alpha_0 + \left(\overline{\alpha\beta}\right)_{0\bullet} - \left(\alpha_1 + \left(\overline{\alpha\beta}\right)_{1\bullet}\right)$$

$$= \alpha_0 - \alpha_1 + \left(\overline{\alpha\beta}\right)_{0\bullet} - \left(\overline{\alpha\beta}\right)_{1\bullet}.$$

SLICEDIFF and LSMESTIMATE are available *only* in the GLIMMIX procedure. For SAS PROC GLM, MIXED, or GENMOD, these options can be accessed, although somewhat less conveniently, via PROC PLM.

Finally, the E options appear throughout the program. SAS's internal logic handles the location effects to achieve estimability. The E option allows you to see what GLIMMIX is doing with location effects "inside the black box."

The selected output is as follows:

Coefficients for Estimate with Multiple Rows

Effect	Loc	a	b	Row1	Row2	Row3	Row4
Intercept				1	1		
a		0		1	0.5	1	
a		1			0.5	−1	
b			0	0.5	1		1
b			1	0.5			−1
a*b		0	0	0.5	0.5	1	1
a*b		0	1	0.5			−1
a*b		1	0		0.5	−1	
a*b		1	1				
Loc	1			0.125	0.125		
Loc	2			0.125	0.125		
Loc	3			0.125	0.125		
Loc	4			0.125	0.125		
Loc	5			0.125	0.125		
Loc	6			0.125	0.125		
Loc	7			0.125	0.125		
Loc	8			0.125	0.125		

Estimates

Label	Estimate	Standard Error	DF	*t* Value	Pr > |*t*|
a_0 marginal mean	24.5438	0.4240	21	57.88	<.0001
b_0 marginal mean	26.1625	0.4240	21	61.70	<.0001
$a \mid b_0$ simple effect	−2.4250	0.8481	21	−2.86	0.0094
$b \mid a_0$ simple effect	0.8125	0.8481	21	0.96	0.3489

These are the results of the ESTIMATE statement. Notice that for the marginal means of A_0 and B_0, you need to give all of the coefficients for treatment-related effects, η, α_i, β_j, and $(\alpha\beta)_{ij}$, but the program's internal logic averages over the location effects—that is, adds $\bar{L}_{\bullet}$—which are required for estimability.

To compare, LSMESTIMATE gives the following results for the A_0 marginal mean:

Coefficients for a Least Squares Means Estimates

Effect	Loc	a	b	Row1
Intercept				1
a		0		1
a		1		
b			0	0.5
b			1	0.5
a*b		0	0	0.5
a*b		0	1	0.5
a*b		1	0	
a*b		1	1	
Loc	1			0.125
Loc	2			0.125
Loc	3			0.125
Loc	4			0.125
Loc	5			0.125
Loc	6			0.125
Loc	7			0.125
Loc	8			0.125

The other LSMESTIMATE coefficients are not shown here in the interest of space. Notice that even though the LSMESTIMATE statement only mentions the model effect A and assigns coefficients of 1 and 0 to A_0 and A_1, respectively, the coefficients GLIMMIX "understands" from this statement, in terms of the model effects, are identical to the ESTIMATE statement you write out explicitly. The estimates are also identical. For example, for the A_0 marginal mean, you get

Least Squares Means Estimates

Effect	Label	Estimate	Standard Error	DF	*t* Value	Pr > |*t*|
a	a_0 marg mean	24.5438	0.4240	21	57.88	<.0001

This is identical to the ESTIMATE listing shown earlier.

The SLICEDIFF output is as follows:

a*b Least Squares Means

a	b	Estimate	Standard Error	DF	*t* Value	Pr > |*t*|
0	0	24.9500	0.5997	21	41.60	<.0001
0	1	24.1375	0.5997	21	40.25	<.0001
1	0	27.3750	0.5997	21	45.65	<.0001
1	1	27.0875	0.5997	21	45.17	<.0001

Simple Effect Comparisons of a*b Least Squares Means By a							
Simple Effect Level	b	_b	Estimate	Standard Error	DF	*t* Value	Pr > |*t*|
a_0	0	1	0.8125	0.8481	21	0.96	0.3489
a_1	0	1	0.2875	0.8481	21	0.34	0.7380

Simple Effect Comparisons of a*b Least Squares Means By b							
Simple Effect Level	a	_a	Estimate	Standard Error	DF	*t* Value	Pr > |*t*|
b_0	0	1	−2.4250	0.8481	21	−2.86	0.0094
b_1	0	1	−2.9500	0.8481	21	−3.48	0.0022

Other output is similar to what we have seen before in previous example. It is not shown here in the interest of space, but left as an exercise for the reader.

Just as a final note, the *p*-value for the test of A × B interaction is 0.6661. Also, in the SLICEDIFF listing provided earlier, the differences in the simple effects appear to be small. In practice, we would need to know what the response variable is and seek the subject matter specialist's judgment of what magnitude of difference is considered important or negligible. But it would appear that with what we know about the problem and the lack of statistically compelling evidence of an interaction, our analysis should focus on main effects for these data.

KEY IDEAS: *model or link scale, data or inverse link scale, inverse link, delta rule, measure of location a.k.a. mean, measure of difference, how data/model scale considerations affect estimate of mean vs. estimate of location*

3.4 Issue II: Inference Space

In Section 3.2, we introduced the terms *broad* and *narrow* inference. The distinction only exists for mixed models, models with the linear predictor $\mathbf{X\beta}+\mathbf{Zb}$. For these models, we can base inference on estimable functions *only*—$\mathbf{K'\beta}$—or on predictable functions, $\mathbf{K'\beta}+\mathbf{M'b}$. Notice that the estimable function can be regarded as a predictable function with $\mathbf{M}=\mathbf{0}$. In an estimable function, all of the information for the estimate comes from the fixed effects and the uncertainty derives from *all* of the random effects. Put another way, uncertainty for estimable functions comes from the combined contribution of all of the random effects. In a predictable function, the nonzero components of $\mathbf{M'b}$ remove those effects from uncertainty. As a result, they narrow inference from the entire population represented by the random effects to just those effects included in the predictable function. Inference on estimable functions, $\mathbf{K'\beta}$ only, is called *broad-space inference*. Inference on predictable functions, $\mathbf{K'\beta}+\mathbf{M'b}$, is called *narrow-space inference*.

To illustrate the distinction between broad and narrow inference, we consider two examples in this section. The first revisits the four-treatment multilocation data used in the last three examples of Section 3.3. The second is a multi-batch regression where the batch effects are random. We start with a review of essential ideas.

3.4.1 Broad Inference

In mixed models, the levels of a random effect represent the population from which they were sampled. These could be clinics sampled from a state or region, schools from a district, individual workers representing a larger work force, farms representing an agricultural region, and so forth. Ideally, selection occurs via random sampling so that all members of the population have an equal and independent chance of being represented. The random effect levels should be exchangeable, meaning, for example, in our eight location example, that we could just as well use any eight randomly selected locations—there is nothing special about the eight we happened to observe in this study. In practice, random effect levels are included in studies through less than ideal sampling. Nonetheless, if these levels (clinics, schools, workers, farms, etc.) reasonably represent the population, then the main ideas underlying random model effects (their defining properties) hold. These are as follows:

1. Random effect levels represent a target population.
2. Random effect levels have a probability distribution.

Implicit in random effects' two defining properties is broad inference. Informally, broad inference refers to point estimates, interval estimates, and hypothesis tests meant to be applicable to the entire population represented by the random model effects. Formally, broad inference refers to a mixed-model-based estimate or hypothesis test defined exclusively by an estimable function, $\mathbf{K}'\boldsymbol{\beta}$. If there is a nonzero coefficient in $\mathbf{M}$, then the estimate or test is defined by a predictable function $\mathbf{K}'\boldsymbol{\beta} + \mathbf{M}'\mathbf{b}$. At that point, we no longer have broad inference.

Broad inference often confuses data analysts, because, at first glance, all inference with fixed effects models has to be based on $\mathbf{K}'\boldsymbol{\beta}$ so all inference with fixed effects models must be broad inference. Right? The reality is more nuanced.

The multilocation study we have been following provides a perfect example. The question "are location effects fixed or random?" has a long history of controversy in statistics. The crux of the dilemma is "how broadly can we justifiably apply inference as a consequence of defining location effects as fixed or random?"

Example 3.6 Four-treatment multilocation, Gaussian data, random location effects

Example 3.3 introduced Data Set 3.2 with a model that assumed fixed location effects. We did this with a caveat that regarding location effects as fixed was questionable and should be revisited. We do so in this example. We know that if locations represent a population and the observed locations constitute a representative sample of this population, then the case for fixed location effects is weak. If, in addition, locations were selected by random sampling, and *any* eight locations could just as well have been included, then location effects are, by definition, random. In this example, suppose that the locations were indeed selected by random sampling. The resulting model is

1. Linear predictor: $\eta_{ij} = \eta + \tau_i + L_j$, where τ_i:i = 0,1,2,3 denotes the ith treatment effect, and L_j:j = 1,2,…,8 denotes the jth location effect
2. Distributions:
   a. $y_{ij}|L_j \sim NI(\mu_{ij}, \sigma^2)$
   b. L_j i.i.d. $N\left(0, \sigma_L^2\right)$
3. Link: identity—$\eta_{ij} = \mu_{ij}$

Notice the impact of this change on the estimable function for treatment means. In the location-effects-fixed model, $\eta + \tau_i$ was not estimable. Instead, the function defined by $E(\bar{y}_{i\bullet}) = \eta + \tau_i + \bar{L}_{\bullet}$ was required to satisfy estimability requirements. In the

```
proc glimmix data=MultiLoc_4Trt;
 class loc trt;
 model Y=trt /solution;
 Random Loc / solution;
 lsmeans trt / diff e;
 estimate 'BLUP Trt 0' intercept 8 Trt 8 0 0 0 | Loc 1 1 1 1 1 1 1 1 / divisor=8;
 estimate 'LSM Trt 0' intercept 1 Trt 1 0 0 | Loc 0 0 0 0 0 0 0 0;
```

FIGURE 3.6
GLIMMIX program for location-effects-random Gaussian model.

location-effects-random model, $E(\bar{y}_{i\bullet}) = \eta + \tau_i$ because the L_j are now random variables with expected value equal to zero. Thus, $\eta + \tau_i$ is the estimable function for treatment means in the location-effects-random model. Two questions arise as a result:

1. How do the estimates of treatment least squares means in the location-effects-random model differ from those in the location-effects-fixed case?
2. How would a *predictable* function defined by the same coefficients that the location-effects-fixed model uses for least squares means compare? In other words, the location-effects-fixed least squares mean is $\eta + \tau_i + \bar{L}_\bullet$. In the location-effects-random model $\eta + \tau_i + \bar{L}_\bullet$ is a predictable function. How do these compare?

Figure 3.6 shows GLIMMIX statements for the location-effects-random model.

Notice the ESTIMATE statements. Coefficients associated with the **K** matrix, for η and τ_i, appear before the vertical bar (|); coefficients associated with the **M** matrix, for L_j, appear after the vertical bar.

The solutions for the fixed and random effects are

Solutions for Fixed Effects

	Effect	Trt	Estimate	Standard Error
$\hat{\eta}$	Intercept		27.0875	0.7593
$\hat{\tau}_i$	Trt	0	−2.1375	0.8481
	Trt	1	−2.9500	0.8481
	Trt	2	0.2875	0.8481
	Trt	3	0	

Solution for Random Effects

	Effect	Loc	Estimate	Std Err Pred
$\hat{L}_j$	Loc	1	0.1326	0.8135
	Loc	2	−1.4758	0.8135
	Loc	3	−0.3623	0.8135
	Loc	4	0.5744	0.8135
	Loc	5	1.4051	0.8135
	Loc	6	0.6098	0.8135
	Loc	7	−1.7055	0.8135
	Loc	8	0.8218	0.8135

Comparing these solutions to the fixed location model in Example 3.3, notice that in the fixed part of the model the last-effect-zero generalized inverse convention holds, but this is not the case for the location effects. In Chapter 4, we will develop the formal theory of best linear unbiased prediction for random effects. For location effects in this example, estimation goes as follows. Use location 1 to illustrate. Define the unadjusted estimate of the effect of location 1 as the difference between the location 1 mean and the overall mean, $\bar{y}_{\bullet 1} - \bar{y}_{\bullet\bullet} = 26.075 - 25.8875 = 0.1875$. The adjusted random effects estimate, also known as the *best linear unbiased predictor* (BLUP), is

$$\hat{L}_1 = E(L_j \mid \mathbf{y}) = E(L_j) + Cov(L_j, \bar{y}_{\bullet j})\left[Var(\bar{y}_{\bullet j})\right]^{-1}(\bar{y}_{\bullet j} - \bar{y}_{\bullet\bullet})$$

Noting that $E(L_j) = 0$, using methods presented in Chapter 4, one can show that the location 1 BLUP is

$$\hat{L}_j = 0 + \hat{\sigma}_L^2\left(\frac{\sigma^2 + 4\sigma_L^2}{4}\right)^{-1}(\bar{y}_{\bullet j} - \bar{y}_{\bullet\bullet})$$

The variance component estimates are

Covariance Parameter Estimates

	Cov Parm	Estimate
$\hat{\sigma}_L^2$	Loc	1.7352
$\hat{\sigma}^2$	Residual	2.8770

Substituting, $\hat{L}_j = 0 + 1.7352\left(\frac{2.877 + 4 \times 1.7352}{4}\right)^{-1}(26.075 - 25.8875) = 0.1326$

BLUPs are also called *shrinkage estimators* because they use information about the distribution, in this case of the location effect, to attenuate extreme values—in essence to "shrink" the estimate toward zero. The amount of shrinkage depends on the variance components. If σ_L^2 is small, shrinkage is large, because the spread of location effects is known to be very tight around zero. If σ_L^2 is large, shrinkage is small, because the location effect distribution is relatively diffuse.

Now returning to the treatment least squares means, their estimates and standard errors are

Trt Least Squares Means

Trt	Estimate	Standard Error
0	24.9500	0.7593
1	24.1375	0.7593
2	27.3750	0.7593
3	27.0875	0.7593

Comparing these to the least squares means of the location-effects-fixed model, the estimates are identical but the standard errors differ: 0.7593 here vs. 0.5997 in the location-effects-fixed model. Why the discrepancy? For the two models under consideration here, the standard error is the estimate of the square root of $Var(\bar{y}_{i\bullet})$.

For both models, $Var(\bar{y}_{i\bullet}) = Var\left\{1/8\sum_j (\eta + \tau_i + L_j + w_{ij})\right\}$, where w_{ij} denotes random variation among units of observation (over and above treatment effects) within locations. In the location-effects-fixed model, only w_{ij} is a random variable, so $Var(\bar{y}_{i\bullet}) = Var\left\{1/8\sum_j w_{ij}\right\} = (1/8)^2 \sum_j Var(w_{ij}) = \sigma^2/8$. In the location-effects-random model, L_j is also a random variable, so

$$Var(\bar{y}_{i\bullet}) = Var\left\{\frac{1}{8}\sum_j (L_j + w_{ij})\right\} = Var\left(\frac{1}{8}\right)^2\left[\sum_j Var(L_j) + \sum_j Var(w_{ij})\right] = \frac{\sigma^2 + \sigma_L^2}{8}.$$

Now, consider the results of the ESTIMATE statements:

Estimates

Label	**Estimate**	**Standard Error**
BLUP Trt 0	24.9500	0.5997
LSM Trt 0	24.9500	0.7593

Notice that the standard error of the BLUP for treatment 0, which uses the same coefficients as the location-effects-fixed least squares mean, has exactly the same standard error as we obtained for the least squares mean with the location-effects-fixed model.

This gives us a window into interpretation. Imagine that we want to take our estimate of the mean of treatment 0 based on the eight locations we have observed in this data set and use it to anticipate what would happen if we apply treatment 0 at *our* location, not one of the eight locations in the study. Clearly, the best estimate of the expected response is 24.95—all of the estimates agree on this number. If we want lower and upper confidence limits on our estimate, however, we need to decide whether to use 0.5997 or 0.7593 as the standard error for our confidence interval. The answer hinges on what we expect to be the sources of uncertainty as we try to anticipate how treatment 0 will perform. Does uncertainty come exclusively from variation among units of observation *within* each location, or does variation *among* locations over and above unit-of-observation variance also contribute?

The location-effects-random least squares mean standard error includes both sources of uncertainty. The location-effects-random BLUP includes only variance among units-of-observation within locations. In effect, the BLUP limits the scope of inference from statements applicable to the entire population of locations that could have been included in the study (including ours) to only the locations with nonzero coefficients in the predictable function for the treatment 0 BLUP.

The fact that the location-effects-random BLUP and the location-effects-fixed least squares mean share identical coefficients provides an insight into the implicit consequence of regarding location effects as fixed. In effect, defining location effects to be fixed narrows the scope of inference: confidence intervals for the treatment means apply *only* to the locations observed in the study. They do not apply—and cannot be interpreted to apply—to the population of locations supposedly represented by the locations observed in this study.

The least squares mean from the location-effects-random model is a *broad*-inference space estimate. The BLUP (and by implication, the least squares mean from the location-effects-fixed model) is a *narrow* inference space estimate. The former applies broadly to the entire population. The latter applies narrowly to only the eight locations observed. This example illustrates an important aspect of the inference space issue. The next example illustrates an important application of narrow inference space—SS inference.

3.4.2 Narrow Inference

The last section focused on broad inference. Understanding what broad inference is—and what it is not—is central to working with contemporary linear models. In this section, we turn our attention to narrow inference—often called **SS** inference. Narrow inference has its roots in many areas.

In animal genetics, mating trials are conducted to obtain estimates of variance among sires and dams. These estimates are used to identify for traits with selective breeding potential—susceptibility to disease, increased milk production, etc. Viewed this way, sire and dam effects clearly must be random. In the same trial, however, sires with exceptional breeding or commercial potential have to be identified. In the classical understanding of fixed and random effects, this seems like trying to have it both ways. Either the effect was included in the model to estimate variance (random effect) or to estimate or compare a measure of location (fixed). Animal breeders want to estimate the variance *and* estimate a measure of location on the same effect! This is the problem that motivated Henderson's development of *BLUP*, seminal work in mixed model literature (Henderson, 1950, 1963, 1975, 1984).

In medicine, clinical trials typically involve a random sample of patients. A common objective of these trials is to determine patient response to drug dosage. Broad inference—also called PA inference—on response to dosage is one objective, but following individual patient response is often equally important. The individual patient response—also called the SS response—is a form of narrow inference. It is defined by a linear combination of fixed effects (dosage) and random effects (patients).

There are other examples. In multilocation studies, location-specific treatment effects can be used to identify unanticipated local aberrations in a treatment's performance. Recently, much interest has focused on "value-added" models to estimate teacher effectiveness. In essence, teacher effects are defined as random in these models and estimated using teacher-specific predictable functions.

Example 3.7 illustrates a common example of narrow inference.

Example 3.7 Random coefficient regression model

In this example, there are four batches randomly sampled from production runs of a perishable product. The products are stored under recommended conditions and measured at specified storage times for a product quality characteristic denoted hereafter as Y. Suppose that past experience with this type of study allows us to assume that Y deteriorates linearly over time and has an approximate Gaussian distribution. The data appear in the SAS Data and Program Library as Data Set 3.3.

In Chapter 2, we developed a plausible model for this type of study—the random coefficient regression model. For this study, the model is

1. Linear predictor: $\eta_{ij} = \beta_0 + b_{0i} + (\beta_1 + b_{1i})X_j$, where β_0 and β_1 denote fixed intercept and slope coefficients; b_{0i} and b_{1i} denote the ith batch effects—both random—on intercept and slope, respectively; and X_j denotes the jth time
2. Distributions: there are two, one for y_{ij}, the observation on the ith batch at the jth time, and one for the joint distribution of b_{0i} and b_{1i}
   a. $y_{ij} \mid b_{0i}, b_{1i} \sim NI(\mu_{ij}, \sigma^2)$
   b. $\begin{bmatrix} b_{0i} \\ b_{1i} \end{bmatrix} \sim N\left(\begin{bmatrix} 0 \\ 0 \end{bmatrix}, \begin{bmatrix} \sigma_0^2 & \sigma_{01} \\ & \sigma_1^2 \end{bmatrix} \right)$
3. Link function: identity—$\eta_{ij} = \mu_{ij}$

With regard to inference space, two sets of estimates are of interest. The first set involves fixed effects only: the intercept, β_0; the slope, β_1; and the expected response at time T, $\beta_0 + \beta_1 T$. The intercept can be interpreted as the measure of product quality at time 0, when the product is first stored. The slope measures the rate of product deterioration. In random coefficient language, these are called PA estimates. Notice that they depend exclusively on fixed effects. All are forms of $\mathbf{K}'\boldsymbol{\beta}$. Thus, PA estimates are a form of broad inference.

The second set involves predictable functions $\beta_0 + b_{0i}$, $\beta_1 + b_{1i}$, and $\beta_0 + b_{0i} + (\beta_1 + b_{1i})T$. These are called SS "estimates"—the intercept, slope, and expected response at time T for the ith batch. SS "estimates" are, in fact, *predictors*, not *estimates*. They are all forms of $\mathbf{K}'\boldsymbol{\beta} + \mathbf{M}'\mathbf{b}$. SS estimators are BLUPs and, thus, a form of narrow inference.

3.4.2.1 GLIMMIX Implementation

Figure 3.7 shows GLIMMIX statements to estimate the parameters of the random coefficient model for Data Set 3.3.

These statements were discussed in Appendix 2.B. Notice that the variable X denotes the storage time. The GCORR option allows you to examine the estimated **G** matrix, the variance-covariance matrix of the random intercept, and slope effects. A look at this matrix shows why we need to look at it before proceeding:

Covariance Parameter Estimates

	Cov Parm	Subject	Estimate	Standard Error
$\hat{\sigma}_0^2$	UN(1,1)	Batch	0.07515	0.07379
$\hat{\sigma}_{01}$	UN(2,1)	Batch	−0.04620	0.03970
$\hat{\sigma}_1^2$	UN(2,2)	Batch	0.02818	0.02303
$\hat{\sigma}^2$	Residual		0.1182	0.02090

Estimated G Correlation Matrix

Effect	Row	Col1	Col2
Intercept	1	1.0000	−1.0037
X	2	−1.0037	1.0000

A warning in the SAS LOG appears: "`NOTE: Estimated G matrix is not positive definite.`" Looking at the covariance parameter estimates, shown in the first table, nothing stands out, but a look at the **G** correlation matrix reveals that we have nonsense estimates. This indicates that we are over-modeling the covariance of the random intercept and slope coefficients. We could speculate that no consequential correlation between random intercept and slope exists or we simply have too few batches on which to base a reasonable estimate

```
proc glimmix data=multi_batch_1;
 class batch;
 model y = X / solution;
 random intercept X / solution subject=batch type=un gcorr;
```

FIGURE 3.7
GLIMMIX statements for random coefficient model.

of the covariance. Either way, we need to simplify the model by deleting σ_{01} from $Var\begin{bmatrix} b_{0i} \\ b_{1i} \end{bmatrix}$.

Figure 3.8 shows the revised random statement and the ESTIMATE statements to obtain PA and SS estimates. In this illustration, we use time $T = 10$ to show how to obtain expected responses at a given time. In practice, you can set T to any value of interest in the study.

Eliminating the TYPE=UN option from the RANDOM statement deletes σ_{01} from the model. The PA ESTIMATE statements are defined on fixed effects only. The syntax for the SS estimates uses the structure of the RANDOM statement to help define the predictable function. Consider the first statement, "SS intercept for batch 1." INTERCEPT 1 before the vertical bar identifies the fixed effect, β_0; INTERCEPT 1 after the vertical bar identifies the random effect, b_{0i}; and SUBJECT 1 0 0 0 refers back to the RANDOM statement where SUBJECT was defined as BATCH. With these specifications, we have $\beta_0 + b_{01}$ the batch-specific (SS) BLUP of the intercept for batch 1. In "SS intercept for batch 2," everything is the same except SUBJECT 0 1 0 0, so the predictable function is defined on the second batch, that is, $\beta_0 + b_{02}$.

The statements for the batch-specific slope predictors appear in a single ESTIMATE statement. Notice the syntax for identifying the subject associated with each line of the statement. These predictors are defined as a group, and a multiplicity adjustment appears because testing the null hypothesis that each SS slope is 0 may be of interest. If it is, we want the option of controlling for experiment-wise type I error.

The last two statements show how to define differences and averages involving SS predictors. In the first statement, nothing appears before the vertical bar, so this predictable function involves no fixed effects, that is, $\mathbf{k} = \mathbf{0}$. INTERCEPT 1 X 10 appear after the vertical bar. This defines $b_{0i} + b_{1i} \times 10$. SUBJECT 1 –1 0 0 says multiply all coefficients from batch 1 by 1 and all coefficients from batch 2 by –1. The result is $\mathbf{m}'\mathbf{b} = b_{01} - b_{02} + (b_{11} - b_{12})10$, the difference between the batch 1 and batch 2 SS predictors at time $T = 10$. The fixed effects do not appear in this statement because the fixed effect component, $\beta_0 + \beta_1 \times 10$, cancels out in

```
proc glimmix data=multi_batch_1;
 class batch;
 model y = X / solution;
 random intercept X / solution subject=batch;
 estimate 'PA intercept' intercept 1;
 estimate 'PA slope'    X 1;
 estimate 'PA y-hat at X=10' intercept 1 x 10;
 estimate 'SS intercept for batch 1' intercept 1 | intercept 1 / subject 1 0 0 0;
 estimate 'SS intercept for batch 2' intercept 1 | intercept 1 / subject 0 1 0 0;
 estimate 'SS slope for batch 1' x 1 | x 1,
          'SS slope for batch 2' x 1 | x 1,
          'SS slope for batch 3' x 1 | x 1,
          'SS slope for batch 3' x 1 | x 1 / subject 1 0 0 0, 0 1 0 0,
             0 0 1 0, 0 0 0 1 adjust=bonferroni;
 estimate 'SS y-hat @ X=10, batch 1 ' intercept 1 x 10 | intercept 1 x 10
      / subject 1 0 0 0;
 estimate 'SS y-hat @ X=10, batch 2 ' intercept 1 x 10 | intercept 1 x 10
      / subject 0 1 0 0;
 estimate 'SS diff @ x=10, batch 1 v 2' | intercept 1 x 10 / subject 1 -1 0 0;
 estimate 'SS y-hat @ x=10, avg batch 1+2' intercept 2 x 20 | intercept 1 x 10
      / subject 1 1 0 0 divisor=2;
```

FIGURE 3.8
GLIMMIX program for random coefficient model, independent random effects, with ESTIMATE statements for PA and SS estimates.

the full SS statement of the difference: $\beta_0 + b_{01} + (\beta_1 + b_{11})10 - [\beta_0 + b_{02} + (\beta_1 + b_{12})10]$. The SS prediction averaged over batches 1 and 2 is

$$\frac{1}{2}\{[\beta_0 + b_{01} + (\beta_1 + b_{11})10] + [\beta_0 + b_{02} + (\beta_1 + b_{12})10]\} = \beta_0 + \beta_1 \times 10 + \frac{1}{2}[b_{01} + b_{02} + (b_{11} + b_{12})10].$$

Literally as implemented in the ESTIMATE statement, $\mathbf{k}'\boldsymbol{\beta} + \mathbf{m}'\mathbf{b} = 2\beta_0 + \beta_1 \times 20 + [b_{01} + b_{02} + (b_{11} + b_{12})10]/2$.

The listing appears as

Solutions for Fixed Effects

	Effect	Estimate	Standard Error
$\hat{\beta}_0$	Intercept	96.4827	0.1489
$\hat{\beta}_1$	X	0.1245	0.08321

Solution for Random Effects

	Effect	Subject	Estimate	Std Err Pred
$\hat{b}_{01}$	Intercept	Batch 1	−0.3308	0.1667
$\hat{b}_{11}$	X	Batch 1	0.1986	0.08328
$\hat{b}_{02}$	Intercept	Batch 2	0.2660	0.1667
$\hat{b}_{12}$	X	Batch 2	−0.1665	0.08328
$\hat{b}_{03}$	Intercept	Batch 3	0.01828	0.1667
$\hat{b}_{13}$	X	Batch 3	0.07132	0.08328
$\hat{b}_{04}$	Intercept	Batch 4	0.04652	0.1667
$\hat{b}_{14}$	X	Batch 4	−0.1033	0.08328

Estimates

Label	Estimate	Standard Error	DF	*t* Value	Pr > |*t*|
PA intercept	96.4827	0.1489	3	648.05	<.0001
PA slope	0.1245	0.08321	3	1.50	0.2316
PA y-hat at X = 10	97.7273	0.8438	3	115.81	<.0001
SS intercept for batch 1	96.1519	0.1150	3	836.13	<.0001
SS intercept for batch 2	96.7487	0.1150	3	841.32	<.0001
SS y-hat @ X = 10, batch 1	99.3820	0.08503	3	1168.75	<.0001
SS y-hat @ X = 10, batch 2	96.3280	0.08503	3	1132.84	<.0001
SS diff @ X = 10, batch 1 v 2	3.0539	0.1178	64	25.93	<.0001
SS y-hat @ X = 10, avg batch 1+2	97.8550	0.06134	3	1595.32	<.0001

Estimates Adjustment for Multiplicity: Bonferroni						
Label	Estimate	Standard Error	DF	*t* Value	Pr > |*t*|	*Adj P*
SS slope for batch 1	0.3230	0.005151	3	62.71	<.0001	<.0001
SS slope for batch 2	−0.04207	0.005151	3	−8.17	0.0038	0.0154
SS slope for batch 3	0.1958	0.005151	3	38.01	<.0001	0.0002
SS slope for batch 3	0.02112	0.005151	3	4.10	0.0263	0.1050

The estimates/predictors result from straightforward application of the estimable and predictable functions described earlier. Unlike the examples up to this point, the standard errors do not arise from easily described and intuitive formulas. In Chapter 6, we will develop the theory needed to compute the standard errors of these estimates and predictors. Note that the standard errors of the batch-specific slopes are especially small. This is not an error, just a reflection of how much uncertainty is removed by this particular predictable function.

KEY TOPICS: *broad inference, narrow inference, BLUP, PA inference, and SS inference*

3.5 Issue III: Conditional and Marginal Models

In Section 3.3, we considered the model/data scale issue arising in non-Gaussian models. In Section 3.4, we considered inference space issues that arise when we add random effects to the model. What happens when we have a non-Gaussian response *and* random model effects?

In addition to the model/data scale and inference space issues, we have a third issue: we can now define two different kinds of models—the *conditional* and the *marginal*. For non-Gaussian data, they are not equivalent; they do not estimate the same thing. Worse, conditional-marginal issues are embedded in multi-block and clustered non-Gaussian data even if one uses methodology thought to be free of GLMM issues. To paraphrase the boxer, Joe Louis, "you can run but you can't hide." To illustrate, we continue with the four-treatment multilocation study that has been our running example through this chapter.

Example 3.8 Binomial multilocation with location effects random

As in Example 3.6, assume random location effects. However, instead of Gaussian data, suppose we have a binomial response with $N = 100$ observations on each treatment at each location. That is, for treatment $i = 1,2,3,4$ and location $j = 1,2,\ldots,8$, $y_{ij}|L_j \sim Binomial(100,\pi_{ij})$.

3.5.1 Normal Approximation

At first glance these data appear to be a candidate for a normal approximation. Even with probabilities, π_{ij}, as close to 0 as 0.05 or as close to 1 as 0.95, with $N = 100$ the normal approximates the binomial plausibly well, as one can readily demonstrate via simulation—a common activity in an intro stat course. One approach—arguably the most common approach—for data of this kind is to fit a Gaussian model to the sample proportion, $p_{ij} = y_{ij}/N_{ij}$. Here, all $N_{ij} = 100$. Formally, the model is

1. Linear predictor: $\eta_{ij} = \eta + \tau_i + L_j$
2. Distributions:
   a. Data: $p_{ij}|L_j \sim N(\mu_{ij}, \sigma^2)$
   b. Location effects: L_j i.i.d. $N(0, \sigma_L^2)$
3. Link: identity—$\eta_{ij} = \mu_{ij}$

In this approach, the estimate $\hat{\mu}_{ij}$ would typically be interpreted as an estimate of $\hat{\pi}_{ij}$. Interest would presumably focus on treatment least squares means, which would be interpreted as $\hat{\pi}_0$, $\hat{\pi}_1$, $\hat{\pi}_2$, and $\hat{\pi}_3$, the probability of a favorable outcome for each of the treatments, respectively. This model is essentially identical to the Gaussian models for this design discussed in the previous two sections. We have seen that both the fixed and random location effect versions of these models yield identical results for broad inference on treatment differences. Both, in fact, are equivalent to results one would obtain from analysis of variance. This is why data analysts who say GLMMs make them uneasy believe they avoid GLMM issues when they do analysis of variance on the sample proportion—or on a variance-stabilizing transformation such as the arc sine square root, $\sin^{-1}\left(\sqrt{p_{ij}}\right)$. This is where "you can run but you can't hide" makes its appearance. Let us see why.

Using the same methods shown in Sections 3.3 and 3.4 for Gaussian multi-block models, the pertinent results here are

Test of treatment effect	$H_0: \tau_0 = \tau_1 = \tau_2 = \tau_3$	F-value = 40.01; p-value <0.0001	
Variance component estimates	Location	$\hat{\sigma}_L^2 = 0.0172$	
	Residual	$\hat{\sigma}^2 = 0.00452$	
Treatment mean estimates	Treatment 0	$\hat{\mu}_{0\bullet} = 0.11$	
	Treatment 1	$\hat{\mu}_{1\bullet} = 0.12$	Std error$\left(\hat{\mu}_{i\bullet}\right) = 0.0521$
	Treatment 2	$\hat{\mu}_{2\bullet} = 0.34$	
	Treatment 3	$\hat{\mu}_{3\bullet} = 0.40$	

There are obvious problems with these results. Notably, if the treatment means can indeed be interpreted as estimates of the corresponding π_i, should not the variance of their estimates, and hence their standard errors, show some dependence on $\pi_i(1 - \pi_i)$, the variance of a binomial? We leave these problems for now, but we will return to them.

3.5.2 Binomial GLMM

How does the normal approximation compare to a true GLMM? For these data, a GLMM can be described as

1. Linear predictor: $\eta_{ij} = \eta + \tau_i + L_j$
2. Distributions:
   a. Data: $y_{ij}|L_j \sim Binomial(N_{ij}, \pi_{ij})$
   b. Location effects: L_j i.i.d. $N(0, \sigma_L^2)$
3. Link: logit—$\eta_{ij} = \log[\pi_{ij}/(1 - \pi_{ij})]$

Notice that this model is identical to the normal approximation except for the distribution of the data—now explicitly binomial—and the link function. Figure 3.9 shows a GLIMMIX program for this model.

The selected results are as follows:

Covariance Parameter Estimates

Cov Parm	Subject	Estimate	Standard Error
Intercept	Loc	0.9462	0.5342

Type III Tests of Fixed Effects

Effect	Num DF	Den DF	*F* Value	Pr > *F*
Trt	3	21	57.99	<.0001

Trt Least Squares Means

Trt	Estimate	Standard Error	Mean	Standard Error Mean
0	−2.3558	0.3770	0.08661	0.02983
1	−2.2694	0.3753	0.09369	0.03186
2	−0.7826	0.3599	0.3138	0.07748
3	−0.5045	0.3589	0.3765	0.08426
	Data Scale (logits)		Model Scale—$\hat{\pi}_i$	

We see several differences between these results and the normal approximation. First, there is only one variance component estimate: $\hat{\sigma}_L^2 = 0.946$. This can be interpreted as the variance of the log odds among locations. There is no residual variance because within a treatment-location combination, observations have a binomial distribution. Their variance is $\pi_{ij}(1 - \pi_{ij})$. Once you have an estimate of π_{ij}, by definition you also have an estimate of the variance. No further variance component estimation should be necessary.

```
proc glimmix data=MultiLoc_4Trt;
 class loc trt;
 model S_1/N_Bin=trt;
 random intercept / subject=loc;
 lsmeans trt / ilink;
```

FIGURE 3.9
GLIMMIX program to implement GLMM for binomial multilocation data.

The F-value is 57.99 vs. 40.01 for the normal approximation. That these are different is no surprise. However, we might ask if there is anything to the fact that the F-value for the GLMM is greater (it turns out there is—we will pursue that later). The estimated probabilities for the four treatments have different standard errors, reflecting the mean–variance relationship for the binomial. As expected, standard errors increase as the estimated probability approaches 0.5.

Question for students: If we have estimated probabilities greater than 0.5, what should happen to their standard errors as $\hat{\pi}_i$ approaches 1?

Comment: Readers will notice that the data-scale standard errors in the aforementioned table cannot be obtained directly from applying the variance function to $\hat{\pi}_i$. Partly, this is because the variance function applies to π_{ij}, whereas the $\hat{\pi}_i$ are averaged over the locations *and* averaging occurs before applying the inverse link. There are additional subtleties in determining standard errors. We will consider the "big picture" of standard error computation in Chapter 5.

Finally, comparing the GLMM estimated probabilities to those for the normal approximation, we have

Treatment	Normal Approximation	GLMM
0	0.110	0.087
1	0.119	0.093
2	0.348	0.314
3	0.400	0.377

Two things stand out. First, the probabilities are different—no surprise; the normal approximation is just that, an approximation. Second, the GLMM-estimated probabilities are *all* less than their normal approximation counterparts. Is there something to that? If so, what? Pursuing this point, if we take the arithmetic mean of the sample proportions for each treatment, they equal the estimates for the normal approximation. Does this mean the GLMM estimates are wrong? If not—if in fact the GLMM actually yields *better* estimates of $\hat{\pi}_i$—then we need a convincing explanation to counter the cognitive dissonance. We are used to arithmetic means being *best* linear *unbiased* estimates, at least for balanced data.

3.5.3 Conditional and Marginal Distribution

To understand the discrepancy between the normal approximation and GLMM results, we need to return to modeling first principles and apply them meticulously to this example. Recall the first sentence of Chapter 1: A statistical model is a mathematical and probabilistic description of a plausible process giving rise to the observed data. Both the normal approximation and the GLMM in Example 3.8 assume the following process. For the ith treatment at the jth location, y_{ij} successes are observed. The number of successes has a binomial distribution: formally $y_{ij}|L_j \sim Binomial(N_{ij}, \pi_{ij})$. In our example, all $N_{ij} = 100$, but the distributional assumption about $y_{ij}|L_j$ would apply to any $N_{ij} \geq 1$. Note that the probability statement refers to the number of successes observed *conditional* on a given location. The word *conditional* is important. It means that π_{ij} itself behaves like a random variable. Specifically, $\pi_{ij} = 1\Big/\left(1+e^{-(\eta+\tau_i+L_j)}\right)$, which means π_{ij} varies through the random location effects L_j which we assume are i.i.d. $N\left(0,\sigma_L^2\right)$. It helps to visualize the distribution of π_{ij} conditional on L_j by reexpressing it as $\text{logit}\left(\pi_{ij}\right)|L_j \sim N\left(\eta+\tau_i,\sigma_L^2\right)$.

In general, a mixed model describes a process with two stochastic elements: the distribution of $\mathbf{y}|\mathbf{b}$, the observations conditional on the random model effects; and $\mathbf{b}$, the random model effects. However, we *cannot* observe either of these random processes directly, in isolation from each other. The only random variable we ever observe is, $\mathbf{y}$, the response variable. For non-Gaussian data, the conditional distribution of $\mathbf{y}|\mathbf{b}$ and the distribution of the observed data, $\mathbf{y}$, are not the same. Most importantly, their expected values are not the same. This is where confusion about how to interpret the results from a model—or an analysis of variance—begins.

The distribution of the observed data is the *marginal* distribution. We obtain it by applying well known probability theory. Let $f(\mathbf{y}|\mathbf{b})$ denote the p.d.f. of the data conditional on the random effects and $f(\mathbf{b})$ denote the p.d.f. of the random effects. The marginal distribution of the data is

$$f(\mathbf{y})=\int_{\mathbf{b}} f(\mathbf{y}\,|\,\mathbf{b})f(\mathbf{b})d\mathbf{b}$$

The marginal p.d.f. $f(\mathbf{y})$ describes the probabilistic behavior of the data as we actually observe them. We must infer everything about the model from the marginal distribution.

In the Gaussian case, the relationships between the probability processes assumed to give rise to the data—$f(\mathbf{y}|\mathbf{b})$ and $f(\mathbf{b})$—and the probability distribution of the data as we observe them—$f(\mathbf{y})$—are straightforward: $\mathbf{y}|\mathbf{b} \sim N(\mathbf{X\beta}+\mathbf{Zb},\mathbf{R})$, $\mathbf{b} \sim N(\mathbf{0},\mathbf{G})$, and $\mathbf{y} \sim N(\mathbf{X\beta},\mathbf{ZGZ}'+\mathbf{R})$. Importantly, there is no ambiguity about the meaning of the fixed effects vector $\boldsymbol{\beta}$ and, hence, no ambiguity about how to interpret estimable functions $\mathbf{K}'\boldsymbol{\beta}$. This is not true in the non-Gaussian case.

Our binomial example illustrates the interplay among these distributions for non-Gaussian data. The p.d.f. of the conditional distribution of the data given the location effects is

$$f\left(y_{ij}\,|\,L_j\right)=\binom{100}{y_{ij}}\left(\frac{1}{1+e^{-\eta_{ij}}}\right)^{y_{ij}}\left[1-\left(\frac{1}{1+e^{-\eta_{ij}}}\right)\right]^{100-y_{ij}}, \quad \text{where } \eta_{ij}=\eta+\tau_i+L_j$$

The distribution of the random model effects can be expressed alternatively starting with the assumption L_j i.i.d. $N\left(0,\sigma_L^2\right)$ or its implication: η_{ij} i.i.d. $N\left(\eta_i,\sigma_L^2\right)$ where $\eta_i=\eta+\tau_i$. The latter form allows easier development of the marginal p.d.f. Specifically, the marginal p.d.f. is

$$f\left(y_{ij}\right)=\int_{-\infty}^{\infty}\binom{100}{y_{ij}}\left(\frac{1}{1+e^{-\eta_{ij}}}\right)^{y_{ij}}\left[1-\left(\frac{1}{1+e^{-\eta_{ij}}}\right)\right]^{100-y_{ij}}\frac{1}{\sqrt{2\pi\sigma_L^2}}e^{-\frac{\left(\eta_{ij}-\eta_i\right)^2}{2\sigma_L^2}}d\eta_{ij}$$

This is a typical form of a marginal p.d.f. for data arising from the processes defined by non-Gaussian mixed models. With some exceptions, marginal p.d.f.'s arising from GLMMs defy direct, analytic evaluation. However, they can be visualized via simulation and they can be evaluated via numeric approximation. We will discuss approximation methods for GLMMs in Chapter 5.

For this section, it suffices to visualize this distribution and determine two measures of central tendency—the expected value and the median—of y_{ij}. This will shed light on our dilemma regarding the estimators from the normal approximation and the GLMM in Example 3.8. Taking a larger view, we need to understand how the characteristics of the marginal distribution relate to the processes we describe for GLMMs in order to fully understand what our modeling options are and what these models actually estimate (and, just as importantly in many cases, what they *do not* estimate).

3.5.4 Visualizing the Marginal p.d.f.

In our multilocation binomial example, treatments 0 and 1 had estimates of π_i in the vicinity of 0.1, and, in the GLMM, the estimate of location variance was ~ $\hat{\sigma}_L^2 = 1$. Note that this, unlike the variance component estimates from the normal approximation, is expressed on the scale used in the integral that defines the marginal distribution. Figure 3.10 shows

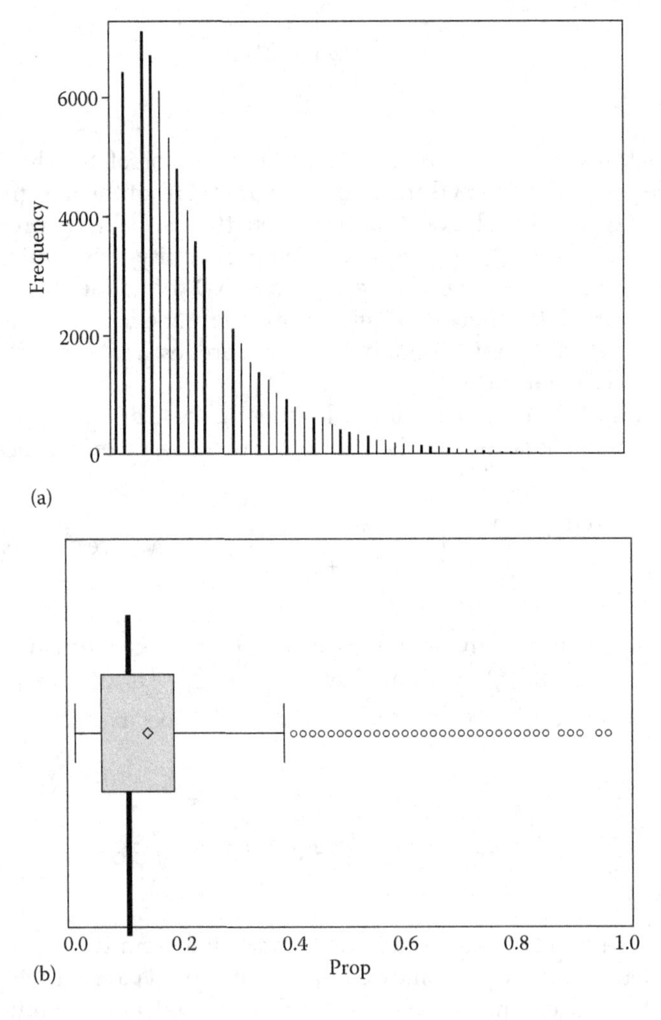

FIGURE 3.10
Plot of simulated data with proportion (prop) conditional on location binomial and location effects normal. (a) Histogram. (b) Box and Whisker plot.

the histogram of 1,000,000 observations generated from random number simulation of the processes involved in this example. The random numbers were generated using the following sequence:

- Set η = logit(prob = 0.1) = log[0.1/(1−0.1)] = −2.19722.
- Generate a standard normal deviate, z_i, for the *i*th simulated observation.
- $\sqrt{\sigma_L^2} z_i$ is thus $\sim N\left(0, \sigma_L^2\right)$. For this simulation, $\sigma_L^2 = 1$. Up to this point, we have simulated the process giving rise to the random model effects. That is, $\sqrt{\sigma_L^2} z_i$ simulates the process giving rise to the location effects.
- Thus, for the *i*th simulated observation, the binomial probability is $p_i = 1/\left(1+e^{-(\eta+z_i)}\right)$. This mimics the impact of the location effect on the binomial probability.
- Generate y_i as a random *Binomial*(100, p_i) deviate. This simulates the process giving rise to the data conditional on the location effect

From the histogram, it is evident that the marginal distribution is strongly right skewed. From the box and whisker plot, it is clear that the *median* of this distribution is 0.1. This makes sense: The only time the probability generating the observations is *exactly* 0.1 is when the simulated location effect is 0, that is, the *center* of location effect distribution. Because the translation of the location distribution to the generating probability is nonlinear, any location effect not at the exact center of its distribution will translate asymmetrically to the distribution of binomial probabilities. Thus, the highly skewed marginal distribution. The marginal distribution will always be skewed unless the conditional distribution has a binomial p.d.f. with a probability of exactly 0.5—that is, the probability that determines η is 0.5. If it is less than 0.5, the marginal distribution will be right skewed; if it is greater than 0.5, the marginal distribution will be left skewed. Skewness increases as the probability approaches 0 or 1. Skewness also increases as the variance, in this case σ_L^2, increases.

What about the mean of the marginal distribution? Following McCulloch, et al. (2008), we can approximate this using Gauss–Hermite quadrature. The conditional distribution of the sample proportion, $y_i/100$, is proportional to *Binomial*(100,p_i). The binomial proportion $p_i = 1/1+e^{-\eta_i}$ and $\eta_i \sim N\left(\eta, \sigma_L^2\right)$, where η = logit(0.1) and $\sigma_L^2 = 1$. The expected value of the sample proportion can be expressed as

$$E_{p_i}\left[E\left(\frac{y_i}{100}\middle| p_i\right)\right] = E\left(p_i\right) = \int_{-\infty}^{\infty}\left[\frac{1}{1+e^{-\eta_i}}\right]\frac{1}{\sqrt{2\pi\sigma_L^2}}e^{-\frac{(\eta_i-\eta)^2}{2\sigma_L^2}}d\eta_i.$$

Letting $x_i^2 = \left(\eta_i - \eta\right)^2/2\sigma_L^2$ and substituting yields $\int_{-\infty}^{\infty}\frac{1}{\left[1+e^{-\left(\eta+\sqrt{2\sigma_L x_i}\right)}\right]\sqrt{\pi}}e^{-x_i^2}dx_i$. Using Gauss–Hermite quadrature, this can be approximated by $\sum_{k=1}^{q} w_k\left\{\frac{1}{\left[1+e^{-\left(\eta+\sqrt{2\sigma_L x_k}\right)}\right]\sqrt{\pi}}\right\}$,

where w_k denotes the quadrature weights, x_k denotes the quadrature nodes, and q denotes the number of quadrature points. Using nine quadrature points, the approximate expected value of the sample proportion is 0.1339. The sample mean of the simulated data shown in Figure 3.10 is 0.1338.

3.5.5 What Do the Normal Approximation and the GLMM Estimate?

We can now see, via the marginal distribution, what the analysis of variance defined on the normal approximation and the GLMM estimate and why they differ. The normal approximation uses the sample proportion of the marginal distribution as the response variable. We know that the analysis of variance estimates of treatment means are unbiased estimates of the expected values of the response variable for each treatment. That is, the analysis of variance provides unbiased estimates of the mean of the marginal distribution for each treatment.

On the other hand, the GLMM obtains treatment means by estimating the parameters of the linear predictor, determining the model scale estimate $\hat{\eta}+\hat{\tau}_i$ and finally the data scale estimate $\hat{\pi}_i = 1/\left(1+e^{-(\hat{\eta}+\hat{\tau}_i)}\right)$. Note that these are broad inference space estimates because they are based on estimable functions $\mathbf{K}'\boldsymbol{\beta}$. Also, because $\mathbf{M} = \mathbf{0}$ in estimable functions—for this model all $L_j = 0$—these estimates occur at the center of the distribution of the binomial probabilities assumed to be giving rise to these data. In other words, $\hat{\pi}_i$ estimates the *median* of the marginal distribution for the ith treatment.

This resolves the question "why do the normal approximation and the GLMM produce different results?" It also frames the question "which estimate should one use?" in an intelligible way.

One way to think about the question "which estimate should one use?" is to recall discussions about measures of location in introductory statistics classes. Skewed distributions like the one shown in Figure 3.10 are routinely used to illustrate cases where the median is the preferred measure of location and the mean is considered ill advised.

A second way to think about this dilemma is to return to the definition of a statistical model. It should describe a "plausible process giving rise to the data." The conditional model clearly provides a coherent and believable probability mechanism for the data in this example. The normal approximation does not. In the conditional GLMM, the meaning of π_i, the probability of the outcome of interest occurring when the ith treatment is applied, is clear and well defined. For the normal approximation, it is muddled.

Another way to think about this question is to consider how investigators conceptualize data of this kind. Each treatment has a probability that the outcome of interest occurs. This is the "π_i" that the investigator wants to determine. The investigator also understands that locations perturb the treatment probabilities. What the investigator understands as "*the* probability for the ith treatment" we can think of as the "Goldilocks" probability—the probability of the outcome of interest occurring at a location that is "just right"—not below average, not above average. In formal terms, this is the probability at the center of the distribution of locations—that is, the estimable function $\eta + \tau_i$ from the GLMM. This is what investigators *think* they are estimating regardless of whether they compute ANOVA on the normal approximation or use the GLMM. This is the "you can run but you can't hide" aspect of GLMM-related issues. ANOVA users do not avoid the issue; they are simply oblivious to the fact that there *is* an issue.

A final way to think about this question is to ask "what is the investigator trying to estimate?" more carefully. Is the objective to estimate the probability of the outcome of interest for a typical member of the population—in this case, at a typical location? If so, the "Goldilocks" approach seems appropriate. If you drew a location at random from the population, its expected value would be neither above nor below average. On the other hand, if you truly do want to estimate the mean over the entire population and you understand the implications of using the mean to characterize a skewed distribution, then the estimates produced by the normal approximation may in fact be more appropriate. For example, if you are estimating disease incidence, the GLMM estimates would give you expected incidence in a typical community; the *estimates* produced by the normal approximation would give you a picture of *aggregate* disease incidence over the entire population.

The word *estimate* in the last sentence appears in italics because, as we know, in statistical inference, it is not *all* about the estimate. If the marginal mean does emerge as the focus of your study, there is still the issue of accurate standard errors and hence accurate test statistics. As noted earlier, not only does the normal approximation estimate the mean of the marginal distribution but it also assumes homogeneity of variance—which we *know* is not true for binomial data except in the (usually) uninteresting case that there is no treatment effect on the probabilities. To fully address these issues, we now introduce the ultimate focus of this section—conditional and marginal mixed models.

3.5.6 Gaussian Conditional and Marginal Models

The standard specification of a LMM includes

- The distribution of $\mathbf{y}|\mathbf{b}$, the data conditional on the random model effects
- The distribution of $\mathbf{b}$, the random model effects
- The link function, $\boldsymbol{\eta} = g[E(\mathbf{y}|\mathbf{b})]$
- The linear predictor, $\boldsymbol{\eta} = \mathbf{X}\boldsymbol{\beta} + \mathbf{Z}\mathbf{b}$

Alternatively, models that describe the same processes as those defined by the mixed model can be specified in terms of the marginal distribution and a linear predictor $\mathbf{X}\boldsymbol{\beta}$ without any explicit reference to the random effects. This is done by embedding variance associated with the random effects in the variance of the marginal distribution.

For Gaussian data, the mixed model defines $\mathbf{y}|\mathbf{b} \sim N(\mathbf{X}\boldsymbol{\beta} + \mathbf{Z}\mathbf{b},\mathbf{R})$ and $\mathbf{b} \sim N(\mathbf{0},\mathbf{G})$. The marginal distribution of $\mathbf{y}$ is $N(\mathbf{X}\boldsymbol{\beta},\mathbf{Z}\mathbf{G}\mathbf{Z}' + \mathbf{R})$. A fixed-effects-only model defined by $\boldsymbol{\eta} = \mathbf{X}\boldsymbol{\beta}$ and variance $\mathbf{V} = Var(\mathbf{y}) = \mathbf{Z}\mathbf{G}\mathbf{Z}' + \mathbf{R}$ yields the same inference *on the fixed effects*, as inference based on estimable functions from a mixed model.

For non-Gaussian data, this equivalence does not hold. Models defined in terms of the distributions of $\mathbf{y}|\mathbf{b}$ and $\mathbf{b}$ and the linear predictor $\boldsymbol{\eta} = \mathbf{X}\boldsymbol{\beta} + \mathbf{Z}\mathbf{b}$ yield fixed effects inference along the lines of the GLMM in the binomial multilocation example we considered in the last section, Example 3.8. A GLM defined exclusively in terms of $\boldsymbol{\eta} = \mathbf{X}\boldsymbol{\beta}$ yields inference whose target is the *mean* of the marginal distribution. As we saw in the last section, these two approaches yield different estimators with different meanings.

Models defined in terms of $\mathbf{y}|\mathbf{b}$ and $\mathbf{b}$ and the linear predictor $\boldsymbol{\eta} = \mathbf{X}\boldsymbol{\beta} + \mathbf{Z}\mathbf{b}$ are called *conditional models;* models defined in terms of the marginal distribution of $\mathbf{y}$ and the linear predictor $\boldsymbol{\eta} = \mathbf{X}\boldsymbol{\beta}$ are called *marginal models.*

To see how this works in the Gaussian case, we return the multilocation data assuming random location effects, Example 3.6. Recall that the conditional distribution of the data was $y_{ij}|L_j \sim NI(\mu_{ij}, \sigma^2)$, the linear predictor was $\eta_{ij} = \eta + \tau_i + L_j$, and L_j i.i.d. $N\left(0,\sigma_L^2\right)$. This is the conditional model. We can express this in model equation form as $y_{ij} = \eta + \tau_i + L_j + w_{ij}$, where w_{ij} denotes within location variation, assumed i.i.d. $N\left(0,\sigma_W^2\right)$. Note that $\sigma^2 = \sigma_W^2$.

The model equation can be reexpressed as $y_{ij} = \eta + \tau_i + e_{ij}$, where e_{ij} is defined as $L_j + w_{ij}$. Notice that what we are doing is removing all random model effects from the linear predictor—it is now $\eta + \tau_i$—and embedding all of the random variation in the variance of y_{ij}. The location effect variation is still in the model, it just is not explicitly in the linear predictor. The variance structure of the model is now

- $Var(y_{ij}) = Var(e_{ij}) = Var(L_j + w_{ij}) = \sigma_L^2 + \sigma_W^2$
- $Cov(y_{ij}, y_{i'j}) = Cov(e_{ij}, e_{i'j}) = Cov(L_j + w_{ij}, L_j + w_{i'j}) = \sigma_L^2$
- $Cov(y_{ij}, y_{i'j'}) = Cov(e_{ij}, e_{i'j'}) = 0$

Thus, for the jth location,

$$Var\begin{bmatrix} e_{1j} \\ e_{2j} \\ e_{3j} \\ e_{4j} \end{bmatrix} = \begin{bmatrix} \sigma_L^2+\sigma_W^2 & \sigma_L^2 & \sigma_L^2 & \sigma_L^2 \\ & \sigma_L^2+\sigma_W^2 & \sigma_L^2 & \sigma_L^2 \\ & & \sigma_L^2+\sigma_W^2 & \sigma_L^2 \\ & & & \sigma_L^2+\sigma_W^2 \end{bmatrix} = \sigma_E^2 \begin{bmatrix} 1 & \rho & \rho & \rho \\ & 1 & \rho & \rho \\ & & 1 & \rho \\ & & & 1 \end{bmatrix}$$

where

$$\sigma_E^2 = \sigma_L^2 + \sigma_W^2$$

$$\rho = \frac{\sigma_L^2}{(\sigma_L^2 + \sigma_W^2)}$$

This is called a *compound symmetry* covariance model and in many contexts $\rho = [\sigma_L^2/(\sigma_L^2+\sigma_W^2)]$ is called the *intra class correlation*. Note that it is equivalent to the mixed model, provided $\rho \geq 0$. The only difference is that $\hat{L}_j$ and predictable functions involving them cannot be obtained from the compound symmetry model.

For the multilocation data, the compound symmetry model is the marginal model. Location effects do not appear in the model, but the impact of random location effects is embedded in the variance structure.

Figure 3.11 shows the GLIMMIX statements to implement the marginal multilocation model.

Notice that the only RANDOM statement that appears uses the _RESIDUAL_ specification. GLIMMIX understands this as defining a response Y whose distribution is $\mathbf{y} \sim N(\mathbf{X\beta},\mathbf{V})$—normal because it is the default distribution, $\mathbf{X\beta}$ described by the right-hand side of the MODEL statement, and the structure of $\mathbf{V}$ defined by the RANDOM _RESIDUAL_ statement. In the RANDOM _RESIDUAL_ statement, SUBJECT=LOC sets up a block diagonal matrix; TYPE=CS defines each block (LOC) as having a compound symmetry structure as shown earlier.

```
proc glimmix data=MultiLoc_4Trt;
 class loc trt;
 model Y= trt / solution;
 random _residual_  / subject=loc type=cs;
 lsmeans trt / diff;
```

FIGURE 3.11
GLIMMIX for Gaussian compound symmetry marginal model.

The selected results are as follows:

Covariance Parameter Estimates			
Cov Parm	**Subject**	**Estimate**	**Standard Error**
CS	Loc	1.7352	1.3306
Residual		2.8770	0.8879

Type III Tests of Fixed Effects				
Effect	**Num DF**	**Den DF**	**F Value**	**Pr > F**
Trt	3	21	7.04	0.0019

Trt Least Squares Means		
Trt	**Estimate**	**Standard Error**
0	24.9500	0.7593
1	24.1375	0.7593
2	27.3750	0.7593
3	27.0875	0.7593

Differences of Trt Least Squares Means						
Trt	**_Trt**	**Estimate**	**Standard Error**	**DF**	***t* Value**	**Pr > |*t*|**
0	1	0.8125	0.8481	21	0.96	0.3489
0	2	−2.4250	0.8481	21	−2.86	0.0094
0	3	−2.1375	0.8481	21	−2.52	0.0199
1	2	−3.2375	0.8481	21	−3.82	0.0010
1	3	−2.9500	0.8481	21	−3.48	0.0022
2	3	0.2875	0.8481	21	0.34	0.7380

Aside from the labeling of the variance component estimates, as far as inference on the fixed effects—treatment means, differences, and any contrasts that might be of interest—the results here are identical to those shown earlier in the conditional location-effects-random Gaussian model of Example 3.8.

3.5.7 Non-Gaussian Marginal vs. Conditional Model

Now, consider the binomial case. Earlier in this section, we considered the conditional model—the binomial GLMM.

1. Linear predictor: $\eta_{ij} = \eta + \tau_i + L_j$
2. Distributions:
   a. $y_{ij}|L_j \sim Binomial(N_{ij}, \pi_{ij})$
   b. L_j i.i.d. $N\left(0, \sigma_L^2\right)$
3. Link function: $\text{logit}(\pi_{ij})$

What about the marginal model?

First, we eliminate random effects from the linear predictor. This leaves us with $\eta_{ij} = \eta + \tau_i$.

The "distribution" is "Binomial"$(N_{ij}, \tilde{\pi}_{ij})$. Note that the distribution is actually not binomial. It is a marginal distribution, the result of integrating over random location effects. The binomial conditional distribution given the location effects is a building block, but, as we saw earlier, the distribution in this model is not binomial. Nonetheless, GLM convention is to label the model "binomial," so we will adhere to this convention here. Key notation here, to avoid confusion with the conditional model: we refer to the probability parameter as $\tilde{\pi}_{ij}$ not π_{ij}. $\tilde{\pi}_{ij}$ denotes the marginal expectation; π_{ij} denotes the probability of the binomial distribution in the conditional model.

Finally, we need to account for location effects in the covariance structure. In the non-Gaussian marginal GLMM, this is done using a *working covariance matrix*. The working covariance matrix mimics the structure of the Gaussian covariance model, but it is not a true covariance. Here, the working covariance matrix for the jth location is

$$P_W \begin{bmatrix} y_{1j} \\ y_{2j} \\ y_{3j} \\ y_{4j} \end{bmatrix} = \phi \begin{bmatrix} 1 & \rho & \rho & \rho \\ & 1 & \rho & \rho \\ & & 1 & \rho \\ & & & 1 \end{bmatrix}$$

where

P_W denotes the working covariance
ϕ is a scale parameter
ρ denotes the working correlation

The scale parameter and working correlation are interpreted in the same spirit as σ_E^2 and ρ in the Gaussian case, but interpretation should not be too literal. The working covariance model is mainly a device to obtain marginal estimates and allow the impact of location variance on standard errors and test statistics to be taken into account.

> *Quasi-likelihood*: When we introduce the working covariance structure, we have a structure whose basic form derives from the binomial p.d.f. but with the working covariance grafted in. The result is a mathematical form that is not a probability distribution and, in fact, does not correspond to any feasible probability mechanism. In Chapter 4, we will introduce *quasi-likelihood*. Quasi-likelihood, introduced by Wedderburn (1974), refers to mathematical forms that share characteristics essential for GLM estimation with the likelihoods of known probability distributions, but whose expectation and covariance structure mean that they are not true likelihoods. All non-Gaussian marginal GLMMs depend on quasi-likelihood theory.

Figure 3.12 shows the GLIMMIX statements for the binomial marginal model.

Note that this program is identical to the program for the Gaussian marginal model except for the response variable and the use of the ILINK option in the LSMEANS statement. For non-Gaussian data, the RANDOM _RESIDUAL_ statement defines the working covariance structure.

```
proc glimmix data=MultiLoc_4Trt;
 class loc trt;
 model S_1/N_Bin = trt / solution;
 random _residual_  / subject=loc type=cs;
 lsmeans trt / diff ilink;
```

FIGURE 3.12
Marginal model for binomial multilocation data.

The selected results are as follows:

Covariance Parameter Estimates			
Cov Parm	**Subject**	**Estimate**	**Standard Error**
CS	Loc	6.3508	3.5321
Residual		1.0222	0.3155

Type III Tests of Fixed Effects				
Effect	**Num DF**	**Den DF**	***F* Value**	**Pr > *F***
Trt	3	21	26.31	<.0001

Trt Least Squares Means				
Trt	**Estimate**	**Standard Error**	**Mean**	**Standard Error Mean**
0	−2.0865	0.3954	0.1104	0.03884
1	−2.0043	0.3831	0.1188	0.04009
2	−0.6466	0.2609	0.3438	0.05887
3	−0.4055	0.2530	0.4000	0.06072

In the context of this discussion, the most noteworthy results are the data scale estimates and standards errors in the "Trt Least Squares Means" table. Notice that the estimates are identical to the normal approximation: these are estimates of the marginal means. They are unbiased estimates of the $\bar{\pi}_i$—not the probabilities of the binomial distribution but the expected values of the marginal distribution.

As a final comment on the marginal GLMM, recall that one of our defining criteria for a statistical model says that, ideally, it should describe a plausible mechanism giving rise to the observed data. By this standard, marginal GLMMs fail. Once we define the model in terms of the working covariance, we no longer have a true probability distribution. Instead, we have a quasi-likelihood, which we will define formally in Chapter 4. There is no known probability mechanism that could give rise to data as described by the marginal model. Among the models we have considered here, the conditional GLMM is the only one that actually describes a process that is even possible. Any legitimate alternative would still have a two-step process—vary the binomial probability according to some distribution and then generate binomial observations given the binomial probability conditional on that location. Like it or not, the generating mechanism describes a conditional model. The marginal GLMM does not describe a process; it simply allows marginal means to be estimated when they are deemed to be the objective.

3.5.8 One Last Aspect of the Conditional Model: The Role of "Residual" in Gaussian LMM vs. One-Parameter Non-Gaussian GLMM

In Chapter 1, we established the "probability distribution form" as the preferred way of presenting a linear model. This discussion included the realization that the "error" or "residual" term in the traditional "model equation" form of the model should not appear in the linear predictor of the linear model expressed in "probability distribution" form. In Chapter 1, we justified this for models with non-Gaussian data, such as the binomial and Poisson, by noting that the "residual" has no meaning—once you know the mean, you also know the variance. For Gaussian data, the mean and variance of the conditional distribution of $\mathbf{y}|\mathbf{b}$ are independent parameters and must be estimated separately. The linear predictor estimates the mean. Information needed to estimate the $Var(\mathbf{y}|\mathbf{b})$ has to come from elsewhere. We saw in Chapter 2 that excluding "residual" from the linear predictor amounts to reserving the information from the residual component of the skeleton ANOVA for estimating $Var(\mathbf{y}|\mathbf{b})$.

Taking this reasoning one step further, we pose the following question. In models for data with a one-parameter non-Gaussian distribution, such as the binomial or Poisson, since we do *not* need the "last term" in the skeleton ANOVA to estimate the variance, might this term have another role? The answer, in some cases, is "yes." Here, we explore why, using the multilocation data to illustrate.

Recall from Chapter 2 the WWFD skeleton ANOVA exercise we used to help construct the linear predictor. Table 3.3 summarizes this exercise for the multilocation data.

Three versions of the combined skeleton ANOVA appear in Table 3.3. All three agree in the way location and treatment main effects are represented. Any resulting linear predictor must contain at least $\eta + \tau_i + L_j$—the linear predictor we have used for all examples with the multilocation data throughout this chapter. The skeleton ANOVAs differ in their understanding of the final 21-d.f. term. Models with Gaussian data must use "Combined v.3." The data distribution is $y_{ij} \sim NI(\mu_{ij}, \sigma^2)$ for fixed-effects-only models, $y_{ij}|L_j \sim NI(\mu_{ij}, \sigma^2)$ for location-effects-random models. Either way, "residual" must be reserved for estimating σ^2.

Without multiple observations per location-treatment combination, models with Gaussian data cannot use the "Combined v.2" ANOVA. Location × treatment interaction cannot be estimated in a Gaussian model without leaving 0 d.f. to estimate σ^2.

For models with binomial data, this is not true. In all of our examples so far in this chapter, we used the same linear predictor as we did for the Gaussian models. However, there is no reason in principle why we could not have included the location × treatment interaction in the linear predictor.

TABLE 3.3

Skeleton ANOVAs Used to Construct Linear Predictor for Multilocation Data

Topographical		Treatment		Combined v.1		Combined v.2		Combined v.3	
Source	d.f.	Source	d.f.	Source	d.f.	Source	d.f.	Source	d.f.
Location	7			Location	7	Location	7	Location	7
		Treatment	3	Treatment	3	Treatment	3	Treatment	3
Obs(Loc)	24			Obs(Loc)	21				
						Loc × Trt	21		
		"Parallels"	28					Residual	21
Total	31	Total	31	Total	31	Total	31	Total	31

```
proc glimmix data=MultiLoc_4Trt;
 class loc trt;
 model s_1/N_Bin= trt / solution;
 random intercept trt / subject=loc;
 /* or alternatively random loc loc*trt; */
 lsmeans trt / ilink;
```

FIGURE 3.13
GLIMMIX program for binomial multilocation model with interaction.

What happens if we do? We could change the linear predictor to $\eta_{ij} = \eta + \tau_i + L_j + (\tau L_{ij})$, where $(\tau L)_{ij}$ denotes the ijth location × treatment effect, assumed i.i.d. $N\left(0, \sigma_{TL}^2\right)$, and compute the analysis using the GLIMMIX program shown in Figure 3.13.

The variance component estimates are $\hat{\sigma}_L^2 = 0.948$ and $\hat{\sigma}_{TL}^2 = 0.003$. This compared to $\hat{\sigma}_L^2 = 0.946$ for the model with no interaction. Other estimates and test statistics are also changed only negligibly. For this response variable, there is no evidence of a nonnegligible location × treatment interaction. It does not matter if you include the term or leave it out—you get essentially the same results.

However, this is not always the case. We now consider the same data but with a different response variable, denoted S_2 in Data Set 3.2. Four analyses will be compared:

1. Binomial conditional GLMM without interaction. Linear predictor: $\eta_{ij} = \eta + \tau_i + L_j$. Location effects random. Logit link. Same model used to introduce conditional GLMM.
2. Normal approximation. Response variable: sample proportion. Linear predictor: $\eta_{ij} = \eta + \tau_i + L_j$. Same model used to introduce marginal distribution.
3. Binomial conditional GLMM with interaction. Linear predictor: $\eta_{ij} = \eta + \tau_i + L_j + (\tau L)_{ij}$. Logit link. Same model used in Figure 3.13.
4. Binomial marginal GLMM. Linear predictor $\eta_{ij} = \eta + \tau_i$. Logit link. Compound symmetry working covariance. Same model used in Figure 3.12.

Table 3.4 contains a summary of the pertinent results.

TABLE 3.4
Comparison of Four Models for Multilocation Four-Treatment Binomial Data

		Conditional GLMM no Interaction		Normal Approximation		Conditional GLMM with Interaction		Marginal GLMM	
F- and p-values H_0: all $\tau_i = 0$		$F = 19.63$ $p < 0.0001$		$F = 2.08$ $p = 0.1334$		$F = 2.89$ $p = 0.0594$		$F = 2.10$ $p = 0.1307$	
		Estimate	**Std Err**	**Estimate**	**Std Err**	**Estimate**	**Std Err**	**Estimate**	**Std Err**
Data scale estimates of treatment expected values[a]	Trt 0	0.124	0.046	0.160	0.073	0.106	0.051	0.160	0.061
	Trt 1	0.159	0.056	0.200	0.073	0.141	0.064	0.200	0.067
	Trt 2	0.260	0.080	0.304	0.073	0.263	0.101	0.304	0.077
	Trt 3	0.298	0.087	0.340	0.073	0.288	0.108	0.340	0.079

[a] Estimates for conditional GLMM are conditional $\hat{\pi}_i$; estimates for normal approximation and marginal GLMM are marginal $\hat{\bar{\pi}}_i$.

The most striking result in Table 3.4 is the large difference between the Conditional GLMM without interaction and the other three models regarding the *F*-test for H_0: all $\tau_i = 0$. Why the discrepancy? In the no-interaction GLMM, the linear predictor is $\eta_{ij} = \eta + \tau_i + L_j$, and all variation among observations within a given location are assumed to result from π_{ij} through the variance relationship $\pi_{ij}(1 - \pi_{ij})$. In each of the other models, there is a provision for relative differences among the treatments within each location to be perturbed by random processes. This occurs in the normal approximation through the residual variance, σ^2; in the marginal GLMM, it occurs through the scale parameter, ϕ. In the conditional GLMM with interaction, by definition location × treatment interaction is random variation in log odds ratios between treatments from location to location.

The conditional GLMM without interaction makes the potentially unrealistic assumption that odds ratios between treatment remain absolutely constant from location to location, while at the same time assuming that log odds averaged over all treatments at each location do vary randomly from location to location. Failure to account for random variation among locations with respect to the odds ratios has the effect of inflating the *F*-value, distorting the estimates of treatment probabilities, and underestimating their standard errors. This is a preview of one way in which GLMMs can display symptoms of overdispersion. Overdispersion occurs when the observed variance is greater than the variance accounted for by the proposed model.

As in the previous examples, the normal approximation and the marginal GLMM yield identical estimates of treatment expectation, the marginal $\tilde{\pi}_i$ discussed earlier. The standard errors of the normal approximation do not account for heterogeneity of variance known to occur by definition with binomial data. The marginal GLMM does. The *F*-values for these two analyses essentially agree.

The conditional GLMM with interaction yields accurate estimates of the π_i as defined in the conditional model. Recall these are the estimated "Goldilocks" probabilities—what we would expect at a typical location. The standard errors reflect the full impact of the mean–variance relationship of the binomial distribution. The *F*-value for the test of treatment effect reflects the impact of the random location × treatment effect so it is not inflated like the conditional GLMM without interaction.

At the same time, the *F*-value is noticeably greater than the *F*-value for the normal approximation and the marginal GLMM. This is not an artifact of this data set. It results from the skewness of the marginal distribution. Recall that marginal "probabilities" are shifted toward 0.5 relative to the probabilities of the actual process giving rise to the data, the π_i. This has the effect of attenuating differences among the marginal $\tilde{\pi}_i$ relative to differences among the π_i. This in turn reduces the power of tests for treatment effect: assuming type I error is appropriately controlled, as it is here by the conditional GLMM *with* interaction, marginal models will always have lower power than their conditional analogs.

KEY TOPICS: *conditional distribution of the data given the random effects, marginal distribution of the observations, what is the distribution of the data we actually see?, what happens when ANOVA is applied to non-normal data?, two measures of location for marginal distribution and which models estimate what?, compound symmetry, conditional model, marginal model, working correlation, working covariance, GEE, role of the residual in Gaussian data, possible alternative understanding of "residual" in non-Gaussian data*

3.6 Summary

In many ways, the material in this chapter is the most important in this textbook. The ideas are complex and include some of the most widely misunderstood and misapplied concepts in statistical modeling. In particular, the distinction between inference with conditional and marginal models is just beginning to be recognized and appreciated in the modeling world. Understanding this distinction and its implications for how models are written and how inference proceeds is a crucial prerequisite for working effectively with contemporary statistical models.

Four overarching ideas summarize the essence of this chapter:

1. Inference with statistical models occurs via *estimable* and *predictable* functions. They form the foundation of hypothesis testing and interval estimation with linear models. For statistical inference to proceed, one must be able to express research objectives in terms of estimable or predictable functions.
2. For GLMs with nonidentity link functions, all estimation and inference occurs on the *model* scale—that is, in terms of the parameter vector $\boldsymbol{\beta}$ for all GLMs and, additionally, the random vector $\mathbf{b}$ for GLMMs. However, for reporting purposes, it is often desirable to convert model scale estimates to the *data* scale.
3. For mixed models, inference may be broadly applicable to the entire population represented by the random model effects or narrowly applicable to a subset of the random effects observed. Estimable functions define *broad* inference; predictable functions define *narrow* inference. We refer to these as *inference space* issues.
4. There are two kinds of broad inference—*conditional* and *marginal*. With Gaussian data, they are equivalent; *for non-Gaussian data, they are not.* Conditional inference targets the expected value of the assumed distribution of the observations given the random model effects. Marginal inference targets the expected value of the marginal distribution of the observations. Except for Gaussian data, the marginal distribution of the observations is *not* the same as the assumed distribution of the response variable. For example, with binomial data, conditional broad inference estimates a binomial probability; marginal inference does not. Marginal broad inference is also called—somewhat deceptively—PA inference. In general, marginal distributions of GLMMs are highly skewed. With marginal broad inference, a.k.a. PA inference, estimable functions depend on the mean of the marginal distribution as the measure of location. With conditional broad inference, also called "random-effects inference" (Molenbergh and Verbeke, 2006), estimable functions depend on the mean of the conditional distribution at $\mathbf{M}'\mathbf{b} = \mathbf{0}$, which translates to the *median* of the marginal distribution as the measure of location. The choice of marginal vs. conditional broad inference is essentially a choice between using the mean or the median as the preferred measure of location with a highly skewed distribution.

The conditional-marginal issue only occurs when we have random model effects *and* non-normal data. It does not occur with Gaussian data or with fixed effects only non-Gaussian data. For this reason, it is not an issue that would have occurred to anyone in the modeling world until the advent of GLMMs. Given that GLMM theory is relatively recent and useable GLMM software has only appeared during the past decade, it is not surprising that the conditional-marginal issue does not appear in traditional linear model literature and is only now beginning to be appreciated. However, whenever non-Gaussian data and mixed model structure coexist, for example, split-plot experiments with binomial or count data, the conditional-marginal issue exists and cannot be avoided. *In this sense, learning the material in the rest of this textbook is to some extent a moot point unless one understands the issues in this chapter.*

Exercises

There are four data sets on SAS data and programs for this chapter:

- 4trt_blocked_1 a randomized complete block with four treatments
- 4trt_blocked_2 a balanced incomplete block (BIB) with four treatments
- 4trt_blocked_3 an incomplete block design—not balanced in BIB sense but augmented so all treatments replicated four times
- 4trt_blocked_4 an incomplete block design with four treatments; Note: you could call this a "disconnected design", doing so misses the point of what it really is Why?

All "4trt_" data sets have two response variables—*Y* may be assumed normal (Gaussian); *Count* should be assumed to be Poisson

- regression_y_count quantitative levels applied in multiple "reps" (which could be locations, batches, blocks, etc.)—the response variable *Y* can be assumed normal (Gaussian); assumed (unless instructed otherwise) response variable COUNT has Poisson distribution
- regression_binomial quantitative levels applied in multiple "reps" (which could be locations, batches, blocks, etc.)—the response variable is binomial (*Y* "successes" out of *N* observations per rep x level of *X*)

Treat this as an exploration of inference issues—model vs. data scale, narrow vs. broad, and conditional vs. marginal—using SAS linear model software focusing on PROC GLIMMIX. The following are suggested questions you should answer. In addition, you should go above and beyond and pose your own "what happens when I do this?" explorations.

3.1 For 4trt_blocked_1 data set—*Y* variable

a. Write the model with blocks fixed, then with blocks random.

b. Write the required SAS statements for the fixed block and random block case.

c. Verify that `random block` and `random intercept/subject=block` yield identical results (and make sure you understand why).

d. Add the `e` option to the `LSMEANS` statement. Compare and contrast the results for the fixed and random block models. What do these tell you? Why are they different?

e. Compare and contrast the standard errors of treatment means and differences computed by the `LSMEANS` statement. Account for similarities and differences.

3.2 For 4trt_blocked_2 data set—*Y* variable

a. Repeat parts a-e from (3.1).

b. Run a second LSMEANS statement using the `bylevel` and `e` options. How do these compare with the default LSMEANS? What are the LSMEANS BYLEVEL estimating?

c. Write an ESTIMATE statement to compute the difference between the Trt 1 and Trt 2 LSMEANS BYLEVEL estimates—what are these estimating?

d. Ordinarily, would you use results from the LSMEANS default or LSMEANS BYLEVEL? explain—*briefly*!

e. Reexpress the blocks random model as a marginal model with a compound symmetry covariance structure. Verify that the block-effects-random and compound symmetry models produce identical results.

3.3 Repeat steps in part (3.2) for 4trt_blocked_3.

3.4 Repeat steps in part (3.2) for 4trt_blocked_4. Pay particular attention to how the fixed and random block model results compare?

We can make a compelling argument that calling this design "disconnected" misses the point. What is a disconnected design? Why call this design disconnected? Why does this miss the point? (Hint: How would Fisher look at this design? Following Fisher, what is another name [commonly discussed in design of experiments courses] for this design?) This is an example of where the limitations of fixed block thinking are on full display—where it backs you into a corner with no good escape.

3.5 For data set regression_Y_count. This problem refers to response variable *Y*.

a. Write the model assuming "reps" fixed; then assuming "reps" random.

b. Write the required SAS statements assuming the fixed "rep" model then repeat for the random "rep" model. Here, assume the random effects are mutually independent.

c. Compare and contrast the results from each analysis.

d. Include in (c) the estimate of the "population average" regression equation. Note that the random "rep" model obtains this easily; the fixed "rep" model requires literate `ESTIMATE` statements.

e. Rewrite the random "rep" model assuming that the random intercept and random slope terms are correlated within each "rep." Compare and contrast these results with the results in (b).

3.6 Repeat (3.5) but use the data set regression_binomial.

3.7 Repeat 3.1 through 3.4 for the *Count* response variable in data set regression_y_count.

3.8 Repeat (3.1) with the *count* variable, but run the analysis assuming count is approximately normal.

3.9 Repeat (3.1) with the *count* variable, but use the log transformation and assume log_count is approximately normal.

3.10 Compare and contrast your results from (3.7) to (3.9), paying particular attention to marginal vs. conditional inference issues.

Part II

Estimation and Inference Essentials

4

Estimation

4.1 Introduction

This chapter begins the second major part of this textbook: the theory and methodology of linear model estimation and inference. In the first part—the first three chapters—we explored the essential structure of Linear models, the interplay between study design and model construction, and the major inference issues, including those introduced by random model effects and non-Gaussian response variables. Until now, we have been intentionally vague about what goes on inside the black box. How do we estimate parameters and construct inference statistics? What theory justifies these approaches?

Our focus in this chapter is on estimation. Recall that a fully specified linear model consists of

- A linear predictor $\mathbf{X\beta}+\mathbf{Zb}$ or simply $\mathbf{X\beta}$ if there are no random effects
- The distribution of $\mathbf{y}|\mathbf{b}$—the observations, conditional on the random effects if there are random effects
- The link function, $\boldsymbol{\eta}=g(\boldsymbol{\mu}|\mathbf{b})$, where $\boldsymbol{\mu}|\mathbf{b}=E(\mathbf{y}|\mathbf{b})$
- The random effects distribution, $\mathbf{b}\sim N(\mathbf{0},\mathbf{G})$

Our task includes estimating $\boldsymbol{\beta}$, any components of $\text{Var}(\mathbf{y}|\mathbf{b})$ that do not depend on $\boldsymbol{\mu}|\mathbf{b}$ and, if random effects are present, $\mathbf{b}$ and $\mathbf{G}$.

Unlike classical LM theory, which is least squares based, generalized linear mixed model (GLMM) estimation uses maximum likelihood. We approach estimation one piece at a time, starting with $\boldsymbol{\beta}$ for fixed-effects-only models, then $\mathbf{b}$ and its components variance and covariance for Gaussian-only models, and then finally integrating these into full-fledged GLMM estimation procedures. Along the way, prior to the each increasing level of elaboration, we introduce essential background. After each step, we will pause to show the relationship between likelihood-based GLMM estimation and classical, least-squared-based estimation.

4.2 Essential Background

Nelder and Wedderburn (1972), in their foundational generalized linear model (GLM) paper, extended the fixed effects LM from response variables assumed to have a Gaussian distribution to response variables belonging to the exponential family. Our first stop will be the exponential family. We then review the basic principles of maximum likelihood estimation, including the Newton–Raphson and Fisher scoring algorithms. These methods extend to *quasi-likelihood*, which allows GLMs to be used with response variables whose expected value and variance can be specified even though a full likelihood cannot.

4.2.1 Exponential Family

Consider a random variable Y whose expected value we denote by $E(Y)=\mu$ and whose variance we denote by $\text{Var}(Y)=\sigma^2$. As presented in most mathematical statistics textbooks, the distribution of Y is said to be a member of the exponential family if its p.d.f. can be written as

$$f(y\mid\theta)=m(y)r(\theta)e^{s(\theta)t(y)}$$

where θ denotes a canonical parameter. Following Nelder et al., this can alternatively, and generally more conveniently, be written as

$$f(y\mid\theta)=\exp\left[\frac{y\theta-b(\theta)}{a(\phi)}+c(y,\phi)\right]$$

or

$$\log\left[f(y\mid\theta)\right]=\ell(\theta;y,\phi)=\frac{y\theta-b(\theta)}{a(\phi)}+c(y,\phi) \tag{4.1}$$

where ϕ denotes a scale parameter. This form has the advantages of explicitly accounting for the scale parameter and lending itself naturally to maximum likelihood estimation.

As examples, consider the Gaussian, binomial, and Poisson probability density/distribution functions, distributions we saw in Chapters 1 through 3.

- *Gaussian*:

$$\frac{1}{\sigma\sqrt{2\pi}}e^{-(1/2\sigma^2)(y-\mu)^2}$$

- *Binomial*:

$$\binom{n}{y}p^y(1-p)^{n-y}$$

- *Poisson*:

$$\frac{e^{-\lambda}\lambda^y}{y!}$$

Their log-likelihoods are, respectively,

- *Gaussian*:

$$-\log\left(\sigma\sqrt{2\pi}\right)-\frac{y^2-2y\mu+\mu^2}{2\sigma^2}=\frac{y\mu-\left(\mu^2/2\right)}{\sigma^2}-\left\{\frac{y^2}{2\sigma^2}+\log\left(\sigma\sqrt{2\pi}\right)\right\}$$

- *Binomial*:

$$\log\binom{n}{y}+y\log(p)+(n-y)\log(1-p)=y\log\left(\frac{p}{1-p}\right)+n\log(1-p)+\log\binom{n}{y}$$

- *Poisson*:

$$-\lambda+y\log(\lambda)-\log(y!)=y\log(\lambda)-\lambda-\log(y!)$$

From the log-likelihood, we can see that each of these distributions belongs to the exponential family by identifying its elements.

Distribution	Canonical Parameter, θ	$b(\theta)$	$a(\phi)$	$c(y,\phi)$
Gaussian	μ	$\mu^2/2$	σ^2	$\frac{y^2}{2\sigma^2}+\log\left(\sigma\sqrt{2\pi}\right)$
Binomial	$\log\left(\frac{p}{1-p}\right)$	$-n\log(1-p)$	1	$\log\binom{n}{y}$
Poisson	$\log(\lambda)$	λ	1	$-\log(y!)$

Most of the distributions that we will encounter in this textbook, including the negative binomial, multinomial, and gamma distributions, can be similarly represented. Some, for example, the beta distribution, can also be shown belong to the exponential family, though you must use the standard math stat textbook representation rather than the simpler form shown in (4.1).

Important: The canonical parameter is always a function of the expected value of the distribution. Also the Gaussian, binomial, and Poisson are examples of distributions for which the canonical parameter happens to be the standard link function for GLMs involving these distributions.

4.2.1.1 Essential Terminology and Results

The following properties of the exponential family are crucial for estimation and inference.

The *score function*, denoted $S(\theta)$, is defined as the derivative of the log-likelihood, that is, $S(\theta)=\partial\ell(\theta;y,\phi)/\partial\theta$. The variance of the score function is called the *information* and is denoted

Var[S(θ)] = I(θ). The score function has the important property that its expected value is zero. Several important results follow:

$$E(Y) = \mu = \frac{\partial b(\theta)}{\partial \theta} \tag{4.2}$$

$$E\left[S(\theta)\right] = E\left[\frac{y - (\partial b(\theta)/\partial \theta)}{a(\phi)}\right] = 0$$

Therefore, $E(y) - (\partial b(\theta)/\partial\theta) = 0$ and the result follows:

$$E\left(\frac{\partial S}{\partial \theta}\right) = -E\left[S(\theta)\right]^2 \tag{4.3}$$

By definition,

$$E\left(\frac{\partial S}{\partial \theta}\right) = E\left(\frac{\partial^2 \log f(y;\theta)}{\partial \theta^2}\right) = \int \left[\partial \frac{(\partial \log f(y;\theta)/\partial\theta)}{\partial \theta}\right] f(y;\theta)dy$$

This can be written

$$\int \left[\partial\left(\frac{\partial f(y;\theta)/\partial\theta}{f(y;\theta)}\right)\Big/\partial\theta\right] f(y;\theta)dy = \int -\frac{\partial f(y;\theta)/\partial\theta}{f(y;\theta)f(y;0)} \partial f(y;\theta)/\partial\theta\, f(y;\theta)dy$$

Rearranging yields

$$\int -\left(\frac{\partial f(y;\theta)/\partial\theta}{f(y;\theta)}\right)\left(\frac{\partial f(y;\theta)/\partial\theta}{f(y;\theta)}\right) f(y;\theta)dy$$

or equivalently

$$-\int \left(\frac{\partial \log f(y;\theta)}{\partial \theta}\right)\left(\frac{\partial \log f(y;\theta)}{\partial \theta}\right) f(y;\theta)dy$$

which we recognize as $-E[S(\theta)]^2$

$$\text{Var}\left[S(\theta)\right] = -E\left(\frac{\partial S}{\partial \theta}\right) \tag{4.4}$$

By definition, $\text{Var}[S(\theta)] = E[S(\theta)]^2 - \{E[S(\theta)]\}^2$, which reduces to $E[S(\theta)]^2$ since $E[S(\theta)] = 0$. From the previous result, $E[S(\theta)]^2 = -E(\partial S/\partial\theta)$, which completes the proof.

$$\text{Var}(Y) = a(\phi)\left[\frac{\partial^2 b(\theta)}{\partial\theta^2}\right] \tag{4.5}$$

By definition, $\text{Var}(Y) = E(Y-\mu)^2$. By (4.2), we can write this as

$$E\left\{[a(\phi)]\frac{y-[\partial b(\theta)/\partial(\theta)]}{a(\phi)}\right\}^2 = [a(\phi)]^2 E[S(\theta)]^2$$

By (4.3) this is

$$-[a(\phi)]^2 E\left[\frac{\partial S}{\partial\theta}\right] = [a(\phi)]^2\left[\frac{\partial^2 b(\theta)/\partial\theta^2}{a(\phi)}\right] = a(\phi)\left[\frac{\partial^2 b(\theta)}{\partial\theta^2}\right]$$

which completes the proof.

The second derivative $\partial^2 b(\theta)/\partial\theta^2$ is called the *variance function* and denoted $V(\mu)$. This is because $V(\mu) = \partial^2 b(\theta)/\partial\theta^2$ describes the dependence of the variance of Y on its expected value. For example, for the normal, or Gaussian, distribution, $V(\mu) = 1$, meaning that the variance does not depend on the mean at all. For the binomial and Poisson distributions, the variance functions are $V(p) = p(1-p)$ and $V(\lambda) = \lambda$; this, in conjunction with the fact that the scale parameter is a known constant for both the binomial and Poisson distributions, implies that the variance is completely determined by the expected value. For other distributions, for example, the beta, gamma, and negative binomial, the variance depends on both the mean and the scale parameter.

4.2.2 Maximum Likelihood Estimation

The essential idea of maximum likelihood estimation is to determine the model parameter values that maximize the likelihood. In the context of the exponential family, we do this by maximizing the log-likelihood function, $\ell(\theta; y, \phi)$, with respect to the canonical parameter θ given the observation, y, and the scale parameter, ϕ.

At first glance, this maximization appears a bit removed from the GLM, $\boldsymbol{\eta} = \mathbf{X}\boldsymbol{\beta}$. However, upon reflection, we can see a logical flow. We need to keep this flow in mind as we work through each elaboration of the estimation process for each elaboration of the model, culminating with the GLMM. The flow: by model definition, the parameter vector $\boldsymbol{\beta}$ determines the link function, $\boldsymbol{\eta}$; the link function in turn determines the mean, that is, $\boldsymbol{\mu} = g^{-1}(\boldsymbol{\eta})$, where $g^{-1}(\ )$ is the inverse link. Finally, recall that the canonical parameter is a function of the mean—a more informative notation for the canonical parameter would be $\theta(\boldsymbol{\mu})$. Putting this together, we see that the general form of the exponential family log-likelihood (4.1) can be reexpressed as

$$\ell(\beta; y, \phi) = \frac{y\{\theta[g^{-1}(X\beta)]\} - b(\{\theta[g^{-1}(X\beta)]\})}{a(\phi)} + c(y, \phi) \tag{4.6}$$

Now it is clear that maximization with respect to $\boldsymbol{\beta}$ makes sense.

While the expression (4.6) captures the essential *form* of the log-likelihood, we need to state it more precisely to reflect that fact that we will observe a data *vector* and upon these data base an estimate of a parameter *vector*. Therefore, the log-likelihood needs to be stated in matrix language. Specifically

$$\ell(\boldsymbol{\theta};\mathbf{y},\phi)=\mathbf{y}'\mathbf{A}\boldsymbol{\theta}-\mathbf{1}'\mathbf{A}b(\boldsymbol{\theta})+c(\mathbf{y},\phi) \tag{4.7}$$

where

$\mathbf{A}=diag[1/a(\phi_i)]$, a $n\times n$ diagonal matrix (assuming n observations) whose ith element is the value of the scale parameter for the ith observation y_i

$\boldsymbol{\theta}$ is the vector of canonical parameters

$\mathbf{1}'$ is a $1\times n$ vector of ones

$b(\boldsymbol{\theta})$ is an $n\times 1$ vector defined by applying the function $b(\,)$ to each element of $\boldsymbol{\theta}$

$c(\mathbf{y},\phi)$ is the remaining term that does not depend on $\boldsymbol{\theta}$

While not shown here, we could express $\boldsymbol{\theta}$ in terms of $\boldsymbol{\beta}$ as in (4.6).

Maximum likelihood estimation for the fixed effects GLM involves of setting the derivative $\partial\ell(\boldsymbol{\theta};\mathbf{y},\phi)/\partial\boldsymbol{\beta}'$ equal to zero (understanding that $\boldsymbol{\theta}$ depends on $\boldsymbol{\beta}$) and solving for $\boldsymbol{\beta}$.

4.2.3 Newton–Raphson and Fisher Scoring

Beginning in Section 4.3, we will see that estimation for generalized and mixed LMs require iterative procedures. To accomplish this, Newton–Raphson and Fisher scoring are two approaches that figure prominently in LM methodology.

Both procedures are based on second-order Taylor series approximations of the log-likelihood. For convenience, we use the shorthand notation $\ell(\theta)$ for the log-likelihood. The Taylor series approximation, evaluated at $\tilde{\theta}$, is

$$\ell(\theta)\cong\ell(\tilde{\theta})+\left.\frac{\partial\ell(\theta)}{\partial\theta}\right|_{\theta=\tilde{\theta}}(\theta-\tilde{\theta})+\left(\frac{1}{2}\right)\left.\frac{\partial^2\ell(\theta)}{\partial\theta^2}\right|_{\theta=\tilde{\theta}}(\theta-\tilde{\theta})^2 \tag{4.8}$$

Differentiating with respect to θ yields

$$\frac{\partial\ell(\theta)}{\partial\theta}\cong\frac{\partial\left\{\ell(\tilde{\theta})+(\partial\ell(\theta)/\partial\theta)\big|_{\theta=\tilde{\theta}}(\theta-\tilde{\theta})+(\partial^2\ell(\theta)/\partial\theta^2)\big|_{\theta=\tilde{\theta}}(\theta-\tilde{\theta})^2\right\}}{\partial\theta}=0$$

$$+\left.\frac{\partial\ell(\theta)}{\partial\theta}\right|_{\theta=\tilde{\theta}}+\left.\frac{\partial^2\ell(\theta)}{\partial\theta^2}\right|_{\theta=\tilde{\theta}}(\theta-\tilde{\theta})$$

Setting $\partial\ell(\theta)/\partial\theta=0$ and rearranging terms yields

$$\theta\cong\tilde{\theta}-\left[\left.\frac{\partial\ell(\theta)}{\partial\theta}\right|_{\theta=\tilde{\theta}}\right]\left[\left.\frac{\partial^2\ell(\theta)}{\partial\theta^2}\right|_{\theta=\tilde{\theta}}\right]^{-1}$$

The aforementioned derivation can be written in matrix terms (not shown here—left as an exercise for the reader). Doing so, we recognize $\left.\frac{\partial\ell(\boldsymbol{\theta})}{\partial\boldsymbol{\theta}}\right|_{\boldsymbol{\theta}=\tilde{\boldsymbol{\theta}}}$ as the *score vector* evaluated at

$\tilde{\boldsymbol{\theta}}$ and $\left.\frac{\partial \ell^2(\boldsymbol{\theta})}{\partial\boldsymbol{\theta}\partial\boldsymbol{\theta}'}\right|_{\boldsymbol{\theta}=\tilde{\boldsymbol{\theta}}}$ as the *Hessian matrix* evaluated at $\tilde{\boldsymbol{\theta}}$. Denoting these as $\mathbf{s}(\tilde{\boldsymbol{\theta}})$ and $\mathbf{H}(\tilde{\boldsymbol{\theta}})$, respectively, we have

$$\boldsymbol{\theta} \cong \tilde{\boldsymbol{\theta}} - \left[\mathbf{H}(\tilde{\boldsymbol{\theta}})\right]^{-1} \mathbf{s}(\tilde{\boldsymbol{\theta}}) \tag{4.9}$$

This defines the basic Newton–Raphson algorithm. To implement it, take a current estimate of $\boldsymbol{\theta}$, denoted by $\tilde{\boldsymbol{\theta}}$; use it to update the score vector and Hessian matrix; compute a new value of $\boldsymbol{\theta}$ using Equation 4.9; and use the new value as $\tilde{\boldsymbol{\theta}}$ in the next update. This proceeds until the difference $(\boldsymbol{\theta} - \tilde{\boldsymbol{\theta}})$ is acceptably small.

Fisher scoring replaces the Hessian matrix with the *information matrix*. From (4.4), the expected value of the Hessian matrix $E[\mathbf{H}(\boldsymbol{\theta})] = -\text{Var}[\mathbf{s}(\boldsymbol{\theta})] = -\text{I}(\theta)$. Substituting in (4.9) yields

$$\boldsymbol{\theta} \cong \tilde{\boldsymbol{\theta}} + \left[\mathbf{I}(\tilde{\boldsymbol{\theta}})\right]^{-1} \mathbf{s}(\tilde{\boldsymbol{\theta}}) \tag{4.10}$$

This defines the Fisher scoring algorithm. It proceeds exactly as the Newton–Raphson algorithm except that we use the information matrix instead of the Hessian matrix.

Neither algorithm is best for all situations. In most of the examples we will consider in this text book, the two procedures would be indistinguishable in terms of speed and results. Both eventually yield the same solution, but one may converge to the solution more quickly that the other for a given model or data set. Computer software programs typically use one as the default, but allow you the option of choosing the other or even allow you to switch at some point during the iteration process. Which algorithm works best in any given situation is often a matter of trial and error.

4.2.4 Quasi-Likelihood

Maximum likelihood estimates depend on $\partial\ell(\theta;\, y,\, \phi)/\partial\theta$. We have seen that, in its scalar form, this equals $[(y-\boldsymbol{\mu})/a(\phi)]$. Thus, while the term $c(y,\phi)$ is required for a complete specification of the exponential family likelihood, it is not required for maximum likelihood estimation. Only $[(y-b(\theta))/a(\phi)]$ is required. Wedderburn (1974) formalized this idea though *quasi-likelihood* theory.

Quasi-likelihood is defined as $\int^{\mu}(y-t)/v(t)a(\phi)dt$, where $v(t)$ corresponds to the form of the variance function and $a(\phi)$ denotes the scale parameter function. For example, setting $v(t)=t$ and $a(\phi)=1$ yields $\int^{\mu}(y-t/t)dt = y\log(\mu)-\mu$, the Poisson log-likelihood without $c(y,\phi) = -\log(y!)$. Similarly, setting $v(t)=1$ and $a(\phi)=\phi^2$ yields $\int^{\mu}((y-t)/\phi^2)dt = y\mu-(\mu^2/2)/\phi^2$, the "quasi-likelihood" part of the Gaussian log-likelihood.

Wedderburn showed that GLM estimation and inference theory, developed for the exponential family, also applied to models with response variables whose mean and variance could be specified even though they were not associated with a known likelihood. Important applications of quasi-likelihood include *overdispersion* and correlated error models for non-Gaussian repeated measures, both of which are discussed in Part III of this textbook.

As an example, count data are often assumed to have a Poisson distribution. The Poisson assumption requires expected value and variance to be equal, that is, $E(y) = \lambda = \text{Var}(y)$. In practice, however, we often observe overdispersion, that is, sample variance is substantially greater than the mean and hence substantially greater than what theory would suggest. A common approach to modeling overdispersion with count data is to multiply the assumed variance by a scale parameter. In quasi-likelihood terms, we define $v(t) = t$, as we would with the Poisson distribution, but change the scale parameter to $a(\phi) = \phi$. The resulting quasi-likelihood is $\int^{\mu} (y - t)/t\phi = (y \log(\mu) - \mu)/\phi$. Thus, $E(y) = \mu$ but $\text{Var}(y) = \phi\mu$. No actual probability distribution exists with this structure, but in many cases it adequately models the distribution of the count data, and the quasi-likelihood is perfectly well defined for GLM estimation and inference purposes.

4.3 Fixed Effects Only

The estimating equations developed in this section apply to *all* models with linear predictor $\boldsymbol{\eta} = \mathbf{X}\boldsymbol{\beta}$. While we refer to these as the GLM estimating equations, they apply equally to the classical "general" LM. Recall that the latter is simply a special case of the GLM with independent, homoscedastic Gaussian data. Similarly, the classical, ordinary least squares—the so-called normal equations—is simply a special case of the GLM estimating equations.

Scalar form

Ultimately, we want a set of matrix equations to estimate $\boldsymbol{\beta}$, but for simplicity, we start with the scalar form of the log-likelihood, Equation 4.1, or, better yet, Equation 4.6, which shows how the model parameters are embedded in the canonical parameter. Using the chain rule, we can write the derivative to be used for maximum likelihood estimation as

$$\frac{\partial \ell\left(\theta\left[g^{-1}(X\beta)\right]; y, \phi\right)}{\partial \beta} = \frac{\partial \ell(\theta)}{\partial \theta} \frac{\partial \theta}{\partial \mu} \frac{\partial \mu}{\partial \eta} \frac{\partial \eta}{\partial \beta}$$

Now we note the following:

- $\dfrac{\partial \ell(\theta)}{\partial \theta} = \dfrac{y - [\partial b(\theta)/\partial \theta]}{a(\phi)} = \dfrac{y - \mu}{a(\phi)}$
- $\dfrac{\partial \theta}{\partial \mu} = \left(\dfrac{\partial \mu}{\partial \theta}\right)^{-1} = \left(\dfrac{\partial[\partial b(\theta)/\partial \theta]}{\partial \theta}\right)^{-1} = \left(\dfrac{\partial^2 b(\theta)}{\partial \theta^2}\right)^{-1} = \dfrac{1}{V(\mu)}$
- $\dfrac{\partial \eta}{\partial \beta} = \dfrac{\partial (X\beta)}{\partial \beta} = X$

Note that these results apply identically to quasi-likelihood. This gives us

$$\frac{\partial \ell\left(\theta\left[g^{-1}(X\beta)\right]; y, \phi\right)}{\partial \beta} = \frac{y - \mu}{a(\phi)} \left(\frac{1}{V(\mu)}\right) \left(\frac{\partial \mu}{\partial \eta}\right) X$$

Now, recall (4.5): $a(\phi)V(\mu) = \text{Var}(y)$ giving us

$$\frac{\partial \ell\left(\theta\left[g^{-1}(X\beta)\right];y,\phi\right)}{\partial \beta} = (y-\mu)\left(\frac{1}{V(y)}\right)\left(\frac{\partial \mu}{\partial \eta}\right)X \tag{4.11}$$

Matrix form

We are now ready to develop the estimating equations in matrix form. We can write (4.11) as

$$\frac{\partial \ell(\boldsymbol{\theta})}{\partial \boldsymbol{\beta}'} = \mathbf{X}'\mathbf{D}^{-1}\mathbf{V}^{-1}(\mathbf{y}-\boldsymbol{\mu}) \tag{4.12}$$

where $\mathbf{y}$ is the $n \times 1$ vector of observations, $\ell(\boldsymbol{\theta})$ is the $n \times 1$ vector of log-likelihood values associated with the observations, $\mathbf{V} = diag[\text{Var}(y_i)]$ is the $n \times n$ variance matrix of the observations, $\mathbf{D} = diag[\partial\eta_i/\partial\mu_i]$ is the $n \times n$ matrix of derivatives, and $\boldsymbol{\mu}$ is the $n \times 1$ mean vector.

Let $\mathbf{W} = (\mathbf{DVD})^{-1}$. We can rewrite (4.12) as follows, also noting that it is the score vector:

$$\mathbf{s}(\boldsymbol{\theta}) = \frac{\partial \ell(\boldsymbol{\theta})}{\partial \boldsymbol{\theta}'} = \mathbf{X}'\mathbf{D}^{-1}\mathbf{V}^{-1}\left(\mathbf{D}^{-1}\mathbf{D}\right)(\mathbf{y}-\boldsymbol{\mu}) = \mathbf{X}'\mathbf{W}\mathbf{D}(\mathbf{y}-\boldsymbol{\mu}) \tag{4.13}$$

We complete the estimating equations using Fisher scoring. To do this, we need the information matrix. By definition, the information matrix is

$$\text{Var}\left[\mathbf{s}(\boldsymbol{\theta})\right] = \mathbf{X}'\mathbf{W}\mathbf{D}\left[\text{Var}(\mathbf{y}-\boldsymbol{\mu})\right]\mathbf{D}\mathbf{W}\mathbf{X} = \mathbf{X}'\mathbf{W}\mathbf{D}\mathbf{V}\mathbf{D}\mathbf{W}\mathbf{X} = \mathbf{X}'\mathbf{W}\mathbf{W}^{-1}\mathbf{W}\mathbf{X} = \mathbf{X}'\mathbf{W}\mathbf{X} \tag{4.14}$$

Substituting into the Fisher scoring equation (4.10) yields $\boldsymbol{\beta} = \tilde{\boldsymbol{\beta}} + \left(\mathbf{X}'\tilde{\mathbf{W}}\mathbf{X}\right)^{-1}\mathbf{X}'\tilde{\mathbf{W}}\tilde{\mathbf{D}}(\mathbf{y}-\tilde{\boldsymbol{\mu}})$, where $\tilde{\mathbf{W}}$, $\tilde{\mathbf{D}}$, and $\tilde{\boldsymbol{\mu}}$ denote $\mathbf{W}$, $\mathbf{D}$, and $\boldsymbol{\mu}$ evaluated at $\tilde{\boldsymbol{\beta}}$. Multiplying both sides of the equation by $\mathbf{X}'\tilde{\mathbf{W}}\mathbf{X}$ we have $\mathbf{X}'\tilde{\mathbf{W}}\mathbf{X}\boldsymbol{\beta} = \mathbf{X}'\tilde{\mathbf{W}}\mathbf{X}\tilde{\boldsymbol{\beta}} + \mathbf{X}'\tilde{\mathbf{W}}\tilde{\mathbf{D}}(\mathbf{y}-\tilde{\boldsymbol{\mu}})$. This gives us the GLM estimating equations:

$$\mathbf{X}'\tilde{\mathbf{W}}\mathbf{X}\boldsymbol{\beta} = \mathbf{X}'\tilde{\mathbf{W}}\mathbf{y}* \tag{4.15}$$

where $\mathbf{y}^* = \mathbf{X}\tilde{\boldsymbol{\beta}} + \tilde{\mathbf{D}}(\mathbf{y}-\tilde{\boldsymbol{\mu}}) = \tilde{\boldsymbol{\eta}} + \tilde{\mathbf{D}}(\mathbf{y}-\tilde{\boldsymbol{\mu}})$. In GLM terminology, $\mathbf{y}^*$ is called the *pseudo-variable*.

We implement the GLM estimating equations (4.15) by iteration. A starting value of $\tilde{\boldsymbol{\beta}}$ allows us to compute the equation, yielding a new estimate of $\boldsymbol{\beta}$. We use this as an updated $\tilde{\boldsymbol{\beta}}$ and continue in this manner until the difference $(\boldsymbol{\beta} - \tilde{\boldsymbol{\beta}})$ is acceptably negligible, at which point we say that the estimation has converged.

4.3.1 Relation to Least Squares Estimation

There are two relationships to least squares estimation of importance here. The first anticipates the pseudo-likelihood estimating equations we will develop in Section 4.5 for

GLMMs. The second notes the fact that classical least squares estimation for Gaussian LMs is in fact a special case of GLM estimation.

4.3.1.1 Pseudo-Likelihood for GLM

Consider the pseudo-variable $\mathbf{y}^*$. Its expectation and variance are, respectively,

- $E(\mathbf{y}^*) = E\left[\mathbf{X}\tilde{\boldsymbol{\beta}} + \tilde{\mathbf{D}}(\mathbf{y} - \tilde{\boldsymbol{\mu}})\right] = \mathbf{X}\boldsymbol{\beta}$
- $\text{Var}(\mathbf{y}^*) = \text{Var}\left[\mathbf{X}\tilde{\boldsymbol{\beta}} + \tilde{\mathbf{D}}(\mathbf{y} - \tilde{\boldsymbol{\mu}})\right] = \tilde{\mathbf{D}}\left[\text{Var}(\mathbf{y} - \tilde{\boldsymbol{\mu}})\right]\tilde{\mathbf{D}} = \tilde{\mathbf{D}}\tilde{\mathbf{V}}\tilde{\mathbf{D}} = \tilde{\mathbf{W}}^{-1}$

It follows that a generalized least squares estimator of $\boldsymbol{\beta}$, weighted by the inverse variance, can be obtained from the estimating equation

$$\mathbf{X}'\left[\text{Var}(\mathbf{y}^*)\right]^{-1}\mathbf{X}\boldsymbol{\beta} = \mathbf{X}'\left[\text{Var}(\mathbf{y}^*)\right]^{-1}\mathbf{y}^* \Rightarrow \mathbf{X}'\mathbf{W}\mathbf{X}\boldsymbol{\beta} = \mathbf{X}'\mathbf{W}\mathbf{y}^*$$

This is identical to the GLM estimating equations in (4.15).

4.3.1.2 Gaussian Linear Models and Ordinary Least Squares

The development of classical LM theory was based on least squares. The LM with independent, homoscedastic data, that is,

- Linear predictor: $\boldsymbol{\eta} = \mathbf{X}\boldsymbol{\beta}$
- Link: $\boldsymbol{\eta} = \boldsymbol{\mu}$
- Distribution: $\mathbf{y} \sim N(\boldsymbol{\mu}, \mathbf{I}\sigma^2)$

was called, in classical LM literature, the "general" LM. Estimation uses *ordinary least squares*, that is, minimizing the sum of squared differences between $\mathbf{y}$ and $\hat{\mathbf{y}}$. Because the Gaussian model uses the identity link (although the idea of a link function would never have occurred to anyone in pre-GLM days), $\hat{\mathbf{y}} = \mathbf{X}\hat{\boldsymbol{\beta}}$. The ordinary least squares estimator is obtained by minimizing $(\mathbf{y} - \mathbf{X}\boldsymbol{\beta})'(\mathbf{y} - \mathbf{X}\boldsymbol{\beta})$, yielding

$$\mathbf{X}'\mathbf{X}\boldsymbol{\beta} = \mathbf{X}'\mathbf{y} \tag{4.16}$$

The ordinary least squares equations in (4.16) are also called the *normal equations* in LM literature. Notice that under the identity link, the derivative matrix $\mathbf{D}$ reduces to the identity matrix. Thus, $\mathbf{W}$ reduces to $(\mathbf{I}\sigma^2)^{-1}$, $\mathbf{y}^*$ reduces to $\mathbf{y}$, and the GLM estimating equations become $\mathbf{X}'(\mathbf{I}\sigma^2)^{-1}\mathbf{X}\boldsymbol{\beta} = \mathbf{X}'(\mathbf{I}\sigma^2)^{-1}\mathbf{y}$, that is, $\mathbf{X}'\mathbf{X}\boldsymbol{\beta} = \mathbf{X}'\mathbf{y}$.

For Gaussian models with nontrivial variance structure, that is, with distribution $\mathbf{y} \sim N(\boldsymbol{\mu}, \mathbf{V})$, we know that the ordinary least squares estimator is less efficient than the generalized least squares estimator. Because the Gaussian model uses the identity link, $\mathbf{W}$ reduces to $\mathbf{V}^{-1}$, $\mathbf{y}^* = \mathbf{y}$, and GLM equations reduce to $\mathbf{X}'\mathbf{V}^{-1}\mathbf{X}\boldsymbol{\beta} = \mathbf{X}'\mathbf{V}^{-1}\mathbf{y}$, Gaussian generalized least squares. Thus, we see that Gaussian least squares is a special case of the more general GLM estimating equations.

4.4 Gaussian Mixed Models

In this section we develop estimating equations for the linear mixed model (LMM)—that is, models with Gaussian data and random effects in the linear predictor. Recall the essential features of the LMM:

- Linear predictor: $\boldsymbol{\eta} = \mathbf{X}\boldsymbol{\beta} + \mathbf{Z}\mathbf{b}$
- Distributions: $\mathbf{y}|\mathbf{b} \sim N(\boldsymbol{\mu}|\mathbf{b},\mathbf{R})$; $\mathbf{b} \sim N(\mathbf{0},\mathbf{G})$
- Link: $\boldsymbol{\eta} = \boldsymbol{\mu}|\mathbf{b}$

As we saw in Chapter 3, it follows from the distributional assumptions of the LMM that the marginal distribution of $\mathbf{y}$ is $N(\boldsymbol{\mu},\mathbf{V})$, where $\mathbf{V} = \mathbf{ZGZ}' + \mathbf{R}$ and $\mathbf{X}\boldsymbol{\beta}$ models $\boldsymbol{\mu}$. Recall from Chapter 3 that the LMM is the only mixed model for which the fully specified conditional model and marginal model have the same distribution (Gaussian) and the estimates of $\boldsymbol{\beta}$ from the conditional and marginal model are identical.

Our tasks in this section are to (1) develop the estimating equations for the linear predictor effects, $\boldsymbol{\beta}$ and $\mathbf{b}$, and (2) develop estimating equations for the components of the variance–covariance matrices $\mathbf{G}$ and $\mathbf{R}$. Along the way, we will formally demonstrate the equivalence of the estimates of the fixed model effects, $\boldsymbol{\beta}$, obtained from the conditional and marginal forms of the LMM.

4.4.1 Mixed Model Equations for β and b

From the distributional assumptions (and taking liberties with notational conventions of $\ell(\,)$ in the interest of clarity), the log-likelihood equations for $\mathbf{y}|\mathbf{b}$ and $\mathbf{b}$, respectively, are

$$\ell(\mathbf{y}\,|\,\mathbf{b}) = -\left(\frac{n}{2}\right)\log(2\pi) - \left(\frac{1}{2}\right)\log(|\mathbf{R}|) - \left(\frac{1}{2}\right)(\mathbf{y} - \mathbf{X}\boldsymbol{\beta} - \mathbf{Z}\mathbf{b})'\,\mathbf{R}^{-1}(\mathbf{y} - \mathbf{X}\boldsymbol{\beta} - \mathbf{Z}\mathbf{b})$$

and

$$\ell(\mathbf{b}) = -\left(\frac{b}{2}\right)\log(2\pi) - \left(\frac{1}{2}\right)\log(|\mathbf{G}|) - \left(\frac{1}{2}\right)\mathbf{b}'\mathbf{G}^{-1}\mathbf{b}$$

where b denotes the total number of levels of the random effects. Focusing on the "quasi-likelihood" part, that is, ignoring $c(y,\phi)$, we can write the joint log-likelihood as

$$\ell(\mathbf{y},\mathbf{b}) = -\left(\frac{1}{2}\right)(\mathbf{y} - \mathbf{X}\boldsymbol{\beta} - \mathbf{Z}\mathbf{b})'\,\mathbf{R}^{-1}(\mathbf{y} - \mathbf{X}\boldsymbol{\beta} - \mathbf{Z}\mathbf{b}) - \left(\frac{1}{2}\right)\mathbf{b}'\mathbf{G}^{-1}\mathbf{b}$$

We now obtain the maximum likelihood estimator by setting $\partial\ell(\mathbf{y},\mathbf{b})/\partial\boldsymbol{\beta}'$ and $\partial\ell(\mathbf{y},\mathbf{b})/\partial\mathbf{b}'$ equal to zero and solving the resulting set of equations for $\boldsymbol{\beta}$ and $\mathbf{b}$. The derivatives are, respectively,

$$\frac{\partial[\ell(\mathbf{y},\mathbf{b})]}{\partial\boldsymbol{\beta}'} = \mathbf{X}'\mathbf{R}^{-1}\mathbf{y} - \mathbf{X}'\mathbf{R}^{-1}\mathbf{X}\boldsymbol{\beta} - \mathbf{Z}'\mathbf{R}^{-1}\mathbf{X}\mathbf{b}$$

$$\frac{\partial[\ell(\mathbf{y},\mathbf{b})]}{\partial\mathbf{b}'} = \mathbf{Z}'\mathbf{R}^{-1}\mathbf{y} - \mathbf{X}'\mathbf{R}^{-1}\mathbf{Z}\boldsymbol{\beta} - \mathbf{Z}'\mathbf{R}^{-1}\mathbf{Z}\mathbf{b} - \mathbf{G}^{-1}\mathbf{b}$$

Setting these equal to zero and solving for $\boldsymbol{\beta}$ and $\mathbf{b}$ yields the LMM estimating equations, often referred to as the *mixed model equations* in LMM literature:

$$\begin{bmatrix} \mathbf{X}'\mathbf{R}^{-1}\mathbf{X} & \mathbf{X}'\mathbf{R}^{-1}\mathbf{Z} \\ \mathbf{Z}'\mathbf{R}^{-1}\mathbf{X} & \mathbf{Z}'\mathbf{R}^{-1}\mathbf{Z}+\mathbf{G}^{-1} \end{bmatrix} \begin{bmatrix} \boldsymbol{\beta} \\ \mathbf{b} \end{bmatrix} = \begin{bmatrix} \mathbf{X}'\mathbf{R}^{-1}\mathbf{y} \\ \mathbf{Z}'\mathbf{R}^{-1}\mathbf{y} \end{bmatrix} \tag{4.17}$$

If the components of $\mathbf{G}$ and $\mathbf{R}$ are known, then we can obtain exact estimates of $\boldsymbol{\beta}$ and $\mathbf{b}$ with a single calculation of (4.17). Rarely is this the case, meaning that solving (4.17) entails estimating $\mathbf{G}$ and $\mathbf{R}$ as well. This turns solving the mixed model equations into an iterative process: Starting values of $\mathbf{G}$ and $\mathbf{R}$ enable an initial solution of (4.17); we use the resulting estimates of $\boldsymbol{\beta}$ and $\mathbf{b}$ to update $\mathbf{G}$ and $\mathbf{R}$, and the process continues until we reach convergence. We discuss estimation of $\mathbf{G}$ and $\mathbf{R}$ in Section 4.4.3.

4.4.2 Relation to Least Squares

The LM (fixed-effects-only model assuming Gaussian data) least squares estimating equations are special cases of (4.17). We readily see this, recalling that the linear predictor for the LM is $\boldsymbol{\eta}=\mathbf{X}\boldsymbol{\beta}$ and the general form of the form of the distribution is $\mathbf{y}\sim N(\boldsymbol{\mu},\mathbf{V})$. Note that in the LMM $\text{Var}(\mathbf{y})=\mathbf{V}=\mathbf{Z}\mathbf{G}\mathbf{Z}'+\mathbf{R}$. The LM is simply the LMM with $\mathbf{Z}\mathbf{b}$ and $\mathbf{G}$ removed; hence, for the LM $\mathbf{V}=\mathbf{R}$. Removing $\mathbf{Z}\mathbf{b}$ and $\mathbf{G}$ from (4.17) leaves $\mathbf{X}'\mathbf{R}^{-1}\mathbf{X}\boldsymbol{\beta}=\mathbf{X}'\mathbf{R}^{-1}\mathbf{y}$ or, equivalently, $\mathbf{X}'\mathbf{V}^{-1}\mathbf{X}\boldsymbol{\beta}=\mathbf{X}'\mathbf{V}^{-1}\mathbf{y}$, the generalized least squares estimator. If we assume independent, homoscedastic observations, the classical "general" LM, that is, $\mathbf{V}=\mathbf{R}=\mathbf{I}\sigma^2$, then (4.17) reduces to the "normal equations" $(\mathbf{X}'\mathbf{X})\boldsymbol{\beta}=\mathbf{X}'\mathbf{y}$.

For the marginal form of the LMM, the linear predictor is $\boldsymbol{\eta}=\mathbf{X}\boldsymbol{\beta}$ and the assumed distribution is $\mathbf{V}=\mathbf{Z}\mathbf{G}\mathbf{Z}'+\mathbf{R}$. It follows that the estimating equations for $\boldsymbol{\beta}$ using the marginal form of the LMM are given by the generalized least squares estimate

$$\mathbf{X}'\mathbf{V}^{-1}\mathbf{X}\boldsymbol{\beta} = \mathbf{X}\mathbf{V}^{-1}\mathbf{y} \text{ or equivalently } \mathbf{X}'(\mathbf{Z}\mathbf{G}\mathbf{Z}'+\mathbf{R})^{-1}\mathbf{X}\boldsymbol{\beta} = \mathbf{X}(\mathbf{Z}\mathbf{G}\mathbf{Z}'+\mathbf{R})^{-1}\mathbf{y} \tag{4.18}$$

The important result here is that the estimates of $\boldsymbol{\beta}$ from both the mixed model estimating equations (4.17) and generalized least squares for the marginal LMM (4.18) are identical. Searle (1971) gives the proof. The essential flow of Searle's proof is given here for the reader's convenience and owing to the importance of the result.

We can write the mixed model equations (4.17) as two equalities:

$$\mathbf{X}'\mathbf{R}^{-1}\mathbf{X}\boldsymbol{\beta} + \mathbf{X}'\mathbf{R}^{-1}\mathbf{Z}\mathbf{b} = \mathbf{X}'\mathbf{R}^{-1}\mathbf{y} \tag{4.19}$$

$$\mathbf{Z}'\mathbf{R}^{-1}\mathbf{X}\boldsymbol{\beta} + (\mathbf{Z}'\mathbf{R}^{-1}\mathbf{Z}+\mathbf{G}^{-1})\mathbf{b} = \mathbf{Z}'\mathbf{R}^{-1}\mathbf{y} \tag{4.20}$$

Solving for $\mathbf{b}$ in (4.20) gives $\mathbf{b}=(\mathbf{Z}'\mathbf{R}^{-1}\mathbf{Z}+\mathbf{G}^{-1})^{-1}(\mathbf{Z}'\mathbf{R}^{-1}\mathbf{y}-\mathbf{Z}'\mathbf{R}^{-1}\mathbf{X}\boldsymbol{\beta})$. Substituting into (4.19), we have $\mathbf{X}'\mathbf{R}^{-1}\mathbf{X}\boldsymbol{\beta}+\mathbf{X}'\mathbf{R}^{-1}\mathbf{Z}[(\mathbf{Z}'\mathbf{R}^{-1}\mathbf{Z}+\mathbf{G}^{-1})^{-1}(\mathbf{Z}'\mathbf{R}^{-1}\mathbf{y}-\mathbf{Z}'\mathbf{R}^{-1}\mathbf{X}\boldsymbol{\beta})]=\mathbf{X}'\mathbf{R}^{-1}\mathbf{y}$. Rearranging terms, we rewrite this as

$$\mathbf{X}'\mathbf{R}^{-1}\mathbf{X}\boldsymbol{\beta}-\mathbf{X}'\mathbf{R}^{-1}\mathbf{Z}\left(\mathbf{Z}'\mathbf{R}^{-1}\mathbf{Z}+\mathbf{G}^{-1}\right)^{-1}\mathbf{Z}'\mathbf{R}^{-1}\mathbf{X}\boldsymbol{\beta}=\mathbf{X}'\mathbf{R}^{-1}\mathbf{y}-\mathbf{X}'\mathbf{R}^{-1}\mathbf{Z}\left(\mathbf{Z}'\mathbf{R}^{-1}\mathbf{Z}+\mathbf{G}^{-1}\right)^{-1}\mathbf{Z}'\mathbf{R}^{-1}\mathbf{y}$$

or

$$\mathbf{X}'\left[\mathbf{R}^{-1}-\mathbf{R}^{-1}\mathbf{Z}\left(\mathbf{Z}'\mathbf{R}^{-1}\mathbf{Z}+\mathbf{G}^{-1}\right)^{-1}\mathbf{Z}'\mathbf{R}^{-1}\right]\mathbf{X}\boldsymbol{\beta}=\mathbf{X}'\left[\mathbf{R}^{-1}-\mathbf{R}^{-1}\mathbf{Z}\left(\mathbf{Z}'\mathbf{R}^{-1}\mathbf{Z}+\mathbf{G}^{-1}\right)^{-1}\mathbf{Z}'\mathbf{R}^{-1}\right]\mathbf{y}$$

Notice that this last expression is a form of generalized least squares with weight matrix $\mathbf{R}^{-1}-\mathbf{R}^{-1}\mathbf{Z}(\mathbf{Z}'\mathbf{R}^{-1}\mathbf{Z}+\mathbf{G}^{-1})^{-1}\mathbf{Z}'\mathbf{R}^{-1}$. It now remains to show that this weight matrix is $\mathbf{V}^{-1}$. We do this by demonstrating that the product of the weight matrix and $\mathbf{V}$ is the identity matrix. Noting that $\mathbf{V}=\mathbf{ZGZ}'+\mathbf{R}$, the product is

$$\begin{aligned}
&\left[\mathbf{R}^{-1}-\mathbf{R}^{-1}\mathbf{Z}\left(\mathbf{Z}'\mathbf{R}^{-1}\mathbf{Z}+\mathbf{G}^{-1}\right)^{-1}\mathbf{Z}'\mathbf{R}^{-1}\right](\mathbf{ZGZ}'+\mathbf{R})\\
&=\mathbf{R}^{-1}\mathbf{ZGZ}'+\mathbf{R}^{-1}\mathbf{R}-\mathbf{R}^{-1}\mathbf{Z}\left(\mathbf{Z}'\mathbf{R}^{-1}\mathbf{Z}+\mathbf{G}^{-1}\right)^{-1}\left(\mathbf{Z}'\mathbf{R}^{-1}\mathbf{ZGZ}'+\mathbf{Z}'\mathbf{R}^{-1}\mathbf{R}\right)\\
&=\mathbf{R}^{-1}\mathbf{ZGZ}'+\mathbf{I}-\mathbf{R}^{-1}\mathbf{Z}\left(\mathbf{Z}'\mathbf{R}^{-1}\mathbf{Z}+\mathbf{G}^{-1}\right)^{-1}\left(\mathbf{Z}'\mathbf{R}^{-1}\mathbf{ZGZ}'+\mathbf{Z}'\right)\\
&=\mathbf{R}^{-1}\mathbf{ZGZ}'+\mathbf{I}-\mathbf{R}^{-1}\mathbf{Z}\left(\mathbf{Z}'\mathbf{R}^{-1}\mathbf{Z}+\mathbf{G}^{-1}\right)^{-1}\left(\mathbf{Z}'\mathbf{R}^{-1}\mathbf{ZG}+\mathbf{I}\right)\mathbf{Z}'\\
&=\mathbf{R}^{-1}\mathbf{ZGZ}'+\mathbf{I}-\mathbf{R}^{-1}\mathbf{Z}\left(\mathbf{Z}'\mathbf{R}^{-1}\mathbf{Z}+\mathbf{G}^{-1}\right)^{-1}\left(\mathbf{Z}'\mathbf{R}^{-1}\mathbf{ZGG}^{-1}+\mathbf{G}^{-1}\right)\mathbf{GZ}'\\
&=\mathbf{R}^{-1}\mathbf{ZGZ}'+\mathbf{I}-\mathbf{R}^{-1}\mathbf{Z}\left(\mathbf{Z}'\mathbf{R}^{-1}\mathbf{Z}+\mathbf{G}^{-1}\right)^{-1}\left(\mathbf{Z}'\mathbf{R}^{-1}\mathbf{Z}+\mathbf{G}^{-1}\right)\mathbf{GZ}'\\
&=\mathbf{R}^{-1}\mathbf{ZGZ}'+\mathbf{I}-\mathbf{R}^{-1}\mathbf{ZGZ}'=\mathbf{I}
\end{aligned}$$

which completes the proof.

A similar derivation can be applied to the estimating equation for $\mathbf{b}$. From (4.19)

$$\mathbf{b}=\left(\mathbf{Z}'\mathbf{R}^{-1}\mathbf{Z}+\mathbf{G}^{-1}\right)^{-1}\left(\mathbf{Z}'\mathbf{R}^{-1}\mathbf{y}-\mathbf{Z}'\mathbf{R}^{-1}\mathbf{X}\boldsymbol{\beta}\right)$$

rearranging terms and inserting $\mathbf{VV}^{-1}=(\mathbf{ZGZ}'+\mathbf{R})\mathbf{V}^{-1}$ yields

$$\begin{aligned}
&\left(\mathbf{Z}'\mathbf{R}^{-1}\mathbf{Z}+\mathbf{G}^{-1}\right)^{-1}\mathbf{Z}'\mathbf{R}^{-1}(\mathbf{ZGZ}'+\mathbf{R})\mathbf{V}^{-1}(\mathbf{y}-\mathbf{X}\boldsymbol{\beta})\\
&=\left(\mathbf{Z}'\mathbf{R}^{-1}\mathbf{Z}+\mathbf{G}^{-1}\right)^{-1}\left(\mathbf{Z}'\mathbf{R}^{-1}\mathbf{ZGZ}'+\mathbf{G}^{-1}\mathbf{GZ}'\mathbf{I}\right)\mathbf{V}^{-1}(\mathbf{y}-\mathbf{X}\boldsymbol{\beta})\\
&=\left(\mathbf{Z}'\mathbf{R}^{-1}\mathbf{Z}+\mathbf{G}^{-1}\right)^{-1}\left(\mathbf{Z}'\mathbf{R}^{-1}\mathbf{Z}+\mathbf{G}^{-1}\right)\mathbf{GZ}'\mathbf{V}^{-1}(\mathbf{y}-\mathbf{X}\boldsymbol{\beta})\\
&=\mathbf{GZ}'\mathbf{V}^{-1}(\mathbf{y}-\mathbf{X}\boldsymbol{\beta})
\end{aligned}\tag{4.21}$$

Searle et al. (1992), Robinson (1991), and others have shown that the best linear predictor of $\mathbf{b}$ is

$$E(\mathbf{b} \mid \mathbf{y}) = E(\mathbf{b}) + \mathrm{Cov}(\mathbf{b}, \mathbf{y}')\left[\mathrm{Var}(\mathbf{y})\right]^{-1}(\mathbf{y} - \mathbf{X}\boldsymbol{\beta})$$

For the LMM, the joint distribution of $\mathbf{y}$ and $\mathbf{b}$ is $\begin{bmatrix} \mathbf{y} \\ \mathbf{b} \end{bmatrix} \sim N\left(\begin{bmatrix} \mathbf{X}\boldsymbol{\beta} \\ \mathbf{0} \end{bmatrix}, \begin{bmatrix} \mathbf{V} & \mathbf{ZG}' \\ \mathbf{GZ}' & \mathbf{G} \end{bmatrix}\right)$. It follows that the best linear predictor of $\mathbf{b}$ is $\mathbf{GZ}'\mathbf{V}^{-1}(\mathbf{y} - \mathbf{X}\boldsymbol{\beta})$. Goldberger (1962) gave the mixed model solution to $\mathbf{b}$ the name "best linear unbiased predictor" and Henderson used the acronym BLUP, an acronym which is widely used in mixed model circles. We will explore BLUPs in detail in Chapter 9.

The difference between the estimation of $\boldsymbol{\beta}$ and $\mathbf{b}$ stems from the fact that the elements of $\boldsymbol{\beta}$ are fixed parameters, whereas $\mathbf{b}$ is a random vector. For the former, the goal is to produce a minimum variance estimator about a fixed target; for the latter, the goal is to minimize the prediction error about the realized value of a random variable. Robinson (1991) gives an excellent discussion of the issues involved in best linear unbiased prediction.

4.4.3 Unknown G and R: ML and REML Variance–Covariance Component Estimation

Typically, the components of $\mathbf{G}$ and $\mathbf{R}$ must be estimated in order to obtain a solution to the mixed model equations. As examples, consider the two LMMs from Example 3.1. with four treatments (fixed effects) and eight locations (random effects). The linear predictor was $\eta_{ij} = \eta + \tau_i + L_j$ and the location effect distribution was L_j i.i.d. $N(0, \sigma_L^2)$. Example 3.2 was a random coefficient regression with linear predictor $\eta_{ij} = \beta_0 + b_{0i} + (\beta_1 + b_{1i})X_j$ and random regression coefficient distribution $\begin{bmatrix} b_{0i} \\ b_{1i} \end{bmatrix} \sim N\left(\begin{bmatrix} 0 \\ 0 \end{bmatrix}, \begin{bmatrix} \sigma_0^2 & \sigma_{01} \\ & \sigma_1^2 \end{bmatrix}\right)$. For Example 3.1, $\mathbf{G} = \mathbf{I}_8\sigma_L^2$ and $\mathbf{R} = \mathbf{I}_{32}\sigma^2$. The variance components we need to estimate are σ_L^2 and σ^2. In Example 3.2, $\mathbf{G} = \mathbf{I}_4 \otimes \begin{bmatrix} \sigma_0^2 & \sigma_{01} \\ & \sigma_1^2 \end{bmatrix}$ and $\mathbf{R} = \mathbf{I}_{72}\sigma^2$. The components of $\mathbf{G}$ and $\mathbf{R}$ include variance and covariance terms: σ_0^2, σ_{01}, σ_1^2, and σ^2.

The best-known variance component estimation procedure—because it is widely taught in introductory statistical methods courses—uses analysis of variance based expected mean squares (EMS). These are called *ANOVA estimators*. They can be used with *variance-component-only* LMMs: LMMs whose random effects are all i.i.d. The multilocation Example 3.4.1 is an example of a variance-component-only model. If the multilocation data had subsampling, making it possible to include a random treatment-by-location term, then the linear predictor would be $\eta_{ij} = \eta + \tau_i + L_j + (tL)_{ij}$, where L_j i.i.d. $N(0, \sigma_L^2)$ and $(tL)_{ij}$ i.i.d. $N(0, \sigma_{TL}^2)$. This would still be a variance-component-only model because both random model effects are i.i.d. with one variance component per effect. On the other hand, the random coefficient LMM is not a variance-component-only model because there is a covariance term, σ_{01}, to account for correlation between the random intercept and slope coefficients.

We cannot use the ANOVA method for Example 3.2, or any other LMM with covariance or correlation terms in $\mathbf{G}$ or $\mathbf{R}$. Clearly, we need a method that can be used for all LMMs. Ideally, this method would be amendable to extension for use with GLMMs as well. As with estimation of $\boldsymbol{\beta}$ and $\mathbf{b}$, we turn to maximum likelihood—but with a twist. Maximum likelihood estimates of the variance components, as we shall see, are biased. This is a problem in and of itself when the goal is inference on the variance components per se. Often, however, inference focuses on estimable functions, $\mathbf{K}'\boldsymbol{\beta}$, or predictable functions, $\mathbf{K}'\boldsymbol{\beta}+\mathbf{M}'\mathbf{b}$, and the variance–covariance components are just means to an end—that is, to get interval estimates, test hypotheses, etc. In this case, biased estimates of $\mathbf{G}$ and $\mathbf{R}$ have a ripple effect, biasing standard errors and test statistics. The solution to this problem is *restricted*—or *residual—maximum likelihood*, usually referred to by its acronym *REML*.

We first work through the ANOVA and pure ML estimation, using the multilocation data from Example 3.1 as an example, to illustrate. Then we introduce REML and show how it is implemented in practice. One caveat: there are excellent textbooks—and stand-alone courses—on variance component estimation. Our goal here is not a comprehensive exposition. We focus on just enough background and methodology to allow us to get on with the task at hand—learning to work with linear statistical models. Readers interested in more detail and depth are referred to texts such as Searle et al. (1992) or Demidenko (2004).

4.4.3.1 ANOVA Estimator

The analysis of variance table for the four treatment multilocation data used in Example 3.4.1 is

Source of Variation	Degrees of Freedom	Sum of Squares	Mean Square	Expected Mean Square
Treatment	3	60.7525	20.2508	
Location	7	68.7250	9.8179	$\sigma^2+4\sigma_L^2$
Residual	21	60.4175	2.8770	σ^2
Total	31	189.8950		

For variance component estimation, we only need the EMS for location and residual, so these are the only EMS shown. The expected value of the mean square for residual is σ^2, hence the estimate of the residual variance is $\hat{\sigma}^2 = MS(\text{residual}) = 2.877$. The mean square for location estimates $\sigma^2+4\sigma_L^2$. Solving for σ_L^2 yields $\hat{\sigma}_L^2 = (1/4)\left[MS(\text{location}) - MS(\text{residual})\right] = 1.7532$. These are the ANOVA estimates of the variance components. The coefficients of the variance components in the EMS are straightforward for variance-component-only LMMs with balanced data with no missing values. Rules for these coefficients appear in most introductory statistical methods textbooks, and the underlying justification appears in variance component textbooks referenced earlier. With unbalanced or missing data, the coefficients become messier, but in principle the ANOVA method can be used with any variance-component-only LMM. But it can be used *only* with this subset of LMMs. We move on, therefore, to likelihood-based variance–covariance estimation.

4.4.3.2 Maximum Likelihood

As a first step, we establish notation. First, for convenience, in this section, the terms "covariance component" and "covariance matrix" generically refer to variance and well as covariance. Let $\boldsymbol{\sigma}$ denote the vector of covariance components to be estimated. For the multilocation data, we just considered $\boldsymbol{\sigma}' = \begin{bmatrix} \sigma_L^2 & \sigma^2 \end{bmatrix}$. For the random coefficient model, $\boldsymbol{\sigma}' = \begin{bmatrix} \sigma_0^2 & \sigma_{01} & \sigma_1^2 & \sigma^2 \end{bmatrix}$. Let $\hat{\boldsymbol{\sigma}}$ denote the vector of covariance component estimates, for example, $\hat{\boldsymbol{\sigma}}' = \begin{bmatrix} \hat{\sigma}_L^2 & \hat{\sigma}^2 \end{bmatrix}$. Recognizing that each covariance matrix—**G**, **R**, and **V**—depends on components of $\boldsymbol{\sigma}$, use the notation $\mathbf{G}(\boldsymbol{\sigma})$, $\mathbf{R}(\boldsymbol{\sigma})$, and $\mathbf{V}(\boldsymbol{\sigma})$ to denote **G**, **R**, and **V** with known covariance components. Recognize that $\mathbf{G}(\boldsymbol{\sigma})$ and $\mathbf{R}(\boldsymbol{\sigma})$ depend on some but not all of the elements of $\boldsymbol{\sigma}$ but the notation is clear enough so we use it for convenience. With estimated covariance components, we use the notation $\hat{\mathbf{G}} = \mathbf{G}(\hat{\boldsymbol{\sigma}})$, $\hat{\mathbf{R}} = \mathbf{R}(\hat{\boldsymbol{\sigma}})$, and $\hat{\mathbf{V}} = \mathbf{V}(\hat{\boldsymbol{\sigma}})$, sticking with the $\boldsymbol{\sigma}$ notation, for example, $\mathbf{G}(\hat{\boldsymbol{\sigma}})$ unless the context makes using **G** possible without ambiguity or confusion.

Using this notation, we write the marginal log-likelihood for the LMM as

$$\ell(\boldsymbol{\sigma};\boldsymbol{\beta},\mathbf{y}) = -\left(\frac{n}{2}\right)\log(2\pi) - \left(\frac{1}{2}\right)\log\left(\left|\mathbf{V}(\boldsymbol{\sigma})\right|\right) - \left(\frac{1}{2}\right)(\mathbf{y}-\mathbf{X}\boldsymbol{\beta})'\left[\mathbf{V}(\boldsymbol{\sigma})\right]^{-1}(\mathbf{y}-\mathbf{X}\boldsymbol{\beta}) \quad (4.22)$$

We obtain maximum likelihood estimates of the components of $\boldsymbol{\sigma}$ by setting the score vector $\partial\ell(\boldsymbol{\sigma};\boldsymbol{\beta},y)/\partial\boldsymbol{\sigma}$ equal to zero and solving. Note that $\partial\ell(\boldsymbol{\sigma};\boldsymbol{\beta},y)/\partial\boldsymbol{\sigma}$ is a vector whose dimension follows from the total number of covariance components to be estimated. For example, for the random coefficient LMM,

$$\mathbf{s}(\boldsymbol{\sigma}) = \frac{\partial\ell(\boldsymbol{\sigma};\boldsymbol{\beta},y)}{\partial\boldsymbol{\sigma}} = \begin{bmatrix} \dfrac{\partial\ell(\boldsymbol{\sigma};\boldsymbol{\beta},y)}{\partial\sigma_0^2} \\ \dfrac{\partial\ell(\boldsymbol{\sigma};\boldsymbol{\beta},y)}{\partial\sigma_{01}} \\ \dfrac{\partial\ell(\boldsymbol{\sigma};\boldsymbol{\beta},y)}{\partial\sigma_1^2} \\ \dfrac{\partial\ell(\boldsymbol{\sigma};\boldsymbol{\beta},y)}{\partial\sigma^2} \end{bmatrix} \quad (4.23)$$

In general, $\mathbf{s}(\boldsymbol{\sigma})$ denotes the score vector with respect to covariance estimation and the ith element of the score vector is $\partial\ell(\boldsymbol{\sigma};\boldsymbol{\beta},y)/\partial\sigma$, where σ_i denotes the ith component of the $\boldsymbol{\sigma}$ vector. For example, for the random coefficient model, $\sigma_1 \equiv \sigma_0^2$, $\sigma_2 \equiv \sigma_{01}$, $\sigma_3 \equiv \sigma_1^2$, and $\sigma_4 \equiv \sigma^2$.

While there are specific forms for the maximum likelihood estimating equations for variance-component-only LMMs, our goal here is to present a method that can be used for LMMs *in general*. To do this, we use the Newton–Raphson and Fisher scoring approaches. This means we need the Hessian matrix—$\mathbf{H}(\boldsymbol{\sigma})$—for Newton–Raphson and the information matrix—$\mathrm{I}(\boldsymbol{\sigma})$—for Fisher scoring. Harville (1977) derived the elements of these matrices. Linstrom and Bates (1988) presented Newton-Raphson algorithms for LMM estimation. We also need the specific form of the score vector. All have two forms, one for maximum likelihood (shown here) and one for REML (shown later).

The ith element of the score vector for maximum likelihood estimation of $\boldsymbol{\sigma}$ is

$$s_i(\boldsymbol{\sigma}) = \frac{\partial\ell(\boldsymbol{\sigma};\mathbf{y},\boldsymbol{\beta})}{\partial\sigma_i} = -\left(\frac{1}{2}\right)tr\left[\mathbf{V}^{-1}\left(\frac{\partial\mathbf{V}(\boldsymbol{\sigma})}{\partial\sigma_i}\right)\right] + \left(\frac{1}{2}\right)(\mathbf{y}-\mathbf{X}\boldsymbol{\beta})'\mathbf{V}^{-1}\left(\frac{\partial\mathbf{V}(\boldsymbol{\sigma})}{\partial\sigma_i}\right)\mathbf{V}^{-1}(\mathbf{y}-\mathbf{X}\boldsymbol{\beta}) \quad (4.24)$$

The *ij*th element of the Hessian matrix for maximum likelihood estimation of $\boldsymbol{\sigma}$ is

$$\frac{\partial^2 \ell(\boldsymbol{\sigma};\mathbf{y},\boldsymbol{\beta})}{\partial\sigma_i\partial\sigma_j} = -\left(\frac{1}{2}\right) tr\left[\mathbf{V}^{-1}\left(\frac{\partial^2\mathbf{V}(\boldsymbol{\sigma})}{\partial\sigma_i\partial\sigma_j}\right) - \mathbf{V}^{-1}\left(\frac{\partial\mathbf{V}(\boldsymbol{\sigma})}{\partial\sigma_i}\right)\mathbf{V}^{-1}\left(\frac{\partial\mathbf{V}(\boldsymbol{\sigma})}{\partial\sigma_j}\right)\right]$$

$$+\left(\frac{1}{2}\right)(\mathbf{y}-\mathbf{X}\boldsymbol{\beta})'\mathbf{V}^{-1}\left[\left(\frac{\partial^2\mathbf{V}(\boldsymbol{\sigma})}{\partial\sigma_i\partial\sigma_j}\right) - 2\left(\frac{\partial\mathbf{V}(\boldsymbol{\sigma})}{\partial\sigma_i}\right)\mathbf{V}^{-1}\left(\frac{\partial\mathbf{V}(\boldsymbol{\sigma})}{\partial\sigma_j}\right)\right]\mathbf{V}^{-1}(\mathbf{y}-\mathbf{X}\boldsymbol{\beta}) \quad (4.25)$$

For convenience, we can denote this element $H_{ij}(\boldsymbol{\sigma})$. The *ij*th element of the information matrix for maximum likelihood estimation of $\boldsymbol{\sigma}$ is

$$\mathrm{I}_{ij}(\boldsymbol{\sigma}) = -E\left(\frac{\partial^2 \ell(\boldsymbol{\sigma},\mathbf{y},\boldsymbol{\beta})}{\partial\sigma_i\partial\sigma_j}\right) = \left(\frac{1}{2}\right) tr\left[\mathbf{V}^{-1}\left(\frac{\partial\mathbf{V}(\boldsymbol{\sigma})}{\partial\sigma_i}\right)\mathbf{V}^{-1}\left(\frac{\partial\mathbf{V}(\boldsymbol{\sigma})}{\partial\sigma_j}\right)\right] \quad (4.26)$$

If we implement the Newton–Raphson procedure with the random coefficient model, the estimating equations are

$$\begin{bmatrix}\sigma_1\\ \sigma_2\\ \sigma_3\\ \sigma_4\end{bmatrix} = \begin{bmatrix}\tilde{\sigma}_1\\ \tilde{\sigma}_2\\ \tilde{\sigma}_3\\ \tilde{\sigma}_4\end{bmatrix} - \begin{bmatrix} H_{11}(\tilde{\boldsymbol{\sigma}}) & H_{12}(\tilde{\boldsymbol{\sigma}}) & H_{13}(\tilde{\boldsymbol{\sigma}}) & H_{14}(\tilde{\boldsymbol{\sigma}}) \\ & H_{22}(\tilde{\boldsymbol{\sigma}}) & H_{23}(\tilde{\boldsymbol{\sigma}}) & H_{24}(\tilde{\boldsymbol{\sigma}}) \\ & & H_{33}(\tilde{\boldsymbol{\sigma}}) & H_{34}(\tilde{\boldsymbol{\sigma}}) \\ & & & H_{44}(\tilde{\boldsymbol{\sigma}}) \end{bmatrix}^{-1} \begin{bmatrix} s_1(\tilde{\boldsymbol{\sigma}})\\ s_2(\tilde{\boldsymbol{\sigma}})\\ s_3(\tilde{\boldsymbol{\sigma}})\\ s_4(\tilde{\boldsymbol{\sigma}})\end{bmatrix} \quad (4.27)$$

$\tilde{\boldsymbol{\sigma}}$ and its elements $\tilde{\sigma}_i$ denote the value of the covariance component from the previous iteration. For Fisher scoring, replace each $H_{ij}(\boldsymbol{\sigma})$ by $\mathrm{I}_{ij}(\boldsymbol{\sigma})$ and change the sign from minus to plus. The key to setting up the procedure is determining the derivatives. Start by writing $\mathbf{V}(\boldsymbol{\sigma})$ explicitly in matrix terms. For the random coefficient model, $\mathbf{V}(\boldsymbol{\sigma}) = \begin{bmatrix}\mathbf{Z}_0 & \mathbf{Z}_1\end{bmatrix}\left(I_4 \otimes \begin{bmatrix}\sigma_0^2 & \sigma_{01}\\ \sigma_{01} & \sigma_1^2\end{bmatrix}\right)\begin{bmatrix}\mathbf{Z}_0'\\ \mathbf{Z}_1'\end{bmatrix} + \mathbf{I}\sigma^2$, where $\mathbf{Z}_0$ and $\mathbf{Z}_1$ are submatrices of $\mathbf{Z}$ associated with the random intercept and random slope effects, respectively (these were derived in the matrix section of Chapter 1). The resulting derivatives are $\partial\mathbf{V}(\boldsymbol{\sigma})/\partial\sigma_1 = \mathbf{Z}_0\mathbf{Z}_0'$, $\partial\mathbf{V}(\boldsymbol{\sigma})/\partial\sigma_2 = \mathbf{Z}_1\mathbf{Z}_0' + \mathbf{Z}_0\mathbf{Z}_1'$, $\partial\mathbf{V}(\boldsymbol{\sigma})/\partial\sigma_3 = \mathbf{Z}_1\mathbf{Z}_1'$, and $\partial\mathbf{V}(\boldsymbol{\sigma})/\partial\sigma_4 = \mathbf{I}$. All second derivatives are zero.

Continuing the four treatment multilocation example, the maximum likelihood estimators are $\hat{\sigma}_L^2 = 1.52$ and $\hat{\sigma}^2 = 2.52$ compared with the ANOVA estimators 1.74 and 2.88, respectively. Recall from Chapter 3 that PROC GLIMMIX, on default, produced covariance parameter estimates equal to the ANOVA estimates (this happens for variance-component-only LMMs with balanced data, but not in general). This is our first illustration of the downward bias associated with maximum likelihood. Recall what was likely your first encounter with a variance estimate: the usual formula for the sample variance of a random sample from a single population, $S^2 = \sum_i (y_i - \bar{y})^2/(n-1)$. Some of you probably asked, "If the sample mean is $\bar{y} = \sum_i y/n$, why don't we also divide the sample variance by n?" It happens that $\sum_i (y_i - \bar{y})^2/n$ is the maximum likelihood estimate when y_i i.i.d.

$N(\mu,\sigma^2)$—and the reason you divide by $n-1$ instead is that if you divide by n you get a downward biased estimate.

4.4.3.3 Restricted Maximum Likelihood

The sample variance, it turns out, was your first encounter with a REML estimate.

What impact does ML variance estimate's downward bias have? Confidence intervals and test statistics both depend on the standard error, which in turn depends on the variance component estimates. Downward biased variance estimates mean downward biased standard errors which in turn mean confidence intervals that are too narrow (inadequate coverage) and upwardly-biased test statistics (inflated type I error rate). For instance, in the multilocation example, the F-value for the overall test of equal treatment effects is 7.04 using the ANOVA or REML variance estimates, but it is 8.04 if you use the maximum likelihood variance estimates. While these both have p-values that would be declared statistically significant by most α-criteria, it is clearly possible for ANOVA or REML estimates to produce different conclusions than ML-based results.

As an example of the seriousness of the bias' impact, consider an example Littell et al. (2006) present at the start of *SAS for Mixed Models*, 2nd edn. Littell et al. apply mixed analysis to an incomplete block design that appears in Cochran and Cox (1957), using a model essentially identical to the multilocation example in this section: Gaussian data, random block effects in place of random location effects, and fixed treatment. The following table shows the variance estimates and resulting overall treatment F- and p-values. The data and the GLIMMIX programs that duplicate these results appear in the SAS Data and Program Library.

Variance Estimation Method	Block Variance $\hat{\sigma}_B^2$	Residual Variance $\hat{\sigma}_B^2$	F_{Trt}	p-Value
ANOVA	4.73	8.62	1.52	0.1622
REML	4.63	8.56	1.53	0.1576
ML	4.50	6.04	2.04	0.0484

A simulation using the design from the Cochran and Cox example with all treatment means equal and variance ratios of $\sigma_B^2/\sigma^2 = 0.5$ and $\sigma_B^2/\sigma^2 = 5$ showed that the variance ratio had little impact, but the observed rejection rates, assuming $\alpha = 0.05$, for runs of 1000 simulated experiments ranged from 4% to 6% for F tests computed using the ANOVA and REML variance estimates (there was little to choose between ANOVA and REML) but the rejection rates were always greater than 20% (as high as 28%) when ML variance estimates were used. This is the typical impact of the downward bias of ML variance estimates for LMMs and why REML is the de facto gold standard for LMM variance estimation.

REML ideas first appeared in statistical literature in the 1950s. Patterson and Thompson (1971) presented comprehensive REML theory for LMMs. The fundamental idea of REML is to maximize the likelihood *after* accounting for the model's fixed effects. Instead of maximizing the likelihood of $\mathbf{y} \sim N(\mathbf{X}\boldsymbol{\beta},\mathbf{V})$, we obtain estimators from the likelihood of $\mathbf{K}'\mathbf{y}$ where $\mathbf{K}$ is any matrix such that $E(\mathbf{K}'\mathbf{y}) = 0$ and hence $\mathbf{K}'\mathbf{y} \sim N(\mathbf{0},\mathbf{K}'\mathbf{V}\mathbf{K})$. This effectively removes the fixed effects from the estimation of $\boldsymbol{\sigma}$. The likelihood of $\mathbf{K}'\mathbf{y}$ is called the REML likelihood. Typically, the ordinary least squares residual operator, $\mathbf{I} - \mathbf{X}(\mathbf{X}'\mathbf{X})^{-}\mathbf{X}'$, is used, noting that $E\{[\mathbf{I}-\mathbf{X}(\mathbf{X}'\mathbf{X})^{-}\mathbf{X}']\mathbf{y}\}=[\mathbf{I}-\mathbf{X}(\mathbf{X}'\mathbf{X})^{-}\mathbf{X}']\mathbf{X}\boldsymbol{\beta}=0$ and assuming $(\mathbf{X}'\mathbf{X})^{-}$ is a $G^{(2)}$ (reflexive) generalized inverse of $\mathbf{X}'\mathbf{X}$. The REML log-likelihood is

$$\ell_R(\boldsymbol{\sigma};\mathbf{y})=-\left(\frac{n-p}{2}\right)\log(2\pi)-\left(\frac{1}{2}\right)\log\left(\left|\mathbf{V}(\boldsymbol{\sigma})\right|\right)-\left(\frac{1}{2}\right)\log\left(\left|\mathbf{X}'\left[\mathbf{V}(\boldsymbol{\sigma})\right]^{-1}\mathbf{X}\right|\right)-\left(\frac{1}{2}\right)\mathbf{r}'\left[\mathbf{V}(\boldsymbol{\sigma})\right]^{-1}\mathbf{r} \tag{4.28}$$

where $p=rank(\mathbf{X})$ and $\mathbf{r}=\mathbf{y}-\mathbf{X}(\mathbf{X}'[\mathbf{V}(\boldsymbol{\sigma})]^{-1}\mathbf{X})^{-}\mathbf{X}'[\mathbf{V}(\boldsymbol{\sigma})]^{-1}\mathbf{y}$. Note that $\mathbf{r}$ could be viewed as $\mathbf{y}-\mathbf{X}\hat{\boldsymbol{\beta}}_{ML}$, where $\hat{\boldsymbol{\beta}}_{ML}$ is the maximum likelihood estimate of $\boldsymbol{\beta}$.

As with maximum likelihood, we implement REML using Newton–Raphson or Fisher scoring. The score vector and Hessian and information matrices derived from the REML log-likelihood (again following Harville, 1977) are

Score (*i*th element)

$$s_i(\boldsymbol{\sigma})=\frac{\partial \ell_R(\boldsymbol{\sigma};\mathbf{y})}{\partial\sigma_i}=-\left(\frac{1}{2}\right)tr\left[\mathbf{P}\left(\frac{\partial\mathbf{V}(\boldsymbol{\sigma})}{\partial\sigma_i}\right)\right]+\left(\frac{1}{2}\right)(\mathbf{y}-\mathbf{X}\boldsymbol{\beta})'\,\mathbf{V}^{-1}\left(\frac{\partial\mathbf{V}(\boldsymbol{\sigma})}{\partial\sigma_i}\right)\mathbf{V}^{-1}(\mathbf{y}-\mathbf{X}\boldsymbol{\beta}) \tag{4.29}$$

where

$$\mathbf{P}=\left[\mathbf{V}(\boldsymbol{\sigma})\right]^{-1}-\left[\mathbf{V}(\boldsymbol{\sigma})\right]^{-1}\mathbf{X}\left(\mathbf{X}'\left[\mathbf{V}(\boldsymbol{\sigma})\right]^{-1}\mathbf{X}\right)^{-}\mathbf{X}'\left[\mathbf{V}(\boldsymbol{\sigma})\right]^{-1}$$

Hessian (*ij*th element)

$$\begin{aligned}\frac{\partial^2\ell_R(\boldsymbol{\sigma};\mathbf{y})}{\partial\sigma_i\partial\sigma_j}&=-\left(\frac{1}{2}\right)tr\left[\mathbf{P}\left(\frac{\partial^2\mathbf{V}(\boldsymbol{\sigma})}{\partial\sigma_i\partial\sigma_j}\right)-\mathbf{P}\left(\frac{\partial\mathbf{V}(\boldsymbol{\sigma})}{\partial\sigma_i}\right)\mathbf{P}\left(\frac{\partial\mathbf{V}(\boldsymbol{\sigma})}{\partial\sigma_j}\right)\right]\\&\quad+\left(\frac{1}{2}\right)(\mathbf{y}-\mathbf{X}\boldsymbol{\beta})'\,\mathbf{V}^{-1}\left[\left(\frac{\partial^2\mathbf{V}(\boldsymbol{\sigma})}{\partial\sigma_i\partial\sigma_j}\right)-2\left(\frac{\partial\mathbf{V}(\boldsymbol{\sigma})}{\partial\sigma_i}\right)\mathbf{P}\left(\frac{\partial\mathbf{V}(\boldsymbol{\sigma})}{\partial\sigma_j}\right)\right]\mathbf{V}^{-1}(\mathbf{y}-\mathbf{X}\boldsymbol{\beta})\end{aligned} \tag{4.30}$$

Information (*ij*th element)

$$\mathrm{I}_{ij}(\boldsymbol{\sigma})=-E\left(\frac{\partial^2\ell_R(\boldsymbol{\sigma};\mathbf{y})}{\partial\sigma_i\partial\sigma_j}\right)=\left(\frac{1}{2}\right)tr\left[\mathbf{P}\left(\frac{\partial\mathbf{V}(\boldsymbol{\sigma})}{\partial\sigma_i}\right)\mathbf{P}\left(\frac{\partial\mathbf{V}(\boldsymbol{\sigma})}{\partial\sigma_j}\right)\right] \tag{4.31}$$

Aside from replacing $\mathbf{V}^{-1}$ by $\mathbf{P}$ where indicated, estimation proceeds exactly as described for Newton–Raphson or Fisher scoring for maximum likelihood.

To summarize, LMM estimation proceeds iteratively. The process starts with initial values of $\boldsymbol{\sigma}$, denoted $\tilde{\boldsymbol{\sigma}}$. From these, compute $\mathbf{G}(\tilde{\boldsymbol{\sigma}})$ and $\mathbf{R}(\tilde{\boldsymbol{\sigma}})$, which allows us the complete the mixed model equations and compute solutions for $\boldsymbol{\beta}$ and $\mathbf{b}$. We use these to determine new values of the score vector and the Hessian or information matrix, allowing us to update $\tilde{\boldsymbol{\sigma}}$. Estimation continues in this manner until convergence.

4.5 Generalized Linear Mixed Models

We come now to the most ambitious generalization of the LM to be covered in this textbook—the GLMM. Here, we combine the essential features of the generalized and mixed LMs. From the LMM, we allow random model effects in the linear predictor. From the GLM, we drop classical LM assumption of normality—we only assume that the observations, conditional on the random effects, have a distribution belonging to the exponential family or at least have a quasi-likelihood. In summary, the essential features of the GLMM are

- Linear predictor: $\boldsymbol{\eta} = \mathbf{X}\boldsymbol{\beta} + \mathbf{Zb}$
- Distribution: $\mathbf{b} \sim N(\mathbf{0}, \mathbf{G})$
- Distribution or quasi-likelihood: $E(\mathbf{y}|\mathbf{b}) = \boldsymbol{\mu}|\mathbf{b}$; $\text{Var}(\mathbf{y} \mid \mathbf{b}) = \mathbf{V}_\mu^{1/2}\mathbf{A}\mathbf{V}_\mu^{1/2}$, where $\mathbf{V}_\mu^{1/2} = diag\left[\sqrt{V(\mu)}\right] = diag\left[\sqrt{\partial^2 b(\theta)/\partial\theta^2}\right]$ and $\mathbf{A} = diag[1/a(\phi)]$ and $\mathbf{y}|\mathbf{b}$ either has a distribution that belongs to the exponential family or a quasi-likelihood
- Link: $\boldsymbol{\eta} = g(\boldsymbol{\mu}|\mathbf{b})$, or alternatively, inverse link: $\mathbf{X}\boldsymbol{\beta} + \mathbf{Zb} = h(\boldsymbol{\eta})$. Typically, $h(\cdot) = g^{-1}(\cdot)$

The quasi-likelihood of the observations conditional on the random effects is $ql(\mathbf{y}|\mathbf{b}) = \mathbf{y}'\mathbf{A}\theta - \mathbf{1}'\mathbf{A}b(\boldsymbol{\theta})$ and the log-likelihood of the random effects distribution is $\ell(\mathbf{b}) = -(b/2)\log(2\pi) - (1/2)\log(|\mathbf{G}|) - (1/2)\mathbf{b}'\mathbf{G}^{-1}\mathbf{b}$. The joint log(quasi)-likelihood is thus $\ell(\mathbf{b}) + ql(\mathbf{y}|\mathbf{b})$ and the marginal (quasi-) likelihood is

$$\iint_{\mathbf{b}} \left[ql(\mathbf{y} \mid \mathbf{b}) + \ell(\mathbf{b})\right] d\mathbf{b} = ql(\mathbf{y})$$

$$= \iint_{\mathbf{b}} \left[\mathbf{y}'\mathbf{A}\boldsymbol{\theta} - \mathbf{1}'\mathbf{A}b(\boldsymbol{\theta}) - \left(\frac{b}{2}\right)\log(2\pi) - \left(\frac{1}{2}\right)\log(|\mathbf{G}|) - \left(\frac{1}{2}\right)\mathbf{b}'\mathbf{G}^{-1}\mathbf{b}\right] d\mathbf{b} \tag{4.32}$$

If $\mathbf{y}|\mathbf{b}$ has a *bona fide* distribution, we can add the $c(\phi,\mathbf{y})$ term; this gives us a true log-likelihood.

While $q\ell(\mathbf{y})$, or $\ell(\mathbf{y})$, is well defined, further simplification, with few exceptions, is not possible. Thus, we face the problem of trying to minimize an essentially intractable function. The LMM is one such exception—because $\mathbf{y}|\mathbf{b}$ is Gaussian, the joint distribution and, hence, the marginal distribution are also Gaussian, as we saw in Section 4.4. In general, however, we need to use some form of approximation. Two alternatives we will use in this text are as follows:

- *Linearization*: Specifically, the *pseudo-likelihood* method used by PROC GLIMMIX.
- *Integral approximation*: Two methods implemented by PROC GLIMMIX are the Laplace approximation and adaptive Gauss–Hermite quadrature.

In the rest of this section, we present the pseudo-likelihood method in some detail. The integral approximation methods are presented in considerably less detail. Instead, subsequent chapters focus extensively on the use (and informed *non*use) of these methods in practical modeling situations.

4.5.1 Pseudo-Likelihood for GLMM

Schall (1991) and Breslow and Clayton (1993) proposed quasi-likelihood-based linearization for GLMM estimation, known in the literature as penalized-quasi-likelihood (PQL). Wolfinger and O'Connell (1993) developed the pseudo-likelihood approach motivated from the perspective of a Laplace approximation. Both are similar in that they extend the idea of a pseudo-variable—$\mathbf{y}^*$ from Section 4.3—and end up with a GLM version of the mixed model equations (4.17). *Pseudo-likelihood* is preferred here because when the conditional distribution belongs to the exponential family, this distributional assumption is explicit (we do not have a quasi-likelihood) and—perhaps more importantly—pseudo-likelihood conveys the idea that the approximating function has much of the structure of the Gaussian log-likelihood, allowing us to use LMM-like estimating equations for the covariance components as well as the linear predictor effects.

We start with a Taylor series expansion of the inverse link function evaluated at $\tilde{\boldsymbol{\eta}}$. This follows from the objective of a GLMM: to model $E(\mathbf{y}|\mathbf{b})=\boldsymbol{\mu}|\mathbf{b}$ by $h(\mathbf{X}\boldsymbol{\beta}+\mathbf{Z}\mathbf{b})=h(\boldsymbol{\eta})$. The Taylor series expansion is $h(\boldsymbol{\eta})\cong h(\tilde{\boldsymbol{\eta}})+\left.\frac{\partial h(\boldsymbol{\eta})}{\partial\boldsymbol{\eta}}\right|_{\boldsymbol{\eta}=\tilde{\boldsymbol{\eta}}}(\boldsymbol{\eta}-\tilde{\boldsymbol{\eta}})$. Following notation introduced in (4.12), define $\mathbf{D}=diag[\partial g(\boldsymbol{\mu}|\mathbf{b})/\partial\boldsymbol{\mu}]$ or, alternatively, $\mathbf{D}^{-1}=diag[\partial h(\boldsymbol{\eta})/\partial\boldsymbol{\eta}]$. We can reexpress the Taylor series expansion as $h(\boldsymbol{\eta})\cong h(\tilde{\boldsymbol{\eta}})+\tilde{\mathbf{D}}^{-1}\left(\mathbf{X}\boldsymbol{\beta}+\mathbf{Z}\mathbf{b}-\mathbf{X}\tilde{\boldsymbol{\beta}}-\mathbf{Z}\tilde{\mathbf{b}}\right)$, where $\tilde{\mathbf{D}}$ denotes $\mathbf{D}$ evaluated at $\tilde{\boldsymbol{\eta}}=\mathbf{X}\tilde{\boldsymbol{\beta}}+\mathbf{Z}\tilde{\mathbf{b}}$. Rearranging terms yields $\tilde{\mathbf{D}}\left[h(\boldsymbol{\eta})-h(\tilde{\boldsymbol{\eta}})\right]+\mathbf{X}\tilde{\boldsymbol{\beta}}+\mathbf{Z}\tilde{\mathbf{b}}\cong\mathbf{X}\boldsymbol{\beta}+\mathbf{Z}\mathbf{b}$.

Now recalling the pseudo-variable $\mathbf{y}^*$ from the GLM estimating equations (4.15), the mixed model version is $\mathbf{y}^*=\tilde{\boldsymbol{\eta}}+\tilde{\mathbf{D}}^{-1}\left[\mathbf{y}-\left(\tilde{\boldsymbol{\mu}}\,|\,\tilde{\mathbf{b}}\right)\right]=\mathbf{X}\tilde{\boldsymbol{\beta}}+\mathbf{Z}\tilde{\mathbf{b}}+\tilde{\mathbf{D}}^{-1}\left[\mathbf{y}-h(\tilde{\boldsymbol{\eta}})\right]$. It follows that

$$E(\mathbf{y}^*\,|\,\mathbf{b})=\tilde{\mathbf{D}}\left[h(\boldsymbol{\eta})-h(\tilde{\boldsymbol{\eta}})\right]+\mathbf{X}\tilde{\boldsymbol{\beta}}+\mathbf{Z}\tilde{\mathbf{b}}\cong\mathbf{X}\boldsymbol{\beta}+\mathbf{Z}\mathbf{b} \tag{4.33}$$

$$\mathrm{Var}(\mathbf{y}^*\,|\,\mathbf{b})=\mathbf{D}\mathbf{V}_{\mu}^{1/2}\mathbf{A}\mathbf{V}_{\mu}^{1/2}\mathbf{D} \tag{4.34}$$

At this point, you can either use a generalized least squares argument with $Var(\mathbf{y}^*|\mathbf{b})$ playing the role of $\mathbf{R}$ in the LMM, or you can simply look at $\mathbf{y}^*$ as the response variable in a LMM. Either way, the pseudo-likelihood estimating equations, also called the generalized mixed model equations, are the mixed model equations with $\mathbf{y}^*$ replacing $\mathbf{y}$ and $\mathbf{D}\mathbf{V}_{\mu}^{1/2}\mathbf{A}\mathbf{V}_{\mu}^{1/2}\mathbf{D}$ replacing $\mathbf{R}$. Recalling that we defined $\mathbf{W}=(\mathbf{DVD})^{-1}=\left(\mathbf{D}\mathbf{V}_{\mu}^{1/2}\mathbf{A}\mathbf{V}_{\mu}^{1/2}\mathbf{D}\right)^{-1}$ in the GLM estimating equations, and hence $\mathbf{W}$ replaces $\mathbf{R}^{-1}$, the pseudo-likelihood GLMM equation is

$$\begin{bmatrix}\mathbf{X}'\mathbf{W}\mathbf{X} & \mathbf{X}'\mathbf{W}\mathbf{Z}\\ \mathbf{Z}'\mathbf{W}\mathbf{X} & \mathbf{Z}'\mathbf{W}\mathbf{Z}+\mathbf{G}^{-1}\end{bmatrix}\begin{bmatrix}\boldsymbol{\beta}\\ \mathbf{b}\end{bmatrix}=\begin{bmatrix}\mathbf{X}'\mathbf{W}\mathbf{y}^*\\ \mathbf{Z}'\mathbf{W}\mathbf{y}^*\end{bmatrix} \tag{4.35}$$

4.5.2 Variance–Covariance Estimation with Pseudo-Likelihood

Several results follow that are useful for covariance component estimation.

- The "marginal" pseudo-variance is

$$\mathrm{Var}(\mathbf{y}^*)=\mathbf{Z}\mathbf{G}\mathbf{Z}'+\mathbf{W}^{-1}\text{or, alternatively}$$

$$\mathbf{Z}\mathbf{G}\mathbf{Z}'+\mathbf{D}\mathbf{V}_{\mu}^{1/2}\mathbf{A}\mathbf{V}_{\mu}^{1/2}\mathbf{D}.\text{ Denote this term }\mathbf{V}^*(\boldsymbol{\sigma}). \tag{4.36}$$

- The pseudo-log-likelihood function is

$$p\ell(\boldsymbol{\beta},\boldsymbol{\sigma};\mathbf{y}^*) = -\left(\frac{n}{2}\right)\log(2\pi) - \left(\frac{1}{2}\right)\log\left(\left|\mathbf{V}^*(\boldsymbol{\sigma})\right|\right) - \left(\frac{1}{2}\right)(\mathbf{y}^* - \mathbf{X}\boldsymbol{\beta})'\left[\mathbf{V}^*(\boldsymbol{\sigma})\right]^{-1}(\mathbf{y}^* - \mathbf{X}\boldsymbol{\beta}) \tag{4.37}$$

- The REML pseudo-log-likelihood is

$$\begin{aligned} pl_R(\boldsymbol{\sigma};\mathbf{y}^*) = &-\left(\frac{n-p}{2}\right)\log(2\pi) - \left(\frac{1}{2}\right)\log\left(\left|\mathbf{V}^*(\boldsymbol{\sigma})\right|\right) \\ &-\left(\frac{1}{2}\right)\log\left(\left|\mathbf{X}'\left[\mathbf{V}^*(\boldsymbol{\sigma})\right]^{-1}\mathbf{X}\right|\right) - \left(\frac{1}{2}\right)(\mathbf{r}^*)'\left[\mathbf{V}^*(\boldsymbol{\sigma})\right]^{-1}\mathbf{r}^* \end{aligned} \tag{4.38}$$

It follows that the ML and REML score vector terms and the ML and REML Hessian and information matrix terms given in Equations 4.25 through 4.30 can be used for GLMM covariance component estimation. As with the ML and REML pseudo-log-likelihood, simply replace $\mathbf{y}$ by $\mathbf{y}^*$ and $\mathbf{V}(\boldsymbol{\sigma})$ by $\mathbf{V}^*(\boldsymbol{\sigma})$.

PROC GLIMMIX uses the acronym RSPL to refer to pseudo-likelihood (PL) implemented using the REML version of the score vector and Hessian or information matrix for covariance estimation. The full acronym stands for Restricted Subject-specific Pseudo-Likelihood. The acronym MSPL—Maximum Subject-specific Pseudo-Likelihood—designates PL implemented using the ML version of the score vector and Hessian or information matrix for covariance estimation. We will use the acronym PL from this point forward. The GLIMMIX acronyms we will use only when referring specifically to the GLIMMIX procedure.

PL's similarity to LMM estimating gives it several desirable features. These include ease of implementation and the fact that it accommodates all non-Gaussian analogs of LMMs. PL can be viewed as the truly general statement of estimating equations for LMs. For Gaussian mixed models, PL reduces to the mixed model equations (4.17). For non-Gaussian fixed-effects-only models, PL reduces to the GLM estimating equations (4.15). For Gaussian fixed-effects-only models, PL reduces to the GLS estimating equations and further reduces to ordinary least squares normal equations if $\mathbf{V} = \mathbf{I}\sigma^2$. In other words, just as the LMM, GLM, and LM are all special cases of the GLMM, their estimating equations are all special cases of PL.

In Chapter 5 we will also see that PL allows us to use GLMM analogs of most well-known inference techniques associated with LMMs. In subsequent chapters, we explore these capabilities in greater detail.

Unfortunately, PL is not universally applicable. The linearization upon which PL depends does not always approximate the likelihood well enough to yield useable estimates. One known problem exists for binomial GLMMs when the number of Bernoulli trials per unit of observation is small. Another includes two parameter exponential distributions (e.g., the beta and negative binomial). As we proceed through subsequent chapters, we will add more specifics. The point here: PL cannot be used as a one-size-fits-all estimation method for all GLMMs. Integral approximation methods provide alternatives for the problem models.

4.5.3 Integral Approximation: Laplace and Quadrature

For problem models, direct maximization of the likelihood provides an alternative that often addresses the problems associated with PL-based estimation. Recall from (4.32) that the marginal likelihood involves integration over a (generally) messy product of Gaussian and exponential family (or quasi-) likelihoods that defy further simplification. While direct maximization, therefore, is generally not possible, we can come very close using integral approximation. Two useful approaches are Gauss–Hermite quadrature and the Laplace approximation. These follow McCulloch (1997) and Pinheiro and Bates (1995).

The basic ideas of these approximations are as follows:

- *Gauss–Hermite quadrature*: $\int f(x)e^{-x^2}dx \cong \sum_k w_k f(x_k)$, where w_k are weights and x_k are nodes. For given k, values of w_k and x_k are given in standard reference books of mathematical tables, for example, Zwillinger (1996).
- *Laplace approximation*: $\int e^{h(x)}dx \cong (2\pi)^{-1/2} e^{h(\tilde{x})} \left| \frac{\partial^2 h(x)}{\partial x^2}\Big|_{x=\tilde{x}} \right|^{-1/2}$, where $\tilde{x}$ denotes the value of x that maximizes $h(x)$.

In Chapters 1 through 3, we looked at several example GLMMs with $y|b \sim$ Binomial and $b \sim$ Gaussian. Let us use a very simple version of this set up, with $y|b \sim \text{Binary}(\pi)$ and $b \sim N(0,\sigma^2)$ to illustrate each of these approximate. Each approximation can get very intricate and computationally intense as the complexity of the GLMM increases, but this simple illustration will serve our purpose—which is simply to have a conceptual feel for how the approximations work.

We start with Gauss–Hermite quadrature. The conditional p.d.f. of the observation is $f(y|b) = p^y(1-p)^{1-y}$, where $y=0$ or 1 and $p=\Pr\{Y=1\}$. The p.d.f. of the random effect is $f(b) = (2\pi)^{-1/2}\sigma^{-1}e^{-b^2/2}$. Here, we use the notation p for the binary probability rather than the usual notation, π, to avoid confusion with the mathematical constant π. Define the GLMM by linear predictor $\eta + b$ and logit link. Letting w denote the random variable corresponding to the link, we have $w \sim N(\eta,\sigma^2)$ and $p = e^w/(1+e^w)$. We write the resulting joint p.d.f. as

$$\left(\frac{e^w}{1+e^w}\right)^y \left(\frac{1}{1+e^w}\right)^{1-y} (2\pi)^{-1/2}\sigma^{-1}e^{-(w-\eta)^2/2\sigma^2}$$

Now let $z = (w-\eta)/\sigma$. Substituting, we now have

$$\left(\frac{e^{\eta+\sigma z}}{1+e^{\eta-\sigma z}}\right)^y \left(\frac{1}{1+e^{\eta-\sigma z}}\right)^{1-y} (2\pi)^{-1/2}\sigma^{-1}e^{-z^2/2}$$

We can now write the marginal likelihood as

$$\int_{-\infty}^{\infty} \left(\frac{e^{(\eta+\sigma z)y}}{1+e^{\eta+\sigma z}}\right)(2\pi)^{-1/2} e^{-z^2/2}\,dz \tag{4.39}$$

Finally, let $x^2 = z^2/2$. This yields,

$$\int_{-\infty}^{\infty} \left(\frac{e^{(\eta+\sigma\sqrt{2}x)y}}{1+e^{\eta+\sigma\sqrt{2}x}}\right)(\pi)^{-1/2} e^{-x^2}\,dx$$

We now recognize

$$f(x) = \left(\frac{e^{(\eta+\sigma\sqrt{2}x)y}}{1+e^{\eta+\sigma\sqrt{2}x}} \right) (\pi)^{-1/2}$$

as the required function for Gauss–Hermite quadrature: The marginal likelihood is approximately equal to

$$\sum_k w_k \left(\frac{e^{(\eta+\sigma\sqrt{2}x_k)y}}{1+e^{\eta+\sigma\sqrt{2}x_k}} \right) (\pi)^{-1/2}$$

For the Laplace approximation, we return to the marginal distribution given in (4.39). Letting $h(z) = (\eta + \sigma z)y - \log(1 + e^{\eta} + \sigma z) - (1/2)z^2$ gives us the desired form for the Laplace approximation, with

$$\frac{\partial^2 h(z)}{\partial z^2} = -\left(\frac{\sigma^2 e^{\eta+\sigma z}}{\left(1+e^{\eta+\sigma z}\right)^2} + 1 \right)$$

The approximation with Gauss–Hermite quadrature becomes more accurate as k, the number of quadrature points, increases. However, increasing k also increases the procedure's computational burden, to the point where it becomes prohibitive. Also, there is a point of diminishing returns that depends on the data and model being fit. Many software packages that implement quadrature, including the GLIMMIX procedure, are *adaptive*, meaning they have data-driven decision rules to select a nominally optimal number of quadrature points. The adaptive rules can be overridden—for example, in some situations the complexity of the model renders the adaptive procedure itself computationally prohibitive. In other cases, it simply makes sense to see what happens for a specified number of quadrature points.

The Laplace approximation can be shown to be equivalent to the Gauss–Hermite procedure with 1 quadrature point. The Laplace procedure is less computationally intensive than quadrature and is considerably more flexible in terms of the models with which it can be used.

Both Gauss–Hermite quadrature and Laplace approximate the marginal likelihood. Aside from the REML version of PL, there is no analog to residual likelihood for the GLMM—residual likelihood only has meaning in conjunction with Gaussian data. As a result, quadrature and Laplace compute only maximum likelihood covariance component estimates. This is generally not a problem with GLMMs, but it does suggest that neither method should be used for LMs, mixed or otherwise, with Gaussian data.

One major advantage of quadrature and Laplace will become evident when we explore comparative model-fitting, beginning in Chapter 6. Because these procedures focus on the actual likelihood, rather than a PL linearization, statistics derived from the likelihood, for example, fit statistics and likelihood ratio tests, are well defined whereas they are not for PL. This is only an issue for non-Gaussian GLMMs; for LMM, GLM, and LM *all* likelihoods are well defined. For the *non-Gaussian* GLMM, however, quadrature and Laplace compute reasonable approximations of the likelihood (or at least they are intended to do so) whereas PL does not. As we will see when we consider correlated error

GLMMs, this makes the Laplace approximation especially useful because it can accommodate appropriately defined correlated error models for certain GLMMs (models that are generally too complex for quadrature) *and* allow comparison of competing models using fit statistics or likelihood ratio statistics.

Interested readers are referred to Nelder and Lee (1982), McCulloch et al. (2008) and Wu (2010) for additional background.

4.6 Summary

The general approach to LM estimation is maximum likelihood. Historically, least squares has been the dominant approach. In the case of the LM, least squares and the maximum likelihood yield identical estimates of $\boldsymbol{\beta}$, the vector of model parameters. However, least squares becomes progressively less applicable as modeling complexity increases.

Maximum likelihood can be divided into four categories, corresponding to the four LM classifications:

1. *LM*: exact solutions can be obtained analytically for maximum likelihood estimators. If $\mathbf{V} = \mathbf{I}\sigma^2$, ML solutions are equivalent to ordinary least squares. For general $\mathbf{V}$, ML solutions are equivalent to generalized least squares.
2. *GLM*: estimating equations that follow from likelihood-based estimation can be written in exact form. Analytic solutions, however, do not exist—estimating equations must be solved iteratively.
3. *LMM*: there are two sets of estimating equations, one for the model effects $\boldsymbol{\beta}$ and $\mathbf{b}$—commonly referred to as the "mixed model equations"—and one for the vector of covariance components, $\boldsymbol{\sigma}$. The mixed model equations are maximum likelihood. For the fixed effects, $\boldsymbol{\beta}$, the mixed model equation solutions are equivalent to generalized least squares. Estimating equations for $\boldsymbol{\sigma}$ have two forms, one based on the full likelihood, the other based on the residual likelihood. The latter yield essentially unbiased estimates of the covariance components; the former yield biased estimates and are generally to be avoided. The latter yield REML estimates of $\boldsymbol{\sigma}$. The distinction between ML and REML applies only to estimates of $\boldsymbol{\sigma}$. Solutions for $\boldsymbol{\beta}$ and $\mathbf{b}$ are always maximum likelihood. The LMM estimating equations can be written exactly, but their solution requires iteration.
4. *GLMM*: estimation requires maximizing the marginal likelihood, which is generally intractable. Therefore, the estimating equations cannot be written exactly. The two approaches to doing this discussed in this chapter are pseudo-likelihood and integral approximation. Pseudo-likelihood essentially applies the LMM estimating equations to the pseudo-variate $\mathbf{y}^*$. Pseudo-likelihood therefore has ML and REML forms for estimation of $\boldsymbol{\sigma}$. Two integral approximation methods were discussed: Laplace and Gauss–Hermite quadrature. The former allows more flexibility and is less computationally intense; the latter is more accurate but complexity and computing demands limit its use for certain applications.

All true likelihood procedures yield a value of the likelihood that can be used for likelihood ratio tests and fit criteria, which are both discussed in Chapter 6. Thus, all LM, GLM, and LMM procedures yield a likelihood. For GLMMs, integral approximation methods yield a likelihood; pseudo-likelihood procedures do not.

All LM, GLM, and LMM estimating equations are special cases of the pseudo-likelihood estimating equations. However, these estimating equations are *pseudo*-likelihood *only* in the GLMM case—they are *not* "pseudo" in their LM, GLM, and LMM forms. So, for example, saying "we fit a Gaussian LMM using 'restricted pseudo-likelihood'" is incorrect.

Exercises

4.1 Let Y be a random variable whose p.d.f. is $f(y,\theta)$. Let $\ell(\theta;y)=\log[f(y,\theta)]$ denote the log-likelihood.

a. The score function is defined as $S=\partial\ell/\partial\theta$. Show that $E(S)=0$.

Two hints:
i. Recall that $\int f(y,\theta)dy=1$ and differentiate both sides w.r.t. θ
ii. Use the result $\partial\log[g(x)]/\partial x=g'(x)/g(x)$, where $g'(x)$ denotes $\partial g(x)/\partial x$.

b. For S defined as in (a) show that $\text{Var}(S)=-E(\partial S/\partial\theta)$.

c. Suppose the random variable Y is a member of the exponential family—and thus $\ell(\theta;y)=[(y\theta-b(\theta))/\phi]+c(y,\phi)$. Use (a) and (b), show
i. $b'(\theta)=\mu$, where $\mu=E(Y)$ and $b'(\theta)=\partial b(\theta)/\partial\theta$
ii. $\text{Var}(y)=\phi b''(\theta)$, where $b''(\theta)=\partial^2\ell/\partial\theta^2$

4.2 Let $\underline{\theta}$ be a $n\times 1$ vector and $\ell(\underline{\theta})$ be a $n\times 1$ vector whose ith element is $\ell(\theta_i)$, θ_i being the ith element of $\underline{\theta}$. For convenience, these vectors will be denoted θ and $\ell(\theta)$ for the rest of the problem. In matrix form, the second-order Taylor series expansion of $\ell(\theta)$ about θ^* is given by

$$\ell(\theta)\approx\ell(\theta^*)+\left[\left(\frac{\partial\ell}{\partial\theta'}\right)\Bigg|_{\theta^*}\right](\theta-\theta^*)+\frac{1}{2}(\theta-\theta^*)'\left[\left(\frac{\partial^2\ell(\theta)}{\partial\theta\partial\theta'}\right)\Bigg|_{\theta^*}\right](\theta-\theta^*)$$

From this expansion, derive the Newton–Raphson algorithm for obtaining the maximum likelihood estimate of θ follows. *Hint:* consider θ^* to be a starting value.

4.3 Consider a GLM, $\underline{\eta}=g(\underline{\mu})=x'\underline{\beta}$, for the observation y whose expected value is μ and variance is $\phi V(\mu)$. For notational convenience, assume lowercase Latin and Greek symbols are vectors and uppercase are matrices. The first-order Taylor series expansion of $g(y)$ about μ° is

$$g(y)\approx g(\mu^\circ)+\left[\frac{\partial g(y)}{\partial y'}\right]\Bigg|_{\mu^\circ}(y-\mu^\circ)$$

a. Show that it follows that $g(y)\approx X\beta^\circ+f(\mu^\circ)$, where $f(\mu^\circ)=\Delta(y-\mu^\circ)$, where $\Delta=[\partial g(y)/\partial y']\mu^\circ$. Note that $[\partial g(y)/\partial y']|\mu^\circ$ can alternatively be represented as $\Delta=[\partial\eta/\partial\mu']|\mu^\circ$

b. Show that $\text{Var}[(f(\mu^{\circ})] = \Delta \text{Var}(y) \Delta$
c. From (a) and (b) show that
   i. $E[g(y)] \approx X\beta$
   ii. $\text{Var}[(g(y)] \approx \Delta V \Delta$ where $V = \text{Var}(Y)$
d. Recall that for the LM $y = X\beta + e$, where $\text{Var}(e) = V$, the *GLS* (Generalized Least Squares, a.k.a., *Weighted* Least Squares) estimate of β is obtained by minimizing $(y - X\beta)' V(y - X\beta)$. By analogy, obtain the GLS estimate of β in the *GLM* defined in this problem using the Taylor series "pseudo-response variable" $g(y)$ in place of y.
e. How does your estimator in (d) compare to the maximum likelihood estimator developed from Fisher scoring?
f. From (e) derive the standard error of $k'\hat{\beta}$, assuming $k'\boldsymbol{\beta}$ is estimable.
g. How would you test $H_0{:}k'\boldsymbol{\beta} = 0$? (Give the form of the test statistic you would use and describe how you would evaluate statistical significance.)

4.4 Consider the estimate $\hat{\eta} = X\hat{\beta}$. The Delta rule states that $\text{Var}\left(\left[h(\hat{\eta})\right]\right) = [\partial\eta / \partial\mu']\text{Var}(\hat{\eta})[\partial\eta / \partial\mu]$, where $h(\eta)$ is the inverse link.

a. If the *link function*, $g(\boldsymbol{\mu})$, is the *natural parameter*, derive a simplified form of $\text{Var}\left[h(\hat{\eta})\right]$.
b. When would estimates of the form $h(\hat{\eta})$ be useful?

4.5 To fully understand the estimating equations, students should write programs in a matrix language, such as SAS PROC IML, that implements the estimating equations described in this chapter. Such an exercise can be applied to any of the models discussed in Chapters 1 through 3 for which data and GLIMMIX programs have been provided. Suggested: one GLM, one LMM, and one GLMM programming exercise.

The following problems refer to a SAS IML program that is provided in the SAS Data and Program Library entitled **`TwoWay MM withAB:`**

4.6 Modify the LMM IML program **`TwoWay MM withAB.sas`** so that it obtains REML estimates of the variance components using *Fisher scoring*. In addition to the variance components, use the program to obtain a complete solution of the mixed model equations.

4.7 Stay with the same data set as Problem 4.6 and stay with Fisher scoring. Modify the program so that it uses Fisher scoring to obtain ML rather than REML estimates of the variance components. Also, obtain a complete solution of the mixed model equations based on ML variance components estimates.

4.8 Using the data set **`TwoWayMMnoAB.sas`**, modify the REML IML program (Newton–Raphson or Fisher scoring—your choice) to obtain variance component and model parameter estimates for these data. Verify that your results agree with those obtained using GLIMMIX for the two-way, no-interaction mixed model.

5

Inference, Part I: Model Effects

5.1 Introduction

In Chapter 4, we focused on estimation. In Chapters 5 and 6, we turn our attention to inference. Chapter 5 concerns estimable and predictable functions defined on the linear predictor effects, $\boldsymbol{\beta}$ and $\mathbf{b}$. Chapter 6 concerns elements of $\boldsymbol{\sigma}$ and the variance and covariance components.

By inference, we mean interval estimation—the construction and interpretation of confidence intervals—and hypothesis testing. For the linear predictor effects, all inference begins with estimable and predictable functions. These were introduced in Chapter 3. They include treatment means, differences, odds ratios, contrasts, predicted values from regression models, main, simple, and interaction effects from factorial treatment structures, etc. For mixed models, all have population-averaged, broad-inference forms and subject-specific or narrow-inference forms. All models with nonidentity links are defined, and all inference takes place on the model scale. However, inference *results* can be *expressed* on the model or data scale.

As with estimation, in Chapter 3 we were intentionally vague about formal criteria required for estimability, distribution theory associated with interval estimates and hypothesis tests, and how the building blocks of inference—standard errors and test statistics—are computed. In this chapter, we fill in these details. Our goal in this chapter is to develop a strong, practical-application-oriented sense of what goes on "inside the generalized linear mixed model (GLMM) 'black box.'" In many cases, we will see that alternative approaches exist and literate use of GLMM inference requires knowing what tools go with what application—and when certain tools should *not* be used even though they can be computed.

Our goal in this chapter is to develop the essential thought processes that justify interval estimation and hypothesis testing procedures used in working with linear models. However, we will stop short of rigorous, formal theorem-proof mathematical statistics and probability theory. At several points, we draw on important matrix theory and methodology, especially generalized inverses and distribution theory for quadratic forms. Readers will find essential background in Appendix A: Matrix Operations, and Appendix B: Distribution Theory for Matrices.

5.2 Essential Background

Recall from Chapter 3 that a linear combination of the model effects, denoted $\mathbf{K}'\boldsymbol{\beta}+\mathbf{M}'\mathbf{b}$, is called a *predictable function* if $\mathbf{K}'\boldsymbol{\beta}$ is *estimable*. Also, $\mathbf{K}'\boldsymbol{\beta}$ is called an *estimable function*.

For fixed-effects-only models, we can only construct estimable functions; for mixed models, setting $\mathbf{M}=\mathbf{0}$ yields estimable functions. We use predictable functions for narrow or subject-specific inference and estimable functions for broad or population-averaged inference. In this section, we present formal criteria $\mathbf{K}'\boldsymbol{\beta}$ must satisfy to be considered estimable, general results regarding the distribution of estimators of estimable and predictable functions, and an overview of strategies leading to the two most common forms of inference: interval estimation and hypothesis testing.

5.2.1 Estimable and Predictable Functions

All estimable functions share the general form $\mathbf{K}'\boldsymbol{\beta}$. However, not all $\mathbf{K}'\boldsymbol{\beta}$ are estimable. The concept of estimability originated with ordinary least squares fixed-effects-only linear model theory—LM using our acronym. In principle, for $\mathbf{K}'\boldsymbol{\beta}$ to be estimable, we must be able to write it as a linear combination of $E(\mathbf{y})$. GLMMs use an extension of the same concept. In formal LM theory mathematical terms, estimable functions are defined as follows:

Definition (LM): $\mathbf{K}'\boldsymbol{\beta}$ is estimable if there exists a matrix $\mathbf{A}$ such that

$$\mathbf{A}'E(\mathbf{y}) = \mathbf{K}'\boldsymbol{\beta} \tag{5.1}$$

Noting that in the LM $E(\mathbf{y})=\mathbf{X}\boldsymbol{\beta}$ it follows that $\mathbf{A}'\mathbf{X}\boldsymbol{\beta}=\mathbf{K}'\boldsymbol{\beta}$ and, hence, $\mathbf{K}'$ must equal $\mathbf{A}'\mathbf{X}$. One way of testing candidate matrices for estimability is the use following result:

$$\mathbf{K}'\boldsymbol{\beta} \text{ is estimable if and only if } \mathbf{K}'(\mathbf{X}'\mathbf{X})^{-}\mathbf{X}'\mathbf{X} = \mathbf{K}' \tag{5.2}$$

where $(\mathbf{X}'\mathbf{X})^{-}$ denotes a *generalized inverse* of $\mathbf{X}'\mathbf{X}$. Solving the estimating equations developed in Chapter 4 requires generalized inverses for model with $\mathbf{X}$ less than full rank. See Appendix A, Matrix Operations, for more information about generalized inverses. Following Searle (1971), we prove the result as follows. First, suppose $\mathbf{K}'(\mathbf{X}'\mathbf{X})^{-}\mathbf{X}'\mathbf{X}=\mathbf{K}'$. Setting $\mathbf{A}' = \mathbf{K}'(\mathbf{X}'\mathbf{X})^{-}\mathbf{X}'$ yields $\mathbf{K}'(\mathbf{X}'\mathbf{X})^{-}\mathbf{X}'\mathbf{X} = \mathbf{A}'\mathbf{X} = \mathbf{K}'$, hence by definition $\mathbf{K}'\boldsymbol{\beta}$ is estimable. Now suppose $\mathbf{K}'\boldsymbol{\beta}$ is estimable. Then $\mathbf{K}' = \mathbf{A}'\mathbf{X}$. Multiplying by $(\mathbf{X}'\mathbf{X})^{-}(\mathbf{X}'\mathbf{X})$ we have $\mathbf{K}'(\mathbf{X}'\mathbf{X})^{-}(\mathbf{X}'\mathbf{X}) = \mathbf{A}'\mathbf{X}(\mathbf{X}'\mathbf{X})^{-}(\mathbf{X}'\mathbf{X}) = \mathbf{A}'\mathbf{X} = \mathbf{K}'$, which proves the result.

For example, in the two-treatment model with linear predictor $\mu+\tau_i$, $i=1,2$, the treatment mean and difference, $\mu+\tau_i$ and $\tau_1-\tau_2$, are estimable, but the intercept parameter, μ, and the treatment effect parameters, τ_i, are not by themselves estimable. We can view this in a couple of ways. First, apply the definition. In order to estimate μ we need to satisfy $\mu = \sum_{i,j} a_{ij}E(y_{ij}) = \sum_{i,j} a_{ij}(\mu+\tau_i) = \sum_{i,j} a_{ij}\mu + \sum_{i,j} a_{ij}\tau_i$. This requires $\sum_{i,j} a_{ij}$ to equal 1 (for μ) and 0 (for each τ_i) simultaneously, which is impossible. Second, we could apply the $\mathbf{K}'(\mathbf{X}'\mathbf{X})^{-}\mathbf{X}'\mathbf{X} = \mathbf{K}'$ criterion. For μ, $\mathbf{K}' = [1 \quad 0 \quad 0]$. But

$$\mathbf{K}'(\mathbf{X}'\mathbf{X})^{-}\mathbf{X}'\mathbf{X} = \begin{bmatrix} 2/3 & 1/3 & 1/3 \end{bmatrix} \neq \mathbf{K}'.$$

Why does estimability matter? Simply put, in models that are not of full rank—for example, all ANOVA-type effects models—estimating equation solutions for the effects themselves have no intrinsic meaning. Their solution depends entirely on the generalized inverse used, and theory tells us that there are infinitely many ways to construct a generalized inverse. On the other hand, estimable functions are invariant to choice of generalized inverse, and therefore have an assignable meaning. The effect estimates *per se* do *not* have *any* legitimate interpretation; estimable functions *do*.

5.2.1.1 Estimability and GLMMs

For linear mixed models (LMMs), extension of LM estimability is straightforward. The LMM models the conditional mean, that is, $E(\mathbf{y}|\mathbf{b}) = \mathbf{X\beta} + \mathbf{Zb}$. Therefore, $E(\mathbf{y}) = E[E(\mathbf{y}|\mathbf{b})] = \mathbf{X\beta}$. At this point, LM estimability theory applies directly.

> *Important consequence*: For mixed models (LMMs and GLMMs), estimability depends on *fixed effects* ***only***, that is, only on $\mathbf{K'\beta}$. Linear combinations of the random effects are not relevant to estimability criteria. This becomes very important when we consider blocking, the impact of defining blocks as fixed or random effects, and the extension of these issues to designs with multiple sources of random variation (e.g., split-plot and repeated measures data). We will consider these in detail in Chapters 8 and 14.

Generalized linear models (GLMs) and GLMMs that use nonidentity links do not model the expected value directly. Therefore, the definition of estimability as stated with the LM and LMM cannot be applied literally. Instead, think of the estimability criterion through the pseudo-variable $\mathbf{y}^*$. While $E(\mathbf{y}^*)$ is not an expected value *per se*, conceptually we understand $\mathbf{X\beta} + \mathbf{Zb}$ to be a model of $E(\mathbf{y}^*|\mathbf{b})$ in the sense that the inverse link $h(\mathbf{X\beta} + \mathbf{Zb})$ *does* model $E(\mathbf{y}|\mathbf{b})$. Following this line of reasoning, in the GLMM (and GLM), $\mathbf{K'\beta}$ is defined as estimable if there exists a matrix $\mathbf{A}$ such that $\mathbf{A}'E(\mathbf{y}^*) = \mathbf{K'\beta}$. Applying this definition, estimability depends entirely on the form of the linear predictor and the presence or absence of observations for given effect level combinations. In other words, the same functions $\mathbf{K'\beta}$ that are estimable for the LM and LMM are also estimable for the GLM and GLMM. Similarly, any $\mathbf{K'\beta}$ that is *not* estimable for the LM and LMM is also *not* estimable for the GLM and GLMM.

5.2.2 Basics of Interval Estimates and Test Statistics

Inference on measures of location in GLMM starts with predictable functions $\mathbf{K'\beta} + \mathbf{M'b}$, where $\mathbf{K'\beta}$ satisfies estimability criteria given earlier in Section 5.2.1. For convenience in this discussion, we use the notation $\boldsymbol{\psi} = \mathbf{K'\beta} + \mathbf{M'b}$ as shorthand for the predictable function. By inference we mean either testing a hypothesis, for example, H_0: $\boldsymbol{\psi} = \boldsymbol{\psi}_0$ vs. H_A: $\boldsymbol{\psi} \neq \boldsymbol{\psi}_0$, or obtaining an interval estimate of $\boldsymbol{\psi}$.

First, consider the cases where $\mathbf{K}$ and $\mathbf{M}$ are vectors, and hence $\psi = \mathbf{k'\beta} + \mathbf{m'b}$ is a scalar. The basic approach is essentially no different than the testing and confidence interval methods taught in introductory statistics. Specifically, we require two quantities: the point estimate and the standard error of the point estimate. Denote these as $\hat{\psi}$ and $s.e.(\hat{\psi})$, respectively. From these, we form the statistic $[(\hat{\psi} - E(\hat{\psi}))/s.e.(\hat{\psi})]$ and determine its distribution if an exact result exists or its approximate distribution otherwise. From this basic form, we can do the following:

- Test a hypothesis: Under the null hypothesis, H_0: $\psi = \psi_0$, $E(\hat{\psi}) = \psi_0$. We use the test statistic $[(\hat{\psi} - \psi_0)/s.e.(\hat{\psi})]$ or $[(\hat{\psi} - \psi_0)/s.e.(\hat{\psi})]^2$ as a criterion for assessing H_0. Typically, the reference distribution for $[(\hat{\psi} - \psi_0)/s.e.(\hat{\psi})]$ is $N(0,1)$—the standard

Gaussian (normal)—for known $s.e.(\hat{\psi})$ and the t-distribution for estimated $s.e.(\hat{\psi})$; for $[(\hat{\psi}-\psi_0)/s.e.(\hat{\psi})]^2$, the reference distributions are χ^2 and F, respectively, for known and estimated $s.e.(\hat{\psi})$.

- Construct a confidence interval: We use the generic formula $\hat{\psi} \pm$ ("table value") $\times\, s.e.(\hat{\psi})$, where the "table value" comes from $N(0,1)$ for known standard error and from the t-distribution when the standard error is estimated.

For the general case when $\boldsymbol{\psi}=\mathbf{K}'\boldsymbol{\beta}+\mathbf{M}'\mathbf{b}$ is a vector, we need matrix extensions of the distribution theory associated with $[(\hat{\psi}-E(\hat{\psi}))/s.e.(\hat{\psi})]$ and matrix forms of the test statistic $[(\hat{\psi}-\psi_0)/s.e.(\hat{\psi})]^2$ for testing hypotheses. We consider various forms of the test statistics in Section 5.3. We now develop the necessary distribution theory associated with $\hat{\boldsymbol{\psi}}$.

5.2.3 Approximate Distribution of Estimable and Predictable Functions

For the LM and certain marginal forms of the LMM, we can derive exact distributions of the estimable functions. Distribution results hold approximately for other linear models: LMMs in general and all generalized models—the GLM and GLMM. We now develop these results, starting with the LM.

5.2.3.1 Distribution of $\hat{\boldsymbol{\beta}}$ in the LM with Known V

Recall that we specify the LM by the linear predictor $\boldsymbol{\eta}=\boldsymbol{\mu}=\mathbf{X}\boldsymbol{\beta}$ and the response variable distribution $\mathbf{y} \sim N(\boldsymbol{\mu},\mathbf{V})$. Under the model, $\mathbf{y} \sim N(\mathbf{X}\boldsymbol{\beta},\mathbf{V})$. We estimate $\boldsymbol{\beta}$ using the generalized least squares estimating equations, $\mathbf{X}'\mathbf{V}^{-1}\mathbf{X}\boldsymbol{\beta}=\mathbf{X}'\mathbf{V}^{-1}\mathbf{y}$, hence $\hat{\boldsymbol{\beta}}=(\mathbf{X}'\mathbf{V}^{-1}\mathbf{X})^{-}\mathbf{X}'\mathbf{V}^{-1}\mathbf{y}$, where $(\mathbf{X}'\mathbf{V}^{-1}\mathbf{X})^{-}$ is a generalized inverse of $\mathbf{X}'\mathbf{V}^{-1}\mathbf{X}$. Note that if we assume the classical "general" LM, $\mathbf{V}=\mathbf{I}\sigma^2$ estimation reduces to ordinary least squares, that is, $\hat{\boldsymbol{\beta}}=(\mathbf{X}'\mathbf{X})^{-}\mathbf{X}'\mathbf{y}$. Also note that unless $\mathbf{X}$ is a full-rank matrix, $\hat{\boldsymbol{\beta}}$ is a *solution* not an *estimate*, in the sense that because the generalized inverse is not unique, $\hat{\boldsymbol{\beta}}$ is not unique either. If $\mathbf{X}$ *is* a full-rank matrix (e.g., cell-means and regression models), a true inverse exists: $(\mathbf{X}'\mathbf{V}^{-1}\mathbf{X})^{-}$ becomes $(\mathbf{X}'\mathbf{V}^{-1}\mathbf{X})^{-1}$ making $\hat{\boldsymbol{\beta}}$ unique in this case. We focus this discussion on the more general case where $\mathbf{X}$ is not of full rank. Any results discussed here for generalized inverses also apply as a special case to models for which a true inverse exists.

We estimate the estimable function by $\mathbf{K}'\hat{\boldsymbol{\beta}}=\mathbf{K}'(\mathbf{X}'\mathbf{V}^{-1}\mathbf{X})^{-}\mathbf{X}'\mathbf{V}^{-1}\mathbf{y}$. For known $\mathbf{V}$, we can use the following: If $\mathbf{y} \sim N(\boldsymbol{\mu},\mathbf{V})$ and $\mathbf{A}$ is a matrix of known constants, then $\mathbf{A}\mathbf{y} \sim N(\mathbf{A}\mu, \mathbf{A}\mathbf{V}\mathbf{A}')$. Setting $\mathbf{A}=\mathbf{K}'(\mathbf{X}'\mathbf{V}^{-1}\mathbf{X})^{-}\mathbf{X}'\mathbf{V}^{-1}$, we have $E(\mathbf{K}'\hat{\boldsymbol{\beta}})=E(\mathbf{A}\mathbf{y})=\mathbf{K}'(\mathbf{X}'\mathbf{V}^{-1}\mathbf{X})^{-}\mathbf{X}'\mathbf{V}^{-1}\boldsymbol{\mu}=\mathbf{K}'(\mathbf{X}'\mathbf{V}^{-1}\mathbf{X})^{-}\mathbf{X}'\mathbf{V}^{-1}\mathbf{X}\boldsymbol{\beta}=\mathbf{K}'\hat{\boldsymbol{\beta}}$ assuming $(\mathbf{X}'\mathbf{V}^{-1}\mathbf{X})^{-}$ is a reflexive generalized inverse (see Appendix A). Also, $Var(\mathbf{K}'\mathbf{y})=\mathbf{K}'(\mathbf{X}'\mathbf{V}^{-1}\mathbf{X})^{-}\mathbf{X}'\mathbf{V}^{-1}Var(\mathbf{y})\mathbf{V}^{-1}\mathbf{X}(\mathbf{X}'\mathbf{V}^{-1}\mathbf{X})^{-}\mathbf{K}=\mathbf{K}'(\mathbf{X}'\mathbf{V}^{-1}\mathbf{X})^{-}\mathbf{K}$, again assuming we use a reflexive generalized inverse. Therefore, for known $\mathbf{V}$

$$\mathbf{K}'\hat{\boldsymbol{\beta}} \sim N\left(\mathbf{K}'\boldsymbol{\beta},\mathbf{K}'(\mathbf{X}'\mathbf{V}^{-1}\mathbf{X})^{-}\mathbf{K}\right) \tag{5.3}$$

5.2.3.2 Distribution of the Quadratic Form Defined on $\hat{\boldsymbol{\beta}}$ for the LM with Known V

In Section 5.3, we will see that test statistics typically use some variation of the expression $(\mathbf{K}'\hat{\boldsymbol{\beta}})'\left[Var(\mathbf{K}'\hat{\boldsymbol{\beta}})\right]^{-1}\mathbf{K}'\hat{\boldsymbol{\beta}}$. We call this a *quadratic form*. By definition, for vector $\mathbf{v}$, $\mathbf{v}'\mathbf{A}\mathbf{v}$ is called a quadratic form of $\mathbf{v}$. Letting $A=\left[Var(\mathbf{K}'\hat{\boldsymbol{\beta}})\right]^{-1}$ gives us a quadratic form defined on $\mathbf{K}'\hat{\boldsymbol{\beta}}$. Recognize the logic behind this: in essence, the quadratic form for $\mathbf{K}'\hat{\boldsymbol{\beta}}$ is the matrix version of $[((\hat{\psi} - \psi_0)/s.e.(\hat{\psi}))]^2$. A standard result in matrix distribution theory involves quadratic forms of random vectors with a Gaussian distribution. Specifically, if $\mathbf{y} \sim N(\boldsymbol{\mu}, \mathbf{V})$ and $\mathbf{AV}$ is *idempotent*, meaning $\mathbf{AVAV} = \mathbf{AV}$, then the quadratic form $\mathbf{y}'\mathbf{Ay}$ has a noncentral chi-square distribution, with degrees of freedom equal to the rank of $\mathbf{A}$ and non-centrality parameter equal to $1/2\boldsymbol{\mu}'\mathbf{A}\boldsymbol{\mu}$, that is,

$$\mathbf{y}'\mathbf{Ay} \sim \chi^2_{\text{rank}(\mathbf{A}),1/2\hat{\boldsymbol{\mu}}'\mathbf{A}\boldsymbol{\mu}} \tag{5.4}$$

For the LM, $Var(\mathbf{K}'\hat{\boldsymbol{\beta}}) = \mathbf{K}'(\mathbf{X}'\mathbf{V}^{-1}\mathbf{X})^{-}\mathbf{K}$ and $\mathbf{A} = \left[\mathbf{K}'(\mathbf{X}'\mathbf{V}^{-1}\mathbf{X})^{-}\mathbf{K}\right]^{-1}$, meaning $\mathbf{AV} = \mathbf{I}$ and hence $\mathbf{AV}$ is idempotent. Therefore, for the LM with known $\mathbf{V}$, the quadratic form

$$\left(\mathbf{K}'\hat{\boldsymbol{\beta}}\right)'\left[\mathbf{K}'(\mathbf{X}'\mathbf{V}^{-1}\mathbf{X})^{-}\mathbf{K}\right]^{-1}\mathbf{K}'\hat{\boldsymbol{\beta}} \sim \chi^2_{r(\mathbf{K}),\varphi} \tag{5.5}$$

where

$r(\mathbf{K})$ denotes the rank of $\mathbf{K}$

φ denotes the non-centrality parameter $\varphi = 1/2(\mathbf{K}'\boldsymbol{\beta})'\left[\mathbf{K}'(\mathbf{X}'\mathbf{V}^{-1}\mathbf{X})\mathbf{K}\right]^{-1}\mathbf{K}'\boldsymbol{\beta}$

In practice, we rarely have a known $\mathbf{V}$. Recalling our notation from Chapter 4, we must estimate the vector covariance parameters, $\boldsymbol{\sigma}$, and use the estimated variance, $\hat{\mathbf{V}} = V(\hat{\boldsymbol{\sigma}})$.

5.2.3.3 LM with Unknown V

Following the notation introduced in Chapter 4, we refer to the estimated covariance matrix using the shorthand $\hat{\mathbf{V}}$. We consider two cases. The first: the special case when we can write $\mathbf{V} = \sigma^2\boldsymbol{\Sigma}$ assuming $\boldsymbol{\Sigma}$ is known and the only covariance parameter to be estimated is σ^2. The second: the more general case when $\boldsymbol{\sigma}$ involves parameters in addition to σ^2 and all of the parameters of $\boldsymbol{\sigma}$ must be estimated.

5.2.3.4 LM with Unknown V: Case 1—$\mathbf{V} = \sigma^2\boldsymbol{\Sigma}$

One obvious example of Case 1 occurs when $\boldsymbol{\Sigma} = \mathbf{I}$—the classical "general" LM. When $\boldsymbol{\Sigma} \neq \mathbf{I}$, $\boldsymbol{\Sigma}$ typically describes a correlation structure. If enough is known about the correlation structure to treat it as a constant, Case 1 applies.

For inference, the main results are what we would expect when we replace known σ^2 by its estimate, $\hat{\sigma}^2$. First, consider vector forms of $\mathbf{K}$ only: hence $\psi = \mathbf{k}'\boldsymbol{\beta}$ is a scalar. The generic form $[(\hat{\psi} - E(\hat{\psi}))/s.e.(\hat{\psi})]$ here equals $\mathbf{k}'\left(\hat{\boldsymbol{\beta}} - \boldsymbol{\beta}\right)\Big/\sqrt{\mathbf{k}'\left(\mathbf{X}\left(\boldsymbol{\Sigma}\hat{\sigma}^2\right)^{-1}\mathbf{X}\right)^{-}\mathbf{k}} = \mathbf{k}'\left(\hat{\boldsymbol{\beta}} - \boldsymbol{\beta}\right)\Big/\sqrt{\hat{\sigma}^2\left[\mathbf{k}'\left(\mathbf{X}(\boldsymbol{\Sigma})^{-1}\mathbf{X}\right)^{-}\mathbf{k}\right]}$ where $\hat{\sigma}^2$ is an estimate of σ^2. Because we replace σ^2 by $\hat{\sigma}^2$, we would expect the distribution to be t instead of $N(0, 1)$ and the distribution of its square to be F rather than chi-square, but we need to demonstrate this.

We know that $\mathbf{k}'(\hat{\boldsymbol{\beta}} - \boldsymbol{\beta}) \sim N(0, \sigma^2\mathbf{k}'(\mathbf{X}'\Sigma^{-1}\mathbf{X})^{-}\mathbf{k})$. Therefore $\mathbf{k}'(\hat{\boldsymbol{\beta}} - \boldsymbol{\beta})\Big/\sqrt{\sigma^2\left[\mathbf{k}'\left(\mathbf{X}(\Sigma)^{-1}\mathbf{X}\right)^{-}\mathbf{k}\right]} = \mathbf{k}'(\hat{\boldsymbol{\beta}} - \boldsymbol{\beta})/\sigma\Big/\sqrt{\left[\mathbf{k}'\left(\mathbf{X}(\Sigma)^{-1}\mathbf{X}\right)^{-}\mathbf{k}\right]} \sim N(0,1)$. Also, $\hat{\sigma}^2 = (\mathbf{y} - \mathbf{X}\hat{\boldsymbol{\beta}})'\Sigma^{-1}(\mathbf{y} - \mathbf{X}\hat{\boldsymbol{\beta}})/N - \text{rank}(\mathbf{X})$ is an unbiased estimate of σ^2. We can rewrite $\hat{\sigma}^2$ as $\left[\left(\mathbf{y}'\left(\mathbf{I} - \mathbf{X}(\mathbf{X}'\Sigma^{-1}\mathbf{X})^{-}\mathbf{X}'\Sigma^{-1}\right)\Sigma^{-1}\left(\mathbf{I} - \mathbf{X}(\mathbf{X}'\Sigma^{-1}\mathbf{X})^{-}\mathbf{X}'\Sigma^{-1}\right)\mathbf{y}\right)\Big/(N - \text{rank}(\mathbf{X}))\right]$. Observe now that $[N - \text{rank}(\mathbf{X})]\hat{\sigma}^2/\sigma^2$ is a quadratic form with $\mathbf{A} = \left(\mathbf{I} - \mathbf{X}(\mathbf{X}'\Sigma^{-1}\mathbf{X})^{-}\mathbf{X}'\Sigma^{-1}\right)'\Sigma^{-1}\left(\mathbf{I} - \mathbf{X}(\mathbf{X}'\Sigma^{-1}\mathbf{X})^{-}\mathbf{X}'\Sigma^{-1}\right)\Big/\sigma^2$ and $\mathbf{AV}$ is idempotent, therefore $[N-\text{rank}(\mathbf{X})]\hat{\sigma}^2/\sigma^2 \sim \chi^2_{N-\text{rank}(\mathbf{X})}$. Note that the non-centrality parameter is zero, yielding a central chi-square distribution. Finally, we use the result that if $\mathbf{y} \sim N(\boldsymbol{\mu}, \mathbf{V})$, $\mathbf{y}'\mathbf{A}\mathbf{y}$ and $\mathbf{B}\mathbf{y}$ are independent if $\mathbf{AVB}=\mathbf{0}$. Using $\mathbf{A}$ from $\hat{\sigma}^2$ and $\mathbf{B}=(\mathbf{X}'\Sigma^{-1}\mathbf{X})^{-}\mathbf{X}'\Sigma^{-1}$, it follows (with some matrix algebra) that $\hat{\boldsymbol{\beta}}$ and $\hat{\sigma}^2$ are independent. We now write the statistic $\mathbf{k}'(\hat{\boldsymbol{\beta}}-\boldsymbol{\beta})\Big/\sqrt{\hat{\sigma}^2\left[\mathbf{k}'\left(\mathbf{X}(\Sigma)^{-1}\mathbf{X}\right)^{-}\mathbf{k}\right]} = \left[\mathbf{k}'(\hat{\boldsymbol{\beta}}-\boldsymbol{\beta})/\sigma\Big/\sqrt{\left[\mathbf{k}'\left(\mathbf{X}(\Sigma)^{-1}\mathbf{X}\right)^{-}\mathbf{k}\right]}\right]\Big/$ $\sqrt{\left(\dfrac{[N-r(\mathbf{X})]\hat{\sigma}^2}{\sigma^2}\right)\Big/[N-r(\mathbf{X})]}$, that is, the ratio $Z\big/\sqrt{X^2/df}$, where Z and X^2 are independent random variables with $Z \sim N(0,1)$ and $X^2 \sim \chi^2_{df}$. Hence,

$$\frac{\mathbf{k}'\left(\hat{\boldsymbol{\beta}}-\boldsymbol{\beta}\right)}{\sqrt{\hat{\sigma}^2\left[\mathbf{k}'(\mathbf{X}(\Sigma)^{-1}\mathbf{X})^{-}\mathbf{k}\right]}} \sim t_{N-\text{rank}(\mathbf{X})} \tag{5.6}$$

When $\mathbf{K}$ is not a vector, we focus on the quadratic form from (5.5). Since $\mathbf{V}=\sigma^2\boldsymbol{\Sigma}$, rewrite (5.5) as $(\mathbf{K}'\hat{\boldsymbol{\beta}})'\left[\mathbf{K}'(\mathbf{X}'\Sigma^{-1}\mathbf{X})^{-}\mathbf{K}\right]^{-1}\mathbf{K}'\hat{\boldsymbol{\beta}}\big/\sigma^2$. Also, unknown σ^2 requires us to replace σ^2 with $\hat{\sigma}^2$. If we write the resulting statistic as $\dfrac{(\mathbf{K}'\hat{\boldsymbol{\beta}})'\left[\mathbf{K}'(\mathbf{X}'\Sigma^{-1}\mathbf{X})^{-}\mathbf{K}\right]^{-1}\mathbf{K}'\hat{\boldsymbol{\beta}}\big/\sigma^2}{\hat{\sigma}^2/\sigma^2}$ a little algebra reveals that when we replace σ^2 by $\hat{\sigma}^2$ in (5.5) and divide it by the rank of $\mathbf{K}$ we have the ratio $\dfrac{X_1^2/df_1}{X_2^2/df_2}$, where $X_i^2, i=1,2$ are independent random variables with $X_1^2 \sim$ non-central χ^2 and $X_?^2 \sim$ central χ^2. Therefore,

$$\frac{\left(\mathbf{K}'\hat{\boldsymbol{\beta}}\right)'\left[\mathbf{K}'(\mathbf{X}'\Sigma^{-1}\mathbf{X})^{-}\mathbf{K}\right]^{-1}\mathbf{K}'\hat{\boldsymbol{\beta}}\big/\hat{\sigma}^2}{\text{rank}(\mathbf{K})} \sim F_{\nu_1,\nu_2,\varphi} \tag{5.7}$$

where

$v_1=\text{rank}(\mathbf{K})$ and $\nu_2=N-\text{rank}(\mathbf{X})$ are the numerator and denominator degrees of freedom, respectively

φ is the non-centrality parameter defined earlier for (5.5)

5.2.3.5 LM with Unknown V: Case 2—All Covariance Components Must Be Estimated

Case 2 occurs with fixed-effects-only models whose covariance structure includes at least two unknown terms and mixed models using the marginal form. In the latter case, we saw in Chapter 4 that the linear predictor is $\boldsymbol{\eta}=\boldsymbol{\mu}=\mathbf{X}\boldsymbol{\beta}$ and, under the model, the assumed distribution of the observations is $\mathbf{y} \sim N(\mathbf{X}\boldsymbol{\beta}, \mathbf{ZGZ}'+\mathbf{R})$. For unknown $\mathbf{V}=\mathbf{ZGZ}'+\mathbf{R}$, at least one component from $\mathbf{G}$ and one from $\mathbf{R}$ must be estimated. For fixed-effects-only models, $\mathbf{V}=\mathbf{R}$, so Case 2 implies that $\mathbf{R}$ has at least two covariance components. Examples include compound symmetry (introduced in Chapter 3) and correlated error models—for example, for repeated measures or spatial data—which we will introduce in Chapter 6 and discuss in more detail in Chapter 14.

As we have seen so far, the two key building blocks for inference are

1. The distribution of $\mathbf{K}'\hat{\boldsymbol{\beta}}$
2. The distribution of $(\mathbf{K}'\hat{\boldsymbol{\beta}})'\left[\mathbf{K}'(\mathbf{X}'\hat{\mathbf{V}}^{-1}\mathbf{X})^{-}\mathbf{K}\right]^{-1}\mathbf{K}'\hat{\boldsymbol{\beta}}\,/\text{rank}(\mathbf{K})$

The latter is a generalization of (5.7) with $\hat{\mathbf{V}} = V(\hat{\boldsymbol{\sigma}})$ replacing $\boldsymbol{\Sigma}\hat{\sigma}^2$. We saw in Chapter 4 that we typically obtain $\hat{\boldsymbol{\sigma}}$ via REML estimation. When we do this, the two distributions in question are no longer Gaussian and F, respectively. Standard mixed model practice assumes

- For scalar element of $\mathbf{K}'\boldsymbol{\beta}$ $\dfrac{\mathbf{k}'\left(\hat{\boldsymbol{\beta}}-\boldsymbol{\beta}\right)}{\sqrt{\mathbf{k}'\left(\mathbf{X}'\hat{\mathbf{V}}^{-1}\mathbf{X}\right)^{-}\mathbf{k}}} \sim t_{\nu_2}$ (5.8)
- $\dfrac{\left(\mathbf{K}'\hat{\boldsymbol{\beta}}\right)'\left[\mathbf{K}'\left(\mathbf{X}'\hat{\mathbf{V}}^{-1}\mathbf{X}\right)^{-}\mathbf{K}\right]^{-1}\mathbf{K}'\hat{\boldsymbol{\beta}}}{\text{rank}(\mathbf{K})} \sim F_{\nu_1,\nu_2,\varphi}$ (5.9)

where

ν_1 denotes the numerator degrees of freedom for F and $\nu_1 = \text{rank}(\mathbf{K})$
ν_2 denotes the degrees of freedom for t and the denominator degrees of freedom for F
φ denotes the non-centrality parameter for F, as defined in (5.5)

For certain variance-component-only mixed models, and only for certain estimable functions from these models, we can determine the denominator degrees of freedom ν_2 from the skeleton analysis of variance, for example, by following the WWFD process described in Chapter 2. For all other cases, we must approximate ν_2. In Section 5.3.2, we will see how to do this.

The argument for approximations (5.8) and (5.9) is based on asymptotic theory. As $N \to \infty$, $V(\hat{\boldsymbol{\sigma}}) \to V(\boldsymbol{\sigma})$ (assuming $\hat{\boldsymbol{\sigma}}$ is a consistent estimate of $\boldsymbol{\sigma}$, which it is using ANOVA, ML or REML) and hence the limiting distribution of $\mathbf{K}'\hat{\boldsymbol{\beta}}$ is $N(\mathbf{K}'\boldsymbol{\beta}, \mathbf{K}'(\mathbf{X}'\mathbf{V}^{-1}\mathbf{X})^{-}\mathbf{K})$ and the limiting distribution of $(\mathbf{K}'\hat{\boldsymbol{\beta}})'\left[\mathbf{K}'(\mathbf{X}'\hat{\mathbf{V}}^{-1}\mathbf{X})^{-}\mathbf{K}\right]^{-1}\mathbf{K}'\hat{\boldsymbol{\beta}}\,/\text{rank}(\mathbf{K})$ is $\chi^2_{\nu_1,\varphi}$ and (we assume) it approaches the limit along the lines of $F_{\nu_1,\nu_2,\varphi} \to \chi^2_{\nu_1,\varphi}$ as $\nu_2 \to \infty$. For more detail on the asymptotic theory, see, for example, Vonesh and Chinchilli (1997), Demidenko (2004), and Jiang (2007).

In practice, our interest focuses more on the small sample behavior of these statistics rather that their asymptotic distributions. Why? We generally deal with data sets whose number of observations falls considerably short of the point where asymptotics come into play. Countless simulation studies, most of them unpublished, form something of an oral tradition among LMM practitioners dating from the early 1990s when SAS® introduced PROC MIXED. After the introduction of MIXED, other mixed model software began to proliferate allowing simulation studies to be easily done and become widespread. By the end of the 1990s, a consensus had emerged that approximations (5.8) and (5.9) perform as advertised with one caveat: this stems from Kackar and Harville (1984) and Kenward and Roger (1997) concerning $\mathbf{K}'(\mathbf{X}'\hat{\mathbf{V}}^{-1}\mathbf{X})^{-}\mathbf{K}$ as an approximation of $\mathbf{K}'(\mathbf{X}'\mathbf{V}^{-1}\mathbf{X})^{-}\mathbf{K}$. In general, $\mathbf{K}'(\mathbf{X}'\hat{\mathbf{V}}^{-1}\mathbf{X})^{-}\mathbf{K}$ has a downward bias. We discuss this issue; its consequences, if not addressed; and what to do about it in Section 5.3.3.

5.2.3.6 GLM

As with the LMM, we need to consider two cases. Case 1: the assumed distribution belongs to the one-parameter exponential family (e.g., binomial and Poisson); the mean determines the variance and we need no additional variance estimates. Case 2: we have either a two-parameter exponential family or quasi-likelihood; we must estimate the scale parameter(s).

5.2.3.7 GLM: Case 1—No Scale Parameter to Estimate

We use two asymptotic results:

- $\mathbf{K}'\hat{\boldsymbol{\beta}} \sim N\left(\mathbf{K}'\boldsymbol{\beta}, \mathbf{K}'(\mathbf{X}'\mathbf{W}\mathbf{X})^{-}\mathbf{K}\right)$ (5.10)
- $\left(\mathbf{K}'\hat{\boldsymbol{\beta}}\right)'\left[\mathbf{K}'(\mathbf{X}'\mathbf{W}\mathbf{X})^{-}\mathbf{K}\right]^{-1}\mathbf{K}'\hat{\boldsymbol{\beta}} \sim \chi^2_{r(\mathbf{K}),\varphi}$ (5.11)

where

W denotes the inverse variance of the pseudo-variable as defined by (4.13)
the non-centrality parameter $\varphi = 1/2(\mathbf{K}'\boldsymbol{\beta})'[\mathbf{K}'(\mathbf{X}'\mathbf{W}\mathbf{X})\mathbf{K}]^{-1}\mathbf{K}'\boldsymbol{\beta}$

Note that (5.10) and (5.11) are simply (5.3) and (5.5) replacing $\mathbf{V}^{-1}$ by $\mathbf{W}$.

Interested readers can refer to Wedderburn (1974), Vonesh and Chinchilli (1997), Demidenko (2004), and other textbooks for more in-depth development of the asymptotic theory. It suffices for our purposes that the accumulated body of experience with the small sample behavior of these approximations suggests that they perform as advertised. The rare exceptions will be noted as we discuss GLM applications in subsequent chapters.

Applying these two results, we construct confidence intervals for GLMs using the standard Gaussian [i.e., $N(0,1)$] distribution and test hypotheses using the chi-square distribution. For example, $k'\hat{\beta} \pm Z_{\alpha/2}\sqrt{k'(X'WX)^{-}k}$ gives a two-sided $100(1-\alpha)\%$ confidence interval *on the model scale*. Where meaningful, we obtain confidence limits on the *data scale* by applying the inverse link to the upper and lower model scale confidence limits. We further develop the approach to hypothesis testing with GLMs in Sections 5.3.2 and 5.3.3.

5.2.3.8 GLM: Case 2—Estimated Scale Parameter(s)

These GLMs occur in three contexts: (1) the response variable belongs to the two-parameter exponential family and the scale parameter must be estimated; (2) we define a quasi-likelihood with a scale parameter to account for overdispersion (as is often done with Poisson models); and (3) embed a *working correlation* matrix in **W** (a common strategy for non-Gaussian correlated error models). We discuss overdispersion in Chapter 11 and correlated error GLMs in Chapter 16. Correlated-error GLMs with working correlation matrices are widely known as *generalized estimating equation (GEE) models*. See Zeger et al. (1988) for early developmental work on GEEs and texts such as Diggle et al. (2002) and Hardin and Hilbe (2003) for a more extensive discussion. In general, GEE models are marginal forms of GLMMs—we encountered our first GEE, a binomial model with compound symmetry working correlation, in Chapter 3. GEE models are always quasi-likelihood models.

The basic results involve replacing $\mathbf{W}$ by $\hat{\mathbf{W}}$ in (5.10) and (5.11), which in turn means instead of $N(0, 1)$ we use t and instead of χ^2 we use F. Specifically,

- For scalar, $\dfrac{\mathbf{k}'(\hat{\boldsymbol{\beta}}-\boldsymbol{\beta})}{\sqrt{\mathbf{k}'(\mathbf{X}'\hat{\mathbf{W}}\mathbf{X})^{-}\mathbf{k}}} \sim t_{v_2}$ (5.12)
- $\dfrac{(\mathbf{K}'\hat{\boldsymbol{\beta}})'\left[\mathbf{K}'(\mathbf{X}'\hat{\mathbf{W}}\mathbf{X})^{-}\mathbf{K}\right]^{-1}\mathbf{K}'\hat{\boldsymbol{\beta}}}{\text{rank}(\mathbf{K})} \sim F_{v1,v2,\varphi}$ (5.13)

For models that involve only an estimated scale parameter, $\hat{\mathbf{W}} = \left[\mathbf{D}\mathbf{V}_\mu^{1/2}\hat{\mathbf{A}}\mathbf{V}_\mu^{1/2}\mathbf{D}\right]^{-1}$, where $\hat{\mathbf{A}} = diag\left[1/a(\hat{\phi})\right]$ and $\hat{\phi}$ denotes the estimated scale parameter. For models with working correlation, we replace the scale matrix $\mathbf{A}$ by the working correlation matrix, denoted by $\mathbf{A}_W$. The working correlation model typically depends on multiple parameters and each parameter must be estimated. For example, for the compound symmetry working correlation introduced in Chapter 3, $\mathbf{A}_W = \phi\begin{bmatrix} 1 & \rho & \cdots & \rho \\ & 1 & \cdots & \rho \\ & & \cdots & \cdots \\ & & & 1 \end{bmatrix}$. Follow our notation convention for covariance components, denote the vector of working correlation components by $\boldsymbol{\rho}$—for example, for compound symmetry $\boldsymbol{\rho}' = [\phi, \rho]$. We then fully specify the working correlation matrix as $\mathbf{A}_W(\boldsymbol{\rho})$ and its estimate by $\hat{\mathbf{A}}_W = \mathbf{A}_W(\hat{\boldsymbol{\rho}})$, where $\hat{\boldsymbol{\rho}}$ is the vector of estimated working correlation components, for example, $\hat{\boldsymbol{\rho}}' = [\hat{\phi}\,\hat{\rho}]$. We can estimate the components of $\boldsymbol{\rho}$ by including them as terms in the pseudo-likelihood variance estimation procedures described in Chapter 4.

5.2.3.9 Mixed Models

For the purposes of this discussion, we can consider the Gaussian LMM and the GLMM together. Both use the same asymptotic results although the history of their development differs. Results for LMMs draw on work that traces back at least to Eisenhart (1947). The seminal papers are Harville (1976) and Laird and Ware (1982). Results for GLMMs are more recent and tend to be part asymptotic theory and part *ad hoc* modification of LMM procedure for GLMM applications. As with the LM with unknown variance and the GLM, an accumulating body of simulation work suggests that, with some caveats we will discuss in Sections 5.3 and 5.4, these approximations perform as advertised.

We will present the basic results in their GLMM, pseudo-likelihood form, noting that the LMM equations are special cases resulting from LMMs' use of the identity link and the fact that the LMM likelihood in its conditional and marginal forms are Gaussian and easy to work with using standard likelihood methodology.

Recall the GLMM estimating equations $\begin{bmatrix} \mathbf{X}'\mathbf{W}\mathbf{X} & \mathbf{X}'\mathbf{W}\mathbf{Z} \\ \mathbf{Z}'\mathbf{W}\mathbf{X} & \mathbf{Z}'\mathbf{W}\mathbf{Z}+\mathbf{G}^{-1} \end{bmatrix}\begin{bmatrix} \boldsymbol{\beta} \\ \mathbf{b} \end{bmatrix} = \begin{bmatrix} \mathbf{X}'\mathbf{W}\mathbf{y}^* \\ \mathbf{Z}'\mathbf{W}\mathbf{y}^* \end{bmatrix}$, where $\mathbf{W} = \left(\mathbf{D}\mathbf{V}_\mu^{1/2}\mathbf{A}\mathbf{V}_\mu^{1/2}\mathbf{D}\right)^{-1}$, $\mathbf{D} = \partial\boldsymbol{\mu}/\partial\boldsymbol{\eta}$, and $\mathbf{y}^* = g(\tilde{\boldsymbol{\mu}}) + \mathbf{D}(\mathbf{y} - \tilde{\boldsymbol{\mu}})$. The following are basic results for inference on $\mathbf{K}'\boldsymbol{\beta} + \mathbf{M}'\mathbf{b}$, assuming that $\mathbf{K}'\boldsymbol{\beta}$ satisfies estimability criteria:

- $\mathbf{K}'\hat{\boldsymbol{\beta}}+\mathbf{M}'\hat{\mathbf{b}}$ is the **e-BLUP**—estimated best linear unbiased predictor—of $\boldsymbol{\psi}=\mathbf{K}'\boldsymbol{\beta}+\mathbf{M}'\mathbf{b}$. $\hat{\boldsymbol{\beta}}$ and $\hat{\mathbf{b}}$ are the solutions to the GLMM estimating equations. When $\mathbf{M}=\mathbf{0}$, $\mathbf{K}'\hat{\boldsymbol{\beta}}$ is the best linear unbiased estimate (BLUE) of $\mathbf{K}'\boldsymbol{\beta}$ for Gaussian models and "approximately BLUE" for GLMMs.

The following results apply when $\boldsymbol{\sigma}$*, the vector of components of G and R, is known.*

- Let $\mathbf{C}$ denote the generalized inverse of the left-hand side of the GLMM estimating equations, that is, $\mathbf{C}=\begin{bmatrix}\mathbf{X}'\mathbf{W}\mathbf{X} & \mathbf{X}'\mathbf{W}\mathbf{Z}\\ \mathbf{Z}'\mathbf{W}\mathbf{X} & \mathbf{Z}'\mathbf{W}\mathbf{Z}+\mathbf{G}^{-1}\end{bmatrix}^{-}$. Let $\mathbf{L}'=[\mathbf{K}'\ \mathbf{M}']$.

$$Var\left[\mathbf{K}'\hat{\boldsymbol{\beta}}+\mathbf{M}'\left(\hat{\mathbf{b}}-\mathbf{b}\right)\right]=Var\left(\mathbf{L}'\begin{bmatrix}\hat{\boldsymbol{\beta}}\\ \left(\hat{\mathbf{b}}-\mathbf{b}\right)\end{bmatrix}\right)=\mathbf{L}'\mathbf{C}\mathbf{L} \tag{5.14}$$

where $\mathbf{b}$ is the realized value of the random effect vector. The full derivation of this result appears in the Appendix of Henderson (1975). A summary of the derivation appears in Appendix B.

If $\mathbf{K}$ and $\mathbf{M}$, and hence $\mathbf{L}$, are vectors, then $\mathbf{L}'\mathbf{C}\mathbf{L}$ is a scalar and $\sqrt{\mathbf{L}'\mathbf{C}\mathbf{L}}=s.e.(\hat{\psi})$. The sampling distribution of $\mathbf{K}'\hat{\boldsymbol{\beta}}+\mathbf{M}'(\hat{\mathbf{b}}-\mathbf{b})$ is approximately $N(\mathbf{K}'\boldsymbol{\beta}+\mathbf{M}'\mathbf{b},\ \mathbf{L}'\mathbf{C}\mathbf{L})$ or, equivalently, approximately $N(\boldsymbol{\psi},\ \mathbf{L}'\mathbf{C}\mathbf{L})$, and it follows that

- $[(\hat{\psi}-\psi)/s.e.(\hat{\psi})]$ has an approximate $N(0,1)$ distribution
- $Z=[(\hat{\psi}-\psi_0)/\sqrt{L'CL}]$ can be used as a *z-statistic* for testing hypotheses of the form $H_0\colon \psi=\psi_0$
- The *confidence interval* for ψ is $\hat{\psi}=(\mathbf{K}'\hat{\boldsymbol{\beta}}+\mathbf{M}'\hat{\mathbf{b}})\pm Z_{\alpha/2}\sqrt{\mathbf{L}'\mathbf{C}\mathbf{L}}$, where $1-\alpha$ reflects the level of confidence
- When $\mathbf{K}$ and $\mathbf{M}$, and hence $\mathbf{L}$, are multicolumn matrices, not vectors,

$$\mathbf{L}'\begin{bmatrix}\hat{\boldsymbol{\beta}}\\ \left(\hat{\mathbf{b}}-\mathbf{b}\right)\end{bmatrix}\left(\mathbf{L}'\mathbf{C}\mathbf{L}\right)^{-1}\begin{bmatrix}\hat{\boldsymbol{\beta}}\\ \left(\hat{\mathbf{b}}-\mathbf{b}\right)\end{bmatrix}'\mathbf{L}\sim\chi^2_{\text{rank}(\mathbf{L})} \tag{5.15}$$

To repeat, all the aforementioned results assume that the covariance components are known, and hence $\mathbf{W}$ and $\mathbf{C}$ are known.

Results that apply when the covariance components are NOT known.

This is the more realistic case. Almost without exception in GLMMs, the covariance components are not known and must be estimated. We saw in Chapter 4 that when this happens, in the LMM we replace $\mathbf{G}$ and $\mathbf{R}$ by their estimates $\hat{\mathbf{G}}=\mathbf{G}(\hat{\boldsymbol{\sigma}})$ and $\hat{\mathbf{R}}=\mathbf{R}(\hat{\boldsymbol{\sigma}})$. In the GLMM, instead of replacing $\mathbf{R}$ by $\hat{\mathbf{R}}$ we replace $\mathbf{W}$ by $\hat{\mathbf{W}}$ if the model has unknown scale or working correlation components to estimate. Effectively, this means replacing $\mathbf{C}$ by $\hat{\mathbf{C}}$, using $\hat{\mathbf{G}}$, $\hat{\mathbf{R}}$, and $\hat{\mathbf{W}}$ as needed. Doing so yields the following:

- $$Var\left[\mathbf{K}'\hat{\boldsymbol{\beta}} + \mathbf{M}'\left(\hat{\mathbf{b}} - \mathbf{b}\right)\right] = Var\left(\mathbf{L}'\begin{bmatrix}\hat{\boldsymbol{\beta}} \\ \left(\hat{\mathbf{b}} - \mathbf{b}\right)\end{bmatrix}\right) \cong \mathbf{L}'\hat{\mathbf{C}}\mathbf{L} \quad (5.16)$$
- For scalar $\mathbf{L}$, $$t = \frac{\hat{\psi} - \psi_0}{\sqrt{\mathbf{L}'\hat{\mathbf{C}}\,\mathbf{L}}} \sim t_{\nu_2} \quad (5.17)$$

where ν_2 denotes the degrees of freedom involved in estimating $\mathbf{C}$. In general, we must approximate ν_2—see Section 5.4.2. It follows from (5.17) that a two-sided *confidence interval* for scalar ψ is $\hat{\psi} = (\mathbf{K}'\hat{\boldsymbol{\beta}} + \mathbf{M}'\hat{\mathbf{b}}) \pm t_{\nu_2,\alpha/2}\sqrt{\mathbf{L}'\mathbf{C}\mathbf{L}}$, where $1 - \alpha$ defines the level of confidence. The basic result for hypothesis testing is

$$\frac{\mathbf{L}'\begin{bmatrix}\hat{\boldsymbol{\beta}} \\ \left(\hat{\mathbf{b}} - \mathbf{b}\right)\end{bmatrix}\left(\mathbf{L}'\hat{\mathbf{C}}\mathbf{L}\right)^{-1}\begin{bmatrix}\hat{\boldsymbol{\beta}}' \\ \left(\hat{\mathbf{b}} - \mathbf{b}\right)'\end{bmatrix}\mathbf{L}}{\text{rank}(\mathbf{L})} \sim F_{\nu_1,\nu_2,\varphi} \quad (5.18)$$

where

$\nu_1 = \text{rank}(\mathbf{L})$

ν_2 generally requires approximation (see 5.4.2)

φ denotes the non-centrality parameter

Because $\hat{\mathbf{b}} - \mathbf{b}$ has zero expectation, the non-centrality parameter has the same form as the general case of the LM (for Gaussian models and more generally models with identity link) or GLM (for models with nonidentity link).

This concludes the section on distributions of key statistics associated with estimable and predictable functions. We now turn our attention to using these results to test hypotheses.

5.3 Approaches to Testing

Linear models use two primary approaches—likelihood ratio (LR) and Wald-based—for testing hypotheses about estimable and predictable functions, that is, tests whose null hypothesis can be written as H_0: $\boldsymbol{\psi} = \boldsymbol{\psi}_0$. In Section 5.3.1, we discuss LR tests. In Section 5.3.2, we discuss Wald-based tests. For LMs with known variance or where $\mathbf{V} = \sigma^2\boldsymbol{\Sigma}$ and σ^2 is the only unknown variance component, we show that LR and Wald-based tests are identical. For GLMMs, LR tests are defined only when we estimate model effects using integral approximation (e.g., Laplace and Gauss–Hermite quadrature). For pseudo-likelihood estimation, the LR is undefined; you *can* compute a *pseudo*-likelihood ratio, but it has dubious meaning and, therefore, should not be used. Even with integral approximation, LR testing has limited value because of its computational intensity. For this reason, Wald-tests are generally more practical and, thus, preferred for true GLMMs. For GLMs and true LMMs, there are trade-offs. Wald tests are more convenient. Small sample behavior favors LR testing in the sense that when there is a difference between LR and Wald-based tests, the difference favors LR testing. For most cases, no clear evidence favors LR over Wald-based testing or vice-versa. Thus, for most GLM and LMM applications, convenience becomes the overriding consideration. With the exception of testing covariance components (Chapter 6) we will use Wald-based statistics for most of the applications discussion beginning in Chapter 7.

5.3.1 Likelihood Ratio and Deviance

In its simplest form, the LR test involves determining the likelihood under H_0, that is, when the estimable or predictable function equals $\boldsymbol{\psi}_0$ and the likelihood using the maximum likelihood estimate of the linear predictor $\boldsymbol{\eta}$ obtained from the estimating equations developed in Chapter 4, and using their ratio to assess the hypothesis. In LR language, we refer to the likelihood under the null hypothesis as the likelihood under the "reduced model" and the likelihood under the alternative hypothesis as the likelihood under the "full" model. Here, for clarity, we denote the likelihoods for the reduced (H_0) and full (H_A) models, respectively, by $f(\hat{\boldsymbol{\psi}}_0;\mathbf{y})$ and $f(\hat{\boldsymbol{\eta}};\mathbf{y})$. We define the LR as $\Lambda = f(\hat{\boldsymbol{\psi}}_0;\mathbf{y})/f(\hat{\boldsymbol{\eta}};\mathbf{y})$. When $\mathbf{y}$ belongs to the exponential family, the general form of the likelihood follows (4.1). Wedderburn (1974) showed that you can compute a quasi-LR—as the name implies, replace the likelihood by the quasi-likelihood—and proceed as if you have a true likelihood ratio.

For certain distributions, for example, the Gaussian, we can determine an exact distribution for the likelihood ratio. In general we cannot. For most distributions (and quasi-likelihoods) LR tests use the well-known result that $-2\log(\Lambda) \sim \chi^2_v$. For tests of estimable and predictable functions, the chi-square degrees of freedom $v = \text{rank}(\mathbf{K})$. This means that, in practice, the LR statistic is

$$-2\log(\Lambda) = 2\ell(\hat{\boldsymbol{\eta}}) - 2\ell(\hat{\boldsymbol{\psi}}_0) \tag{5.19}$$

where $\ell(\cdot)$ denotes the log-likelihood (or log quasi-likelihood).

A special case of the LR that is important for GLMs is called the *deviance*. The deviance is defined as −2 multiplied by the log ratio of the likelihood under the observation vector, $\mathbf{y}$, and the likelihood under the model, $\hat{\boldsymbol{\eta}} = \mathbf{X}\hat{\boldsymbol{\beta}}$. That is, $\text{deviance}(\hat{\boldsymbol{\eta}}) = -2\log(f(\hat{\boldsymbol{\eta}};\mathbf{y})/f(\mathbf{y};\mathbf{y})) = 2[\ell(\mathbf{y}) - \ell(\hat{\boldsymbol{\eta}})]$.

For GLMs with $\mathbf{y}$ belonging to one-parameter exponential family, the deviance has a clear interpretation as a measure of goodness-of-fit for the model. Because it has an approximate chi-square distribution, the deviance can be used to formally test goodness-of-fit. We determine the degrees of freedom by taking the difference between the number of observations and the number of parameters under the full model, that is, $N - \text{rank}(\mathbf{X})$. For models from two-parameter exponential families, while the deviance is often used to evaluate goodness-of-fit, its meaning is less clear given that an estimated scale parameter is involved.

For Gaussian models, the deviance is *not* a measure of goodness of fit. In fact, we can show that it is the *SS*(residual) in analysis of variance terms.

For GLMs (and for LMs with known variance), we can use the deviance to test H_0: $\boldsymbol{\psi} = \boldsymbol{\psi}_0$. We do this by comparing the deviance for the full model to the deviance for the reduced model. This gives us $\text{deviance}(\boldsymbol{\psi}_0) - \text{deviance}(\boldsymbol{\eta}) = 2[\ell(\mathbf{y}) - \ell(\hat{\boldsymbol{\psi}}_0)] - 2[\ell(\mathbf{y}) - \ell(\hat{\boldsymbol{\eta}})] = 2[\ell(\hat{\boldsymbol{\eta}}) - \ell(\hat{\boldsymbol{\psi}}_0)]$, which is the LR statistic from (5.19).

Because of the relationship between the deviance and the LR statistic, the LR approach tends to encourage sequential testing for models with more than one effect. Sequential testing may be desirable—for example, with multiple regression models whose effects have an obvious hierarchy—but it may produce nonsense results, or at least results that can easily be misinterpreted. We will explore examples to illustrate in Section 5.3.3.

5.3.2 Wald and Approximate F-statistics

Wald statistics are straightforward applications of results developed in Section 5.2. The general form of a Wald statistic is $\hat{\boldsymbol{\psi}}'[Var(\hat{\boldsymbol{\psi}})]^{-1}\hat{\boldsymbol{\psi}}$, where $Var(\hat{\boldsymbol{\psi}})$ is known. The most general form of the Wald statistic for linear models is (5.15). Equation 5.15 is the Wald statistic for the GLMM with known variance. Equations 5.5 and 5.11 are special cases of (5.15) for the LM with known variance and the GLM with known scale parameter, respectively. Wald statistics are assumed to have a chi-square distribution with degrees of freedom determined by the rank of $\boldsymbol{\psi}$, that is, the rank to the $\mathbf{K}$ matrix that defines $\boldsymbol{\psi}$.

In practice, we know that $Var(\hat{\boldsymbol{\psi}})$ is not known but must be estimated. We saw in Section 5.2 that we can replace the variance components that comprise $Var(\hat{\boldsymbol{\psi}})$ by their estimates to obtain an the "estimated" Wald statistic $\hat{\boldsymbol{\psi}}'\left[\widehat{Var}(\hat{\boldsymbol{\psi}})\right]^{-1}\hat{\boldsymbol{\psi}}$, where $\widehat{Var}(\hat{\boldsymbol{\psi}})$ denotes the estimated $Var(\hat{\boldsymbol{\psi}})$, and then divide the "estimated" Wald statistic by the rank of $\boldsymbol{\psi}$. The result, $\left[\hat{\boldsymbol{\psi}}'\left[\widehat{Var}\left(\hat{\boldsymbol{\psi}}\right)\right]^{-1}\hat{\boldsymbol{\psi}}\right]\Big/\text{rank}(\boldsymbol{\psi})$ has an approximate—and in some cases an exact—F distribution. Equation 5.18 gives the most general expression of the approximate $F = Wald/\text{rank}(\mathbf{K})$ statistic for linear models. All of the other forms, Equations 5.7, 5.9, and 5.13, are special cases.

The Wald and approximate F have two major advantages in linear model inference. First, they do not require estimating two different models, whereas LR tests require estimating the full and reduced model (and the likelihood under the data if we use the deviance approach). Also, we define the Wald and approximate F explicitly for the estimable or predictable function of interest, so we know exactly what we are testing. The LR approach (and, by the way, the reduction sum of square approach from classical analysis of variance) encourage tests based on sequential fitting. As we will see in the next section, it is not always clear what we are actually testing when we use a sequential testing approach.

Finally, for GLMMs using pseudo-likelihood estimation, the LR statistic does not exist. You can compute a *pseudo*-LR from the pseudo-likelihood for the full model and the pseudo-likelihood for the reduced model, but each pseudo-likelihood is based on a different $\mathbf{y}^*$, one that is unique to that particular model. Therefore, it is not clear what—if anything—a pseudo-LR means. In any event, it is not a suitable basis for hypothesis testing. On the other hand, the Wald and approximate F are well-defined for the GLMM and the approximate F appears to perform well in simulation studies. Figure 5.1 shows the result of one such simulation for the two-treatment paired design with 10 pairs. The linear predictor is $\text{logit}(\pi_{ij}) = \eta + \tau_i + p_j$ with p_j i.i.d. $N\left(0, \sigma_P^2\right)$—the figure shows the empirical p.d.f. of the approximate F-statistic for H_0: $\tau_1 = \tau_2$ plotted against the actual central $F_{1,9}$ p.d.f.

5.3.2.1 A Special Case: The Gaussian LM with $V = I\sigma^2$

When we introduced the LR test, we said that in most cases we cannot determine an exact distribution for the LR so we depend on result that $-2\log(\Lambda)$ has an approximate χ^2 distribution. The Gaussian LM is one model for which we can determine an exact distribution. In fact, with certain qualifications, we can show that for the Gaussian LM, the LR and Wald tests are equivalent.

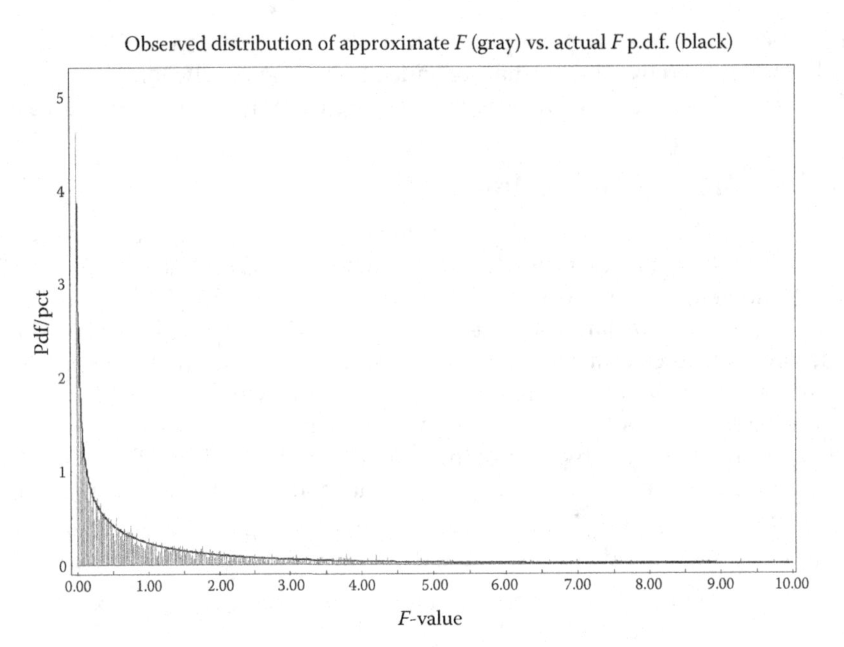

FIGURE 5.1
Plot of approximate F empirical p.d.f. vs. actual F p.d.f. for two-trt paired logistic model.

Suppose we want to test H_0: $\mathbf{K}'\boldsymbol{\beta}=\mathbf{K}'\boldsymbol{\beta}_0$. We first compute the deviance under the full model. Doing so yields deviance($\mathbf{K}'\hat{\boldsymbol{\beta}}$), which for the Gaussian LM is equal to SS(residual) under the full model. Denote this term SSR_F. Then we compute the deviance under the null hypothesis: deviance($\mathbf{K}'\boldsymbol{\beta}_0$) $= SSR_0$, where SSR_0 denotes SS(residual) under the null hypothesis. The LR statistic is the difference between the two deviance terms, which, in sums of squares terms, can be written $SSR_0 - SSR_F = SSH$, where SSH is equal to the sum of squares for the contrast defined by the estimable function $\mathbf{K}'\boldsymbol{\beta}$. We know that SSH is a quadratic form and that $SSH/\sigma^2 \sim \chi^2_{\text{rank}(\mathbf{K})}$. If we knew the variance, we could use it and legitimately perform a chi-square test. If we do not know σ^2, then testing H_0 using the deviance statistic as is obviously cannot be defended (unless, perhaps, σ^2 happens to be close to 1).

Now, it is easy to show that the Wald statistic for H_0: $\mathbf{K}'\boldsymbol{\beta}=\mathbf{K}'\boldsymbol{\beta}_0$ is also SSH/σ^2 (i.e., the Wald and LR statistics are identical for the Gaussian LM), so we face the same quandary that we face with the LR statistic regarding the unknown σ^2. What do we do? Following the process that led to (5.7), we replace σ^2 by $\hat{\sigma}^2$. The REML estimate of σ^2 happens to be $SSR_F/[N - \text{rank}(\mathbf{X})]$, which we also know as MS(residual) under the full model. Replacing σ^2 by $\hat{\sigma}^2$ gives us SSH/MSR_F. We do not know the distribution of this statistic, but if we divide it by rank($\mathbf{K}$), we have MSH/MSR_F, which we immediately recognize as an F-statistic or, in more formal distribution theory terms, as the ratio of two independent chi-square random variables divided by their respective degrees of freedom—that is, a random variable with an F distribution. Notice that SSH/MSR_F is a simple form of *Wald* and the F-statistic is simply *Wald*/rank($\mathbf{K}$).

5.3.3 Multiple Effect Models and Order of Testing

Classical ANOVA-based linear model theory distinguishes between "partial" and "sequential" sums of squares. Partial sums of squares are also referred to as "adjusted" sums of squares. The SAS® linear model procedures (GLM, GENMOD, MIXED, and GLIMMIX) refer to sequential sums of squares (or procedures that use the estimable functions implicit in sequential procedures) as "Type I SS" and partial or adjusted sums of squares (or their GLM, LMM, and GLMM analogs) as "Type II SS" or, more commonly, "Type III SS."

For the classical "general" LM, sequential and partial SS more often serve to mystify and confuse students and practitioners than they serve the cause of sound and appropriate use of linear model methodology. GLMs, LMMs, and GLMMs exacerbate the problem because sums of squares and mean squares have no meaning. The idea of fitting models sequentially does have some meaning when used in the proper context—fitting models for which there is an obvious order of increasing complexity, such as polynomial regression. Sequential and partial lose their meaning when applied to treatment effect models. For these models, it is better to think in terms of estimable functions: What values of **K** define linear combinations of the model parameters that (1) address actual inference objectives in a clear, straightforward, and meaningful way and (2) are indeed estimable?

We will pursue this theme in greater detail, providing additional theoretical background, in Chapter 7. Here, we present two examples to illustrate the issues. Their relevance stems from the fact that LR testing tends to encourage a sequential mindset, whereas Wald and approximate *F* statistics tend to encourage an estimable function mindset. Each has its advantages, and each can lead to utter nonsense. It is important to know the difference.

Example 5.1: Binomial response, polynomial multiple regression

The data for this example appear in the SAS Data and Program Library as Data Set 5.1. The predictor variable is *X*, and the response variable is binomial. There are two binomial observations at each level of *X*: *N* denotes the number of independent Bernoulli trials for a given observation and *F* denotes the number of "successes." Figure 5.2 shows a plot of the observed logits, $\log\left(\frac{F/N}{1-(F/N)}\right)$ by X. We can see that pattern of observed logits over *X* is clearly quadratic.

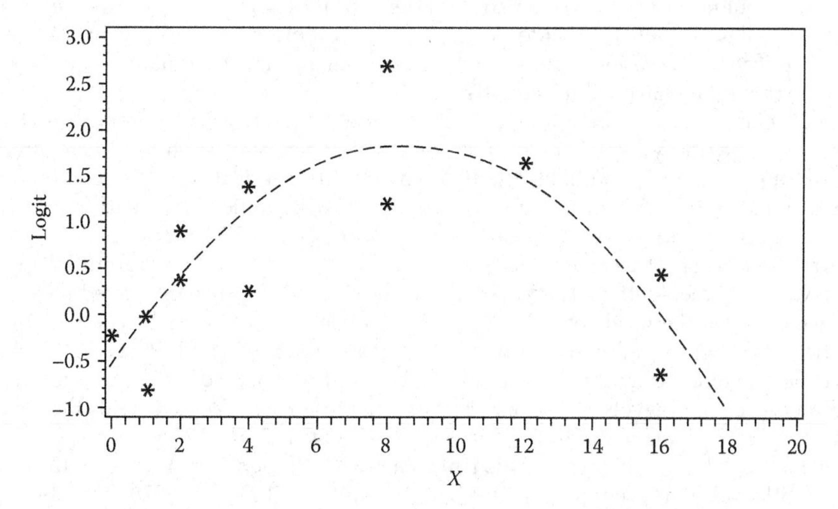

FIGURE 5.2
Plot of observed logit by *X*.

Implementing a sequential fit, we start with the linear predictor $\eta = \beta_0 + \beta_1 X$ and the link $\eta = \text{logit}(\pi)$. For this linear-effect-only logistic model, the LR statistic to test H_0: $\beta_1 = 0$ is 8.49 with a p-value of 0.0036 and the Wald statistic is 8.26 with a p-value of 0.004. However, the deviance is 95.61 with 12 d.f. indicating substantial lack of fit—not surprising given the quadratic pattern observed in Figure 5.2.

Next, we add the quadratic term: the linear predictor is $\eta = \beta_0 + \beta_1 X + \beta_2 X^2$. Now we test H_0: $\beta_2 | \beta_1 = 0$, that is, the additional variability explained by the quadratic effect over and above the linear effect. The LR statistic for this test is 63.85 with a $p < 0.0001$ and the Wald statistic is 58.45 $p < 0.0001$. The deviance is 31.75 with 11 d.f. When we add cubic, quartic, and quintic terms, none have a p-value approaching statistical significance. All this makes sense.

On the other hand, the so-called partial approach involves testing the impact of each potential regression effect after all other regression effects have been fit. For example, we would evaluate the impact of adding the linear term, $\beta_1 X$, *after* fitting the quadratic, cubic, quartic, and quintic terms. This is illogical for two reasons: first, it assumes that fitting the linear predictor $\eta = \beta_0 + \beta_2 X^2 + \beta_3 X^3 + \beta_4 X^4 + \beta_5 X^5$ makes sense (it does not) and, second, it assumes that testing $\beta_1 X$ at this point is a reasonable thing to do (it is not). The LR and Wald statistics for this test are both 0.30 with p-values 0.5822 and 0.5824, respectively (for these data, the LR and Wald statistics differ slightly, but not enough to notice when rounded to the second decimal place). We should immediately recognize the partial approach here as pure nonsense.

The take-home message here is that sequential fitting makes sense and yields sensible results when used with a model for which the effects have an obvious hierarchy, in this case linear, then quadratic, then cubic, etc.

Example 5.2: Three-factor main-effects-only design

The data for this example appear in the SAS Data and Program Library as Data Set 5.2. The data come from a three-factor orthogonal main effects design. Three factors, here called A, B, and C, each observed at two levels, labeled 0 and 1, have observations at only four of the eight possible A × B × C combinations. Orthogonal main effects designs are widely used in applications such as discovery research—especially in its early phases—and quality improvement, so-called "quality by design." Notice that just enough treatment combinations have been observed to allow the main effects of A, B, and C to be estimated, but no interactions—an efficient design to use when interactions are known to be negligible and setting up additional A × B × C combinations is expensive. For these data, replicate observations appear with some, but not all, treatment combinations. Assume these data have a Gaussian distribution.

Suppose we use a sequential approach to assess the effects of the treatment factors. Let us begin by fitting them in alphabetical order. Start with the linear predictor $\eta_{ijk} = \eta + \alpha_i$, where η_{ijkl} denotes the link function for the ijkth A × B × C combination and α_i denotes the effect of factor A. The response distribution for the lth replicate is $y_{ijkl} \sim NI(\mu_{ijk}, \sigma^2)$. We know from the last discussion of Section 5.3.2 that the LR and Wald tests are identical for this model and that we should use the F-test. For our initial model, the F-value for the A-effect, H_0: $\alpha_0 = \alpha_1$, is 3.65 with $p = 0.0978$. The estimated mean difference between the levels of A is 1.62 ± 0.85. Now let us add factor B. The new linear predictor is $\eta_{ijk} = \eta + \alpha_i + \beta_j$, where β_j denotes the jth factor B effect. The test results for H_0: $\beta_0 = \beta_1 | \alpha_i$, the effect of adding B over and above A: $F = 0.86$, $p = 0.4175$, estimated mean difference between levels of B is 0.75 ± 0.86. Finally, we add factor C with the results $F = 36.64$, $p = 0.0018$ mean difference between levels of C is 2.10 ± 0.35.

Now, let us change the order. This time start with B, then add A and finally C. We begin with the linear predictor $\eta_{ijk} = \eta + \beta_j$. The results: $F = 1.71$, $p = 0.2328$, mean B difference is 1.16 ± 0.89. Notice that this is not what we got when we fit B after A. Proceeding, we now fit A. The results: $F=2.26$, $p=0.1831$, mean A difference is 1.37 ± 0.91. Again, this is not what we got before. Also, when we fit the model $\eta_{ijk}=\eta+\beta_j+\alpha_i$ and proceed through LR testing in sequence, we get $\chi^2=1.96$, $p=0.1612$ if we fail to account for fact that σ^2 is unknown, or $F=2.01$, $p=0.2057$ if we do account for estimating σ^2. Clearly we have an issue: the mean differences and significance levels seem to be all over the map. It gets worse: if we fit B first, then add C, the results for B are $F=3.19$, $p=0.1242$, mean B difference is 1.16 ± 0.65. If we fit all three effects using linear predictor $\eta_{ijk}=\eta+\beta_j+\gamma_k+\alpha_i$ and test in sequence, the results for B are $F=13.97$, $p=0.0135$, mean B difference is 1.16 ± 0.31. Depending on the order we fit the effects and how we test, the p-value for B can be anywhere from 0.0135 to 0.8910! Clearly, there is something wrong with this approach.

A better strategy for these data involves thinking clearly through exactly what we want to estimate and exactly what we want to test and framing these objectives in estimable function terms. It turns out that the sequential strategy entails implicit estimable functions that depend on the pattern of the data: the results are artifacts of the treatment combinations used in the design and the pattern of unequal replication. In Chapter 7, we will explore ways to determine the estimable functions implicit in various testing strategies. For now, let us concentrate on what we *should* do.

Presuming that the unequal replication occurred more by accident that by premeditated design, we want estimates of the factor level means that tell us what would have been observed with equal replication. Searle (1987) calls these "adjusted marginal means." SAS® calls them "least squares means" and uses the syntax LSMEANS in its linear model procedures. Whatever we call them, the strategy for estimating the mean involves isolating the factor level of interest and averaging over the levels of the other factors. For example, for factor A, the estimable function that defines the mean of level A_0 is $\eta+\alpha_0+(\beta_0+\beta_1)/2+(\gamma_0+\gamma_1)/2$; similarly, the estimable function for a B_i mean is $\eta+\beta_i+(\alpha_0+\alpha_1)/2+(\gamma_0+\gamma_1)/2$; and so forth. This way, when we estimate the mean difference, we take the difference between the marginal means, yielding $\alpha_0-\alpha_1$, $\beta_0-\beta_1$, and $\gamma_0-\gamma_1$—all of which appear to be the logical focus of the investigation that motivated collecting these data. If we do this, the results are as follows:

Effect Tested	F-Value	p-Value	Estimated Mean Difference
A: $\alpha_0-\alpha_1$	21.27	0.0058	1.60 ± 0.35
B: $\beta_0-\beta_1$	0.02	0.8910	0.05 ± 0.35
C: $\gamma_0-\gamma_1$	36.64	0.0018	2.10 ± 0.35

Summarizing, in general the strategy for testing effects should begin as this chapter began: focus on defining the estimable functions that express your objectives. Thinking in terms of sequential or partial sum of squares or type 1, type 2, or type 3 hypotheses does not get you there. Thinking in terms of $\mathbf{K'\beta}$ does.

5.4 Inference Using Model-Based Statistics

"Inference using model-based statistics" means using (5.16) through (5.18) as the basic tools: interval estimates based on the t-distribution, hypothesis testing using the approximate F, and the defining feature common to both—estimating the variance of estimable and predictable functions by $\mathbf{L}'\hat{\mathbf{C}}\mathbf{L}$. Recall from the discussion toward the end of Section 5.2 that the two main issues involved in implementing this approach are (1) the denominator degrees of freedom for the approximate F—and hence for the t as well—often must be approximated and (2) except for balanced variance-component-only mixed models, $\mathbf{L}'\hat{\mathbf{C}}\mathbf{L}$ is a biased estimate of $\mathbf{L}'\mathbf{C}\mathbf{L}$. In this section, we explore these issues: Section 5.4.1 sets the stage, 5.4.2 specifically discusses degree-of-freedom approximation, and 5.4.3 presents the standard bias-adjustment strategies.

5.4.1 Naive Statistics and Degrees of Freedom

The variance estimate $\mathbf{L}'\hat{\mathbf{C}}\mathbf{L}$, Equation 5.16, is called the "naive" estimate of the variance of $\mathbf{K}'\hat{\boldsymbol{\beta}}+\mathbf{M}'(\hat{\mathbf{b}}-\mathbf{b})$. For estimable or predictable functions for which $\mathbf{K}$ and $\mathbf{M}$ are vectors, $\sqrt{(\mathbf{k}'\mathbf{m}')\mathbf{C}\begin{bmatrix}\mathbf{k}\\ \mathbf{m}\end{bmatrix}}$ is called the *naive standard error*. In this sense, we could call (5.18) the naive F-statistic, although this term is rarely used in mixed model circles. We call the estimates "naive" because Kackar and Harville (1984) showed they have a downward bias (i.e., $E(\mathbf{L}'\hat{\mathbf{C}}\mathbf{L})<\mathbf{L}'\mathbf{C}\mathbf{L}$) for mixed models other than balanced variance-component-only and certain marginal models (e.g., compound symmetry) that have an equivalent variance-component-only conditional model form. Unless we correct for this bias, we have very narrow confidence intervals (hence poor coverage) and inflated test statistics (hence excessive type I error rates).

Denominator degrees of freedom can also be an issue. In Chapter 2, we developed the WWFD skeleton ANOVA strategy for constructing the linear predictor and to distinguish between fixed and random model effects. Writing the skeleton ANOVA entails listing the degrees of freedom for each effect. Doing so provides guidance for determining the denominator degrees of freedom. For some estimable functions of some models, this strategy gives us exact degrees of freedom. In other cases, these degrees of freedom provide reasonable guidelines, but they are not exact. As the complexity of the model's covariance structure increases, the discrepancy between the skeleton ANOVA degrees of freedom and the "true" ν_2 for (5.18) tends to increase. "True" has quotes because (5.18) is an approximate F, so ν_2 must also, by definition, be approximate. We call the degrees of freedom from the skeleton ANOVA the "naive" degrees of freedom. By "true" ν_2 we really mean a high-quality approximation—the Satterthwaite approximation presented in Section 5.4.2.

First, let us work through an example of a simple variance-components only LMM so that we clearly understand the issues motivating the Satterthwaite approximation. Consider a balanced two-factor data set with replicate observations on each A×B combination. Assume that the levels of A are deliberately chosen and the levels of B result from randomly sampling a population of possible levels of B. We can characterize the LMM as follows:

- Linear predictor: $\eta_{ij}=\eta+\alpha_i+b_j+(ab)_{ij}$, where α_i denotes the ith A effect (fixed), b_j denotes the jth B effect (random), and $(ab)_{ij}$ denotes the interaction effect (also random). Assume $i=1,2,\ldots,a$ and $j=1,2,\ldots,b$
- Distributions: b_j i.i.d. $N\left(0,\sigma_B^2\right)$, $(ab)_{ij}$ i.i.d. $N\left(0,\sigma_{AB}^2\right)$, and $y_{ijk}\mid b_j,(ab)_{ij}\sim NI(\mu_{ij},\sigma^2)$ where y_{ijk} denotes the kth observation on the ijth AB combination
- Link: identity

The analysis of variance for this model is

Source of Variation	Degrees of Freedom	Expected Mean Square
A	$a-1$	
B	$b-1$	$\sigma^2 + n\sigma^2_{AB} + na\sigma^2_B$
A×B	$(a-1)(b-1)$	$\sigma^2 + n\sigma^2_{AB}$
Residual	$ab(n-1)$	σ^2
Total	$abn-1$	

Consider the estimable function for estimating an A difference: $\alpha_i - \alpha_{i'}$. We estimate this difference by $\bar{y}_{i\bullet\bullet} - \bar{y}_{i'\bullet\bullet} = 1/nb\sum_{j,k} y_{ijk} - 1/nb\sum_{j,k} y_{i'jk}$. Expressing this in model form, $1/nb\sum_{j,k}(\eta+\alpha_i+b_j+(ab)_{ij}) - 1/nb\sum_{j,k}(\eta+\alpha_{i'}+b_j+(ab)_{i'j})$, we can readily show that $Var(\hat{\alpha}_i - \hat{\alpha}_{i'}) = 2(\sigma^2 + n\sigma^2_{AB})/nb$, which we estimate by $2[MS(AB)]/nb$. $MS(AB)$ is the denominator chi-square in the F-ratio to test $H_0{:}\alpha_i = \alpha_{i'}$ and in the t distribution required to construct a confidence interval. Therefore, $\nu_2 = (a-1)(b-1)$, the degrees of freedom for AB.

Now consider an A mean. The estimable function is $\eta + \alpha_i$ which we estimate by $\bar{y}_{i\bullet\bullet} = 1/nb\sum_{j,k} y_{ijk}$. Again, writing this in model form, we can show that $Var(\eta+\alpha_i) = n(\sigma^2_B + \sigma^2_{AB}) + \sigma^2/nb$. The ANOVA estimates of the variance components are as follows: $\hat{\sigma}^2_B = (1/nb)[MS(B) - MS(AB)]$, $\hat{\sigma}^2_{AB} = (1/n)[MS(AB) - MSR]$, and $\hat{\sigma}^2 = MSR$. A little algebra reveals the estimate of $Var(\eta+\alpha_i)$ to be $(1/b)[MS(B) + (b-1)MS(AB)]$. Now we have a linear combination of two chi-square random variables: which degrees of freedom should we use for ν_2? $b-1$? $(a-1)(b-1)$? A weighted average of the two? We address this in the next section.

5.4.2 Satterthwaite Degrees of Freedom Approximation

The problem we left hanging in the previous section is the problem Satterthwaite (1941, 1946) addressed. Satterthwaite showed that given the ratio $\frac{X^2_{num}/\nu_1}{X_2^*/\nu_2^*}$, where $X^2_{num} \sim \chi^2_{\nu_1}$ and X_2^* is a linear combination of chi-square random variable all independent of X^2_{num}, the $X_2^* \sim \chi^2_{\nu_2^*}$, where

$$\nu_2^* \cong \frac{\left(\sum_m c_m X^2_m\right)^2}{\sum_m \left(c_m X^2_m\right)/df_m} \tag{5.20}$$

where X^2_m denotes the $\chi^2_{df_m}$ random variables.

We call equation (5.20) the *Satterthwaite approximation*. The c_m denote the constants in the linear combination at df_m and the degrees of freedom for the respective X^2_m.

In our example, the Satterthwaite approximate degrees of freedom are

$$\nu_2^* \cong \frac{[MS(\mathrm{B})+(b-1)MS(\mathrm{AB})]^2}{\dfrac{[MS(\mathrm{B})]^2}{b-1}+\dfrac{[MS(\mathrm{AB})]^2}{(a-1)(b-1)}}$$

We could construct a confidence interval for $\eta+\alpha_i$ as $\bar{y}_{i\bullet\bullet} \pm t_{d\nu_2^*,\alpha/2}\sqrt{\dfrac{MS(\mathrm{B})+(b-1)MS(\mathrm{AB})}{b}}$.

We can use the form of Satterthwaite's approximation shown in (5.20) for variance-component-only Gaussian LMMs, but not for more complex LMMs or GLMMs. Geisbrecht and Burns (1985) extended (5.21) to include estimable functions for the general LMM. When $\mathbf{K}$ is a vector, we can write their approximation as $\nu_2 \cong 2\left\{E\left[\mathbf{K}'(\mathbf{X}'\mathbf{V}^{-1}\mathbf{X})^{-}\mathbf{K}\right]\right\}^2 / Var\left[\mathbf{K}'(\mathbf{X}'\mathbf{V}^{-1}\mathbf{X})^{-}\mathbf{K}\right]$.

In practice, we need to replace $\mathbf{V}$ by $\hat{\mathbf{V}}$ because inevitably we have an unknown $\mathbf{V}$. Also, no "nice" expression for the denominator variance term exists. Geisbrecht and Burns use the approximation $Var[\mathbf{K}'(\mathbf{X}'\mathbf{V}^{-1}\mathbf{X})^{-}\mathbf{K}] \cong \mathbf{g}'\mathbf{V}_A(\hat{\boldsymbol{\sigma}})\mathbf{g}$, where $\mathbf{g} = \left.\dfrac{\partial\left[\mathbf{K}'\left(\mathbf{X}'\mathbf{V}(\boldsymbol{\sigma})^{-1}\mathbf{X}\right)\mathbf{K}\right]}{\partial\boldsymbol{\sigma}'}\right|_{\boldsymbol{\sigma}=\hat{\boldsymbol{\sigma}}}$ and $\mathbf{V}_A(\hat{\boldsymbol{\sigma}})$ is the asymptotic covariance matrix of the estimated covariance parameters $\hat{\boldsymbol{\sigma}}$. Using a well-known variance component result (see, e.g., Searle et al., 1992), we write the ijth element of the asymptotic variance as $V_{A,ij} = 2\times[tr\{\mathbf{P}(\partial V(\boldsymbol{\sigma})/\partial\sigma_i)\mathbf{P}(\partial V(\boldsymbol{\sigma})/\partial\sigma_j)\}]^{-1}$, where, recalling from (4.28), $\mathbf{P}=[\mathbf{V}(\boldsymbol{\sigma})]^{-1} - [\mathbf{V}(\boldsymbol{\sigma})]^{-1}\mathbf{X}(\mathbf{X}'[\mathbf{V}(\boldsymbol{\sigma})]^{-1}\mathbf{X})^{-}\mathbf{X}'[\mathbf{V}(\boldsymbol{\sigma})]^{-1}$. The resulting Satterthwaite approximation, as computed in practice, is

$$\nu_2 \cong \frac{2\left[\mathbf{K}'\left(\mathbf{X}'\mathbf{V}(\hat{\boldsymbol{\sigma}})^{-1}\mathbf{X}\right)^{-}\mathbf{K}\right]^2}{\mathbf{g}'\mathbf{V}_A(\hat{\boldsymbol{\sigma}})\mathbf{g}} \tag{5.21}$$

For estimable functions defined on GLMMs, we can use an *ad hoc* version of (5.21). We do this by replacing $[\mathbf{V}(\hat{\boldsymbol{\sigma}})]^{-1}$ with $\mathbf{V}^*$ or, more precisely, $\hat{\mathbf{V}}^*$ as defined in Chapter 4. Note that we can *only* use the Satterthwaite approximation for GLMMs in conjunction with pseudo-likelihood estimation as required elements only have meaning for the GLMM in the context of PL.

For predictable functions, the extension of the Satterthwaite approximation is straightforward: we replace $\mathbf{K}'(\mathbf{X}'\mathbf{V}^{-1}\mathbf{X})^{-}\mathbf{K}$ with $\mathbf{L}'\mathbf{C}\mathbf{L}$ in the numerator of (5.21), using estimated covariance components (and/or working correlation coefficients for GLMMs where they occur). These equations extend to multidimensional $\mathbf{K}$ and $\mathbf{L}$—details are omitted here.

5.4.3 Bias Correction for Model-Based Standard Errors and Test Statistics

For variance-component-only LMMs with balanced data and their compound symmetry marginal model equivalents, the estimated covariance of $\mathbf{K}'\hat{\boldsymbol{\beta}}$ as they appear in (5.8) and (5.9), that is, $\mathbf{K}'(\mathbf{X}'\mathbf{V}(\hat{\boldsymbol{\sigma}})^{-1}\mathbf{X})^{-}\mathbf{K}$, is an unbiased estimate of $Var(\mathbf{K}'\hat{\boldsymbol{\beta}}) = \mathbf{K}'(\mathbf{X}'\mathbf{V}(\boldsymbol{\sigma})^{-1}\mathbf{X})^{-}\mathbf{K}$. Otherwise, for all other LMM and GLMM cases, $E(\mathbf{L}'\hat{\mathbf{C}}\mathbf{L}) < \mathbf{L}'\mathbf{C}\mathbf{L}$. Kackar and Harville (1984) first documented this. Subsequent publications, notably Prasad and Rao (1990) and

Harville and Jeske (1992), pursued this issue, particularly for the estimated variance of predictable functions. Kenward and Rogers (1997) developed a bias correction term that was added as an option to the MIXED procedure and included in the GLIMMIX procedure. Guerin and Stroup (2000) investigated the type I error rates for the LMM approximate F with and without the PROC MIXED Kenward–Roger option with a simulation study involving several correlated error models for longitudinal data. They found that uncorrected type I error rates using "naive" degrees of freedom (as described earlier in this section) were inflated, often in the 15% range for nominal $\alpha=0.05$. The Satterthwaite degree of freedom correction by itself had little impact of type I error rate inflation. The Kenward–Roger correction consistently brought the type I error rate to within the 4%–6% range. Kenward and Roger found similar small sample behavior in their work culminating in their 1997 paper. Subsequent work—largely an oral tradition based on numerous, mostly unpublished, simulation studies—has confirmed the effectiveness of the Kenward–Roger adjustment for LMMs as an established fact. Although the Kenward–Roger correction is technically an option, we may consider it to be the recommended *de facto* standard operating procedure for inference with LMMs.

We briefly describe the correction. Interested readers should refer to original papers for more detail. Kenward and Roger focus on $(\mathbf{X'V^{-1}X})^{-}$. Following Kackar and Harville, they obtain the result we can characterize in estimable function terms as

$$E\left[\mathbf{K}'(\mathbf{X}'\hat{\mathbf{V}}^{-1}\mathbf{X})^{-}\mathbf{K}\right]=\mathbf{K}'(\mathbf{X}'\mathbf{V}^{-1}\mathbf{X})^{-}\mathbf{K}+\frac{1}{2}\sum \text{cov}(\sigma_i,\sigma_j)\mathbf{K}'\frac{\partial^2\left[(\mathbf{X}'\mathbf{V}^{-1}\mathbf{X})^{-}\right]}{\partial\sigma_i\partial\sigma_j}\mathbf{K}$$

where σ_i and σ_j denote the ith and jth elements of the covariance vector $\boldsymbol{\sigma}$ and $\text{cov}(\sigma_i, \sigma_j)$ is the ijth element of the asymptotic covariance matrix—we earlier denoted this as $\mathbf{V}_A(\hat{\boldsymbol{\sigma}})$. Kenward and Roger use this result to obtain the bias adjustment:

$$\left(\hat{\mathbf{V}}_{\hat{\beta}}\right)_{KR-adj}=\left(\hat{\mathbf{V}}_{\hat{\beta}}\right)_N+2\left(\hat{\mathbf{V}}_{\hat{\beta}}\right)_N\left\{\sum_{i,j}\mathbf{V}_{A,ij}\left(\mathbf{Q}_{ij}-\mathbf{X}'\frac{\partial\mathbf{V}^{-1}}{\partial\sigma_i}\mathbf{X}\left[\left(\hat{\mathbf{V}}_{\hat{\beta}}\right)_N\right]\mathbf{X}'\frac{\partial\mathbf{V}^{-1}}{\partial\sigma_j}\mathbf{X}-\frac{1}{4}\mathbf{T}_{ij}\right)\right\}\left(\hat{\mathbf{V}}_{\hat{\beta}}\right)_N \quad (5.22)$$

where

$\hat{\mathbf{V}}_{\hat{\beta}}=(\mathbf{X}'\hat{\mathbf{V}}^{-1}\mathbf{X})^{-}$

$(\hat{\mathbf{V}}_{\hat{\beta}})_{KR-adj}$ denotes the Kenward–Roger adjusted $\hat{\mathbf{V}}_{\hat{\beta}}$

$(\hat{\mathbf{V}}_{\hat{\beta}})_N$ denotes the naive $\hat{\mathbf{V}}_{\hat{\beta}}$

$$\mathbf{Q}_{ij}=\mathbf{X}'\frac{\partial\mathbf{V}^{-1}}{\partial\sigma_i}\mathbf{V}^{-1}\frac{\partial\mathbf{V}^{-1}}{\partial\sigma_j}\mathbf{X}$$

$$\mathbf{T}_{ij}=\mathbf{X}'\mathbf{V}^{-1}\frac{\partial^2 V(\boldsymbol{\sigma})}{\partial\sigma_i\partial\sigma_j}\mathbf{V}^{-1}\mathbf{X}$$

Kackar and Harville have an expression similar to (5.22) but without the $\mathbf{T}_{ij}$ term. For models with $\partial^2 V(\boldsymbol{\sigma})/\partial\sigma_i\partial\sigma_j=0$, the Kackar–Harville and Kenward–Roger corrections are identical.

The Kenward–Roger adjustment was developed for estimable functions using the LMM. However, if we replace $\mathbf{V}(\boldsymbol{\sigma})$ by $\mathbf{V}^*(\boldsymbol{\sigma},\boldsymbol{\rho})$, where $\mathbf{V}^*$ is the variance of the pseudo-variable as defined for PL in Chapter 4 and $\boldsymbol{\sigma}$ and $\boldsymbol{\rho}$ denote the vectors of covariance and working correlation components respectively, then the Kenward–Roger adjustment can be adapted for

GLMM inference. The adaptation is clearly *ad hoc* but informal simulation studies consistently show that it performs well provided the PL linearization is accurate. By "performs well" we mean that the average adjusted standard errors agree with the observed standard deviation of the sampling distribution of $\mathbf{K}'\boldsymbol{\beta}$ being studied and power and type I error characteristics appear to be consistent with theory. The author is familiar with studies for binomial and Poisson GLMMs—these consistently show that the *ad hoc* Kenward–Roger procedure is spot on unless the cluster sizes for the binomial are extremely small. Members of the two-parameter exponential family seem to be more of an issue.

Note that, like the Satterthwaite approximation, the Kenward–Roger procedure cannot be used with integral approximation methods, Laplace or Gauss–Hermite quadrature. In the next section, we discuss alternatives.

5.5 Inference Using Empirical Standard Error

While Kackar and Harville were working on the standard error bias issue in the context of the LMM, Liang and Zeger (1986) encountered analogous issues in the context of GEE theory and methodology they were developing. They proposed a solution they called a "sandwich estimator," also referred to in linear model literature as the "empirical" or "robust" standard error estimate. For LMMs and GLMMs with correlated errors (to be considered in subsequent chapters), the sandwich estimator has the apparent advantage of being less susceptible to misspecification of $\mathbf{V}$ or $\mathbf{V}^*$ than the model-based standard errors discussed in Sections 5.2 and 5.4.

However, the sandwich estimator has a different bias that comes into play for smaller data sets. "Smaller data sets" include experiments typical in disciplines such as agriculture, animal health, and quality improvement—essentially anything smaller than large-scale investigations with thousands of subjects being observed. Uncorrected sandwich estimators are biased downward: the bias is negligible for very large data sets, but as the number of replicate observations decreases the bias increases dramatically. For typical-sized experiments in many disciplines, the uncorrected sandwich estimator is unusable. However, for GLMMs for which PL is unsuitable, some alternative to the Kenward–Roger adjustment is essential. Morel et al. (2003) developed a small sample bias correction for the sandwich estimator.

In Section 5.5.1, we present the basic idea underlying the sandwich estimator and its basic structure. In Section 5.5.2, we describe the Morel et al. bias correction.

5.5.1 Sandwich (a.k.a Robust or Empirical) Estimator

We start with the marginal form of the LMM. We know that $\hat{\boldsymbol{\beta}} = (\mathbf{X}'\mathbf{V}^{-1}\mathbf{X})^{-}\mathbf{X}'\mathbf{V}^{-1}\mathbf{y}$ and therefore

$$Var\left(\hat{\boldsymbol{\beta}}\right) = \left(\mathbf{X}'\mathbf{V}^{-1}\mathbf{X}\right)^{-}\mathbf{X}'\mathbf{V}^{-1}Var(\mathbf{y})\mathbf{X}'\mathbf{V}^{-1}\left(\mathbf{X}'\mathbf{V}^{-1}\mathbf{X}\right)^{-} \tag{5.23}$$

We derive model-based standard errors and the resulting test statistics by replacing $Var(\mathbf{y})$ with its variance, $\mathbf{V}$. For unknown $\mathbf{V}$, we obtain the estimates of the covariance components, determine $\hat{\mathbf{V}}$, and use it in place of $\mathbf{V}$ with all the consequences we have explored in this chapter. For the GLM $\hat{\mathbf{W}}$ replaces $\hat{\mathbf{V}}$ and for the GLMM $\hat{\mathbf{V}}^*$ replaces $\hat{\mathbf{V}}$, but the idea is the same.

An alternative way to estimate $Var(\mathbf{y})$ uses the sum of square and cross-products matrix, $\sum_i (\mathbf{y}_i - \mathbf{X}_i\hat{\boldsymbol{\beta}}_i)(\mathbf{y}_i - \mathbf{X}_i\hat{\boldsymbol{\beta}}_i)'$, where i references subjects, for example, blocks, locations, clusters, etc. Inserting this into (5.23) yields the sandwich estimate

$$Var\left(\hat{\boldsymbol{\beta}}\right) = \left(\mathbf{X}'\mathbf{V}^{-1}\mathbf{X}\right)^{-}\mathbf{X}'\mathbf{V}^{-1}\left[\sum_{i=1}^{m}\left(\mathbf{y}_i - \mathbf{X}\hat{\boldsymbol{\beta}}_i\right)\left(\mathbf{y}_i - \mathbf{X}\hat{\boldsymbol{\beta}}_i\right)'\right]\mathbf{V}^{-1}\mathbf{X}\left(\mathbf{X}'\mathbf{V}^{-1}\mathbf{X}\right)^{-} \quad (5.24)$$

so-called because $\sum_{i=1}^{m} (\mathbf{y}_i - \mathbf{X}_i\hat{\boldsymbol{\beta}}_i)(\mathbf{y}_i - \mathbf{X}_i\hat{\boldsymbol{\beta}}_i)'$ is "sandwiched" between $(\mathbf{X}'\mathbf{V}^{-1}\mathbf{X})^{-}\ \mathbf{X}'\mathbf{V}^{-1}$ terms.

Notice that the sandwich estimator requires us to be able to define the covariance structure in terms of subjects. Recall from Chapter 2, for example, that in a multilocation experiment with random locations, we could define the location variance as a single entity, for example, $Var(\mathbf{L}) = \mathbf{I}_m\sigma_L^2$, where m is the number of locations or as a random intercept structure with each location as a subject $\boldsymbol{I}_m \otimes \mathbf{1}\sigma_L^2$. They are equivalent, but we would need to use the latter form to obtain a sandwich estimate. Putting this in SAS PROC GLIMMIX terms, for a blocked design, the statements

```
proc glimmix empirical;
class treatment block;
model y=treatment;
random block;
```

yield an error message, whereas replacing `random block` by

```
random intercept/subject=block;
```

yields analysis using the sandwich estimator for standard errors, interval estimates, and test statistics. Notice the EMPIRICAL option in the PROC statement invokes the sandwich estimator.

5.5.2 Bias Correction for Sandwich Estimators

Morel et al. (2003) present a refinement of the sandwich estimator. We can regard (5.24) as a "naive" sandwich estimate for two reasons.

First, in a mixed model, the sum of squares and cross-products of the residual should account for the design or sample size: Specifically, $\dfrac{N-1}{N-k}\dfrac{m}{m-1}\sum_{i=1}^{m}(\mathbf{y}_i - \mathbf{X}_i\hat{\boldsymbol{\beta}}_i)(\mathbf{y}_i - \mathbf{X}_i\hat{\boldsymbol{\beta}}_i)'$, where N denotes the total number of observations, m denotes the number of subjects, and $k = \text{rank}(\mathbf{X})$. For example, in the Cochran and Cox (1957) example that appears in Littell et al. (2006) and we used in Chapter 4 to introduce ML and REML variance estimation, there were 15 blocks each with 4 observations, hence $N = 60$ observations, 15 treatments, hence $k = \text{rank}(\mathbf{X}) = 15$ and $m = 15$ subjects, since blocks were the subjects.

Second, the "naive" sandwich estimator implicitly requires us to assume $\sum_i (\mathbf{y}_i - \mathbf{X}_i\hat{\boldsymbol{\beta}}_i) = \mathbf{0}$, which is not necessarily true for all GLMMs. Morel et al. propose a correction term to account for this: $\delta_m\phi\mathbf{V}(\hat{\boldsymbol{\sigma}})$, where $\delta_m = \min(0.5, k/m - k)$ and

$\phi = \max\left[1, trace\left\{(\mathbf{X}'\mathbf{V}^{-1}\mathbf{X})^{-}\sum_i\left[\mathbf{V}_i^{-1}(\mathbf{y}_i - \mathbf{X}_i\boldsymbol{\beta}_i)(\mathbf{y}_i - \mathbf{X}_i\boldsymbol{\beta}_i)'\mathbf{V}_i^{-1}\right]\right\}\right]$. For generalized models (GLM and GLMM), replace $\mathbf{V}$ with $\mathbf{V}^*$, $\mathbf{y}$ with $\mathbf{y}^*$, and $\mathbf{X}_i\boldsymbol{\beta}_i$ with its inverse link, $\tilde{\boldsymbol{\mu}}_i = h(\mathbf{X}_i\boldsymbol{\beta}_i)$.

In the Cochran and Cox data, $\frac{k}{m-k} = \frac{15}{15-15}$, creating a problem evaluating δ_m. To avoid such problems, PROC GLMMIX reexpresses the decision rule as $\delta_m = \begin{cases} \frac{k}{m-k} & \text{if } m > k(d+1) \\ 1/d & \text{otherwise} \end{cases}$.

Also, GLIMMIX sets $d=2$ on default, but allows you to select $d \geq 1$ and to vary ϕ by replacing 1 with $0<r<1$ as the maximum value for ϕ. Identifying the "right" d and r for a given model requires trial and error with simulation.

The corrected sandwich estimator, stated in GLMM marginal model terms, is thus

$$Var\left(\hat{\boldsymbol{\beta}}\right) = \left(\mathbf{X}'\mathbf{V}^{*-1}\mathbf{X}\right)^{-}\mathbf{X}'\mathbf{V}^{*-1}\left[\sum_{i=1}^{m} C\left(\mathbf{y}_i - \tilde{\boldsymbol{\mu}}_i\right)\left(\mathbf{y}_i - \tilde{\boldsymbol{\mu}}_i\right)' + \delta_n\phi\mathbf{V}^*\right]\mathbf{X}'\mathbf{V}^{*-1}\left(\mathbf{X}'\mathbf{V}^{*-1}\mathbf{X}\right)^{-} \quad (5.25)$$

where

$C = \frac{N-1}{N-k}\frac{m}{m-1}$

$\tilde{\boldsymbol{\mu}}_i = h(\mathbf{X}_i\boldsymbol{\beta}_i)$

As an example of how the various covariance estimation and standard error options can affect results, here is a summary of the impact of ML vs. REML, model-based standard errors with and without the Kenward–Roger correction and sandwich (empirical) standard errors with and without the Morel et al. bias correction on the overall F-value for treatment and the average standard error of a treatment difference for the Cochran and Cox data. The SAS programs used to generate these data appear in the SAS Data and Program Library.

Variance Component Estimator	Standard Error Type	Bias Connection	F-value for H_0: all τ_i equal	Average s.e.$(\hat{\tau}_i - \hat{\tau}_{i'})$
REML	Model-based	None	1.53	2.23
		Kenward–Roger	1.48	2.26
	Sandwich	None	428.91	1.77
		Morel et al.	1.44	2.67
ML	Model-based	None	2.07	1.89
		Kenward–Roger	2.02	1.95
	Sandwich	None	395.39	1.77
		Morel et al.	1.72	2.52

Note the extreme bias of the uncorrected sandwich estimators. The REML vs. ML difference ceases to matter for the standard errors—they are both severely biased downward—and the resulting F-values are so upwardly-biased that the numbers themselves cease to mean anything. The model-based and bias-corrected sandwich statistics show the expected

REML vs. ML differences: standard errors show downward-bias and *F*-values show upward bias with ML relative to REML. The ML–REML impact is less pronounced with the bias-corrected sandwich estimate, but still noticeable. The bias-corrected sandwich estimator used in conjunction with REML is too conservative.

These results are more for illustration than to be used as a comprehensive basis for comparison. One-size-fits-all recommendations should be avoided. However, some general guidelines follow.

5.6 Summary of Main Ideas and General Guidelines for Implementation

Inference on measures of location—treatment means, differences, and contrasts—requires defining and working with estimable and predictable functions. Estimable functions concern broad inference as defined in Chapter 3; predictable functions concern narrow inference. Population averaged is an important special case of broad inference; subject specific is an important special case of narrow inference. All inference on estimable and predictable functions occurs on the model scale.

Inference considered in this chapter includes hypothesis testing and confidence interval estimation. There are two general approaches to hypothesis testing: LR and Wald-based. The latter includes true Wald statistics and approximate *F*-statistics obtained by dividing the Wald statistics by the rank of the estimable or predictable function.

We emphasize Wald-based statistics in this chapter. For Gaussian models, Wald and LR statistics can be shown to be equivalent in many important cases making the more computationally demanding LR approach unappealing and unnecessary. For GLMMs, LR statistics are undefined in the context of pseudo-likelihood estimate and often impractical in the context of integral approximation methods. LR statistics are most often used with fixed-effects-only GLMs. For multifactor designs, usual approaches to LR computation use protocols intended for sequential testing. This can yield nonsense results for unbalanced multifactor designs. *Caveat emptor.* LR computation can be modified for computing statistics analogous to partial sums of squares, but not as easily as Wald-type statistics. The partial/sequential testing issue is discussed in more detail in Chapter 7.

LR testing does have an important advantage over Wald-based testing for inference in covariance components. We explore this topic in Chapter 6.

True Wald statistics require the assumption of known scale parameters (including covariance components in Gaussian linear mixed models) and have an approximate χ^2 distribution. Wald-based approximate *F*-statistics use estimated scale parameters and have an approximate *F*-distributions. Exact results can be obtained for the fixed-effects-only Gaussian linear model (LM) $N(\mathbf{X}\boldsymbol{\beta}, \sigma^2\boldsymbol{\Sigma})$ with $\boldsymbol{\Sigma}$ known and certain LMMs—for example, balanced split-plots. We establish exact results using matrix-based distribution results for quadratic forms. For most Gaussian mixed models and all non-Gaussian, generalized models, distribution results are asymptotic.

For approximate *F*-statistics, denominator degrees of freedom often must be approximated, for example, by Satterthwaite's procedure. Satterthwaite's procedure applies to all Gaussian linear models (LM and LMM). GLMMs estimated using pseudo-likelihood use an *ad hoc* adaptation of the Satterthwaite procedure. Satterthwaite's approximation is undefined and therefore unavailable with integral approximation. The WWFD skeleton

ANOVA approach presented in Chapter 2 provides a generally reasonable approximation of denominator degrees of freedom.

All linear models except GLMs with known scale parameter, LMs with a single unknown variance component or variance-component-only LMMs with balanced data, require some form of bias adjustment. For model-based standard errors, this means the Kenward–Roger adjustment. For sandwich estimators, this means the Morel et al. adjustment.

For Gaussian models, covariance estimates obtained via REML should be used. The Kenward–Roger adjustment is well defined in the context of REML estimation and thoroughly tested for standard applications, including correlated error models for repeated measures and spatial data.

For non-Gaussian models, an *ad hoc* form of the Kenward–Roger model-based approach with PL using the REML-like variance estimators seems to work well when the linearization on which PL is based provides a reasonable approximation. This appears to be true of one-parameter models that are not near the limit and even two-parameter exponential family models if the design is not overly complex.

For non-Gaussian models where PL is clearly not indicated—one-parameter models near the limit and two-parameter exponential family models with nontrivial design structure—integral approximation is generally preferable. With integral approximation, the Kenward–Roger approximation is not defined. Thus, with most GLMM and GEE-type models, including those with split-plot or correlated error structure, bias-corrected sandwich estimators are preferred.

For GEE-type models, which we will discuss in detail in Chapter 14, we must use PL. Recall that integral approximation methods can only be used with true likelihoods—GEE models, by definition, are quasi-likelihood models. This means if we use a GEE-type model, we have a choice between the *ad hoc* Kenward Roger adjustment and the bias-corrected sandwich estimator. Morel et al. developed their bias-correction in the context of GEE estimation. Comparative research in this area is a work in progress; as of this textbook's writing, there is much we do not know. We *do* know that if there is an alternative to GEE for which a true likelihood exists—true for many repeated measures GLMMs we consider in Chapter 14—we should use the true likelihood alternative and integral approximation.

Exercises

5.1 This problem refers to data shown in file `Ch _ 5 _ Problem1.sas`. The data are from a study comparing three treatments (denoted `Trt` in the data set). The data were collected from seven locations (denoted `Location` in the data set). For each treatment at each location, data were collected from two plots. Assume that the data are Gaussian with the following model specification:

1. *Linear predictor*: $\eta_{ij} = \eta + \tau_i + L_j + (tL)_{ij}$, where τ_i denotes the ith treatment effect, L_j denotes the jth location effect, and $(tL)_{ij}$ denotes the ijth treatment × location interaction effect
2. *Distributions*:
   a. L_j i.i.d. $N\left(0, \sigma_L^2\right)$
   b. $(tL)_{ij}$ i.i.d. $N\left(0, \sigma_{TL}^2\right)$
   c. $y_{ijk} \sim NI(\mu_{ij}, \sigma^2)$

3. *Identity link*
   a. Define the treatment mean as $\bar{y}_{i\cdot\cdot} = \frac{1}{7 \times 2} \sum_{j,k} y_{ijk}$.
      i. Derive the expectation of the treatment mean consistent with the above model.
      ii. Derive the variance of the treatment mean consistent with the above model. Show relevant work.
   b. "Run 1" in the SAS file is the GLIMMIX program for the model given earlier. Verify that the standard error of the treatment means is consistent with your part (a-ii) derivation by
      i. Giving the estimates of the variance components from the model
      ii. Demonstrating that using these estimates and following your part (a-ii) derivation yield the standard errors shown in the SAS listing
   c. "Run 2" shows the "pre-PROC MIXED" way people implemented this model. The expected mean square was used to determine the "error term" with respect to treatment and this in turn was used as a basis for determining standard errors.
      i. Based on the output from run 2, what is the "error term" with respect to treatment?
      ii. Verify that the LSMEANS statement in "Run 2" follows from your answer in (c-i).
      iii. In terms of the variance component estimates from part (b-i), show how you get the standard error for the treatment means shown in the "Run 2" listing.
      iv. Compare the treatment mean standard errors for the two runs. What does this tell you about use of pre-PROC MIXED software for mixed model analysis?

5.2 This question begins with three GLIMMIX runs on a balanced incomplete block design with seven treatments and seven blocks, each of size 3. The data and GLIMMIX statements are in file `Ch5_problem2.sas`.

   a. Run 1 uses the GLIMMIX default algorithms. State the statistical model in "GMM-appropriate form" that this run fits.
   b. What is the statistical method used to obtain the variance component estimates in run 1?
   c. Introductory textbooks state that the standard error of a treatment mean is $\sqrt{\hat{\sigma}^2/r}$, where $\hat{\sigma}^2$ is the estimate of the residual variance and r is the number of replications per treatment. Compare this to the standard error of a treatment mean in the listing for run 1. Is there a discrepancy? If so, account for it. That is, how does run 1 obtain the standard error of a treatment mean? Which is right—the intro stat methods conventional wisdom or the listing from run 1?
   d. Now look at run 2.
      i. Is the statistical model the same or different from run 1?
      ii. How do the variance component estimates of run 2 compare to run 1?
      iii. What is the statistical method run 2 uses to obtain the variance component estimates?
   e. Now look at run 3.
      i. Is the statistical model the same or different from run 1?
      ii. How do the variance component estimates of run 3 compare to runs 1 and 2?
      iii. What is the statistical method run 3 uses to obtain the variance component estimates?

f. Which run is most appropriate for reporting purposes? Explain, briefly.
g. For the run you selected in (f), are there any adjustments that should be made to the statements given before the analysis is ready for reporting purposes? If so, what and why?

5.3 This problem uses the same data set as problem 2. Questions refer to runs in file `Ch _ 5 _ problem3.sas`.

a. Run 1 uses GLIMMIX default algorithms. State the statistical model in "GMM-appropriate form" that this run fits.
b. Researchers used the generalized chi-square from run 1 as evidence of overdispersion (that is, the generalized chi-square/DF = 5.66 $\gg$1 hence overdispersion). Following the recommendation of 1980s-era GLM textbooks, the researchers used a scale parameter to adjust for overdispersion. Run 2 shows the adjustment. The researchers calculated the difference between the "-2 Res Log Pseudo-Likelihood" in runs 1 and 2 to obtain what they characterized as a LR test of H_0: $\phi = 0$, where ϕ is the standard notation for the scale parameter.
   i. Did the researchers use an appropriate statistic as evidence of overdispersion? If not, what should they have done?
   ii. Verify that that LR statistic the researchers obtained for their test of H_0: $\phi = 0$ was $\chi^2 = 49.92$ with one degree of freedom.
   iii. Were the researchers justified in characterizing their test of H_0: $\phi = 0$ as a LR test? Explain.
   iv. What happens if you attempt to use COVTEST to obtain a test statistic for H_0: $\phi = 0$?
c. Instead of runs 1 and 2 as shown here, show an alternative approach the researchers could have taken that uses integral approximation (use Method = Laplace—in working out the key, the quadrature method took about 90 min to run on a relatively fast 2010-Vintage laptap; Laplace took about 3 s. For this problem, Laplace and quadrature produce almost identical estimates).
   i. In your alternative run 1, what do you use in place of generalized chi-square/DF? What did you get? Does this statistic appear to be evidence of overdispersion?
   ii. In your alternative run 2, how did you account for overdispersion? What was the result of the LR test for this term? How did the *F*-value for treatment you obtained here compare with the researchers' run 2?

5.4 This problem is also based on the same data set as problem 2. Questions refer to runs in file `Ch _ 5 _ problem4.sas`.

There are three runs using the same `model` and `method=` options as in problem 4.

a. Run 1 uses the GLIMMIX default algorithms. State the statistical model in "GMM-appropriate form" that this run fits.
b. What is the statistical method used to obtain the variance component and model estimates in run 1?
c. Now look at run 2.
   i. Is the statistical model the same or different from run 1?
   ii. How do the variance component estimates and *F*-value for treatment of run 2 compare to run 1?
   iii. What is the statistical method run 2 uses to obtain the variance component and model estimates?
e. Now look at run 3.
   i. Is the statistical model the same or different from run 1?

   ii. How do the variance component estimates and *F*-value for treatment of run 3 compare to runs 1 and 2?
   iii. What is the statistical method run 3 uses to obtain the variance component and model estimates?

 f. How do the differences between runs 2 and 3 here compare to differences between run 2 and 3 observed in problem 4? Account for any discrepancies.

5.5 This problem is also based on the same data set as problem 2. Questions refer to runs in file `ch _ 5 _ problem5.sas`.

There are five runs shown in this file. The first two are Gaussian data. The last three assume binomial data.

 a. The first two runs yield identical results. Why?
 b. Run 3 uses the same linear predictor and random statements as run 1. Run 4 uses the same linear predictor and random statements as run 2. Yet runs 3 and 4 do not produce identical results, unlike runs 1 and 2. Why?
 c. Some GLM texts suggest that runs 3 and 4 differ because unlike the Gaussian distribution, the binomial does not have a scale parameter. They suggest that run 4 would be identical to run 3 with a scale parameter added. Run 5 is run 3 with a scale parameter added. Does the result confirm the claim of these texts?
 d. To understand what is going on, recall that $Var(\mathbf{y}^*) = \mathbf{V}^* = \mathbf{ZGZ}' + \mathbf{R}^*$, where $\mathbf{R}^* = \mathbf{\Delta V}_{\mu}^{1/2}\mathbf{P V}_{\mu}^{1/2}\mathbf{\Delta}$. For the set of three observations in a given block, $\mathbf{y}_j = \begin{bmatrix} y_{ij} \\ y_{i'j} \\ y_{i''j} \end{bmatrix}$, give $\mathbf{V}_j^* = Var\left(\mathbf{y}_j^*\right)$ for
   i. Run 1
   ii. Run 2
   iii. Run 3
   iv. Run 4
   v. Run 5
 e. For run 4, show the form of the $\partial \mathbf{V}^*/\partial\sigma_i$ for each covariance term needed to complete the score vector and Hessian (or information) matrix to get PL estimates of the covariance components.
 f. Finally, consider runs 6 and 7. Notice that they yield identical results. The question is why? To answer,
   i. Derive $\mathbf{V}_j^* = Var\left(\mathbf{y}_j^*\right)$ implied by run 6
   ii. Derive $\mathbf{V}_j^* = Var\left(\mathbf{y}_j^*\right)$ implied by run 7
   iii. Verify that $\mathbf{V}_j^* = Var\left(\mathbf{y}_j^*\right)$ implied by runs 6 and 7 are equivalent

5.6 Refer to the IML programs written for Exercise problem 4.5 for two-treatment designs. Augment each program so that it obtains the estimates and standard errors of treatments 1 and 2 and the mean difference between treatments 1 and 2. For non-Gaussian data cases, write statements for both the model and data scale.

5.7 Similar to Exercise 5.6, but apply to the IML program written for Exercise 4.6. In addition, write IML statement(s) that compute the approximate *F*-statistic.

6

Inference, Part II: Covariance Components

6.1 Introduction

In Chapter 5, we focused on inference for estimable and predictable functions. In this chapter, we turn our attention to elements of $\boldsymbol{\sigma}$, the variance and covariance components. Using the convention started in Chapter 4, when we refer to "covariance components," we mean *all* of $\boldsymbol{\sigma}$, variance, covariance, and, in some cases, correlation parameters.

We may be interested in covariance component inference because our objectives target the elements of $\boldsymbol{\sigma}$ *per se*, or we may simply need to evaluate competing covariance models as a means to an end—to fit the "right" covariance model as a prelude to inference on estimable or predictable functions. Common scenarios that focus directly on the covariance components include genetics, where covariance components measure the heritability of traits of interest, and quality improvement, where covariance components identify problem areas. "Means to an end" applications include checking assumptions about the covariance (e.g., can we justify assuming equal variances in a one-way ANOVA?) or choosing from among competing correlated error models for repeated measures or spatial data.

We can divide inference on covariance components into three general categories: (1) formal testing, (2) fit statistics, and (3) interval estimation. We can only do formal tests on covariance components when we have an obvious nesting hierarchy. We present formal covariance component testing in Section 6.2. Formal testing is not always possible or even desirable. We often want to compare covariance models that are not nested. In such cases, we must use fit statistics. We consider examples of non-nested comparisons and introduce fit statistics in Section 6.3. Formal tests and fit statistics require that we have a well-defined likelihood. When we use pseudo-likelihood estimation for non-Gaussian generalized linear mixed models (GLMMs), this is not the case. As a result, non-Gaussian GLMMs present challenges for inference on covariance components that other linear models do not. We consider these issues in Section 6.2.3 for testing and revisit them in Section 6.3 for fit statistics. Finally, Section 6.4 focuses on confidence-interval construction for covariance components.

6.2 Formal Testing of Covariance Components

While there are many ways to test variance components, here we will consider three and concentrate most of our attention on one—the likelihood ratio test. The likelihood ratio test is the one formal test for covariance components flexible enough to use with the

GLMM and all of its special cases—the linear mixed model (LMM), generalized linear model (GLM), and linear model (LM). We also consider analysis of variance (ANOVA)-based tests—briefly—because of their importance in linear model history. Like ANOVA-based variance component estimation, however, ANOVA-based tests can only be used for variance-component-only LMMs, so they are of limited value in contemporary linear modeling. Finally, we consider Wald-based tests. We do this primarily because their small sample behavior is so bad that students of linear models need to be warned.

6.2.1 ANOVA-Based Tests for Variance-Component-Only LMMs

As their name implies, ANOVA-based variance component tests involve applying *F*-tests from the ANOVA table to the sources of variation that represent random model effects. We can illustrate the ANOVA-based test by example. Consider the 4-treatment, 8-location data used in Chapter 3, Data Set 3.2. The ANOVA for these data is

Source of Variation	d.f.	SS	MS	*F*-Value	*p*-Value	EMS
Treatment	3	60.75	20.25	7.04	0.0019	$\sigma^2+4\sigma_L^2+Q_T$
Location	7	68.73	9.82	3.41	0.0135	$\sigma^2+4\sigma_L^2$
Residual	21	60.42	2.88			σ^2

By inspection of the expected mean squares, we see that the *F*-value for location can be used as a test statistic for $H_0\colon \sigma_L^2=0$. Given that $F=3.41$ and $p=0.0135$, sufficient evidence exists to conclude $H_1\colon\sigma_L^2>0$ for any α exceeding the *p*-value. As Self and Liang (1987) pointed out, we should divide the ANOVA *p*-value by 2 because, while the ANOVA *F*-test is, by definition, two-sided, tests for variance components must be one-sided tests. That is, $H_1\colon\sigma_L^2\neq 0$ cannot be the alternative hypothesis since σ_L^2 cannot be negative. The correct *p*-value for this test is $0.0135/2\cong 0.0068$.

All ANOVA tests proceed in this fashion. We construct the *F*-test from inspection of the expected mean squares. For more complex variance-component-only models, the denominator may be a linear combination of mean squares. For example, suppose we have a split-plot experiment with two fixed effect factors, A and C, and one random effect factor, B. Factorial combinations of A and B are randomly assigned to whole-plot units and levels of C are randomly assigned to split-plot units. We can write the resulting LMM as

- Linear predictor: $\eta_{ijk}=\eta+\alpha_i+b_j+(ab)_{ij}+\gamma_k+(\alpha\gamma)_{ik}+(bc)_{ij}+(abc)_{ijk}+w_{ijk}$, where α denotes A effects, b denotes B effects, γ denotes C effects, and w denotes whole-plot effects
- Distributions: $y_{ijkl}\mid b_j,(ab)_{ij},(ac)_{jk},(abc)_{ijk},w_{ijl}$ i.i.d. $N\left(\mu_{ijkl},\sigma_S^2\right)$, where σ_S^2 denotes the variance among split-plot units

$$b_j \text{ i.i.d. } N\left(0,\sigma_B^2\right),\ (ab)_{ij} \text{ i.i.d. } N\left(0,\sigma_{AB}^2\right),\ (bc)_{jk} \text{ i.i.d. } N\left(0,\sigma_{BC}^2\right),$$

$$(abc)_{ijk} \text{ i.i.d. } N\left(0,\sigma_{ABC}^2\right), \text{ and } w_{ijk} \text{ i.i.d. } N\left(0,\sigma_W^2\right)$$

- Link: $\eta_{ijkl}=\mu_{ijkl}$

The analysis of variance table is

Source of Variation	d.f.	EMS
A	1	$\sigma_S^2 + 4\sigma_{ABC}^2 + 2\sigma_W^2 + 8\sigma_{AB}^2 + Q_A$
B	9	$\sigma_S^2 + 4\sigma_{ABC}^2 + 8\sigma_{BC}^2 + 2\sigma_W^2 + 8\sigma_{AB}^2 + 16\sigma_B^2$
A × B	9	$\sigma_S^2 + 4\sigma_{ABC}^2 + 2\sigma_W^2 + 8\sigma_{AB}^2$
Whole-plot	60	$\sigma_S^2 + 2\sigma_W^2$
C	1	$\sigma_S^2 + 4\sigma_{ABC}^2 + 8\sigma_{BC}^2 + Q_C$
A × C	1	$\sigma_S^2 + 4\sigma_{ABC}^2 + Q_{AC}$
B × C	9	$\sigma_S^2 + 4\sigma_{ABC}^2 + 8\sigma_{BC}^2$
A × B × C	9	$\sigma_S^2 + 4\sigma_{ABC}^2$
Residual	60	σ^2
Total	159	

Sources of variation representing random treatment-design effects appear in **bold**. The F-values to test σ_{BC}^2 and σ_{ABC}^2 are simple MS ratios. On the other hand, the ratios needed to isolate σ_{AB}^2 and σ_B^2 require more complicated terms. Specifically, the denominator term for σ_{AB}^2 must estimate $\sigma_S^2 + 4\sigma_{ABC}^2 + 2\sigma_W^2$, the elements of EMS(AB) other than σ_{AB}^2. The resulting test statistic is $F_{AB} = \dfrac{MS(AB)}{MS(ABC) + MSW - MSR}$, where MSW and MSR denote the whole-plot and residual (split-plot) MS, respectively. For σ_B^2, $F_B = \dfrac{MS(B)}{MS(AB) + MS(BC) - MS(ABC)}$ gives the required test statistic. These two tests require Satterthwaite's approximation for the denominator degrees of freedom in order to determine the p-value.

This is as far as we can go using the ANOVA EMS approach. If we have anything other than a variance-component-only LMM, we need to look elsewhere to test covariance components.

6.2.2 Wald Statistics for Covariance Component Testing and Why They Should Not Be Used

Wald statistics for covariance component testing use the diagonal elements of $\mathbf{V}_A(\hat{\boldsymbol{\sigma}})$, the asymptotic covariance matrix of $\hat{\boldsymbol{\sigma}}$, and the estimated covariance components, given in Chapter 5, Section 5.4.2. For the ith covariance component, the asymptotic variance is

$$V_{A,ii} = 2 \times trace\left[\mathbf{P}\left(\frac{\partial V(\boldsymbol{\sigma})}{\partial \sigma_i}\right)\mathbf{P}\left(\frac{\partial V(\boldsymbol{\sigma})}{\partial \sigma_i}\right)\right]^{-1} \tag{6.1}$$

where

$$\mathbf{P} = \left[\mathbf{V}(\hat{\boldsymbol{\sigma}})\right]^{-1} - \left[\mathbf{V}(\hat{\boldsymbol{\sigma}})\right]^{-1}\mathbf{X}\left(\mathbf{X}'\left[\mathbf{V}(\hat{\boldsymbol{\sigma}})\right]^{-1}\mathbf{X}\right)^{-}\mathbf{X}'\left[\mathbf{V}(\hat{\boldsymbol{\sigma}})\right]^{-1}.$$

The Wald statistic for testing H_0: $\sigma_i = 0$ is $Wald_{\sigma_i} = \frac{\hat{\sigma}_i}{\sqrt{V_{A,ii}}}$. $Wald_{\sigma_i}$ is evaluated using the standard Gaussian $N(0,1)$ distribution. For example, using the 4-treatment, 8-location Data Set 3.2, we have the following:

- Variance estimates: $\hat{\boldsymbol{\sigma}} = \begin{bmatrix} \hat{\sigma}_L^2 \\ \hat{\sigma}^2 \end{bmatrix} = \begin{bmatrix} 1.735 \\ 2.877 \end{bmatrix}$
- Asymptotic covariance: $\mathbf{V}_A(\hat{\boldsymbol{\sigma}}) = \begin{bmatrix} 1.771 & -0.197 \\ -0.197 & 0.788 \end{bmatrix}$

The resulting Wald statistic for $H_0\colon \sigma_L^2 = 0$: $Wald\left(\sigma_L^2\right) = \frac{1.735}{\sqrt{1.771}} = 1.30$. The p-value is 0.1922. Recall that for the ANOVA-based test, $F = 3.41$ and $p = 0.0135$.

This is typical of the Wald test for covariance components: a nonsense—usually off-the-charts-conservative—result. The reason this occurs stems from the distribution of the variance component estimator. For variance-components-only LMMs, the elements of $\hat{\boldsymbol{\sigma}}$ depend on linear combinations of χ^2 random variables, distributions that are strongly right skewed. The central limit theorem does hold for the Wald statistic, but for variance estimates convergence to normality is extremely slow. As a result, the Wald statistic is useless unless you have a sample size in the thousands—better yet the tens or hundreds of thousands. For smaller samples, the sampling distribution of the covariance component estimates is markedly non-Gaussian.

Bottom line: No literate data analyst should use the Wald covariance component test, with the possible exception of extremely large studies with tens of thousands or more observations.

6.2.3 Likelihood Ratio Tests for Covariance Components

The likelihood ratio test for covariance components is similar to the likelihood ratio test for estimable functions defined by (5.19) in Chapter 5. Holding the fixed effects portion of the linear predictor constant, we fit the model under H_0: $\boldsymbol{\sigma} = \boldsymbol{\sigma}_0$, determine the likelihood, and then fit the full model. We use the resulting likelihood ratio statistic

$$LR(\boldsymbol{\sigma}_0) = 2\ell\left(\widehat{\boldsymbol{\sigma}}\right) - 2\ell\left(\widehat{\boldsymbol{\sigma}_0}\right) \tag{6.2}$$

Note that $LR(\boldsymbol{\sigma}_0) = -2\log\left[\frac{f(\widehat{\boldsymbol{\sigma}_0})}{f(\widehat{\boldsymbol{\sigma}})}\right]$, where $f(\hat{\boldsymbol{\sigma}})$ denotes the likelihood evaluated at the covariance vector estimate $\hat{\boldsymbol{\sigma}}$. If we fit a model with Gaussian data, an LMM, we should estimate $\boldsymbol{\sigma}$ using REML and hence use the REML likelihood yielding the likelihood ratio statistic $-2\log\left[\frac{f_R(\widehat{\boldsymbol{\sigma}_0})}{f_R(\widehat{\boldsymbol{\sigma}})}\right]$, where $f_R(\hat{\boldsymbol{\sigma}})$ denotes the REML likelihood as given in (4.27). Whether we compute the likelihood ratio from the full likelihood or the REML likelihood, the result that $LR(\boldsymbol{\sigma}_0)$ has an approximate χ_ν^2 distribution holds. The degrees of freedom, ν, equal the difference between the number of covariance parameters in $\boldsymbol{\sigma}$ and the number of covariance parameters in $\boldsymbol{\sigma}_0$. Because of the chi-square approximation, an alternative notation for the likelihood ratio statistics is χ^2_{LR}.

To illustrate, let us return to the 4-treatment, 8-location Data Set 3.2 and the test $H_0: \sigma_L^2 = 0$. Under the full model, $\boldsymbol{\sigma} = \begin{bmatrix} \sigma_L^2 \\ \sigma^2 \end{bmatrix}$. Under H_0, $\boldsymbol{\sigma}_0 = [\sigma^2]$, that is, under the null hypothesis the model is reduced by eliminating the random location effects. Under the full model, the REML likelihood is 4.4477×10^{-28} and $-2 \times$ the REML log-likelihood is thus 125.95. Under the reduced model, the REML likelihood is 4.4148×10^{-29} and $-2 \times$ the REML log-likelihood is 130.58. The resulting likelihood ratio statistic is $LR\left(\sigma_L^2 = 0\right) = 4.62$. Evaluating LR using χ_1^2—because the full model has two variance components and the reduced model has one, hence $\nu = 1$—we obtain $p = 0.0158$. This is consistent with the ANOVA-based test.

6.2.3.1 One-Way ANOVA: Test for Homogeneity of Variance

Let us consider another example. Data Set 6.1 contains data from a completely randomized design with four treatments. Assuming the data are Gaussian, we could consider the following model:

- Linear predictor: $\eta_i = \eta + \tau_i$
- Distribution: usual homoscedasticity assumption: y_{ij} i.i.d. $N(\mu_i, \sigma^2)$
- Unequal variance assumption: y_{ij} i.i.d. $N(\mu_i, \sigma_i^2)$
- Link: $\eta_i = \mu_i$

Before proceeding with inference on the treatment effects (τ_i), we need to resolve the variance question: Do we have equal variance for all treatments or unequal variances?

We cannot address this question via ANOVA-based tests, but we can use the likelihood ratio test. Here, the test has 3 degrees of freedom because the unequal variance model has 4 variance parameters—$\boldsymbol{\sigma}' = \begin{bmatrix} \sigma_1^2 & \sigma_2^2 & \sigma_3^2 & \sigma_4^2 \end{bmatrix}$—and the equal variance model has 1 variance parameter—$\boldsymbol{\sigma}_0 = \sigma^2$. Hence, $\nu = 4 - 1 = 3$. Under the null hypothesis, H_0: all $\sigma_i^2 = \sigma^2$; $i = 1,2,3,4$, the common variance estimate is $\hat{\sigma}^2 = 0.90$ and $-2\ell(\hat{\boldsymbol{\sigma}}_0) = 38.25$. Under the full model, $-2\ell(\hat{\boldsymbol{\sigma}}) = 35.49$, giving us a likelihood ratio statistic $\chi_{LR}^2 = 2.86$, $p = 0.4133$. The likelihood ratio test provides no evidence for rejecting the equal variance model.

> *Aside*: In the "general" linear model days, when a premium was placed on the i.i.d. errors paradigm, if we did reject H_0 it would set off a minor crisis in the form of a hunt for a variance stabilizing transformation. In contemporary modeling, we simply proceed with inference on estimable functions using the unequal variance model.

6.2.3.2 Repeated Measures Example: Selecting a Parsimonious Covariance Model

Our primary discussion of repeated measures data and, more generally, correlated error models occurs in Chapters 14 and 16. However, one of the most important applications of covariance component testing is its use for selecting among competing covariance models.

With that in mind, let us look at an example that previews Chapters 14 and 15 and provides an illustration of how we use likelihood ratio testing for covariance model selection.

Data Set 6.2 consists of repeated measures data with six times of measurement on a total of 12 subjects; 6 of them randomly assigned to one of two treatments, the other six to the other treatment. By "repeated measures data" we mean data from any study where units of replication with respect to treatments are observed at planned times over course of the study. These observations may be hourly, daily, monthly, or they may be unequally spaced, strategically planned to occur at times deemed to be important in the context of the study. Thus, another way to describe Data Set 6.2 is as a completely randomized design with two treatments and six replicates randomly assigned to each treatment. Each subject is observed at six times that are planned in advance, say, 1, 2, 4, 8, 16, and 24 weeks following the initiation of the treatments.

Assuming Gaussian data, we can write the generic form of a model as follows:

- Linear predictor: $\eta_{ijk} = \alpha_{ij} + s_{ik}$, where α_{ij} is the combined fixed effect of time i and treatment j on the linear predictor and s_{ik} is the random effect of the kth subject assigned to the ith treatment. We can partition α_{ij} in any number of ways, depending on the objectives of the study. The two most common are the factorial effects partition, $\alpha_{ij} = \alpha + \tau_i + \gamma_j + (\tau\gamma)_{ij}$, where α denotes the intercept, τ_i, the treatment main effect, γ_j, the time main effect, and $(\tau\gamma)_{ij}$, the time-by-treatment interaction, and the polynomial regression partition, $\alpha_{ij} = \beta_{0i} + \sum \beta_{mi} X_j^m$, where β_{0i} denotes the intercept for the ith treatment, β_{mi} denotes the mth order regression coefficient for the ith treatment, and T_j^m denotes the mth power of the jth time. Thus, $m = 1$ means the linear regression terms, $m = 2$ the quadratic, etc.
- Distributions: s_{ik} i.i.d. $N\left(0, \sigma_S^2\right)$.
- Let $\mathbf{y}'_{ik} \mid s_{ik} = \left(\begin{bmatrix} y_{i1k} & y_{i2k} & y_{i3k} & y_{i4k} & y_{i5k} & y_{i6k} \end{bmatrix} \mid s_{ik}\right)$. $(\mathbf{y} \mid \mathbf{s}) \sim N(\boldsymbol{\mu} \mid \mathbf{s}, \Sigma)$, where $\boldsymbol{\mu} = \mathbf{1}_{ts} \otimes E(\mathbf{y}_{ik} \mid s_{ik})$ and $\Sigma = \mathbf{I}_{ts} \otimes Var(\mathbf{y}_{ik} \mid s_{ik})$ and ts equals the number of treatments $(t) \times$ number of subjects (s).
- Link: Identity, $\boldsymbol{\eta} = \boldsymbol{\mu}$.

For convenience, denote the block diagonal element of the covariance matrix $\boldsymbol{\Sigma}$ corresponding to the ikth subject by $\boldsymbol{\Sigma}_{ij}$. Why all the elaboration for the variance of $y_{ijk} \mid s_{ik}$? Why are they not i.i.d. $N(\mu_{ijk}, \sigma^2)$ like other Gaussian linear models in previous examples? Because when we take observations over time on the same subject, we must allow for the possibility that these observations are correlated, that is, within a given subject, $Cov\left(y_{ijk}, y_{ij'k} \mid s_{ik}\right)$ for times $j \neq j'$ is not necessarily zero.

We write the general form of the covariance matrix $\boldsymbol{\Sigma}_{ik}$ as

$$\sum_{ik} = \begin{bmatrix} \sigma_1^2 & \sigma_{12} & \sigma_{13} & \sigma_{14} & \sigma_{15} & \sigma_{16} \\ & \sigma_2^2 & \sigma_{23} & \sigma_{24} & \sigma_{25} & \sigma_{26} \\ & & \sigma_3^2 & \sigma_{34} & \sigma_{35} & \sigma_{36} \\ & & & \sigma_4^2 & \sigma_{45} & \sigma_{46} \\ & & & & \sigma_5^2 & \sigma_{56} \\ & & & & & \sigma_6^2 \end{bmatrix} \quad (6.3)$$

where $\sigma_j^2 = Var\left(y_{ijk}\,|\,s_{ik}\right)$, the variance among observation taken at time j, and $\sigma_{ij} = Cov(y_{ijk}, y_{ij'k}\,|\,s_{ik})$. This is called the *unstructured* covariance matrix. Potentially, observations at every time have a unique variance and observations within the same subject at every pair of times have a unique covariance. For J times of observation, there are $J + [J(J-1)/2]$ covariance parameters forming $\boldsymbol{\sigma}$ in the unstructured covariance matrix. Not a parsimonious model by any stretch of the imagination.

At the other extreme, if we assume all $y_{ijk}\,|\,s_{ik}$ to be $NI(\mu_{ijk}, \sigma^2)$ and hence $\boldsymbol{\Sigma}_{ik} = \mathbf{I}\sigma^2$, then we have a *split-plot-in-time* covariance model, so-called because it is the same model as a split-plot experiment (we will discuss split-plot and other multilevel mixed models in more detail in Chapter 8). The split-plot-in-time model is also called the *independence* model. In Chapter 3, we saw that the split-plot-in-time or independence model can alternatively be expressed as a compound symmetry (CS) model.

Neither the unstructured nor the independence/CS models are appropriate for most repeated measures analysis. Using the unstructured model is generally wasteful and costly in terms of the statistical power for testing hypotheses. On the other hand, the independence/CS model fails to account for nontrivial correlation among repeated measurements. This results in inflated type I error rates when non-negligible correlation does in fact exist. We can usually find middle ground, a covariance model that adequately accounts for correlation but is more parsimonious than the unstructured model. Doing so allows us full control over type I error rates without needlessly sacrificing power.

We will discuss covariance models in greater detail in Chapters 14 and 16. For now, two common "middle-ground" models are the first-order ante-dependence model—often referred to by the shorthand ANTE(1)—and the first-order auto-regressive model—shorthand AR(1). The ANTE(1) covariance matrix is

$$\sum_{ik} = \begin{bmatrix} \sigma_1^2 & \sigma_1\sigma_2\rho_1 & \sigma_1\sigma_3\rho_1\rho_2 & \sigma_1\sigma_4\rho_1\rho_2\rho_3 & \sigma_1\sigma_5\rho_1\rho_2\rho_3\rho_4 & \sigma_1\sigma_6\rho_1\rho_2\rho_3\rho_4\rho_5 \\ & \sigma_2^2 & \sigma_2\sigma_3\rho_2 & \sigma_2\sigma_4\rho_2\rho_3 & \sigma_2\sigma_5\rho_2\rho_3\rho_4 & \sigma_2\sigma_6\rho_2\rho_3\rho_4\rho_5 \\ & & \sigma_3^2 & \sigma_3\sigma_4\rho_3 & \sigma_3\sigma_5\rho_3\rho_4 & \sigma_3\sigma_6\rho_3\rho_4\rho_5 \\ & & & \sigma_4^2 & \sigma_4\sigma_5\rho_4 & \sigma_4\sigma_6\rho_4\rho_5 \\ & & & & \sigma_5^2 & \sigma_5\sigma_6\rho_5 \\ & & & & & \sigma_6^2 \end{bmatrix} \tag{6.4}$$

where ρ_j denotes the correlation between $y_{ijk}\,|\,s_{ik}$ and $y_{i,j+1,k}\,|\,s_{ik}$ for $j = 1,2, \ldots, 5$. Correlation between nonadjacent observations on the same subject is modeled by the product of the ρ_j contained in the interval. In this example, with 6 repeated measurements, the ANTE(1) allows us to reduce the number of covariance parameters relative to the unstructured model from 21 to 11.

If we further assume that all variances and all adjacent-observation correlations are equal, that is, all $\sigma_j^2 = \sigma^2$ and all $\rho_j = \rho$, then we have an AR(1) covariance model. The AR(1) model has only two covariance parameters. In many applications, the AR(1) provides an adequate model of within-subject correlation, thus maximizing power without sacrificing type I error control.

For Data Set 6.2, the –2 REML log-likelihood values for the competing models are

Covariance Model	$-2\ell_R(\hat{\sigma})$	Covariance Parameters	χ^2_{LR}	*p*-Value
Unstructured	140.26	21	12.17	0.2765
ANTE(1)	152.43	11	8.48	0.4866
AR(1)[a]	160.91	2		
AR(1) + subject[b]	156.69	3	5.42	0.0199
compound symmetry/ independence	162.11	2		

[a] AR(1) covariance model without including s_{ij} in linear predictor.
[b] model includes AR(1) covariance model + s_{ik} effect ⇒ 2 + 1 = 3 covariance parameters.

The likelihood ratio tests compare the model versus the next simpler model in the row immediately below. For example, the likelihood ratio 12.17 tests the null hypothesis that the additional 10 covariance parameters needed to specify the unstructured model rather than the ANTE(1) model are all zero. The *p*-value of 0.2765 means we fail to reject this hypothesis—relative to the ANTE(1), the unstructured model adds nothing of value, but is wasteful and compromises power.

From the table, we see that we fail to reject the null hypothesis that all adjacent-observation correlations are equal and all variances are equal—that is, the extra complication of the ANTE(1) models adds nothing of value for these data—but we do reject the hypothesis that the AR(1) correlation coefficient is zero. Thus, for these data, we would conclude that there is nontrivial within-subject correlation and that, if limited to the covariance model choices shown here, the AR(1) model provides the most parsimonious way to adequately model that correlation. The remaining step—not shown here—would use the likelihood-ratio test to decide between the AR(1) model with and without the extra subject effect, s_{ik}, in the linear predictor.

The likelihood ratio approach works well provided we satisfy two conditions. First, we must have a legitimate likelihood, REML or full. We cannot use the likelihood ratio test with pseudo-likelihood. Second, the models under comparison must be nested. Here, the independence model is a subset of the AR(1) + random subject effect model ($\rho = 0$), which in turn is a subset of the ANTE(1) model (with the variance and correlation equalities shown earlier), which in turn is a subset of the unstructured model. If the models are not nested, we cannot use the likelihood ratio approach. For example, the AR(1) model without the additional s_{ik} effect, often used in repeated measures analysis, cannot be compared to the CS model using likelihood ratio testing because both models have two covariance parameters. In the next section, we discuss the pseudo-likelihood issue. In Section 6.3, we discuss the non-nested model comparison issue.

6.2.4 Consequences of PL versus Integral Approximation for GLMMs

When we fit correlated error models to Gaussian data, within-subject correlation naturally becomes part of the **R** matrix, recalling that $\mathbf{R} = Var(\mathbf{y}|\mathbf{b})$, the conditional variance of the observations given the random effects. For all of the repeated measures models described in the previous section, $\mathbf{R} = \mathbf{I}_{ts} \otimes \boldsymbol{\Sigma}_{ik}$.

How does this work when we fit repeated measures models to non-Gaussian data? With non-Gaussian members of the exponential family and all quasi-likelihoods, $Var(\mathbf{y}|\mathbf{b})$

depends at least partly on $E(\mathbf{y}|\mathbf{b})$. This means we cannot just write an **R** matrix with a correlation structure like we did with Gaussian data, ignoring the variance function.

There are two ways to address this: The "LMM analogy" and WWFD. The former leads to working correlation matrix and GEE models, also called **R-side** modeling. The later leads to modeling correlation through the linear predictor rather than $Var(\mathbf{y}|\mathbf{b})$. This is called **G-side** modeling. The AR(1) + random subject effect model from the previous section provides the best illustration of the difference.

6.2.4.1 R-Side or Working Correlation Model

For non-Gaussian GLMMs, recall that, in matrix form, $Var(\mathbf{y}\,|\,\mathbf{b}) = \mathbf{V}_\mu^{1/2}\mathbf{A}\mathbf{V}_\mu^{1/2}$, where $\mathbf{V}_\mu$ denotes the variance function matrix and **A** denotes the scale parameter matrix. We saw in Chapter 5 that one way to model covariance in the conditional distribution of $\mathbf{y}|\mathbf{b}$ is to replace the scale matrix with a *working correlation,* a.k.a., *working covariance* matrix. For the AR(1) repeated measures model, we write the working covariance matrix as

$$\mathbf{A}_W = \phi\begin{bmatrix} 1 & \rho & \rho^2 & \rho^3 & \rho^4 & \rho^5 \\ & 1 & \rho & \rho^2 & \rho^3 & \rho^4 \\ & & 1 & \rho & \rho^2 & \rho^3 \\ & & & 1 & \rho & \rho^2 \\ & & & & 1 & \rho \\ & & & & & 1 \end{bmatrix} \tag{6.5}$$

For example, suppose we have binomial data. Data Set 6.3 gives one such data set, with design structure identical to Data Set 6.2, but with a binomial rather than Gaussian response variable. The *R-side AR(1) GLMM* for these data is as follows:

1. Linear predictor: identical to Gaussian model, that is, $\eta_{ijk} = \alpha_{ij} + s_{ik}$, where α_{ij} denotes the treatment-time component and can be partitioned as described for the Gaussian model.
2. Distributions:
   a. s_{ik} i.i.d. $N(0, \sigma_S^2)$. This is called *G-side* because it describes the distribution of random effects in the linear predictor and, generically, $Var(\mathbf{b}) = \mathbf{G}$.
   b. $y_{ijk}|s_{ik} \sim \text{Binomial}(n_{ijk}, \pi_{ijk})$, but with quasi-likelihood modification to variance.
   c. $Var(\mathbf{y}_{ik}\,|\,s_{ik}) = \mathbf{V}_\mu^{1/2}\mathbf{A}_W\mathbf{V}_\mu^{1/2}$ where $\mathbf{y}_{ik}$ denotes the vector of observations at six times on the *ik*th subject (as described earlier for the Gaussian model) and $\mathbf{A}_W$ is the working covariance given in (6.5). Thus, $Var(\mathbf{y}\,|\,\mathbf{s}) = \mathbf{I}_{ts} \otimes \left(\mathbf{V}_\mu^{1/2}\mathbf{A}_W\mathbf{V}_\mu^{1/2}\right)$.
3. Link: logit (π_{ijk}).

GLMM practitioners often call this a "GEE-type" model. Because the model contains a random subject effect in the linear predictor, it is not a true GEE model. True GEE models are fully marginal: The linear predictor contains only fixed effects, that is, $\boldsymbol{\eta} = \mathbf{X}\boldsymbol{\beta}$, and all of the variance structure is contained in $\mathbf{V}_\mu^{1/2}\mathbf{A}_W\mathbf{V}_\mu^{1/2}$. *GEE-type models* have working covariance matrices but may contain other random effects in the linear predictor.

Estimation for GEE-type models must use some form of linearization: pseudo-likelihood, as described in Chapter 4; penalized quasi-likelihood, as described by Breslow and Clayton (1993); or explicitly generalized estimating equations, for example, as described by Liang and Zeger (1986). In this discussion, we use pseudo-likelihood.

We now describe the G-side alternative and then compare and contrast the model-selection issues for these two approaches.

6.2.4.2 "What Would Fisher Do?" The G-Side Approach

From Chapter 2, recall that the WWFD process for model building consists of writing the skeleton ANOVAs for the "topographical" (or design) aspect and the treatment aspect and then combining them. Doing this for these data yields the following:

Topographical		**Treatment**		**Combined**	
Source of Variation	**d.f.**	**Source of Variation**	**d.f.**	**Source of Variation**	**d.f.**
		Treatment	1	Treatment	1
Subject	11			Subject(treatment)	10
Meas. occasion	5	Time	5	Time	5
		Treatment × time	5	Treatment × time	5
Subject × occasion	55			Time × subject(trt) a.k.a. residual	50
		"Parallels"	60		
Total	71	Total	71	Total	71

As we saw in Chapter 3, when we use the combined skeleton ANOVA to construct the model for Gaussian data, we must interpret the last line as "residual" in order to account for the Gaussian distribution's scale parameter, σ^2, or in more complex cases like repeated measures, the covariance matrix of $\mathbf{y}|\mathbf{b}$. However, we do not need to do this with one-parameter members of the exponential family, such as the binomial in this example.

This raises the question: At what point does correlation among repeated measurements for the binomial data occur? Conceptually, in the binomial GLMM, we presume that π_{ijk} varies systematically because of the treatment and time effects and that it is randomly perturbed by variation between subjects and by variation within subjects. Correlation among repeated measurements on the same subject must be part of the latter: random perturbation of π_{ijk} within subjects. Another way to put this: Does it make more sense to model correlation in the random perturbation π_{ijk} *before* applying the link or through a working correlation *after* applying the link? The former requires that within-subject variation be part of the linear predictor. Also, the former gives us a true likelihood. The latter gives us a quasi-likelihood and a process that does not correspond to any feasible probability mechanism.

WWFD gives us the following GLMM:

- Linear Predictor: $\eta_{ijk} = \alpha_{ij} + s_{ik} + w_{ijk}$
- Distributions: $y_{ijk}|s_{ik}, w_{ijk} \sim \text{Binomial}(n_{ijk}, \pi_{ijk})$

$$s_{ik} \text{ i.i.d. } N\left(0, \sigma_S^2\right)$$

$$\mathbf{w}_{ik} \sim N\left(\mathbf{0}, \Sigma_{ik}\right),$$

where

$$\mathbf{w}'_{ik} = \begin{bmatrix} w_{i1k} & w_{i2k} & w_{i3k} & w_{i4k} & w_{i5k} & w_{i6k} \end{bmatrix}$$

and

$$\sum_{ik} = \sigma_W^2 \begin{bmatrix} 1 & \rho & \rho^2 & \rho^3 & \rho^4 & \rho^5 \\ & 1 & \rho & \rho^2 & \rho^3 & \rho^4 \\ & & 1 & \rho & \rho^2 & \rho^3 \\ & & & 1 & \rho & \rho^2 \\ & & & & 1 & \rho \\ & & & & & 1 \end{bmatrix} \tag{6.6}$$

This means that for the entire vector of within-subject effects, $\mathbf{W} \sim N(\mathbf{0}, \mathbf{I}_{ts} \otimes \boldsymbol{\Sigma}_{ik})$

- Link: $\eta_{ijk} = \text{logit}(\pi_{ijk})$

This is a G-side model because all of the random variability, aside from the conditional distribution of $y_{ijk}|s_{ik}, w_{ijk}$, occurs in the linear predictor. Note that all of the distributions in this model are true probability distributions. This means it is possible to use either pseudo-likelihood or integral approximation for estimation and inference. This has important consequences for covariance model selection with GLMMs.

6.2.4.3 R-Side versus G-Side: Consequences for Covariance Model Selection

For non-Gaussian models with a working correlation structure, we must use pseudo-likelihood estimation. For non-Gaussian G-side-only models, we can use either pseudo-likelihood or integral approximation. Why does this matter? We can answer this question by examining the likelihood ratio statistic.

We know that in practice we compute the likelihood ratio statistic for testing covariance components as $\chi^2_{LR} = 2\ell(\hat{\sigma}) - 2\ell(\hat{\sigma}_0)$, where $\ell(\hat{\boldsymbol{\sigma}})$ denotes the log-likelihood evaluated at $\hat{\boldsymbol{\sigma}}$. With integral approximation, we obtain $\ell(\hat{\boldsymbol{\sigma}})$ as a part of the estimation process. With pseudo-likelihood, we never evaluate $\ell(\hat{\boldsymbol{\sigma}})$. We construct a pseudo-likelihood, so the closest we can come to $\ell(\hat{\boldsymbol{\sigma}})$ is $p\ell(\hat{\boldsymbol{\sigma}})$, the pseudo-likelihood evaluated at $\hat{\boldsymbol{\sigma}}$. The problem with using this for likelihood ratio testing is that the pseudo-variate, $\mathbf{y}^*$, on which the full-model pseudo-likelihood is based is different than the pseudo-variate on which the reduced model, and hence $p\ell(\hat{\boldsymbol{\sigma}}_0)$, is based. While we can compute a pseudo-likelihood ratio, $2p\ell(\hat{\boldsymbol{\sigma}}) - 2p\ell(\hat{\boldsymbol{\sigma}}_0)$, its interpretation is suspect because of the noncomparability of $\mathbf{y}^*$ under the full model and $\mathbf{y}_0^*$, the pseudo-variate under the reduced model.

For Data Set 6.3, the following table shows the pseudo-log-likelihoods for the R-side models and the log-likelihoods computing using the Laplace method for the G-side models with the covariance structures we considered in the Gaussian example, Data Set 6.2. Note that the R-side CS and independence models are not equivalent (as we saw in Chapter 3), so each has a different pseudo-likelihood. Though not stated explicitly in the table, all log-likelihood and pseudo-log-likelihood statistics have been multiplied by −2.

		R-Side Model		G-Side Model		
(Working) Covariance Model[a]	**Number of Parameters**	**Pseudo-Log-Likelihood**	**Pseudo-LR Statistic**	**Log-Likelihood**	χ^2_{LR}	***p*-Value**
Unstructured	21	154.82	30.12	320.54	10.99	0.3583
ANTE(1)	11	184.94	7.04	331.53	10.39	0.2387
AR(1) + subj	3	191.98	11.17[c]	341.92	4.69	0.0303
CS	2	194.81		346.61		
Independence[b]	2	203.15				

[a] Working covariance for R-side models; true covariance for G-side models.
[b] R-side independence model assumes a scale parameter not usually present in the binomial, that is, $\mathbf{A}_W = \mathbf{I}\phi$ and hence $Var(y_{ijk}|s_{ik}) = \phi n_{ijk}\pi_{ijk}(1-\pi_{ijk})$; G-side independence model assumes w_{ijk} i.i.d. $N\left(0, \sigma_W^2\right)$.
[c] AR(1) pseudo LR computed vs. independence model pseudo-log-likelihood.

If we use the pseudo-likelihood statistics, the p-value testing the unstructured versus the ANTE(1) covariance model is 0.0008, which would (spuriously) indicate that we must use the unstructured working covariance model to adequately account for within-subject correlation. The G-side approach gives a more accurate picture: The AR(1) model with the additional random subject effect, s_{ik} is adequate.

> *How do we know the R-side result is spurious?* Data Set 6.3 is simulated. The data were generated using the AR(1) + random subject model. This is a typical data set from the simulation: R-side pseudo-likelihood ratio testing did not reliably select the AR(1) + random subject model, G-side likelihood ratio testing did.

To summarize, for formal testing of covariance components, likelihood ratio testing is the most versatile and generally applicable tool at our disposal. For Gaussian data, use the REML log-likelihood to construct the likelihood-ratio statistic. For non-Gaussian data, we must use an estimation method that permits evaluating a true likelihood, for example, integral approximation via Laplace or Gauss–Hermite quadrature. This limits formal covariance component testing for non-Gaussian GLMMs to G-side-only models, that is, models with no working correlation structure. For many GLMMs, we have a choice between G-side models and R-side, GEE-type models. If formal covariance testing is a priority, use the G-side model when possible.

The previous paragraph should not be interpreted as a one-size-fits-all condemnation of R-side, GEE-type modeling. There are many situations for which GEE-type models are the only viable approach and others for which the GEE is the method of choice. We will say more about this issue in the Chapters 11 (overdispersion) and 16 (non-Gaussian repeated measures).

Note that *all* of the tests we have discussed require that the fixed effects $\mathbf{X\beta}$ be held constant. The only allowable change between the full model and the reduced model under the null hypothesis is the random part of the model, $\mathbf{Zb}$ and $\mathbf{R}$ for LMMs or $\mathbf{A}_W$ in GLMMs.

We now turn our attention to covariance model comparison for which formal testing is either not possible or simply inappropriate.

6.3 Fit Statistics to Compare Covariance Models

Suppose we want to see how the fit of a CS model compares to the fit of an AR(1) model without the extra subject random effect. Both models have two covariance parameters—σ^2 and ρ, although we interpret these parameters differently in these two models. Neither is a subset of the other. We cannot construct a formal test.

Even when a formal test is possible, many argue that, for many applications, hypothesis testing is beside the point for identifying an adequate, parsimonious model. Some contend that testing is inappropriate; others merely suggest that while testing is not wrong, it is also not informative. See Burnham and Anderson (2002) and Konishi and Kitagawa (2008) for more extensive discussion of these issues.

Information criteria provide an alternative to formal testing. The two most widely used information criteria are the AIC (Akaike [1974] Information Criterion) and BIC (Bayesian Information Criterion proposed by Schwarz [1978]). Burnham and Anderson (1998) recommend a small-sample correction for the Akaike criterion—the amended information criterion goes under the acronym AICC. This section provides a brief overview of the reasoning behind information criteria. We then update the examples using Data Sets 6.2 and 6.3 to illustrate their application.

6.3.1 AIC and AICC

The AIC traces its roots to Kullback and Leibler (1951). They developed a measure of discrepancy between two models: the "truth," which they denoted by the function f; and an "approximating model" denoted by the function g. Following their notation, the "truth" depends only on the data, hence the function $f(\mathbf{y})$ and the approximating model depend on the model and the parameters, thus denoted $g(\mathbf{y}|\boldsymbol{\theta})$. For continuous data, Kullback and Leiber started with a quantity Konishi and Kitagawa refer to as the K-L information, denoted $I(f,g)$ and defined as $E_f\left[\log\left(\frac{f(\mathbf{y})}{g(\mathbf{y}|\boldsymbol{\theta})}\right)\right]$. Note that the K-L information is conceptually akin to a likelihood ratio statistic. Also, a perfect approximating model would yield a ratio $\frac{f(y)}{g(y|\theta)}=1$ and hence a log-ratio equal to 0. We can rewrite $I(f,g)$ as $E_f\{\log[f(\mathbf{y})]\}-E_f\{\log[g(\mathbf{y}|\boldsymbol{\theta})]\}$. Setting $E_f\{\log[f(\mathbf{y})]\}=C$, where C is a constant, yields $-E_f\{\log[g(\mathbf{y}|\boldsymbol{\theta})]\}=I(f,g)-C$. The expression $I(f,g)-C$ is called the K-L distance.

Akaike showed that $-2\ell(\hat{\boldsymbol{\theta}})+2p$, where p denotes the number of parameters in the parameter vector and $\boldsymbol{\theta}$ provides a good approximation to the K-L distance. The expression

$$AIC=-2\ell\left(\hat{\boldsymbol{\theta}}\right)+2p \tag{6.7}$$

defines the Akaike information criterion.

Note that we can view the AIC as −2 multiplied by the log-likelihood minus a penalty equal to the number of model parameters. We know that if we add additional parameters to a model, we can always increase the log-likelihood up to the point where $\ell(\mathbf{y})$ gives the greatest possible value. This is the idea behind the deviance statistic we discussed in Chapter 5. However, we reach a point of diminishing returns, *and* the entire point of fitting a model is to provide a *parsimonious* explanation of variation or prediction of the data's behavior. The AIC exacts a penalty for adding new parameters to the model; unless the increase in the log-likelihood exceeds the number of parameters added to the model, we deem the addition of those parameters to be counterproductive.

Burnham and Anderson (1998) recommend a small-sample correction. Let

$$n^* = \begin{cases} N \text{ if } \ell(\boldsymbol{\theta}) \text{ is a full likelihood} \\ N - rank(\mathbf{X}) \text{ if } \ell(\boldsymbol{\theta}) \text{ is a REML likelihood} \end{cases}$$

where N denotes the total number of observations. The small sample correction is $\frac{2p(p+1)}{n^*-p-1}$. Adding the correction to the AIC yields

$$AICC = AIC + \frac{2p(p+1)}{n^*-p-1} = -2\ell(\boldsymbol{\theta}) + \frac{2pn^*}{n^*-p-1} \tag{6.8}$$

the small sample adjusted, or "corrected" Akaike information criterion.

6.3.2 BIC

Schwarz (1978) proposed an alternative approach using a Bayesian argument. You have a number of competing models, one of which is known to be the "right" model. For each competing model, you have a prior probability that it is the "right" model. Assuming all competing models are equally likely, Schwarz derived the Bayesian information criterion

$$BIC = -2\ell(\hat{\boldsymbol{\theta}}) + p\log(s)$$

where s denotes the number of subjects.

6.3.3 Application to Comparison of Covariance Models

For LMMs using REML estimation, the parameter vector for the log-likelihood is $\boldsymbol{\sigma}$ since the fixed effects parameters do not appear in the REML likelihood. The number of parameters equals the number of covariance parameters. The design size values, n^* and s, remain unchanged. For assessment of LMM covariance models using REML, the three information criteria can be written as

- AIC: $-2\ell_R(\hat{\boldsymbol{\sigma}}) + 2p$
- AICC:

$$AIC + \frac{2p(p+1)}{n^*-p-1} = -2\ell_R\left(\hat{\boldsymbol{\sigma}}\right) + \frac{2pn^*}{n^*-p-1} \tag{6.9}$$

- BIC: $-2\ell_R(\hat{\boldsymbol{\sigma}}) + p\log(s)$

where $\ell_R(\hat{\boldsymbol{\sigma}})$ denotes the REML log-likelihood evaluated at $\hat{\boldsymbol{\sigma}}$.

For GLMMs using integral approximation, the parameter vector includes both the covariance parameters and the fixed effects. If we let p equal the number of *covariance* components—to maintain consistency with (6.9)—and recall that the number of fixed effect parameters equals $rank(\mathbf{X})$, we can write the information criteria for GLMMs using integral approximation as

- AIC: $-2\ell(\hat{\boldsymbol{\sigma}}, \hat{\boldsymbol{\beta}}) + 2[p + rank(\mathbf{X})]$
- AICC:

$$-2\ell\left(\hat{\boldsymbol{\sigma}},\hat{\boldsymbol{\beta}}\right)+\frac{2\left[p+rank(\mathbf{X})\right]n^*}{n^*-\left[p+rank(\mathbf{X})\right]-1} \tag{6.10}$$

- BIC: $-2\ell(\hat{\boldsymbol{\sigma}}, \hat{\boldsymbol{\beta}}) + [p + rank(\mathbf{X})]\log(s)$

where $\ell(\hat{\boldsymbol{\sigma}}, \hat{\boldsymbol{\beta}})$ denotes the full log-likelihood evaluated at the parameter estimates. Again, if we hold the fixed effects portion of the model constant, then differences between information criteria from competing models result from the covariance parameters only, thus allowing us to compare various covariance models.

In Section 6.2, we discussed two examples using likelihood ratio tests to compare covariance models: Data Set 6.2 for Gaussian data with REML and Data Set 6.3 for non-Gaussian data using integral approximation via the Laplace method. Continuing these examples, let us look at the information criteria.

First, we consider the Gaussian data. The REML-based information criteria are as follows:

Covariance Model	**AIC**	**AICC**	**BIC**
Unstructured	182.26	206.58	194.44
ANTE(1)	174.43	179.93	179.77
AR(1) + random subject	162.69	163.12	164.14
Compound symmetry/ independence	166.11	166.33	167.08

Recalling the AIC's origin as an approximation of the K-L distance—that is, the discrepancy between the approximating model and "truth"—the lower the AIC the better. We interpret the other information criteria similarly. By all three criteria, the "winning" model is the AR(1). While the ANTE(1) and unstructured models both have greater log-likelihoods, the improvement does not offset the increased number of covariance parameters. We can see the information criteria's appeal—we can quickly scan the information for all competing models and choose. Notice that the AICC's correction relative to the AIC increases substantially as the number of covariance parameters increases. Burnham and Anderson (2002) suggest that the AICC is the most reliable of the three information criteria shown here as a guide to covariance model selection for these data.

Now, the full-likelihood-based information criteria for the binomial G-side repeated measures GLMM are as follows:

Covariance Model	AIC	AICC	BIC
Unstructured	384.54	438.70	400.06
ANTE(1)	375.53	396.19	386.20
AR(1) + random subject	371.92	380.50	379.20
Compound symmetry/ independence	374.61	381.98	381.40

Again, the results are consistent with what we found using likelihood ratio testing. The AR(1) model provides the most parsimonious fit—we can interpret the inflated information criteria for the unstructured and ANTE(1) models as evidence that they "over-model" the covariance structure and the inflated information criteria for the compound symmetry/ independence model as evidence that it "under-models" the covariance structure. Over-modeling compromises power, under-modeling fails to control type I error.

6.4 Interval Estimation

When the objectives of a study target the covariance components *per se*, we may be more interested in interval estimates of these components than in testing hypotheses about them. We address this briefly. Interested readers can find much more in-depth discussion of this topic in publications specifically devoted to variance components, for example, Searle et al. (1987). Here we consider three methods for constructing confidence intervals. The first, based on the standard Gaussian $N(0,1)$ distribution, we mention mainly because, although many, in their ignorance, use it, they should not, and students of linear models should know this. The second is a Wald procedure based on the χ^2 approximation. The third is based on the profile likelihood.

First: what *not* to do

Earlier, we saw that one test statistic that should never be used for covariance components is the Wald statistic using the normal approximation. This was given as $Wald_{\sigma_i} = \frac{\hat{\sigma}_i}{\sqrt{V_{A,ii}}}$ for testing H_0: $\sigma_i = 0$ and is based on the assumption that $Z = \frac{\hat{\sigma}_i - \sigma_i}{\sqrt{V_{A,ii}}}$, where $V_{A,ii}$ is the asymptotic variance of $\hat{\sigma}_i$ and has an approximate $N(0,1)$ distribution. We did not mince words: "No literate data analyst should use the Wald covariance component test." Unless the sample sizes are *huge*, the normal approximation is very poor.

Related to this is the confidence interval derived from the normal approximation $\hat{\sigma}_i \pm Z_{\alpha/2}\sqrt{V_{A,ii}}$, where $(1 - \alpha)$ denotes the level of confidence. Again, the normal approximation is very poor for sample sizes less than tens or hundreds of thousands. We strongly discourage this approach to confidence-interval construction. To repeat, no literate data analyst should construct confidence intervals this way.

6.4.1 Wald Approach Based on the χ^2

This method is simply an elaboration on the standard interval estimate for the variance taught in introductory statistics classes. Let σ_i be the covariance component of interest. Its estimator, $\hat{\sigma}_i$, has an approximate distribution $\sigma_i \chi_\nu^2/\nu$. Some manipulation yields

$$\frac{\nu \times \hat{\sigma}_i}{\chi^2_{\nu,1-\alpha/2}} < \sigma_i < \frac{\nu \times \hat{\sigma}_i}{\chi^2_{\nu,\alpha/2}}$$

We use this to construct a confidence interval, where $(1 - \alpha)$ denotes the level of confidence.

6.4.2 Likelihood-Based Approach

This procedure is based on the likelihood ratio statistic $\chi^2_{LR} = 2\ell(\hat{\boldsymbol{\sigma}}) - 2\ell(\hat{\boldsymbol{\sigma}}_0)$. Let $\sigma_{0,i}$ be the parameter of interest for the confidence interval, σ_i, under H_0. We can rewrite $\ell(\boldsymbol{\sigma}_0)$ as $\ell(\sigma_{0,i}, \hat{\boldsymbol{\sigma}}_{i'})$, where $\boldsymbol{\sigma}_{i'}$ is defined as the covariance parameter vector with σ_i removed and $\hat{\boldsymbol{\sigma}}_{i'}$ is its maximum likelihood, or REML, estimate, depending on the procedure we use to estimate the covariance components. Following Patiwan (2001), we call this function the profile likelihood.

Define $\ell(\boldsymbol{\sigma}_{i'} \mid \tilde{\sigma}) = \sup_{\boldsymbol{\sigma}_{i'}}\left[\ell(\tilde{\sigma} \mid \boldsymbol{\sigma}_{i'})\right]$, where we hold σ_i at a constant $\tilde{\sigma}$. The *profile likelihood confidence interval* for σ_i is the set of all $\tilde{\sigma}$ such that $2\left[\ell(\hat{\boldsymbol{\sigma}}) - \ell(\boldsymbol{\sigma}_{i'} \mid \tilde{\sigma})\right] < \chi^2_{1,1-\alpha}$, where $(1 - \alpha)$ is the level of confidence.

SAS® PROC GLIMMIX provides options for computing χ^2-based Wald confidence intervals and profile likelihood confidence intervals for covariance components and, if desired, working covariance parameters in GEE-type models. For these models, the profile likelihood uses the pseudo-log-likelihood. See SAS/STAT/PROC GLIMMIX Version 9.22 On-Line Documentation (2010) for additional details.

6.5 Summary

The term "covariance component" refers collectively to variance, covariance, and correlation parameters in the **G** and **R** matrices of the GLMM. For notational purposes, they comprise the vector $\boldsymbol{\sigma}$ referred to in Chapters 5 and 6.

Inference on covariance components takes three basic forms:

1. Formal inference on one or more components of $\boldsymbol{\sigma}$
   a. Primary GLMM tool: likelihood ratio test
   b. Possible reasons include
      i. Verify assumption, for example, homogeneity of variance
      ii. Covariance component of intrinsic interest
      iii. Select among competing covariance models, for example, in repeated measures analysis
      iv. Latter only possible when models are nested

2. Informal inference
   a. Uses fit statistics such as AIC, AICC, and BIC
   b. Fit statistic = log-likelihood–penalty
   c. Used to select from among competing covariance structures
   d. Especially useful when structures are not nested
3. Interval estimation
   a. Wald (easy but not recommended)
   b. Likelihood based—profile likelihood

Inference on covariance components requires keeping the fixed effects portion of the model, $\mathbf{X}\boldsymbol{\beta}$, *unchanged*. Otherwise, tests confound $\boldsymbol{\beta}$ and $\boldsymbol{\sigma}$ and are thus invalid.
For Gaussian data:

- Use restricted likelihood.
- Likelihood ratio tests use REML likelihood.
- Fit statistics are penalized REML likelihood.
- Interval estimates are profile REML likelihood.

For non-Gaussian data:

- Inference *requires conditional* model and *integral* approximation (Laplace or quadrature). Covariance inference on pseudo-likelihood is undefined and invalid.
- Marginal model is by definition a quasi-likelihood model. In GLIMMIX this means estimation by pseudo-likelihood only. Fit statistics are undefined for pseudo-likelihood. Testing requires computing pseudo-likelihood under H_0 and H_A. But H_0 and H_A each imply a different pseudo-variate with means resulting pseudo-likelihoods are not comparable.
- Same issues apply to use of pseudo-likelihood estimation with conditional model.
- This means *students and data analysts new to GLMMs be careful*: pseudo-likelihood is the default GLIMMIX algorithm—you *must* override it with METHOD = LAPLACE or METHOD = QUADRATURE to do inference on covariance components with non-Gaussian data.

Exercises

6.1 This problem refers to data and two GLIMMIX runs given in file **ch6 _ probl.sas**, found in the SAS Data and Program Library.

The data are from any study with three factors: A, B, and storage time. Factors A and B, each with two levels, are manufacturing and finishing processes, respectively. Lots made by each of the four A × B combinations are placed in storage units—there are six storage units per A × B combination. A sample of the lot is removed after 1, 2, 4, 8, and 16 weeks of storage to examine the strength of the product. Sampling is destructive, so you can assume measurements at the various weeks are mutually independent. Also, you can assume "strength" has a Gaussian distribution.

There are two GLIMMIX programs called /* Run 1 */ and /* Run 2 */, respectively.

a. Give a complete description of the model GLIMMIX is processing in these two runs.
b. What is the name of the default method Run 1 uses to estimate the variance components?
c. What is the main difference, in terms of mixed model theory, between these two runs?
d. Which method is preferred (that is, which of the two should you report)? Explain briefly.
e. Formally, in terms of the model parameters you gave in part (a), what does "storage_time*a*b" test in these runs?
f. If you were explaining this to a nonstatistician on the research team for this project, how would you interpret the "storage_time*a*b" test and the result? Hint: an interaction plot in "Run 3" is provided to help you answer this question.

6.2 This problem refers to data and two GLIMMIX runs given in file **ch6 _ prob2.sas** found in the SAS Data and Program Library.

The data are from an incomplete block design with seven treatments and seven blocks. Assume that the data are Gaussian.

a. Use the GLIMMIX COVTEST option to test the hypothesis of homogeneity of variance by treatment for the conditional distribution of the observations given block effects.
b. Depending on your result in (a), write the appropriate linear mixed model for these data.
c. Using (b), what do you conclude about differences among treatment mean response? Assume trt 5 is the "control" or reference treatment. Cite appropriate evidence.

6.3 This problem refers to data and data step statements given in file **ch6 _ prob3.sas** found in the SAS Data and Program Library.

The data are from a randomized complete block design with three treatments and six blocks. The response variable (Y) is a count.

a. Run a standard block-effects-random linear mixed model analysis. Make note of the variance component estimates, *F*-test for treatment, and estimates (and standard errors) of the treatment means.
b. Rerun the analysis using the log count, a standard, pre-GLMM variance stabilizing transformation. Note that "log" means *natural log*. For zero counts, you add a small increment, $\varepsilon > 0$, to the count before taking the log—for example, in SAS, `if y=0 then log _ y=log(y+0.1)`. Obtain the variance component estimates, *F*-value for treatment, and the estimates (and standard errors) of the treatment means *on the data scale*.
c. Rerun the analysis as a GLMM assuming counts given blocks have a Poisson distribution. Use the same linear predictor than you did in parts (a) and (b). Obtain the variance component estimates, *F*-value for treatment, and the estimates (and standard errors) of the treatment means *on the data scale*.
d. Obtain the *Pearson chi-square/DF* fit statistic for the GLMM from part (c).
e. Rerun the GLMM from part (c) but add a random block*trt effect to the linear predictor and obtain the likelihood ratio test of $H_0: \sigma^2_{BT} = 0$. The statistics

from (d) and the test from (e) have essentially the same interpretation. Explain briefly.

f. Characterize the impact of the change in the linear predictor between parts (c) and (e) on the F-test and estimates (and standard errors) of treatment means *on the data scale.*

g. Pick the GLMM (part c or part e) most appropriate to report and compare and contrast the variance component, *F*- and data scale estimate results for the GLMM, log transform LMM, and untransformed LMM.

6.4 This problem refers to data and data step statements given in file **ch6 _ prob4.sas**, found in the SAS Data and Program Library.

This is a set of repeated measures data. Two treatments (coded "0" and "1" in the data set) are randomly assigned to eight subjects each (denoted ID in the data set. Note ID=1 for trt "0" is NOT the same as ID=1 for trt "1"—they are nested, not crossed). Each subject is observed weekly beginning at WEEK=0. The observation at WEEK=0 is NOT a baseline measurement. Assume that it is the first measurement taken AFTER the treatment has been applied to the subject. Also assume the response variable is Gaussian.

a. Complete the following table for the specified plausible covariance models. Assume week is a CLASS variable.

		Test for Trt*Week			
		Without KR		With KR	
Covariance Model	AICC Fit Criterion	*F*-Value	*p*-Value	*F*-Value	*p*-Value
CS					
First-order autoregressive [AR(1)]					
First-order antedependence [ANTE(1)]					
Unstructured (UN)					

Answer (b) and (c) in the following based on information contained in this table.

b. What in the impact of increasing complexity of covariance model on the test of treatment by week interaction? How is this affected by the Kenward-Roger adjustment?

c. Verify that the first-order ante-dependence covariance model provides the best fit (among these models). Why?

d. For the 1st order ante-dependence model, also determine the *F*- and *p*-values for the treatment by week test using the sandwich estimator instead of the model-based statistic. Do this determination for the classical and Morel-adjusted sandwich estimates.

e. Compare the sandwich estimate results for the ANTE(1) model to the sandwich estimate results for the AR(1) and UN models. How does the choice of covariance model affect the *F*- and *p*-values compared to the impact of covariance models on model-based estimates?

6.5 This problem refers to data and data step statements given in file **Ch6 _ prob5.sas** found in the SAS Data and Program Library.

The data are from repeated measures study with a binomial response. There are 12 subjects (coded by variable ID); 6 are assigned—at random—to each of 2 treatments. Each treatment is observed at five equally spaced times (same times for both treatments).

The SAS programs shown in the file show the researcher's attempts to determine the correct covariance structure to use for the analysis. To summarize, the researcher started assuming an unstructured working covariance model. Tests were done (1) to see if evidence of serial correlation exists (the "DiagR" test) and, if so, is there any reason to use an unstructured covariance model or could you simplify the structure to a Toeplitz model. The researcher also started with a first-order ante-dependence model [ANTE(1)] and worked through a series of simplifications ending with an AR(1). Assuming the simplification arrived at Toeplitz and AR(1) working covariance models, then the two models were compared on the basis of the fit statistics provided (e.g., −2 log PL and generalized χ^2/DF) to choose the winner.

a. Suspend disbelief about the researcher's methodology for now. Based on the researcher's logic flow, which of the following covariance structures would you "rule out" and "rule in"?

Covariance Model	Rule In or Out?	Relevant Evidence for Choice
Unstructured (UN)	In Out	
ANTE(1)	In Out	
ARH(1)	In Out	
AR(1) + id(trt)	In Out	
AR(1)	In Out	
Toeplitz	In Out	
CS	In Out	
Indep	In Out	

b. There is a fatal flaw in the researcher's methodology rendering your answers in part (a) unusable. What is it (you should be able to answer this in at most a sentence or two)?

c. As a general principle, how should models be respecified so covariance model selection can actually be done *legitimately*?

d. Redo the model selection consistent with your answer in part (c). What covariance model do you select on this basis? Document your steps and supporting evidence.

e. Without going into detail, what would you conclude about the presence or absence of treatment and/or time effects based on the model you selected in part (d)? Evidence?

Part III

Working with GLMMs

7

Treatment and Explanatory Variable Structure

7.1 Types of Treatment Structures

This chapter begins the third part of this textbook. In the first three chapters, we laid the groundwork for linear modeling and presented the major issues that "generalized" and "mixed" inject into the modeling picture. In the next three chapters, we developed generalized linear modeling's theoretical and methodological underpinning. For the remainder of this book, we turn our attention to the uses of GLMMs and the particulars of applying them in practical situations.

In this chapter, we focus on the types of treatment or, more generally, explanatory variable structures that give rise to $\mathbf{X}\boldsymbol{\beta}$, the fixed effects component of the linear predictor, and how we partition $\mathbf{X}\boldsymbol{\beta}$ meaningfully given the treatment design. There are decompositions of $\mathbf{X}\boldsymbol{\beta}$ that make sense mathematically but are incoherent in the context of the data being modeled. Other decompositions may seem reasonable from the viewpoint of the investigator who collected the data but are not mathematically viable. The modeler's goal is to identify an $\mathbf{X}\boldsymbol{\beta}$ that satisfies both requirements.

Broadly speaking, we can categorize treatment designs and explanatory variables in two different ways:

1. Single factor or multiple factor structures
2. Quantitative or qualitative factors

For single factor structures, the crucial decision we need to make is whether we have a qualitative factor—in which case the linear predictor will be along the lines of $\eta + \alpha_i$, often called an ANOVA-type model—or a quantitative factor—in which case we usually should consider using a regression-based linear predictor, for example, $\beta_0 + \beta_1 X$. When we have more than two levels of a factor, we usually want to test more than one hypothesis about treatment differences. This has an impact on type I error rate that we need to consider.

For multifactor structures, besides deciding how to model each factor depending on whether it is quantitative or qualitative, three additional issues come into play. The first involves partitioning the linear predictor into meaningful effects to capture each factor's impact on the response and the independence—or lack thereof—of their impact. The second involves the order of fitting factor effects—which effect do we fit first, which one do we fit last? Does this order matter? If so, how? Is there a "right" order? If not, how do we remove the impact of order from subsequent inference? Finally, the impact on type I error rate resulting from multiple hypothesis tests becomes more complex with multifactor studies.

In Section 7.2, we consider the order-of-testing issue starting with GLMMs roots in classical ANOVA and regression analysis. In Section 7.2, we also take a brief look at *multiplicity*, a complication that arises when we test multiple hypotheses from a single model. Section 7.3 presents an overview of the various cases of multifactor models arising from the possible combinations of factors. Then we take a more detailed look at each case: all qualitative (Section 7.4), some qualitative, some quantitative (Section 7.5), and all quantitative (Section 7.6).

A final comment: the linear predictors we discuss in this chapter may be used with *any* linear model—the GLMM or any of its special cases, the LMM, GLM, or LM. None of the discussion in this chapter is LMM or GLM or LM specific.

7.2 Types of Estimable Functions

In this section, we focus on two issues. The first three subsections focus on two testing strategies—partial and sequential, also known as type 1, type 2, and type 3—that linear models, generalized, mixed, both, or neither, have inherited from classical ordinary least squares theory. The final subsection concerns alternative ways to control type I error when we test multiple hypotheses in the same analysis.

7.2.1 Relation to Classical ANOVA Reduction Sums of Squares

The theory of estimable functions grew out of classical linear model's ANOVA and regression roots. These roots are steeped in the use of sums of squares and mean squares—quadratic forms in more formal terms—as the building blocks of test statistics. The classical point approach to assessing a multifactor model uses *reduction* sums of squares. Searle (1971) has a comprehensive exposition of this approach, and PROC GLM, SAS's comprehensive ordinary least squares LM procedure, was strongly influenced by Searle. When "generalized" and "mixed" entered the picture, for example, PROC GENMOD, PROC MIXED, and, most recently, PROC GLIMMIX, although sums of squares no longer had any useful meaning, they retained the reduction sum of squares thought process to define *type 1*, *type 2*, and *type 3* estimable functions. We need to understand what these are in order to proceed.

Following Searle's approach, suppose we have a three factor model that results in a linear predictor partition $\mathbf{X\beta}=\mathbf{X}_A\boldsymbol{\beta}_A+\mathbf{X}_B\boldsymbol{\beta}_B+\mathbf{X}_C\boldsymbol{\beta}_C$, that is, $\mathbf{X}=[\mathbf{X}_A\ \mathbf{X}_B\ \mathbf{X}_C]$ and $\boldsymbol{\beta}'=[\boldsymbol{\beta}'_A\ \boldsymbol{\beta}'_B\ \boldsymbol{\beta}'_C]$. If we fit the full model, the sum of squares for the model corresponds to the quadratic form $\mathbf{y}'\mathbf{X}(\mathbf{X}'\mathbf{X})^-\mathbf{X}'\mathbf{y}$. Searle denotes this by $R(\boldsymbol{\beta}_A, \boldsymbol{\beta}_B, \boldsymbol{\beta}_C)$ the reduction sums of squares for the full model. Recall that this is a key building block of the Wald statistics discussed in Chapter 5. Now suppose we fit the submodels in sequence. Starting with factor A, we have the reduction sums of squares for A, $R(\boldsymbol{\beta}_A)=\mathbf{y}'\mathbf{X}_A(\mathbf{X}'_A\mathbf{X}_A)^-\mathbf{X}'_A\mathbf{y}$. Then, fitting B, we define the reduction sums of squares for B over and above fitting A, or B given A, $R(\boldsymbol{\beta}_B\,|\,\boldsymbol{\beta}_A)=R(\boldsymbol{\beta}_B,\boldsymbol{\beta}_A)-R(\boldsymbol{\beta}_A)=\mathbf{y}'\mathbf{X}_{AB}(\mathbf{X}'_{AB}\mathbf{X}_{AB})^-\mathbf{X}'_{AB}\mathbf{y}-\mathbf{y}'\mathbf{X}_A(\mathbf{X}'_A\mathbf{X}_A)^-\mathbf{X}'_A\mathbf{y}$ where $\mathbf{X}_{AB}=[\mathbf{X}_A\ \mathbf{X}_B]$. With some manipulation, we can write $R(\boldsymbol{\beta}_B\,|\,\boldsymbol{\beta}_A)$ as $\mathbf{y}'\mathbf{X}_B\left[\mathbf{X}'_B\left(\mathbf{I}-\mathbf{X}_A(\mathbf{X}'_A\mathbf{X}_A)^-\mathbf{X}'_A\right)\mathbf{X}_B\right]^-\mathbf{X}'_B\mathbf{y}$.

Continuing this process, we end up with two strategies for assessing the parameters associated with each submodel. They are

Sequential tests:

- A: $R(\boldsymbol{\beta}_A)$
- $B|A: R(\boldsymbol{\beta}_B|\boldsymbol{\beta}_A)$ (7.1)
- $C|A, B: R(\boldsymbol{\beta}_C|\boldsymbol{\beta}_A, \boldsymbol{\beta}_B) = R(\boldsymbol{\beta}_A, \boldsymbol{\beta}_B, \boldsymbol{\beta}_C) - R(\boldsymbol{\beta}_A, \boldsymbol{\beta}_B)$

Partial tests:

- $A|B, C: R(\boldsymbol{\beta}_A|\boldsymbol{\beta}_B, \boldsymbol{\beta}_C)$
- $B|A, C: R(\boldsymbol{\beta}_B|\boldsymbol{\beta}_A, \boldsymbol{\beta}_C)$ (7.2)
- $C|A, B: R(\boldsymbol{\beta}_C|\boldsymbol{\beta}_A, \boldsymbol{\beta}_B)$

7.2.2 How Do We Know What We Are Testing?

By taking the expected values of the various reduction sums of squares, we can determine the form of the estimable functions implicit in each testing strategy. These can be summarized as follows.

Define $\mathbf{M}_A = \mathbf{I} - \mathbf{X}_A(\mathbf{X}_A'\mathbf{X}_A)^{-}\mathbf{X}_A$ and $\mathbf{M}_B = \mathbf{M}_A\left[\mathbf{I} - \mathbf{X}_B(\mathbf{X}_B'\mathbf{M}_A\mathbf{X}_B)^{-}\mathbf{X}_B\right]\mathbf{M}_A$. Then the sequential strategy yields the following generators of implicit estimable functions for comparing levels within a factor:

- A: $(\mathbf{X}_A'\mathbf{X}_A)^{-}(\mathbf{X}_A'\mathbf{X}_A)\boldsymbol{\beta}_A + (\mathbf{X}_A'\mathbf{X}_A)^{-}(\mathbf{X}_A'\mathbf{X}_B)\boldsymbol{\beta}_B + (\mathbf{X}_A'\mathbf{X}_A)^{-}(\mathbf{X}_A'\mathbf{X}_C)\boldsymbol{\beta}_C$
- $B|A: (\mathbf{X}_B'\mathbf{M}_A\mathbf{X}_B)^{-}\mathbf{X}_B'\mathbf{M}_A\mathbf{X}_B\boldsymbol{\beta}_B + (\mathbf{X}_B'\mathbf{M}_A\mathbf{X}_B)^{-}\mathbf{X}_B'\mathbf{M}_A\mathbf{X}_C\boldsymbol{\beta}_C$ (7.3)
- $C|A, B: (\mathbf{X}_C'\mathbf{M}_B\mathbf{X}_C)^{-}\mathbf{X}_C'\mathbf{M}_B\mathbf{X}_C\boldsymbol{\beta}_C$

Making appropriate substitutions, we can obtain the implicit estimable functions for the tests of A and B using the partial test strategy, that is, $A|B, C$ and $B|A, C$.

Observe that when we use the sequential approach, with the exception of perfectly balanced data, estimable comparisons among A effects will have some B and C components that we cannot eliminate, and estimable comparisons among levels of factor B include C components that we cannot eliminate.

For example, recall the unbalanced main effects design (Data Set 5.3.1) we discussed in Chapter 5. If we use the sequential approach, the generating functions for factor A comparisons are $\eta + \alpha_1 + (1/3)(\beta_1 + 2\beta_2) + (1/3)(2\gamma_1 + \gamma_2)$ and $\eta + \alpha_2 + (1/3)(2\beta_1 + \beta_2) + (1/3)(2\gamma_1 + \gamma_2)$. Taking the difference, the sequential test of the A effect is $\alpha_1 - \alpha_2 + (1/3)(-\beta_1 + \beta_2)$; the sequential test of the A effect is confounded with the B effect. For each reordering of the model, we redefine the generating equations in (7.3), which is why the results testing the model effects were all over the map depending what order we fit the effects and whether we used a sequential or partial test strategy.

We can apply the generating functions in (7.3) to GLM, LMM, and GLMM as well as ordinary least squares linear models. Their defining feature is the fixed effect component of the linear predictor. *Type 1* estimable functions for all linear models use (7.3) to generate the **K** matrix in the estimable function $\mathbf{K}'\boldsymbol{\beta}$. *Type 2* and *type 3* estimable functions use the generators implicit in the partial strategy—the C|A, B, A|B, C, and B|A, C generators. In many models, type 2 and type 3 strategies are identical. For models with interaction effects, there are differences. We will discuss these in Section 7.3.

7.2.3 How to Decide What to Test Rather than Letting It Be Decided for Us

Notice that only the partial strategy results in hypotheses about factor effects that are not confounded with other factors. This is especially important for testing unbalanced qualitative effect data, such as the main effects example we just considered. You should not interpret this to mean that we should always use partial tests. We saw in Chapter 5 with the regression example (Data Set 5.3.2) that a partial test yields nonsense. Remember the prime operating principle for modeling: one-size-fits-all recipes = poor statistical practice.

A better general rule is as follows: if there is an *obvious* order for fitting models (e.g., polynomial regression on a single factor) use it; otherwise, think in terms of estimable functions that make sense in context. That is a far better statistical practice. Coincidentally, for situations that lack an obvious model-fitting order, when you think in terms of meaningful estimable functions, you usually end up with type 3 tests, but not always. It is much better to end up at the type 3 destination by thinking it through rather than choosing blindly between partial or sequential tests without understanding the implications.

Recalling our discussion of the main effects design from Chapter 5, we concluded by saying that we should define estimable functions that make sense in context. Specifically, use marginal means averaged over the effect estimates of the other factors and the resulting differences not confounded with the other factors. Doing so, we essentially end up using a partial testing strategy. However, we arrive via a more transparent thought process that clearly articulates "why are we doing this?" This, approach becomes a bit more involved when we extend it to multifactor models with interaction terms. We will continue this discussion in Sections 7.4 through 7.6 as we consider the variations on the multifactor model.

7.2.4 Multiplicity

For any treatment design with more than two treatments, we can understand type I error rate in two different ways. The first, called the *comparison-wise* error rate, focuses on the probability of a type I error for each individual test. The other, called the *experiment-wise error rate,* focuses on the probability of *at least one* type I error over the entire set of tests we conduct.

To illustrate, suppose we have three treatments, one a control or reference treatment, the other two experimental treatments we wish to evaluate. Denote their means μ_C for the control and μ_A and μ_B for the two experimental treatments. Let us say we test the equality of the two experimental treatments, H_0: $\mu_A = \mu_B$, and, if the evidence suggests that they are equal, then we test their average against the control, H_0: $\mu_C = (\mu_A + \mu_B)/2$. Now we set an α-level according to our assessment of the seriousness of making a type I error (or, as is more typical, we set it according to the dictates of a mandated protocol). Using the comparison-wise error rate, the α-level defines the criterion for rejecting H_0 for *each test,* one test at a time. If we do this, under H_0, the probability of not rejecting the null hypothesis equals $1-\alpha$. If we do two independent tests (as we are doing here—more about independent, a.k.a. orthogonal, tests in the following), the probability of not making any type I errors equals $(1-\alpha)^2$, and, therefore, the probability of at least one type I error rate is $1-(1-\alpha)^2$; this is the experiment-wise error rate for this example.

In general, if our treatment design has t treatments, we have $t-1$ degrees of freedom. This means we can define up to $t-1$ mutually independent comparisons or, putting it in estimable function terms, we construct $\mathbf{K}'\boldsymbol{\beta}$ so that *rank* $(\mathbf{K}) \le t-1$, with exact equality if we define exactly one comparison per degree of freedom, and the rows of $\mathbf{K}$ define mutually independent treatment differences. If we do this, the experiment-wise error rate equals $1-(1-\alpha)^{t-1}$. Limiting ourselves to $t-1$ comparisons does not guarantee mutual independence. In any event, mutual independence does not stand on its own as a worthy criterion for sets of contrasts. As we shall see later, there may be more scientifically coherent comparisons that are not mutually independent. A more general statement of experiment-wise error rate is as follows: denoting the comparison-wise error rate as α_C, for k comparisons the experiment-wise error rate $\alpha_{EW} \ge 1-(1-\alpha_C)^k$.

How do we know if comparisons are independent? Here, we introduce *orthogonal contrasts*. For a *one-way* treatment design, we have the following definitions:

- *Contrast*: an estimable function is a *contrast* if $\mathbf{k}'\boldsymbol{\beta}$ can be written as $\sum_i k_i\mu_i$ and $\sum_i k_i = 0$. For GLMs and GLMMs, the definition holds if we replace μ_i by μ_i^*, the mean of the pseudo-variate for the ith treatment. Notice that a contrast must refer to an estimable function for which $\mathbf{k}$ is a vector.
- *Orthogonal contrasts*: Let $\mathbf{k}_1'\boldsymbol{\beta}$ and $\mathbf{k}_2'\boldsymbol{\beta}$ be two contrasts. They are orthogonal if and only if

$$\mathbf{k}_1'\mathbf{k}_2 = 0 \tag{7.4}$$

In a one-way treatment design, a *complete set of orthogonal contrasts* must have exactly $t-1$ mutually orthogonal contrasts, one per treatment degree of freedom. This is the only case for which exact equality, $\alpha_{EW} = 1-(1-\alpha_C)^{t-1}$, holds for the experiment-wise error rate. Otherwise $\alpha_{EW} > 1-(1-\alpha_C)^k$ for k comparisons.

As an illustration, let us consider two four-treatment designs. First, the four treatments are

Treatment	Drug	Dose
1	Control	Low
2	Control	High
3	Experimental	Low
4	Experimental	High

Suppose that the objectives call for testing the following three hypotheses:

H_0: $(\mu_1+\mu_2)/2 = (\mu_3+\mu_4)/2$, that is, mean response to control and experimental drugs equal
H_0: $\mu_1 = \mu_2$, that is, no effect of dose for the control drug
H_0: $\mu_3 = \mu_4$, that is, no effect of dose for the experimental drug

This is an example of a complete set of orthogonal contrasts. You can check that

- $\mathbf{k}_1' = (1/2)\begin{bmatrix}1 & 1 & -1 & -1\end{bmatrix}$
- $\mathbf{k}_2' = \begin{bmatrix}1 & -1 & 0 & 0\end{bmatrix}$
- $\mathbf{k}_3' = \begin{bmatrix}0 & 0 & 1 & -1\end{bmatrix}$
- $\mathbf{k}_c'\mathbf{1} = 0$; for all $c = 1,2,3 \Rightarrow$ all three are contrasts
- $\mathbf{k}_c'\mathbf{k}_{c'} = 0$ for all $c \neq c' = 1, 2, 3 \Rightarrow$ the contrasts are mutually orthogonal

Orthogonal contrasts, while mathematically tidy, may also be unsuitable. For many experiments, the objectives defy the restrictions imposed by orthogonal contrasts. Consider our second four-treatment design:

Treatment	Drug
1	Control
2	Experimental 1
3	Experimental 2
4	Experimental 3

Here, an obvious set of mean comparisons is

$$H_0\colon \mu_{control} = \mu_{\exp 1}$$

$$H_0\colon \mu_{control} = \mu_{\exp 2}$$

$$H_0\colon \mu_{control} = \mu_{\exp 3}$$

Each is a contrast, but a quick check reveals that this set of comparisons fails the orthogonality criterion (7.4).

Even in the drug-dose example given earlier, we could reasonably ask "why not compare the two drugs at low (or high) dose?" or "why just compare drugs averaged over doses; why not the other way around?" In many cases, the objective is more general discovery—we simply want to compare all pairs of treatments and see which, if any, are different. An extreme version of this occurs in micro-array testing, where we test thousands of genes to see if any have an effect.

This kind of testing raises the problem of *multiplicity*. If we only test one hypothesis in an experiment, we can set the type I error rate at α. However, only rarely is this the case. Usually, we have $t > 2$ treatments, meaning we can define $t - 1 > 1$ orthogonal contrasts, if they make sense, $t - 1$ meaningful but nonorthogonal contrasts if the treatment design warrants it (e.g., our second four-treatment design), or, if we want to compare all possible pairs of treatments, there are a total of $t(t - 1)/2$ pairs to compare.

Even with orthogonal contrasts, the experiment-wise error rate is greater than the comparison-wise error rate. If we want to make all possible comparisons, it is easy to see that the experiment-wise error rate can increase dramatically.

The question, in practical data analysis, is which type I error rate should we control: the comparison-wise or experiment-wise error rate? There is no one-size-fits-all answer. Here are several strategies. There are others. We list these because they are available with

PROC GLIMMIX, the procedure we will use for the remainder of the textbook to implement most of our examples.

- *Protected LSD*: The LSD test is simply a *t*-test as defined in Chapter 5, with variations for the GLMM, LMM, GLM, and LM, for pairs of treatment means. We implement each test using the comparison-wise error rate, but we precede these tests with a decision rule: if we fail to reject overall *F*-test for treatment, H_0: all μ_i equal, at the specified α-level, we stop. We only do the individual pairwise *t*-tests if we reject the overall treatment hypothesis.
- *Bonferroni*: We set the comparison-wise rejection criterion to α/k, where *k* is the total number of comparisons we plan to do. If we only have one comparison, the unprotected LSD and Bonferroni-adjusted rejection criteria are the same. If we have 10 treatments and do all 45 paired comparisons, the adjusted Bonferroni criterion is $\alpha/45$. For $\alpha = 0.05$, the comparison-wise error rate is 0.05 for the LSD vs. 0.0011 using Bonferroni.
- *Tukey's "Honestly Significant Difference"*: This uses studentized ranges. These are covered in most introductory statistical methods textbooks, for example, Snedecor and Cochran (1989) and Steel et al. (1996).
- *Scheffé's procedure*: This uses a test based on the overall critical value of F. See Westfall et al. (1999) for details.
- *Sidak's procedure*: Essentially starts with the desired experiment-wise error rate and solves for the implied comparison-wise criterion. This is easily done when all comparisons are mutually orthogonal. There is more going on when the comparisons are not independent. See Sidak (1967) for details.

There are two other adjustments for specialized purposes:

1. *Dunnett's procedure*: This is a specific correction intended for a set of comparisons between "control" and each treatment, for example, as shown earlier in the second treatment design. Many applications, including PROC GLIMMIX, use a correction developed by Dunnett (1955) and Hsu (1992). The control versus each treatment comparison set is the only intended application for Dunnett's procedure.
2. *Nelson's procedure*: This results in each treatment being tested against the mean of all levels of a given treatment factor. No means are compared individually directly against one another. See Nelson (1982, 1991, 1993) for details.

7.3 Multiple Factor Models: Overview

We now extend our discussion to multiple-factor treatment or explanatory variable designs. These are studies with two or more factors, which we generically refer to as factors with names from the alphabet, that is, factor A, factor B, etc. These include explanatory variable structures with cross-classification or nesting or possibly both if we have three or more factors. In this section, we review essential concepts, notation, and vocabulary and give an overview of the ways we might meaningfully partition the linear predictor, depending on the context. These partitions can be classified as all qualitative, qualitative–quantitative, and all quantitative. Sections 7.4 through 7.6 discuss each in turn.

For our discussion of factorial structures, we use the following common terminology. We refer to factors by uppercase letters in alphabetical order: factor A and factor B for a two-factor structure, factor C if we add a third factor, etc. Lowercase letters in italic denote the number of levels of a factor: *a* levels of factor A, *b* levels of factor B, etc. Latin letters α, β, and γ refer to model effects corresponding to factors A, B, and C, respectively. When we transition from ANOVA-type effects to regression-type effects, beginning in Section 7.5, we will replace ANOVA-type notation with regression notation, for example, $\beta_0+\beta_1 X$ instead of $\eta+\alpha_i$. A "multifactor" effect refers to any effect that involves more than one factor. A multifactor effect can be an interaction, for example, $(\alpha\beta)_{ij}$, or a nested effect, for example, $\beta(\alpha)_{ij}$.

Most of our discussion will focus on two-factor models. For the most part, working with three factor models involves straightforward extension of methods we use for two-factor models. When extension is not straightforward, there will be some discussion.

We partition multifactor models using two strategies—or three depending on our viewpoint. Despite widespread misconceptions, *there is* no *all-encompassing "right" way to partition the linear predictor in a multifactor model*. The two major strategies are *cross-classification* and *nesting*. We defined both and considered examples in previous chapters. A third strategy—if we choose to look at it this way—is not to partition at all, but simply work with the linear predictor η_{ij} for the *ij*th factor A × factor B combination and define all inference of interest in terms of estimable functions. With the cross-classification strategy, we partition η_{ij} into main effects and interaction, for example, $\eta + \alpha_i + \beta_j + (\alpha\beta)_{ij}$. For the nesting strategy, the general form of the partition is $\eta + \alpha_i + \beta(\alpha)_{ij}$. Notice that this means $\beta(\alpha)_{ij} = \beta_j + (\alpha\beta)_{ij}$.

The cross-classification model makes more sense when we want to start inference by asking if the two factors affect the response independently or does nonnegligible interaction exist between the factors. The nesting model makes sense in two cases: (1) the treatment structure is not cross-classified, and (2) the treatment structure *is* cross-classified but we *know* interaction exists and we want to focus on simple effects. For the latter, we exploit the fact that we can interpret nested effects as simple effects—for example, $\beta(\alpha)_{ij}$ is a model parameterization of the simple effect of B given A.

Important point: For both the cross-classification and nested linear predictors, inference on the main effects, α_i and, where present, β_j, is *contingent* on the multifactor effects, $(\alpha\beta)_{ij}$ or $\beta(\alpha)_{ij}$ being negligible. If the multifactor effect is not negligible—for example, if we reject H_0: $(\alpha\beta)_{ij} = 0$ or H_0: $\beta(\alpha)_{ij} = 0$—then inference on the main effects comprises, to use Nelder's (1968) elegant phrase, a set of "uninteresting hypotheses." To put it more bluntly, inference on main effects is nonsense in the presence of nonnegligible multifactor effects.

There are several variations on the cross-classification and nesting partitions, depending on whether a given factor is qualitative or quantitative and the objectives of a particular study. With both cross-classification and nesting, it may make sense in certain contexts to suppress modeling the intercept explicitly, yielding $\alpha_i + \beta_j + (\alpha\beta)_{ij}$ for the cross-classification and $\alpha_i + \beta(\alpha)_{ij}$ for the nested linear predictor. Further modifications make sense for factors that are quantitative. For qualitative factor A and quantitative factor B, the linear predictors $\eta + \alpha_i + (\beta_1 + \delta_i)X_j$ or $\beta_{0i} + \beta_{1i}X_j$ may be preferable. Observe that these are variations on $\eta + \alpha_i + \beta_j + (\alpha\beta)_{ij}$ and $\alpha_i + \beta(\alpha)_{ij}$ with regression coefficients involving X_j, the quantitative levels of factor B, replacing ANOVA-type model effects. If both factors are quantitative, then we may choose a full multiple regression model, for example, $\beta_0+\beta_1X_1+\beta_2X_2+\beta_{11}X_1^2+\beta_{22}X_2^2+\beta_{12}X_1X_2$, where X_1 and X_2 denote levels of factors A and B, respectively. Observe that $\beta_1X_1+\beta_{11}X_1^2$ replaces α_i, $\beta_2X_2 + \beta_{22}X_2^2$ replaces β_j, and $\beta_{12}X_1X_2$ replaces $(\alpha\beta)_{ij}$.

Finally, a word about type 2 and type 3 tests. We saw earlier that type 1 tests correspond to sequential tests. For main-effects-only models, type 2 and type 3 tests are equivalent—both refer to partial tests. For factorial effects, there is a subtle difference. In terms of reduction sums of squares notation, they are

Type 2 tests

- A: $R(\text{A}|\text{B})$
- B: $R(\text{B}|\text{A})$
- A*B: $R(\text{A*B}|\text{A, B})$

With type 2 tests, main effects are tested after fitting other main effects; interactions are tested given that all main effects involved in the interaction have been fit. This is the classical definition of partial sums of squares for factorial ANOVA models.

Type 3 tests are somewhat unhelpfully defined as

- A: $R(\text{A}|\text{B, A*B})$
- B: $R(\text{B}|\text{A, A*B})$
- A*B: $R(\text{A*B}|\text{ A, B})$

When PROC GLM was introduced, this caused a great deal of confusion, not to mention much consternation in certain quarters. A better way to understand type 3 tests, and type 3 estimable functions in general, is in terms of estimable functions as we described them for the main-effects-only design. Start by writing the estimable functions for the main-effect-adjusted marginal means—what SAS calls least squares means—that is, average over all other model effects to get an idealized statement of what the mean would have been if the data had been perfectly balanced. Assuming missing data occur either by accident or as a result of deliberately incomplete designs such as the orthogonal main effects design, type 3 estimable functions essentially represent the missing-data method of choice, which is why they are the default test for GLMM effects models.

Notice that, for two-factor models, type 2 and type 3 tests differ only for main effects and then only if the data are imbalanced. You can use the methods discussed in Section 7.B to determine exactly what the type 2 approach tests and decide whether it, or the type 3 approach, best addresses your objectives. Again, the best way to resolve such question is *not* which reduction approach makes the most sense but what estimable function do you end up with?

7.4 Multifactor Models with All Factors Qualitative

With all factors qualitative we use effects models only. (Regression models don't make sense here—why?) If we have two factors, denoted factor A and factor B, we may have either a factorial—or cross-classified—structure, here denoted A × B, or a nested structure, denoted B(A). Recall from Chapter 2 the structural difference. We have a factorial structure if all AB combinations either have observations or *could have had* observations; whereas the structure is nested if observing a level of B in conjunction with a given level of factor A precludes observing that same level of B with any other level of A. Visually

	B1	B2	B3	B4
A1	☑	☑	☑	☑
A2	☑	☑	☑	☑

describes a factorial or cross-classified structure, whereas

	B1	B2	B3	B4
A1	☑	☑		
A2			☑	☑

describes a nested structure, presuming that B1 is being observed with A1 means that it cannot also be observed with A2. For example, in a breeding trial, if factor A is sire and factor B is dam, a dam can only mate with a single sire in a given breeding season. We say B is nested within A because a given level of B can only be assigned to one level of A, but not vice versa.

With three factors, all three may be cross-classified, which we denote $A \times B \times C$, or one may be nested within another but crossed with the third, for example, $A \times B(C)$, or we may have one factor nested within a factorial combination of the other two, for example, $C(A \times B)$, or we may have a strictly nested hierarchy, for example, C(B(A)). We continue in this fashion for four and more factors. Beyond three factors, there are too many possibilities to list. The structures take some attention to detail, but they are straightforward extensions of the principles we use for two and three factor structures.

Cross-classified structures use the factorial effects model—$\eta_{ij} = \eta + \alpha_i + \beta_j + (\alpha\beta)_{ij}$ for two factors—or some variation on this form. In certain designs, we may need to delete interaction terms. For example, in the main effects design discussed earlier in this chapter, Data Set 5.3.1, the design was not constructed to allow estimation of interaction effects. Nested structures use the nested effects model $\eta_{ij} = \eta + \alpha_i + \beta(\alpha)_{ij}$ or some variation on this form. We devote the rest of this section to examining in more detail how we work with these models. We illustrate with a two-factor, 2×3 factorial structure with Gaussian data. As we go through the example, it will become clear how we work with nested models. Also, although the example uses Gaussian data, at this point, the illustrated procedures apply to all linear models—GLMs, LMMs, and GLMMs as well as LMs.

7.4.1 Review of Options

Our building blocks are contrasts defined on interaction effects, simple effects, and main effects. Inference proceeds in sequence. First, we determine if nonnegligible interaction exists. The next step depends on the first. If interaction is negligible, inference centers on main effects. In the absence of interaction, inference on simple effects constitutes "uninteresting hypotheses" to use Nelder's characterization. If interaction is nonnegligible, inference centers on simple effects. In the presence of nonnegligible interaction, main effects inference constitutes "uninteresting hypotheses" and in many cases yields outright nonsense.

Inference for three-factor designs follows an extension of the same path. We start with the three-way interaction. If it is negligible, we proceed to two-way interactions. If they are *all* negligible, only then do we proceed to main effects. For any nonnegligible interaction, we direct our focus to simple effects contained within that interaction, choosing the simple effects that best inform us with regard to the objectives that motivated conducting the study.

For factors with more than two levels, we need to either partition variation using contrasts or use pair-wise or multiple comparison tests, described in Section 7.2. This *must* include partitioning interaction. Why? Consider a two-factor model. Factor A has a levels and factor B has b levels. The A × B interaction therefore has $(a-1)(b-1)$ degrees of freedom—that is, it is composed of $(a-1)(b-1)$ independent single degree of freedom contrasts. As a and b increase, the interaction degrees of freedom increase multiplicatively. It becomes possible for very substantial interactions to be concentrated in one or a small number of interaction contrasts, which are likely to be overwhelmed if the remaining contrasts that form the overall interaction effect are essentially zero. Unless we look explicitly at these contrasts, we may miss an important interaction and take subsequent inference down the wrong path.

Finally, with multiple factors, multiplicity can arise in different ways. In the presence of nonnegligible interaction, applying multiplicity adjustments within each set of simple effect comparisons may be appropriate. In other cases, it may be more reasonable to apply the multiplicity adjustment across the board to all treatment combinations simultaneously. In the absence of interaction, multiplicity adjustments, when used, are typically to the main effects comparisons within each factor.

In the next two sections illustrate these ideas. Section 7.4.2 presents the various inference tools. Section 7.4.3 concerns multiplicity adjustments appropriate for this example.

7.4.2 Tools for Qualitative Factorial Inference: "SLICE," "SLICEDIFF," and Other Tools

The data for our working example in this section appear in Data Set 7.1 of the SAS Data Set and Program Library. The treatment structure is a 2×3 factorial. The data are Gaussian and the design structure is completely randomized. We use the standard factorial linear model

- Linear predictor: $\eta_{ij} = \eta + \alpha_i + \beta_j + (\alpha\beta)_{ij}$
- Distribution: $y_{ij} \sim NI(\mu_{ij}, \sigma^2)$
- Link: identity, $\eta_{ij} = \mu_{ij}$

We start with approximate F-tests for the overall model effects, focusing first on the A × B interaction. Using GLIMMIX to do the computations, the required statements are

```
proc glimmix;
  class a b;
  model y=a b a*b;
```

The relevant output is

Type III Tests of Fixed Effects

Effect	Num DF	Den DF	*F*-Value	Pr>*F*
a	1	18	14.94	0.0011
b	2	18	11.50	0.0006
a*b	2	18	6.02	0.0099

The $p = 0.0099$ for the test of A × B interaction. This renders the results for the A and B main effects moot.

> *Attention students*! Do yourself a favor. When there is evidence of a nonnegligible interaction, you are done with the approximate *F*-tests. Many students cannot seem to discipline themselves on homework and exams and proceed to summarize the main effects results in detail even though they are irrelevant and "uninteresting." Spare yourself a bit of writers' cramp and your reader some time and aggravation: if the interaction is nonnegligible, **stop**!

Given the interaction result, we need to characterize it. The means and interaction plot for these data are

a*b Least Squares Means

a	b	Estimate	Standard Error
0	0	50.5000	2.5611
0	1	59.7500	2.5611
0	2	71.5000	2.5611
1	0	67.7500	2.5611
1	1	67.2500	2.5611
1	2	71.0000	2.5611

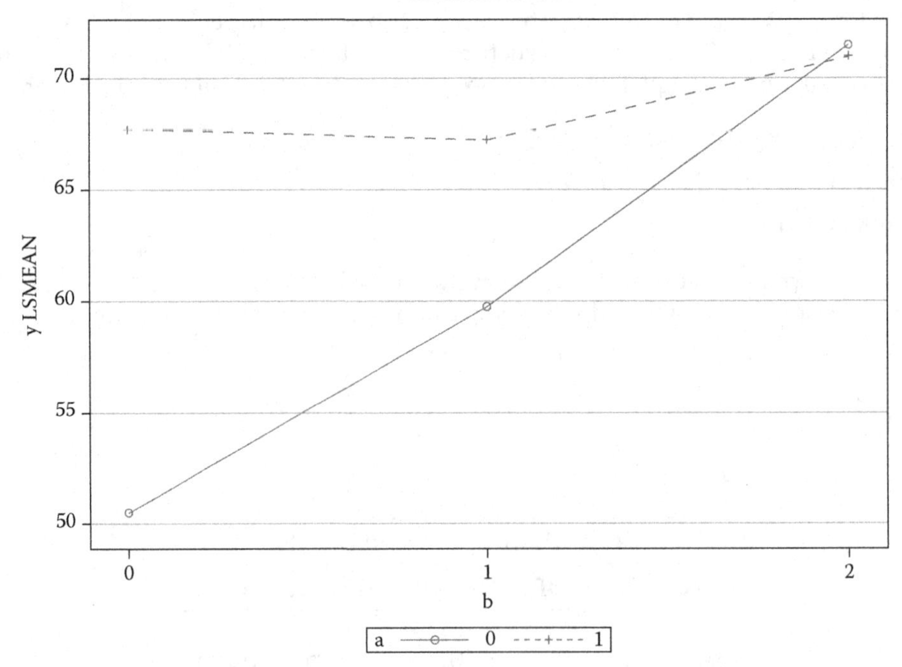

We connect the means of the levels of B for each level of A even though factor B is not quantitative as an aid to help visualize the interaction. There appears to be no mean A effect for B2, but a substantial A effect at B1 and even larger A effect at B0. Taking a different perspective, there appears to be no mean difference among the levels of B within A1 but substantial difference within A0. Both perspectives are correct, but usually one or the other will be more informative with regard to the objectives of the study.

Two tools are useful to decompose the interaction with qualitative factors: the overall simple effect *F*-test and pair-wise simple effect differences. SAS linear model procedures refer to these as SLICE and SLICEDIFF, respectively. SLICEDIFF is only available with GLIMMIX. Suppose we obtain simple effect statistics among the levels of B given A. The required GLIMMIX statement, which we include after the MODEL statement given earlier, is

```
lsmeans a*b/slice=a slicediff=a;
```

The relevant SLICE output is

Tests of Effect Slices for a*b Sliced by a

a	Num DF	Den DF	*F* Value	Pr>*F*
0	2	18	16.89	<0.0001
1	2	18	0.63	0.5429

The SLICEDIFF output:

Simple Effect Comparisons of a*b Least Squares Means by a

Simple Effect Level	b	_b	Estimate	Standard Error	DF	*t* Value	Pr > |*t*|
a0	0	1	−9.2500	3.6219	18	−2.55	0.0199
a0	0	2	−21.0000	3.6219	18	−5.80	<0.0001
a0	1	2	−11.7500	3.6219	18	−3.24	0.0045
a1	0	1	0.5000	3.6219	18	0.14	0.8917
a1	0	2	−3.2500	3.6219	18	−0.90	0.3814
a1	1	2	−3.7500	3.6219	18	−1.04	0.3142

Formally, the two SLICE tests are defined by the null hypothesis H_0: $\mu_{i0} = \mu_{i1} = \mu_{i2}$ for each $i = 0, 1$, and we can construct the estimable functions using any two contrasts that, taken together, imply H_0. For example

$$\mathbf{T}'\boldsymbol{\mu} = \begin{bmatrix} 1 & 0 & -1 & 0 & 0 & 0 \\ 0 & 1 & -1 & 0 & 0 & 0 \end{bmatrix} \begin{bmatrix} \mu_{00} \\ \mu_{01} \\ \mu_{02} \\ \boldsymbol{\mu}_1 \end{bmatrix}$$

where $\boldsymbol{\mu}$ denotes the vector of all μ_{ij} and $\boldsymbol{\mu}_1$ denotes the vector of all μ_{1j}. The SLICEDIFF hypotheses and estimable functions result from all possible pair-wise differences $\mu_{ij} - \mu_{ij'}$, where $j \neq j'$.

The output confirms what we see in the interaction plot. Substantial B effects occur within A0 but not A1.

Given the interaction, an alternative way to focus on the simple effects involves reexpressing the linear predictor as a nested model. If we want to focus on the simple effect of B given A, we use $\eta_{ij} = \eta + \alpha_i + \beta(\alpha)_{ij}$. Using GLIMMIX, we implement this model with the following statements:

```
proc glimmix;
  class a b;
  model y=a b(a);
  contrast 'slice: B | A0' b(a) 1 0 -1,
                           b(a) 0 1 -1;
  contrast 'slice: B | A1' b(a) 0 0 0 1 0 -1,
                           b(a) 0 0 0 0 1 -1;
  lsmeans b(a)/slicediff=a;
```

Output of interest is as follows:

Type III Tests of Fixed Effects				
Effect	**Num DF**	**Den DF**	***F* Value**	**Pr > *F***
a	1	18	14.94	0.0011
b(a)	4	18	8.76	0.0004

Contrasts				
Label	**Num DF**	**Den DF**	***F* Value**	**Pr > *F***
slice of B | A0	2	18	16.89	<0.0001
slice of B | A0	2	18	0.63	0.5429

The overall test of B(A) averages the simple effect slices for B|A0 and B|A1. This is an example of an overly aggregated statistic: By averaging two tests of interest, the resulting statistic does not provide any information we can use. On the other hand, notice that the contrast results are identical to the SLICE tests in using the factorial parameterization of the linear predictor—two different means to the same end. The nested parameterization often helps data analysts see more clearly exactly what they are testing. When there are missing AB combinations, use the nested model to avoid estimability problems.

Exercise for the reader: If you use the factorial parameterization and attempt to compute the contrast defined by the GLIMMIX statement,

```
contrast 'slice: B | A0' a*b 1 0 -1,
                         a*b 0 1 -1;
contrast 'slice: B | A1' a*b 0 0 0 1 0 -1,
                         a*b 0 0 0 0 1 -1;
```

What happens? Why does it happen? If you stay with the factorial parameterization, how must you modify the contrast statements to make them estimable?

7.4.3 Multiplicity Adjustments

We saw in the last section that inference in our working example targets the simple effects. If we focus on the simple effect of B given A, we have three possible pair-wise comparisons between levels of B for each level of A. We saw B given A SLICEDIFF results in the last section that use comparison-wise error rates. They are, in fact, selected LSD tests. If we want to control experiment-wise error rate, then we need to decide how to define it. We could define it to apply over all possible treatment combinations, but seems like overkill. There are 15 possible pair-wise comparisons but we only intend to do six, two sets of three pair-wise comparisons each. An adequate but less draconian approach would apply multiplicity corrections within each set of simple effect tests. For example, we could implement the Bonferroni correction using the following GLIMMIX statement, in conjunction with the factorial parameterization we used first.

```
lsmeans a*b/slicediff=a adjust=bonferroni;
```

The resulting output is as follows:

Simple Effect Comparisons of a*b Least Squares Means by a

Simple Effect Level	b	_b	Estimate	Standard Error	DF	*t* Value	Pr > |*t*|	Adj *P*
a0	0	1	−9.2500	3.6219	18	−2.55	0.0199	0.0598
a0	0	2	−21.0000	3.6219	18	−5.80	<0.0001	<0.0001
a0	1	2	−11.7500	3.6219	18	−3.24	0.0045	0.0135
a1	0	1	0.5000	3.6219	18	0.14	0.8917	1.0000
a1	0	2	−3.2500	3.6219	18	−0.90	0.3814	1.0000
a1	1	2	−3.7500	3.6219	18	−1.04	0.3142	0.9426

The only difference between this output and what we saw before is the last column. These are the Bonferroni-adjusted *p*-values. In this case, they do not affect the simple effect tests within A1—they were already nonsignificant—but they do temper our conclusion regarding the mean difference between B0 and B1 given A0.

In this case, the GLIMMIX algorithm applies the Bonferroni adjustment separately to each set of three simple effects within the levels of A. Compare to an alternative way to define the same simple effects using the LSMESTIMATE statement:

```
lsmestimate a*b 'b1 v b2 | a1' 1 -1 0 0 0 0,
                'b1 v b3 | a1' 1 0 -1 0 0 0,
                'b2 v b3 | a1' 0 1 -1 0 0 0,
                'b1 v b2 | a2' 0 0 0 1 -1 0,
                'b1 v b3 | a2' 0 0 0 1 0 -1,
                'b2 v b3 | a2' 0 0 0 0 1 -1/adjust=bonferroni;
```

This yields the following output:

Least Squares Means Estimates
Adjustment for Multiplicity: Bonferroni

Effect	Label	Estimate	Standard Error	DF	*t* Value	Pr > |*t*|	Adj *P*
a*b	b1 v 2 | a1	−9.2500	3.6219	18	−2.55	0.0199	0.1196
a*b	b1 v 3 | a1	−21.0000	3.6219	18	−5.80	<0.0001	0.0001
a*b	b2 v 3 | a1	−11.7500	3.6219	18	−3.24	0.0045	0.0270
a*b	b1 v 2 | a2	0.5000	3.6219	18	0.14	0.8917	1.0000
a*b	b1 v 3 | a2	−3.2500	3.6219	18	−0.90	0.3814	1.0000
a*b	b2 v 3 | a2	−3.2500	3.6219	18	−0.90	0.3814	1.0000

Here, the GLIMMIX algorithm applies the multiplicity adjustment to the entire set of six mean comparisons. Notice that this results in more conservative adjusted *p*-values, for example, 0.1196 for the simple effect of B1 vs. B2|A1 compared with 0.0598 obtained from the SLICEDIFF option stated earlier. Whether you use GLIMMIX or other software, you need to be aware of the conventions with which multiplicity adjustments are applied.

What if one of the factorial combinations is a "control" or reference treatment?

So far, we have discussed this example with generic identities for factors A and B. Let us suppose that A0 and A1 are actually varieties of a plant or strains of an animal. Suppose A0 is the common variety and is susceptible to a particular disease or pest. Suppose A1 is a newly developed variety that is resistant. Suppose B1 and B2 are two chemicals used to control the pest or disease and B0 is "untreated"—no chemical applied. Finally, the response, Y, is some index of the plant's health—or, more likely, the mean health of plants in the plot that forms the unit of replication.

Now, imagine that under current practice, producers grow variety A0 and control the disease or pest using B2. However, there is accumulating evidence that using B2 may be a problem—for example, an environmental or health hazard. Suppose that B1 is alleged to be safer, although its effectiveness has not been shown. The objective of the study would then be to determine if we can obtain acceptable performance from the susceptible variety, A0, using B1, or, better yet, obtain equivalent performance from the resistant variety, A1, without using any chemical at all (B0).

In this case, it makes sense to disregard the factorial structure of the treatments altogether. Instead, we could define treatment combination A0 × B2 as a "control" or reference treatment and implement Dunnett's test, that is, each treatment combination vs. A0 × B2 and use the Dunnett–Hsu multiplicity correction. The required GLIMMIX statement, used with the factorial parameterization, is

```
lsmeans a*b/diff=control('0' '2') adjust=dunnett;
```

The resulting output is as follows:

Differences of a*b Least Squares Means
Adjustment for Multiple Comparisons: Dunnett

a	b	_a	_b	Estimate	Standard Error	DF	*t* Value	Pr > |*t*|	Adj *P*
0	0	0	2	−21.0000	3.6219	18	−5.80	<0.0001	<0.0001
0	1	0	2	−11.7500	3.6219	18	−3.24	0.0045	0.0184
1	0	0	2	−3.7500	3.6219	18	−1.04	0.3142	0.7519
1	1	0	2	−4.2500	3.6219	18	−1.17	0.2559	0.6602
1	2	0	2	−0.5000	3.6219	18	−0.14	0.8917	1.0000

The last column shows the Dunnett–Hsu adjusted *p*-values. The next-to-last column shows the unadjusted comparison-wise *p*-values. Here, they both show essentially the same thing: no statistically significant evidence of a mean difference between resistant variety and the control treatment, regardless of chemical (level of B) used, but the susceptible variety (A0) without B2 does not compete well.

7.5 Multifactor: Some Factors Qualitative, Some Factors Quantitative

Quantitative factor levels often allow us to take advantage of regression relationships in order to simplify the linear predictor. We saw a simple form of this approach in Chapter 1 and pursued it with the multi-batch examples in Chapter 3 based on Data Set 3.3. In this section, we build on these examples.

Many design texts devote considerable attention to partitioning quantitative factor variation using orthogonal polynomial contrasts. This made sense in the precomputer era (orthogonal polynomials were developed in the 1940s), but it is wasted effort in an era of easy access to contemporary modeling technology. Littell et al. (2002, 2006) present examples illustrating preferable alternatives.

In this section, we tie together three applications of the qualitative–quantitative linear predictor commonly presented in separate and compartmentalized forms. Students and linear modeling practitioners frequently miss the point that these are variations on the same theme.

7.5.1 Generic Form of the Linear Predictor

Suppose we have a two-factor structure with factor A qualitative and factor B quantitative. Examples of such structures abound. Typically, the levels of factor A are different treatments (e.g., control or experimental) or conditions (e.g., exposed or not exposed to an environmental hazard). The levels of factor B can be time, age, dose, temperature, rainfall amount, some measure of the initial condition of an experimental unit, etc.

Often, the relationship between the response and the quantitative level of B can be described by regression. In such cases, we can replace the effects model for factor B with a more parsimonious model that captures the regression relationship. If the relationship between factor B and the response is linear, we can use a model like the multi-batch example from Data Set 3.3. We have seen two useful forms of the linear predictor:

- $\eta_{ij} = \eta + \alpha_i + (\beta_1 + \delta_i)X_j$ (7.5)
- $\eta_{ij} = \beta_{0i} + (\beta_{1i})X_j$ (7.6)

Notice that (7.5) is a special case of $\eta_{ij} = \eta + \alpha_i + \beta_j + (\alpha\beta)_{ij}$ with $\beta_j = \beta_1 X_j$ and $(\alpha\beta)_{ij} = \delta_i X_j$, and (7.6) is a special case of $\eta_{ij} = \alpha_i + \beta(\alpha)_{ij}$ with $\alpha_i = \beta_{0i}$ and $\beta(\alpha)_{ij} = \beta_{1i} X_j$.

When the relationship between the quantitative levels of factor B and the response is more complex than linear regression, we can simply augment (7.5) or (7.6) with additional polynomial regression terms, assuming a polynomial model fits, or substitute an alternative regression model—for example, spline or nonlinear—if these fit. We will look at an example of each in Section 7.6.

The linear predictors (7.5) and (7.6) describe latent growth curve models, analysis of covariance models, and quantitative–qualitative factorial design analysis. While these tend to be compartmentalized, from a modeling viewpoint, they simply represent different ways of interpreting the same model.

7.5.2 Many Uses of the Generic Linear Predictor

7.5.2.1 Latent Growth Curve Models

Disciplines as diverse as medical and pharmaceutical science, sociology and psychology, education, engineering, and agriculture, to name just a few, use variations on this model. In stability testing a drug, for example, we want to understand how stability-limiting characteristics change over time, possibly for different drugs. Letting factor A represent the different drugs and factor B represent time, we clearly have a treatment structure that can be modeled by the linear predictors (7.5) or (7.6) or a polynomial or nonlinear extension of (7.5) and (7.6). In education, we may wish to track student progress over grades and we may further wish to compare this progress for different groups or under different curricula. Again, letting the groups or curricula define factor A and grade level define factor B, we have a structure we can model using (7.5) or (7.6).

The details of latent growth curve analysis follow the examples from Chapter 3 based on Data Set 3.3, so there is no need to repeat them here.

7.5.2.2 Analysis of Covariance

We use analysis of covariance—commonly referred to as ANCOVA models—to account for concomitant variables. We use their relationship with the response variable to improve the accuracy of inference. Usually the concomitant variable, more frequently called the *covariate,* is a characteristic of the unit of replication that we measure at the beginning of the study. Examples include initial soil fertility, pretest score on a competence or achievement test, baseline measurement in a clinical trial, etc.

The most ANCOVA models build on the assumption of a linear relationship between response and covariate. Letting y denote the response and X denote the covariate, we use the linear predictor $\eta_j = \beta_0 + \beta_1 X_j$ to estimate the link $\eta_j = g(\mu_j)$, where $\mu_j = E(y_j)$, or the conditional mean when we extend ANCOVA to mixed models. If we assume the distribution of y is Gaussian, then we proceed as we have in other examples, using the identity link. This simple linear regression becomes an ANCOVA model by adding a treatment effect. ANCOVA distinguishes between two ways this happens: (1) treatment affects the intercept, but has no impact on the slope, and (2) treatment affects both the slope and the intercept. The former is called the *equal slopes ANCOVA* model; the latter is called the *unequal slopes ANCOVA* model. Note that the unequal slopes ANCOVA model is simply (7.5) or (7.6) depending on how we decide to parameterize it. The equal slopes ANCOVA model is (7.5) with all $\delta_i = 0$ or (7.6) with all $\beta_{1i} = \beta_1$, again, depending on how we decide to parameterize. Both parameterizations are equivalent, but (7.5) tends to be more conducive to testing and (7.6) makes estimation easier. Let us now illustrate these ideas with two examples.

7.5.2.2.1 Equal Slopes Example

This example uses Data Set 7.2. There are four treatments observed using a completely randomized design. The response variable is Gaussian. At the beginning of the experiment, a covariate, denoted X, was measured on every unit of replication in the study. At this point, we leave the identity of the treatments, response, and covariate unnamed—readers feel free to use your imagination and supply your own scenario.

All ANCOVA procedures start by testing the equality of slopes using the unequal slopes model. Formally, we start with the linear predictor $\eta_{ij} = \eta + \alpha_i + (\beta_1 + \delta_i)X_{ij}$, where X_{ij} is the covariate measured at the *ij*th unit of replication. We can implement this test using the following GLIMMIX statements:

```
proc glimmix;
  class trt;
  model y=trt x x*trt;
```

where TRT identifies the treatment and X refers to the covariate. The model term `X*TRT` corresponds to $\delta_i X_{ij}$ in the model; the corresponding F-value tests H_0: all $\delta_i = 0$. The output is as follows:

Type III Tests of Fixed Effects

Effect	Num DF	Den DF	*F* Value	Pr>*F*
trt	3	12	3.01	0.0722
x	1	12	60.82	<0.0001
x*trt	3	12	1.41	0.2889

The key output is the p-value for X*TRT, 0.2889; we fail to reject H_0: all $\delta_i = 0$, which means we may assume equal slopes. We proceed with inference on the treatments using the equal slopes model, $\eta_{ij} = \eta + \alpha_i + \beta_1 X_{ij}$. The required GLIIMIX statements are

```
proc glimmix;
  class trt;
  model y=trt x;
  lsmeans trt/diff;
```

We add the LSMEANS statement because, having settled the equal/unequal slopes issue, we want to proceed with inference on the treatment effects. The relevant output is as follows:

Type III Tests of Fixed Effects

Effect	Num DF	Den DF	*F* Value	Pr>*F*
trt	3	15	15.14	<0.0001
x	1	15	81.60	<0.0001

trt Least Squares Means

trt	Estimate	Standard Error
1	47.7120	1.1746
2	54.5984	1.1944
3	44.8822	1.2303
4	53.1131	1.2639

We can see that the p-value for H_0: all α_i equal is <0.0001; we reject the hypothesis of equal treatment means at any reasonable α level. The "least squares means" are adjusted in the sense that they predict what the treatment means would have been if every treatment were observed on completely equivalent units with respect to the covariate. Formally, $LSMean_i = \hat{\eta} + \hat{\alpha}_i + \hat{\beta}_1 \bar{X}_{\bullet\bullet}$, where $\bar{X}_{\bullet\bullet}$ is the overall mean of the covariate.

What would happen if we did not account for the covariate? In other words, if we simply use the usual linear predictor for a completely randomized design, $\eta + \alpha_i$, how would the test for treatment effect and the estimated treatment means be affected? Here are the results deleting the covariate from the model:

Type III Tests of Fixed Effects				
Effect	**Num DF**	**Den DF**	***F* Value**	**Pr>*F***
trt	3	16	3.20	0.0518

trt Least Squares Means		
trt	**Estimate**	**Standard Error**
1	45.7127	2.8344
2	57.3974	2.8344
3	48.7475	2.8344
4	48.4481	2.8344

Notice that the *p*-value for treatment effect is now 0.0518 instead of <0.0001, potentially changing the conclusions regarding the treatment effect. The standard errors of the treatment means, which ranged between 1.17 and 1.26 with the ANCOVA model, are now 2.83, a substantial loss of precision. The relative values of the means change dramatically: for example, with the ANCOVA model, the means for treatment 1 and 3 are 47.7 and 44.9, respectively. They essentially exchange places in the unadjusted model. What accounts for this? The unadjusted means actually estimate unadjusted $LSMean_i = \hat{\eta} + \hat{\alpha}_i + \hat{\beta}_1 \overline{X}_{i\bullet}$, where $\overline{X}_{i\bullet}$ denotes the mean of the covariates observed on only the units that received the *i*th treatment. Unless the randomization results in all $\overline{X}_{i\bullet}$ being exactly equal, the probability of which is essentially zero, the unadjusted means are confounded with the covariate and, hence, estimates of treatment differences are also confounded. Here is the plot of the observed covariates:

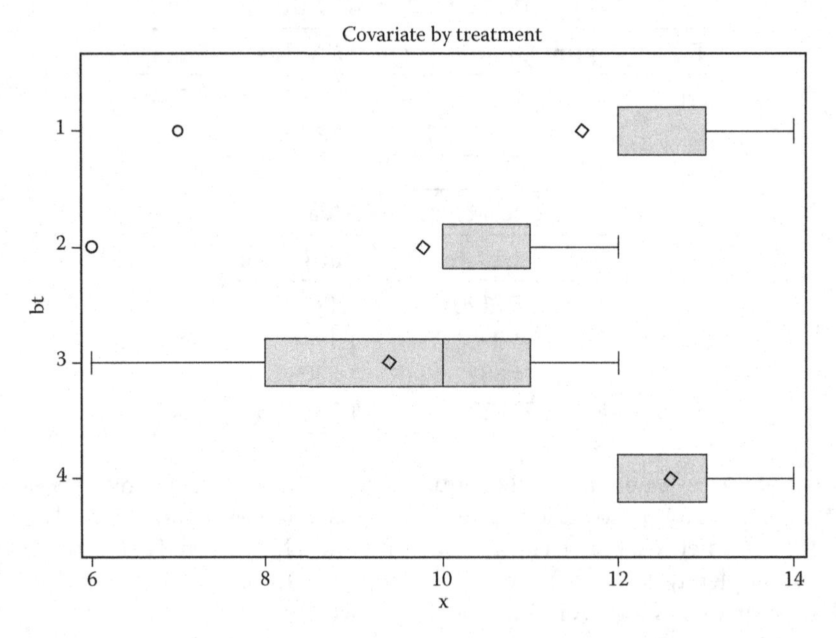

We can easily see that treatments 1 and 4, by the luck of the draw, were observed with distinctly greater average covariates than treatments 2 and 3. The estimated covariate regression coefficient is $\hat{\beta}_1 = -2.67$, meaning that the unadjusted means for treatments 1 and 4 are much lower relative to treatments 2 and 3 than they would have been if the randomization had resulted in an equal distribution of covariate values.

It is important that we stress that the covariates we see in the aforementioned plot *do not* mean a mistake was made in randomization. Randomization necessarily occurs *before* taking data on the covariates. Researchers and statistical scientists need to understand this and be realistic: The randomization we see here goes with the territory; it tends to be more the rule than the exception. Hence, the value of ANCOVA adjustment.

7.5.2.2.2 Unequal Slopes ANCOVA

Let us consider a similar setup but a different data set. The example appears in the SAS Data Set and Program Library as Data Set 7.3. We start with the same model and the same test as we did in the previous example: $\eta_{ij} = \eta + \alpha_i + (\beta_1 + \delta_i)X_{ij}$ and test the equal slopes hypothesis H_0: all $\delta_i = 0$. The results for these data are as follows:

Type III Tests of Fixed Effects

Effect	Num DF	Den DF	*F* Value	Pr>*F*
trt	3	12	6.62	0.0069
x	1	12	94.90	<0.0001
x*trt	3	12	8.40	0.0028

Recalling that *F*-value for x*trt tests H_0: all $\delta_i = 0$, the *p*-value of 0.0028 provides strong evidence that the slopes are *not* equal. Inference on treatment must proceed using the unequal slopes model. This means that for any test of the treatment effects, the estimable function must be $\alpha_i - \alpha_{i'} + (\delta_i - \delta_{i'})X$: there is no alternative that yields estimable functions. All estimates and tests of treatment differences are confounded with the covariate. This means that any conclusions about the treatment are covariate specific. This also raises the question: At what value of *X* was the treatment *F*-value of 6.62, $p = 0.0069$ computed? Two obvious values to try are $X = 0$—if the treatment *F* was computed based on differences among the α_i alone—and the overall mean of the covariates, $\bar{X}_{\bullet\bullet}$. Let us start with $X = 0$. A contrast that tests the equality of all α_i is

```
contrast 'trt at x=0' trt 1 0 0 -1,
                      trt 0 1 0 -1,
                      trt 0 0 1 -1;
```

The *F*-value for this contrast is 6.62, identical to the *F*-value for treatment given previously. On the other hand, for these data, $\bar{X}_{\bullet\bullet} = 9.65$. The estimable function to implement a test among treatments at the overall mean of *X* requires the following contrast statement:

```
contrast 'trt at x=0' trt 1 0 0 -1 x*trt 9.65 0 0 -9.65,
                      trt 0 1 0 -1 x*trt 0 9.65 0 -9.65,
                      trt 0 0 1 -1 x*trt 0 0 9.65 -9.65;
```

The resulting *F*-value is 7.53, with a *p*-value of 0.0043. Which should we use? For these data, the plot of the covariates by treatment is

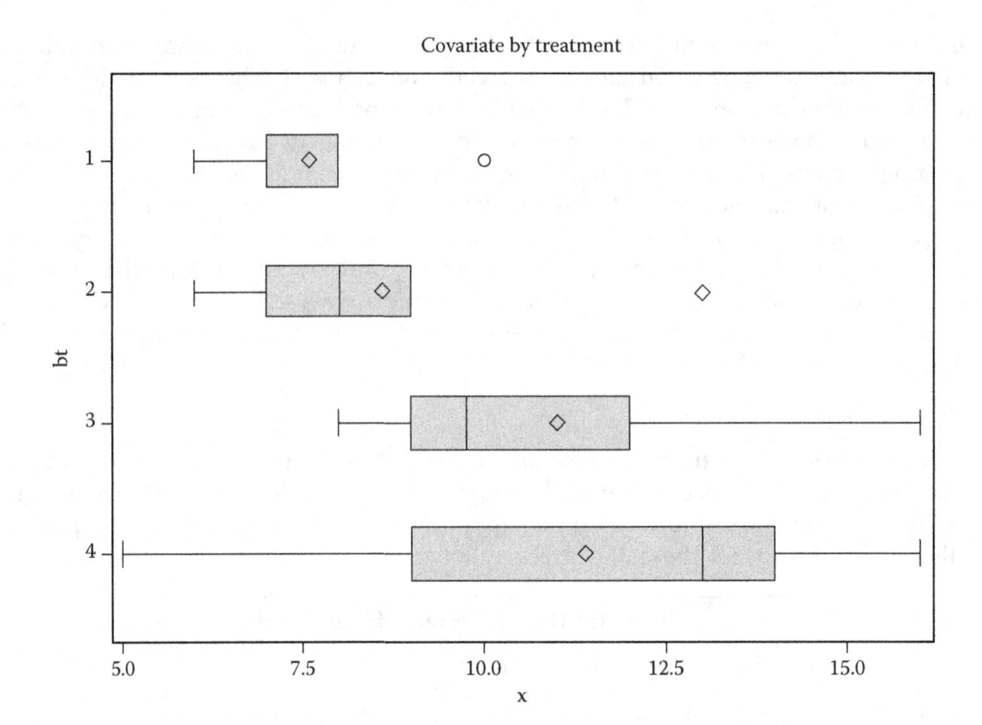

We can see that the covariate ranges predominantly between approximately 6 and 13. If we test at $X=0$, we would be testing at a covariate value that never occurs in this data set. Testing at $\bar{X}_{\bullet\bullet}=9.65$ seems far more reasonable.

> *Note to linear modeling students*: When you work with *any* software, you have to train yourself to ask these kinds of questions.

Some additional information: These data (with some manipulation to blind the data) are from a comparison of treatments to reduce postpartum stress in sows. The covariate is the litter size. Postpartum stress increases with litter size, meaning that we must adjust for litter size to get a fair comparison among the treatments. The inequality of the slopes tells us that the impact of increased litter size differs among the treatments or, to put it in terms more relevant to animal health researchers, some of the treatments do a better job alleviating the negative impact of large litter sizes than other treatments—important information in its own right. This also means that testing the equality of treatments at $X=0$, that is, testing the effectiveness of postpartum stress medication for "mothers" with a litter size of 0 (in other words, mothers who are not really mothers), is nonsense. Modeling is not just about mathematics or statistical theory—you have to pay attention to the subject matter being modeled!

We can use the (7.6) parameterization to obtain estimates of the covariate regression for each treatment. In GLIMMIX language, (7.6) is

```
proc glimmix;
class trt;
model y=trt x(trt)/noint solution;
```

The option NOINT suppresses the intercept η so that TRT corresponds to β_{0i} and X(TRT) corresponds to $\beta_{1i}X_{ij}$. The SOLUTION vector is

Parameter Estimates						
Effect	trt	Estimate	Standard Error	DF	*t* Value	Pr > |*t*|
trt	1	52.2940	6.8869	12	7.59	<0.0001
trt	2	74.2824	4.4731	12	16.61	<0.0001
trt	3	78.2623	4.8592	12	16.11	<0.0001
trt	4	56.8222	3.7134	12	15.30	<0.0001
x(trt)	1	−0.6754	0.8921	12	−0.76	0.4636
x(trt)	2	−3.7459	0.5007	12	−7.48	<0.0001
x(trt)	3	−4.4552	0.4278	12	−10.41	<0.0001
x(trt)	4	−2.3346	0.3080	12	−7.58	<0.0001

For example, the estimated regression equation for treatment 1 is $52.3 - 0.675X$, for treatment 2 is $74.3 - 3.75X$, and so forth. Increasing litter sizes appears to have much less of an impact under treatment 1 than it does under the other treatments.

We can compare treatments at different values of X. In practice, this means having a conversation with the subject matter specialists to find out if there are values for X that are of particular interest. For the sake of illustration, let us determine the adjusted means at $X = 7$ and $X = 12$, near but not quite at the lower and upper limits of the range of X. The GLIIMIX statements to obtain these are

```
lsmeans trt/at x=7;
lsmeans trt/at x=12;
```

The resulting output is as follows:

trt Least Squares Means			
trt	x	Estimate	Standard Error
1	7.00	47.5659	1.3232
2	7.00	48.0614	1.4513
3	7.00	47.0758	2.0959
4	7.00	40.4802	1.8167

trt Least Squares Means			
trt	x	Estimate	Standard Error
1	12.00	44.1887	4.1074
2	12.00	29.3321	2.0887
3	12.00	24.7998	1.2835
4	12.00	28.8073	1.2241

We can see that the differences among treatments 1 through 3 at $X = 7$ are small relative to the differences we see at $X = 12$. Consistent with the estimated slope, treatment 1 is considerably less affected by the change in X than are the other treatments. They are not shown here, but, in practice, we would also obtain mean comparison tests among the treatments at the X values of interest.

7.5.2.3 Factorial Treatment Design

The qualitative–quantitative factorial design is the third major application of linear predictors whose form derives from (7.5) and (7.6). We illustrate the approach with Data Set 7.4. These data are from a 3 × 6 factorial treatment structure. As with the other examples in this chapter, the response variable has a Gaussian distribution and the data were collected using a completely randomized design. The qualitative factor (A) has three levels labeled C, E1, and E2 (for control and experimental treatments 1 and 2). The six levels of the quantitative factor (B) are simply labeled 1 through 6. From past experience, response to the control treatment is known to be approximately linear with a positive slope over the levels of B. Researchers have reason to believe that the experimental treatments increase over levels of B more quickly (i.e., they achieve the target response at lower levels of B)—they want to find out if this in fact happens and if either of the two experimental treatments appears to be better (higher mean response at a given level of B = "better").

The interaction plot for these data is

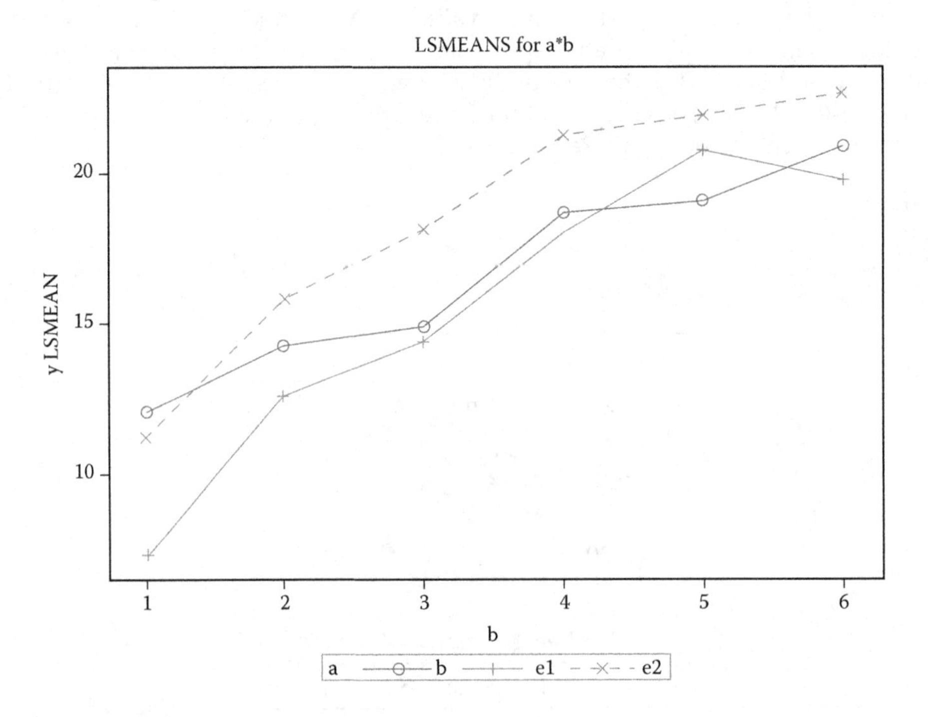

The solid line, level C of factor A, appears to be approximately linear, as expected. The dashed lines, the mean profiles for E1 and E2 over the levels of B, appear to be slightly curvilinear. One profile, E2, appears to be consistently greater than E1.

For a linear predictor to be useful for these data, it must be able to account for curvilinear response to the levels of B. A possible linear predictor is

$$\eta_{ij} = \eta + \alpha_i + (\beta_1 + \delta_i)X_j + (\beta_2 + \gamma_i)X_j^2 + \beta_j + (\alpha\beta)_{ij}$$

where

β_2 is the quadratic regression coefficient
γ_i denotes the *i*th treatment effect on the quadratic regression
β_j is a catchall for factor B main effects over and above linear and quadratic regression
$(\alpha\beta)_{ij}$ is a catchall for AB interaction effects over and above $\delta_i X_j$ and $\gamma_i X_j^2$

You can think of the catchall terms as lack of fit from quadratic regression. The GLIMMIX statements to implement this model are

```
proc glimmix data=a;
  class a b;
  model y=a|x|x a|b/htype=1;
```

The statement A|X|X is SAS modeling shorthand for `A X A*X X*X A*X*X`, the SAS modeling language for $\alpha_i + (\beta_1 + \delta_i)X_j + (\beta_2 + \gamma_i)X_j^2$. The term A|B defines the $\beta_j + (\alpha\beta)_{ij}$ component of the linear predictor. The option HTYPE = 1 requests "type 1" tests as defined earlier in this chapter—that is, sequential tests. Here, they are clearly warranted because we want to test the A main effects terms first, then linear regression, then quadratic, etc. You can try the default type 3 tests if you like, and you will immediately see why they are of no use to us. The type 1 results are

Type I Tests of Fixed Effects

Effect	Num DF	Den DF	*F* Value	Pr>*F*
a	2	35	14.12	<0.0001
x	1	35	261.17	<0.0001
x*a	2	35	2.87	0.0702
x*x	1	35	12.76	0.0011
x*x*a	2	35	2.20	0.1264
B	3	35	1.29	0.2920
a*b	6	35	0.40	0.8753

We see that the two lack-of-fit terms, B and A*B, both have *p*-values nowhere close to statistical significance, so we can safely conclude that the unequal coefficients quadratic regression model provides an adequate fit. The *F*-value for X*X*A tests H_0: all $\gamma_i = 0$, that is, the quadratic component of the regression over levels of B is identical for all three treatments. The *p*-value of 0.1264 suggests that we lack evidence to reject this hypothesis. If we accept it, we would eliminate γ_i from the model and proceed with inference using a model with unequal linear regression components but equal quadratic components. However, this appears to be problematic for two reasons. First, it is not consistent with claims about the treatment's response over levels of B that motivated the study. Second, it contradicts the visual evidence we see in the interaction plot. It may be that the two degrees of freedom for X*X*A decompose into one contrast that is essentially zero and one that is quite substantial. In view of both the interaction plot and prestudy theory, a good bet is that the quadratic components for E1 and E2 are nearly identical, but their average is quite different from quadratic component for C.

To assess this, we need to define contrasts that individually test H_0: $\gamma_{E1} = \gamma_{E2}$ and H_0: $\gamma_C = ((\gamma_{E1} + \gamma_{E2})/2)$. Alternatively, we can reparameterize the model along the lines of (7.6):

$\eta_{ij} = \beta_{0i} + \beta_{1i}X_j + \beta_{2i}X_j^2$ and test H_0: $\beta_{2,E1} = \beta_{2,E2}$ and H_0: $\beta_{2,C} = ((\beta_{2,E1} + \beta_{2,E2})/2)$. The advantage of doing the latter is that it also allows us to estimate the quadratic regression equation directly rather that obtaining estimates in overparameterized, hard-to-interpret form. The GLIMMIX statements are

```
proc glimmix data=a;
  class a b;
  model y=a x(a) x2(a)/noint solution;
  contrast 'quad c vs quad e1+e2' x2(a) 2 -1 -1;
  contrast 'quad e1 v e2' x2(a) 0 1 -1;
```

The term X2 is X*X—you have to define this term in the data step. The terms A and X(A) have the same meaning we saw earlier in the unequal slopes ANCOVA model; X2(A) corresponds to $\beta_{2i}X_j^2$. The output is as follows:

Parameter Estimates

Effect	a	Estimate	Standard Error	DF	*t* Value	Pr > |*t*|
a	c	9.8833	1.6951	45	5.83	<0.0001
a	e1	1.8700	1.6951	45	1.10	0.2758
a	e2	6.4567	1.6951	45	3.81	0.0004
x(a)	c	2.1946	1.1090	45	1.98	0.0540
x(a)	e1	6.0270	1.1090	45	5.43	<0.0001
x(a)	e2	5.4118	1.1090	45	4.88	<0.0001
x2(a)	c	−0.05774	0.1551	45	−0.37	0.7114
x2(a)	e1	−0.4911	0.1551	45	−3.17	0.0028
x2(a)	e2	−0.4530	0.1551	45	−2.92	0.0054

Contrasts

Label	Num DF	Den DF	*F* Value	Pr>*F*
quad c vs quad e1+e2	1	45	4.76	0.0344
quad e1 v e2	1	45	0.03	0.8629

The parameter estimates give the estimated intercept, linear, and quadratic regression coefficients for each treatment. Notice that for the test of the quadratic component for treatment C, $p=0.7114$. For treatments E1 and E2, the p-values are both <0.01. This provides evidence supporting the initial belief that response increases linearly for treatment C but curvilinearly for treatment E1 and E2. The contrast results provide evidence that the quadratic coefficients for E1 and E2 are similar to each other but different from C.

Inference would continue in this manner along the same lines as treatment comparisons in the unequal slopes ANCOVA model. Presumably, we want to determine from the regression estimates at what point on the B axis the mean response for E2 equals or exceeds that of C. We also want to find out if E1 and E2 differ and if so how. We do this by first comparing their linear components and, if they are not significantly different, we compare their intercepts. If the regression components are not different, then the regression lines are parallel; if in addition the intercepts are different, then the lines are parallel and one treatment is consistently higher than the other over all observed levels of B.

7.6 Multifactor: All Factors Quantitative

In structures with all treatments or explanatory factors quantitative, we extend the strategy employed in the previous section. Assuming that we can identify a functional relationship between the explanatory and response variables, we can define the linear predictor in terms of multiple regression.

In many cases, a second-order polynomial regression suffices. A second-order polynomial includes main effects up to quadratic terms and linear-by-linear two-way interaction terms only. This is the classical response surface model, described in depth by Myers et al. (2009), Box et al. (2005), Khuri and Cornell (1996), and Box and Draper (1987). Often, the polynomial approach is inadequate, and we must use alternatives. Important alternatives include splines and nonlinear mean models.

There are entire courses devoted to response surface methodology and other courses devoted entirely to nonlinear models. That is not our purpose here. In this section we introduce these basic approaches via two-factor examples. In Section 7.6.1, we consider the second-order polynomial. In Section 7.6.2, we look at alternative multiple regression models. Again, while all the examples use Gaussian data, our focus is on the linear predictor. Every linear predictor we discuss can be adapted to models with "generalized" and "mixed" features or both.

7.6.1 Second-Order Polynomial, a.k.a. Classical Response Surface Linear Predictors

The data for this example appear in Data Set 7.4. We have five levels each of factors A and B in a factorial arrangement, for a total of 25 treatment combinations. There are three observations per treatment from a completely randomized design. The following shows a three-dimensional plot of the 25-treatment-combination means.

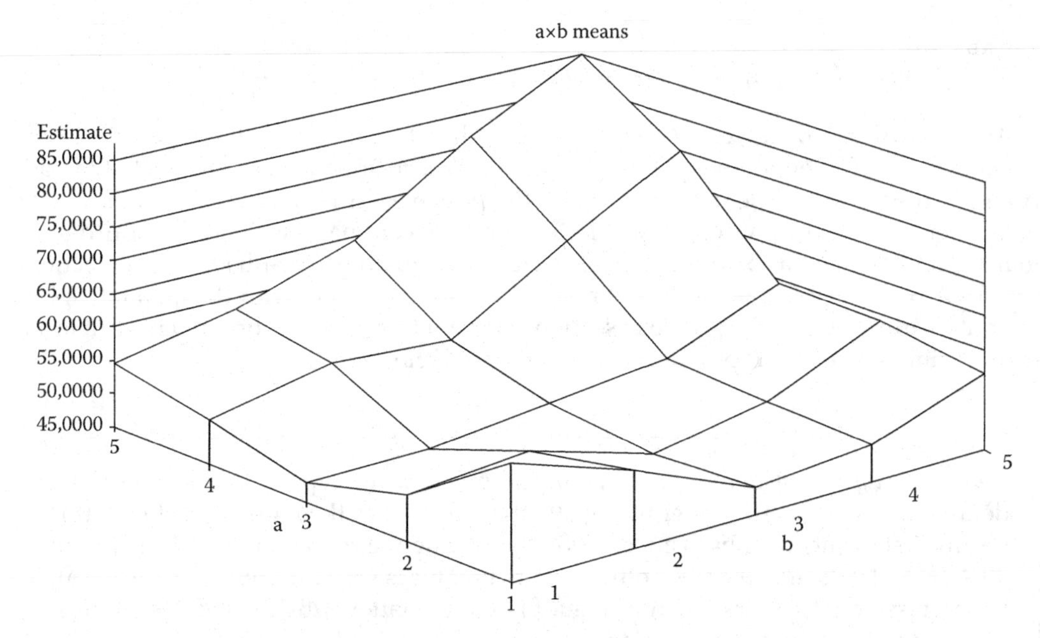

This plot shows the characteristic shape associated with second-order polynomial regression—a dome or inverted dome-like shape with a visually recognizable maximum or minimum response. The quadratic main effects regression accounts for the curvilinear profiles, and the linear-by-linear term accounts for the tilt. Often, estimating the factorial combination that maximizes or minimizes the response is a primary data analysis objective. In this example, assuming lower is better, we appear to have a minimum somewhere in the vicinity of three on both the A and B axes. The second-order polynomial linear predictor allows us to determine the treatment combination that minimizes (or maximizes) the response by ordinary calculus, a distinct advantage assuming the model provides an adequate fit.

We start using the same strategy we saw in the previous section when we fit the quadratic qualitative-quantitative model to Data Set 7.3. Here, we begin with the linear predictor

$$\eta_{ij} = \beta_0 + \beta_{1A}X_A + \beta_{1B}X_B + \beta_{2A}X_A^2 + \beta_{2B}X_B^2 + \beta_{12}X_AX_B + (\alpha\beta)_{ij} \tag{7.7}$$

where
X_k; k = A,B denote the quantitative levels of factors A and B
$\beta_{1,k}$ denote linear regression coefficients
$\beta_{2,k}$ denote quadratic regression coefficients
β_{12} denotes the linear-by-linear interaction coefficient
$(\alpha\beta)_{ij}$ acts as a catchall for all higher-order main effects (in this case cubic and quartic) and all interaction terms other than linear-by-linear

We interpret $(\alpha\beta)_{ij}$ as the lack-of-fit term. Though not shown here, as with the qualitative-by-quantitative example in the previous section, we can partition $(\alpha\beta)_{ij}$.

The GLIMMIX statements required for (7.7) are

```
proc glimmix;
  class a b;
  model response=xa|xa|xb|xb@2 a*b/htype=1;
```

The syntax `xa|xa|xb|xb@2` is SAS modeling shorthand for `xa xb xa*xa xb*xb xa*xb`, where XA and XB denote the quantitative amounts corresponding to the level of factors A and B, respectively. The option `@2` limits the possible XA by XB combinations to second-order terms—otherwise SAS would interpret the command as including third- and fourth-order terms, for example, `xa*xa*xb*xb`, the quadratic-by-quadratic interaction. CLASS A B and `a*b` in the model define the lack-of-fit term. As with the qualitative-by-quantitative example in the previous section, we want to use type 1, or sequential fit tests, here—using the default type 3 approach would yield nonsense.

Why do the type 3 tests give us nonsense in this example? The type 3, or partial tests, by definition, test each effect after fitting all other effects. For the regression effects, this means first fitting all A-by-B effects $(\alpha\beta)_{ij}$ *then* fitting the regression effects. If you fit all A-by-B effects first, there is nothing left to fit! This is very different from fitting all the regression effects first, then fitting $(\alpha\beta)_{ij}$. For the latter, $(\alpha\beta)_{ij}$ fits what is left over after fitting the regression equation.

The type 1 test results are

Type I Tests of Fixed Effects				
Effect	**Num DF**	**Den DF**	***F* Value**	**Pr>*F***
xa	1	75	88.55	<0.0001
xa*xa	1	75	48.94	<0.0001
xb	1	75	82.68	<0.0001
xa*xb	1	75	92.68	<0.0001
xb*xb	1	75	26.29	<0.0001
a*b	19	75	0.83	0.6701

The paramount results here are the *F*- and *p*-values associated with the lack-of-fit term A*B; $p=0.67$ tells us that there is no evidence of lack of fit from the second-order polynomial regression.

We then proceed by dropping $(\alpha\beta)_{ij}$ from the linear predictor, estimating the second-order regression equation and so forth. These are left as exercises.

7.6.2 Other Quantitative-by-Quantitative Models

In many cases, the second-order polynomial does not provide an adequate fit. Moreover, even if we can eventually determine a higher-order polynomial that provides adequate fit *mathematically*, such models usually provide little or no useable insight with regard to objectives that typically motivate studies. The general strategy shown in the previous example still holds: Instead of an effects-based linear predictor, we want to use some form of regression if possible. Alternative regression functions include sine and cosine functions, nonlinear functions, and splines. Sine and cosine functions are very useful when we have regular, cyclic variation of the factor levels, for example, in daily, monthly, seasonal, and other time series. This section presents two examples, one involves a nonlinear form of response surface modeling and the other involves a spline. Readers seeking an introduction to time series are referred to Box and Jenkins (2008); nonlinear models to Bates and Watts (1988), Gallant (1987), or Davidian and Giltinan (1995); and spline regression to Rupert et al. (2003) and Gu (2002).

7.6.2.1 Nonlinear Mean Models

From a GLMM perspective, these models are linear in the sense that we model the link by an additive function of a fixed effects and a random effect part. However, instead of the linear predictor $\boldsymbol{\eta}=\mathbf{X}\boldsymbol{\beta}+\mathbf{Z}\mathbf{b}$, we have $\boldsymbol{\eta}=f(\mathbf{X},\boldsymbol{\beta})+\mathbf{Z}\mathbf{b}$, where $f(\mathbf{X},\boldsymbol{\beta})$ is not necessarily linear.

As an example, Data Set 7.4 contains data from a study whose design structure is similar to Data Set 7.3, but the response profile over the levels of the two factors, shown in the following plot, does not appear to be a good candidate for polynomial regression. These are simulated data, but the response profiles are motivated by nonlinear plant nutrition dose-response profiles discussed by Landes et al. (1999), Paparozzi et al. (2005), Stroup et al. (2006), and Frenzel et al. (2010).

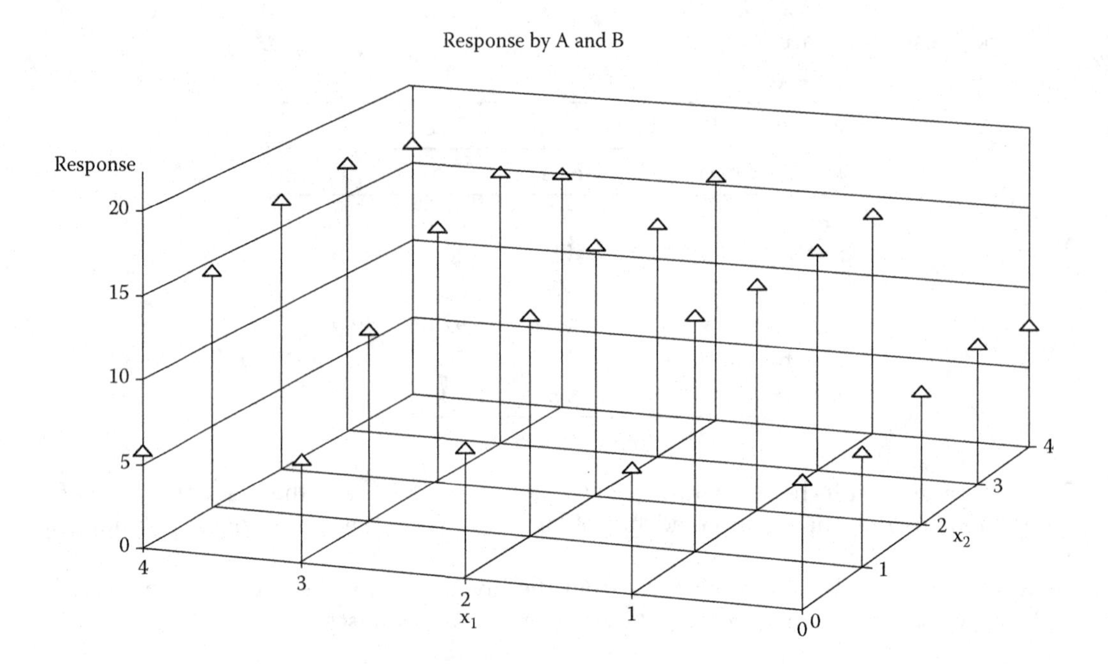
Response by A and B
Response
20
15
10
5
0
4
3
2
x_1
1
0
0
1
2
3
4
x_2

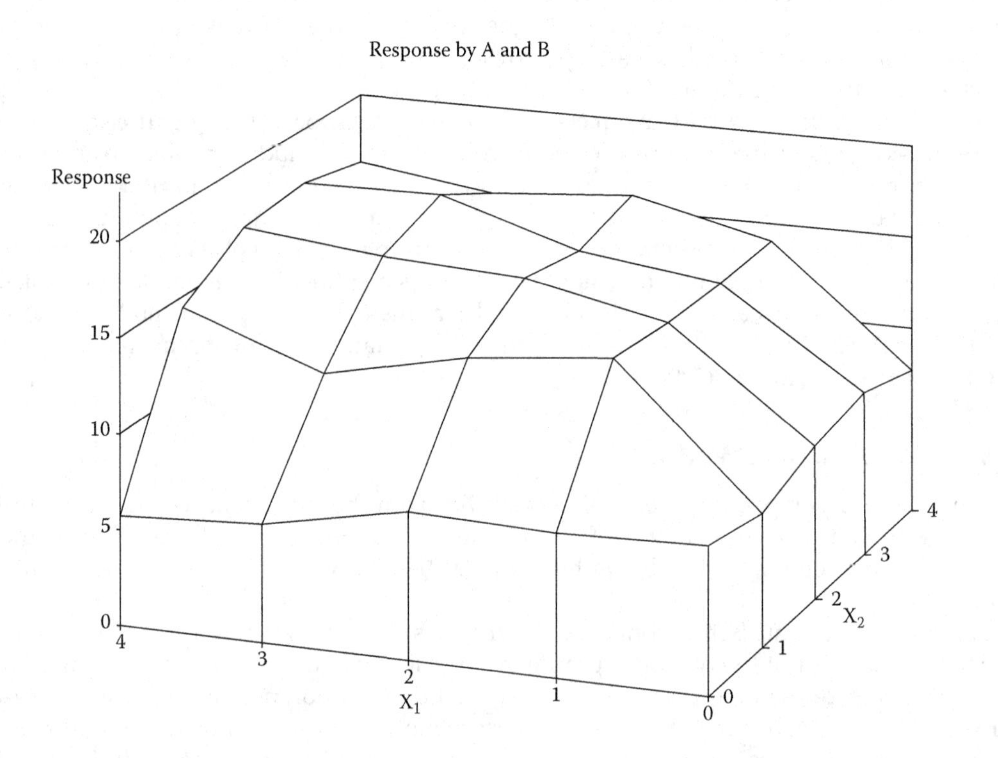
Response by A and B
Response
20
15
10
5
0
4
3
2
X_1
1
0
0
1
2
3
4
X_2

The scatter-plot (top) shows the actual data points; the vertical axis is the response, the two lower axes represent the levels of factors A and B. The bottom plot shows lines connecting the points, making it easier to see the data's response profile. The lower right-hand corner corresponds to the lowest level of both A and B. Along both A and B axes, we see an abrupt rise in the response, followed by a long plateau. If we fit a second-order polynomial, we get an estimated response surface shown by the following plot.

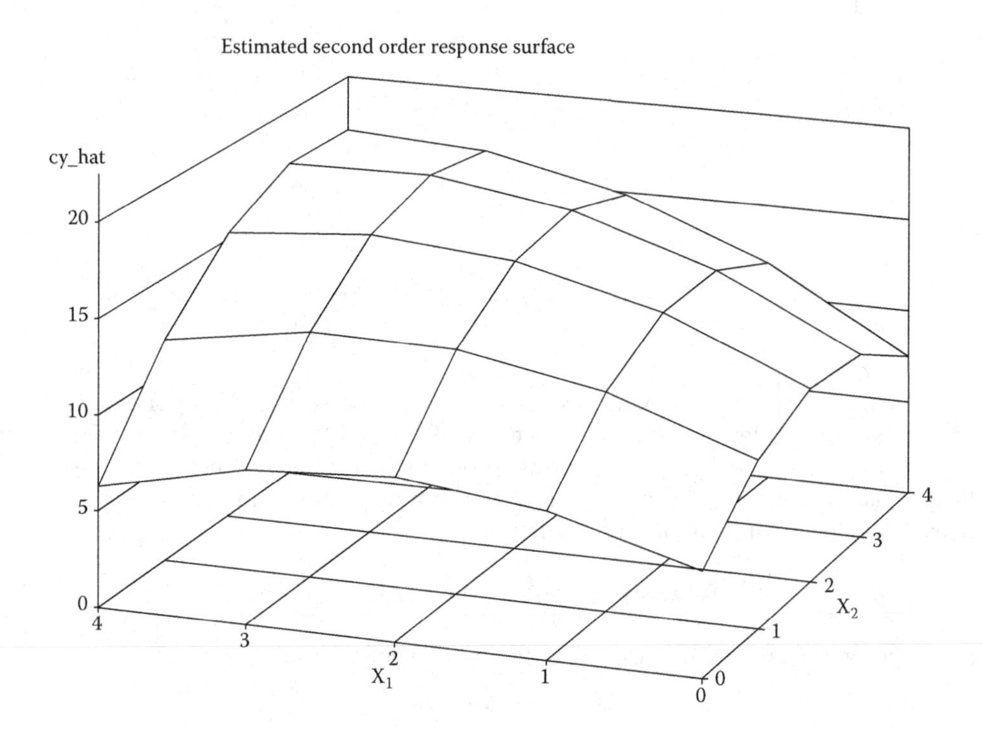

Quadratic regression cannot capture the abrupt rise followed by the long plateau. Instead, it blurs both, producing a poor approximation of the surface. Obviously, we would expect inaccurate conclusions and inappropriate recommendations to result from using the second-order approximation.

Olson et al. (2001) suggested an alternative second-order response surface model using a linearized form of the Hoerl function. In GLMM terms, the linear predictor is

$$\eta_{ij} = \beta_0 + \beta_{1A} X_A + \beta_{1B} X_B + \beta_{11} X_A X_B + \beta_{2A} L_A + \beta_{2B} L_B + \beta_{22} L_A L_B \tag{7.8}$$

where $L_k = \log(X_k)$; $k = \text{A,B}$. With (7.8), we model the curvilinear component of the response surface by log rather than quadratic terms, and we model interaction using the products of the logs as well as the linear-by-linear term. The resulting estimated response surface is

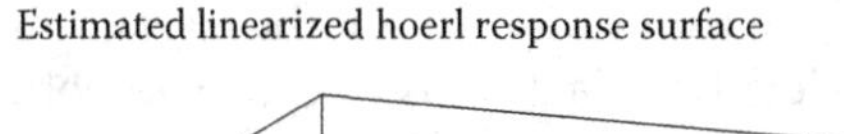

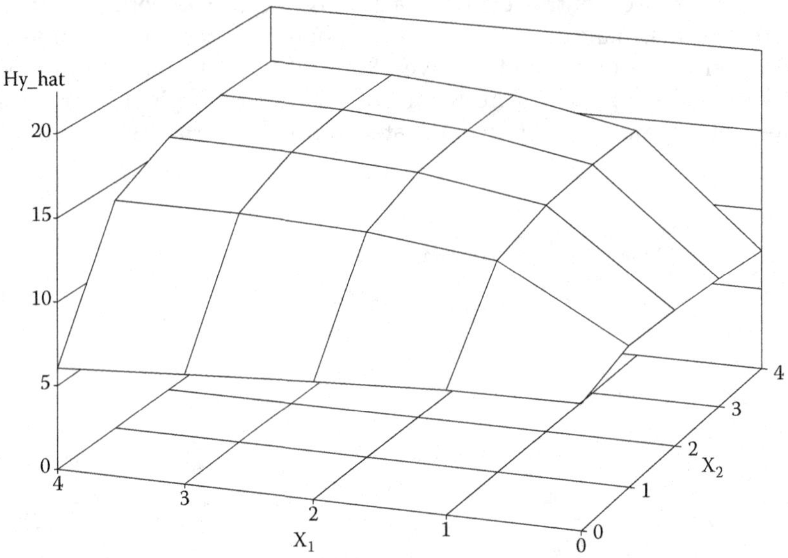

The fit is an obvious improvement. Stroup et al. (2006) showed that this approach allows us to use efficient incomplete factorial designs especially suited to estimating what Nelson and Anderson (1975) call "linear plateau" response profiles when resources are highly constrained.

We can also use explicitly nonlinear η_{ij}. One example, again from Stroup et al. (2006), is a bivariate extension of the Gompertz function:

$$\eta_{ij} = \alpha + \exp\left\{-\beta \times \exp\left[-\gamma_1 X_A - \gamma_2 X_B - \gamma_{12} X_A X_B\right]\right\} \tag{7.9}$$

The Gompertz yields the following estimated response surface:

Estimated Gompertz response surface

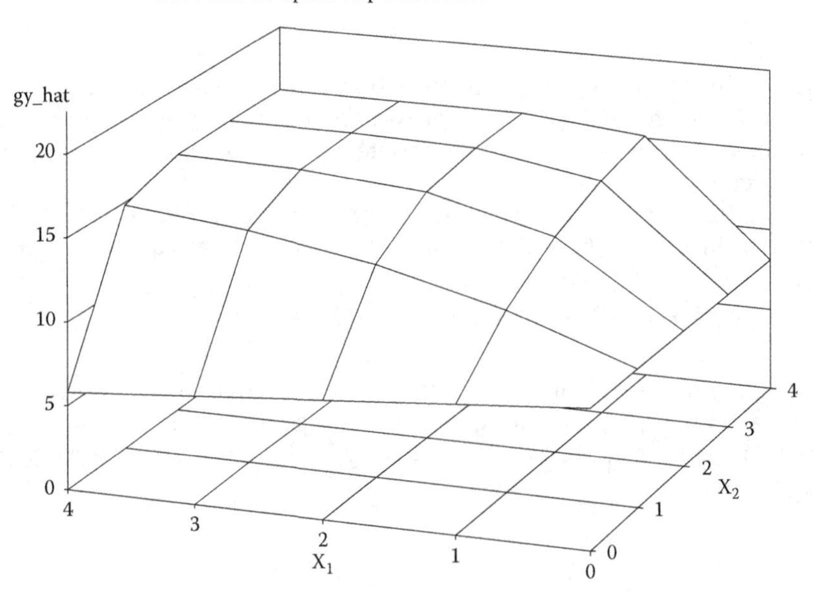

The linearized Hoerl and Gompertz produce similar fits for these data. The main advantages of the Hoerl approach are the ease of identifying a suitable efficient design and the ability to use linear model software for estimation. The advantages of nonlinear models such as the Gompertz lie in greater flexibility to model nonstandard response profiles and the fact that the parameters of nonlinear models lend themselves to more meaningful interpretation. In the Gompertz model, for example, α, β, and γ define the intercept, the asymptote or plateau value of the response, and the rate of increase over the levels of factors A and B.

7.6.2.2 Spline or Segmented Regression

Some response profiles defy characterization by a single functional form. Segmented, or spline, regression partitions the response profile into pieces that are amenable to simpler regression models. Ruppert et al. (2003) present the smoothing spline procedure as a method to address irregular regression profile that is especially well suited to the GLMM. Ruppert et al. give the linear spline function as $\beta_0 + \beta_1 X_i + \sum_{j=1}^{K} \gamma_j I(X_i > t_j)(X_i - t_j)$, where $I(X_i > t_j)$ is an indicator function defined on $X_i > t_j$. The smoothing spline procedure obtains parameter estimates by minimizing

$$(\mathbf{y}^* - \mathbf{X}\boldsymbol{\beta} - \mathbf{Z}\boldsymbol{\gamma})'(\mathbf{y}^* - \mathbf{X}\boldsymbol{\beta} - \mathbf{Z}\boldsymbol{\gamma}) + \lambda^2 \boldsymbol{\gamma}'\boldsymbol{\gamma} \tag{7.10}$$

where

- $\mathbf{y}^*$ corresponds to the pseudo-variate in PL estimation
- $\mathbf{X}\boldsymbol{\beta}$ denotes the $\beta_0 + \beta_1 X$ component of the spline model
- $\mathbf{Z}\boldsymbol{\gamma}$ denotes the rest
- λ is a tuning constant

If we think of the linear spline as a special case of a pseudo model $\mathbf{y}^* = \mathbf{X}\boldsymbol{\beta} + \mathbf{Z}\boldsymbol{\gamma} + \mathbf{e}$ with $\text{Var}(\mathbf{e}) = \phi\mathbf{I}$ and $\text{Var}(\boldsymbol{\gamma}) = \sigma^2\mathbf{I}$, Schabenberger (2008) shows that the tuning constant, $\lambda = \phi/\sigma$, and the PL estimation algorithm can be used to estimate the linear spline.

As an example, Data Set 7.5 contains data from two treatments with observations taken at several points along the X-axis. A plot of the data, along with estimates from the smoothing spline regression, is shown in the following, after the GLIMMIX statements used to estimate the model. The GLIMMIX statements are

```
proc glimmix data=spline;
  class a;
  effect spline_x=spline(x /knotmethod=rangefractions(0.1 0.2 0.3 0.4 0.5
0.6 0.7 0.8 0.9));
  model y=a|spline_x//*noint*/ solution;
    output out=gmxout pred=p;
run;
proc sgplot data=gmxout;
  series y=p x=x/group=a name="fit";
  scatter y=y x=x/group=a;
  keylegend "fit"/title="A";
run;
```

Factor A is the qualitative treatment factor, X is the quantitative factor. The EFFECT statement defines the spline over X, with the RANGEFRACTION option defining the t_j values at which the regression is segmented. With RANGEFRACTION, these points are defined in terms of the percentage in decimal form of the range of the X values. In practice, there is some trial and error involved in selecting range fractions that strike a balance between parsimony and adequate characterization of the response profile.

Here, we show the results of testing the model effects and the plot of the spline regression's fit relative to the observed data.

Type III Tests of Fixed Effects				
Effect	**Num DF**	**Den DF**	***F* Value**	**Pr>*F***
A	1	244	15.16	0.0001
spline_x	12	244	430.68	<0.0001
spline_x*a	12	244	31.18	<0.0001

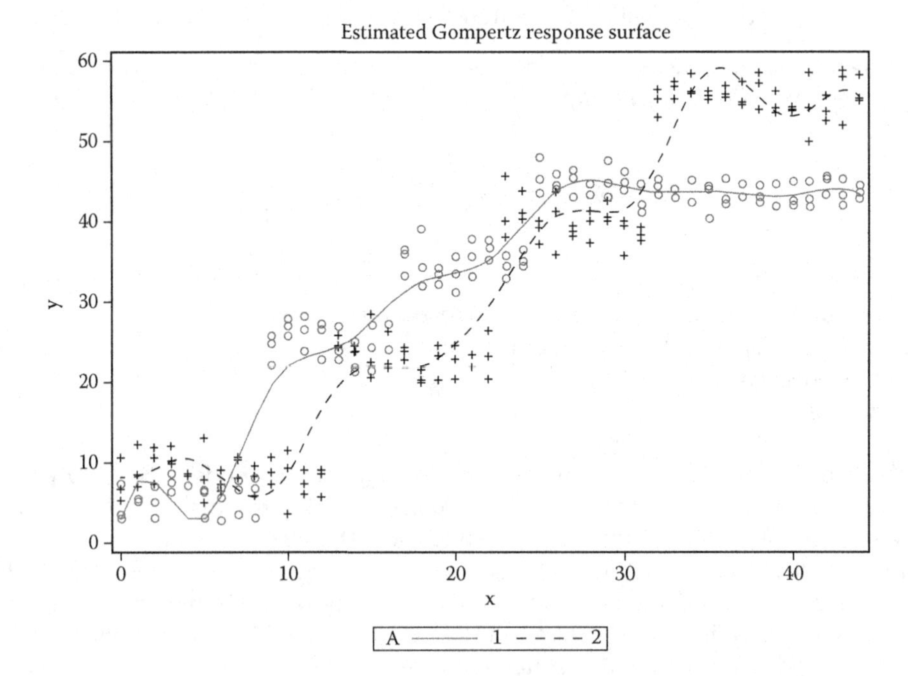

We can see from the plot that the response over levels of X seems to increase in punctuated bursts along the X-axis and that the position and the magnitude of these bursts differ between the two levels of factor A.

7.7 Summary

This concludes our initial consideration of linear predictor and inference strategies for treatment and explanatory variable structures. The main ideas you should take from this chapter are as follows:

1. The form of the $\mathbf{X\beta}$ component of the linear predictor needs to reflect the types of factors in the explanatory variable or treatment design *and* the objectives of the study. A well-chosen $\mathbf{X\beta}$ enables a clear analysis. A poorly chosen $\mathbf{X\beta}$ tends to confuse more than enlighten.
2. Avoid one-size-fits-all rules. Recipes (taken too far) are bad—thinking is more difficult, but much better. You know you are taking recipes too far when they start getting in the way of thinking, or when you are not thinking but just "formula-grabbing."
3. Order of testing can have a drastic impact on what you test. With unbalanced data, what you actually test can be very different from what you think you are testing.
4. Make sure you know what you are testing. If you cannot express what you are testing in estimable function form, stop what you are doing until you can.
5. Multifactor models present the most challenging modeling problems with respect to $\mathbf{X\beta}$.
6. Common themes are as follows:
   a. Test interactions first.
   b. Nonnegligible interaction demands inference on simple effects only.
   c. Absence of interaction demands inference on main effects only.
   d. Multiplicity issues need to be dealt with when they arise.
   e. Sometimes, the specific context of the study overrides all of the above.
7. Points of departure as follows:
   a. In general, qualitative effects call for effects models.
   b. Whenever possible, try to express quantitative factor effects in terms of regression models.
   c. Regression does not just mean polynomial regression.

We will return to most of the model forms in this chapter as we go through the rest of this textbook. Refer back to this chapter for the basic template and motivation of the treatment design component of the linear predictor. The rest of the chapters focus heavily on the various forms of the random component of the linear predictor, $\mathbf{Zb}$; the $\mathbf{R}$ (for LMMs) or working correlation (for GEEs) structures; the distribution of the observations; and various forms of the link.

8

Multilevel Models

8.1 Types of Design Structure: Single- and Multilevel Models Defined

In Chapter 7, we looked at the major strategies for constructing and working with treatment or, more generally, explanatory variable structures. In this chapter, we turn our attention to the design, or what Fisher (1935) called the "topographical" aspect of the linear predictor. To oversimplify somewhat (but not a lot), the treatment aspect primarily determines how we specify $\mathbf{X}\boldsymbol{\beta}$, whereas the design or topographical aspect tends to drive specification and strategies for working with $\mathbf{Zb}$.

As we saw in Chapter 2, the design or topographical aspect—as broadly defined—concerns the units of replication and units of observation and their structure. What are the units? Do we have units that are subsets or subdivisions of larger units? Do we have meaningful groupings of units—strata, clusters, or blocks? How do we apply randomization with these units?

Our primary concern in this chapter is the correct identification of the unit(s) of replication and the specification of $\mathbf{Zb}$ that follows. Do we have one size only or do we have different-sized units of replication? If we have the latter, we have a *multilevel* study. Multilevel studies have different names, depending on whose company you keep. Historically, agricultural statisticians were the first to recognize and name multilevel studies—they called them *split-plot* experiments. In medical centers, biostatisticians call them *nested factorial* or *clustered* designs. Social scientists know them as *hierarchical* designs (Raudenbush and Byrk, 2002). Regardless of the local jargon, all refer to the same design structure.

The components of the model that characterize the design are sometimes called nuisance parameters. This is both unfortunate and unhelpful as "nuisance" is often taken to mean "unimportant." Design components can be complex, and how we define the linear predictor to characterize them can have a substantial influence on subsequent inference. When things go badly wrong in modeling exercises, they tend to go wrong on the design side. The most frequent cause of things going badly is failure to account for multilevel structure when it is present or accounting for it inappropriately.

Much of the language for design or topographical structure comes from the design of experiments. We should not take this to mean that the discussion in this chapter applies only to designed experiments. Surveys, observational studies, and quasi-experiments all have structure. They all have units of observation. They have entities that correspond, at least loosely, to units of replication. Clustering, stratification, and case-control matching are all forms of blocking. To construct a useful and informative model, all of these need to be understood and characterized appropriately. In some respects, this chapter may be *more* important for studies that are *not* formally designed experiments or randomized trials simply because there is a tendency for some involved with nonexperimental studies to operate as if design principles do not apply. So, lest there be any misunderstanding, whenever we do comparative inference, design principles apply. This chapter uses design vocabulary because it is the *lingua franca* of multilevel modeling, but readers should understand the language in this chapter *as broadly defined*.

All of this said, let us establish a precise working definition of what a multilevel study is and what it is not.

We define a study to be multilevel if and only if it has at least two sources of random variation and at least two sizes of units comprising the architecture of the study.

Alternatively, we can define multilevel studies in terms of the number of distinct randomization schemes. The key: more than one size unit or, equivalently, multiple randomization schemes. If we only have one sized unit, meaning the unit of replication and the unit of observation are identical or, equivalently, a single randomization scheme used for all treatment levels or factorial level combinations, then we do not have a multilevel study. One size unit: single level. Two or more sizes of units: multilevel. Single randomization scheme: single level. Two- or more-step randomization: multilevel study.

The simplest multilevel study is a completely randomized design (CRD) with multiple sampling units. In such studies, the unit of replication consists of several individuals, for example, animals in a pen, patients at a clinic, and plants within a plot. Our focus in this chapter will be on more complex multilevel studies, as these are the ones that typically cause modelers and data analysts the greatest difficulty.

In Section 8.2, we will look at how multilevel studies arise in more detail. We will also review the skeleton ANOVA (WWFD) process introduced in Chapter 2 as a way of helping us visualize the distinction between single- and multilevel studies and as a way to help us properly specify the components of the multilevel design.

We cannot discuss multilevel studies without also considering blocking. Issues arise when we use blocking in multilevel studies that do not occur with single-level designs. For single-level designs, deciding whether to regard block effects as fixed or random usually has at most a modest impact on model estimation or inference. Not so with multilevel studies. In particular, defining block effects as fixed can paint you into a corner from which there is no easy escape. The two principle concerns: incorrect inferential statistics and spurious nonestimability. Section 8.3 focuses on blocking.

In Section 8.4, we consider a number of examples to illustrate how multifactor treatment structures and multilevel design structures fit together to form the entire linear predictor. By definition, multifactor models must be mixed models. The material in this chapter through Section 8.4 applies equally to non-Gaussian GLMMs and to LMMs.

In previous chapters, especially Chapters 3, 5, and 6, we discussed marginal and conditional models, G-side and R-side covariance modeling, and GEE-type models. In Section 8.5, we look at how these issues specifically apply to multilevel models.

8.2 Types of Multilevel Models and How They Arise

The unit of replication and the randomization scheme provide the crucial information we need to construct **Zb**. In this section, we review some critical concepts from Chapter 2 and how they manifest themselves in multilevel scenarios.

8.2.1 Units of Replication: Not Just in Designed Experiments

For designed experiments, the unit of replication—the *experimental unit* in design language—figures prominently from the planning stage through data analysis. It is formally defined as the smallest entity to which treatments are *independently* assigned. The word *independent* is crucial. If we have animals in pens and the treatment is assigned to and applied through a common pen feeder, the experimental unit is the *pen*. The animals come with the assignment; we may measure them individually, but their assignment to treatment level is not independent. We cannot assign one animal to one treatment and another animal to a different treatment within the pen. The experimental unit mindset is deeply embedded in the design of experiments thought process. This greatly facilitates constructing a linear predictor from an experiment design. We already have all the necessary information in place.

For surveys, observational studies, and quasi-experiments we have to work harder, but we still need to identify these elements in order to construct a linear predictor capable of yielding an accurate analysis. Every survey, observational study, etc., has the conceptual equivalent of unit(s) of replication. Let us walk through an example of a randomized trial that is almost, but not quite, a designed experiment to reinforce our understanding of the terminology and then consider how these ideas translate to survey and observational studies.

To do this, we use Fisher's term "topographical" as we did in Chapter 2. The word "topographical" is helpful in the sense that it is untainted by design or survey jargon that might get in the way in this discussion, but we need to understand "topographical" as broadly referring to the structure of the study and not narrowly as a geographical or geological concept.

The simplest topographical structure has one kind of unit. We take observations on these units, and we assign—or nature assigns—treatment levels to these units. This structure defines a CRD. The CRD is clearly a single-level structure. In survey terms, a simple random sample has the same structure as a CRD, hence similar modeling considerations.

The CRD is the only design with one unit size only, but the CRD is not the only single-level design. If we have two sizes of units, say classrooms and students within classrooms, and classrooms receive the treatment assignment (e.g., curriculum A1 goes to one classroom, curriculum A2 goes to another classroom, etc.) but we test individual students to see how the treatment performs, then we have a CRD with subsampling, the simplest multilevel structure. Suppose the classrooms are blocked—schools would be an obvious blocking criterion. Now we have three sizes of units in the study. If we still have classrooms receiving the treatment (say at a any given school, one classroom gets curriculum A1 and another gets curriculum A2), then we have a randomized block design—with subsampling if we still take measurements on individual students. Both the CRD with subsampling and the randomized block design—with or without subsampling—are single-factor designs but they nonetheless call for multilevel models. Why? Because, although they have a single treatment factor and require a single randomization scheme that assigns factor levels to units of replication, they still require multiple random variables. At a minimum, to fully account for sources of variation with associated probability processes, we need to specify one random effect in the linear predictor in addition to the distribution of the observations conditional on the random effect.

More complex multilevel studies arise when we add a second factor. Suppose in addition to curriculum (A1 or A2) as a factor, we also have a factor B. If the factorial combinations are assigned to classrooms as before, we still have a single-level experiment if observations are taken at the classroom level or a simple multilevel subsampling model if observations are taken on individual students. We have A × B treatments defined by the factorial combinations, but the unit of replication is still the classroom. However, suppose we assign curriculum to classrooms, as before, but the levels factor B to the entire school, or to individual students within a classroom. Now we have two different-sized units of replication, and we need two randomization schemes, one to assign different curricula (levels of factor A) to the classrooms and another to assign levels of factor B. These are the defining criteria for a more complex multilevel study.

How does this work for observational studies? Suppose we want to find out if exposure to an environmental hazard increases risk of a certain health problem. A possible design would involve identifying people who were exposed to the hazard and an otherwise similar group that was not. We might go further and attempt to create matched pairs—one exposed and one nonexposed person with similar characteristics. Matching translates into linear predictor effects the same way blocking does in a designed experiment. Individual people are clearly the units of observation. What about the unit of replication? The "treatment" factor is hazard, present, or absent. The hazard was "assigned" to the site. If we want to extend inference to all hazards of this type, the unit of replication is site, not individual people. If we only look at one site, we limit the inference space accordingly.

In a survey sample, suppose we want to find out if political orientation of students varies by type of higher education institution. For example, do public universities have a higher proportion of self-identified conservatives? Do private liberal-arts colleges have a higher proportion of self-identified liberals? In this case, the factor is the type of institution, and the smallest entity assigned to type is the entire college or university. The individual students surveyed constitute subsampling. If we added a second factor, say major, then the entire program for a given major at a given institution is the unit of replication with respect to this factor. Now we have a multilevel study: two factors, type of institution and major; two units of replication, college or university and program within a college or university. Individual students are the units of observation.

The skeleton ANOVA (WWFD) process we introduced in Chapter 2 can be a particularly effective way of identifying multilevel studies and translating them into linear predictors.

8.2.2 "What Would Fisher Do?" Revisited: Topographical and Treatment Component

Let us translate the examples given earlier into skeleton ANOVA terms. This will help us visualize the single- vs. multilevel aspects. First, the simplest completely randomized form appears as

Topographical		**Treatment**		**Combined**	
Source	**d.f.**	**Source**	**d.f.**	**Source**	**d.f.**
		trt	$t-1$	trt	$t-1$
Units	$N-1$	"Parallels"	$N-t$	Unit(trt) a.k.a. "residual"	$N-t$
Total	$N-1$	Total	$N-1$	Total	$N-1$

Here, the single-level structure is obvious—"trt" must be the factor and "unit(trt)" is the only possible candidate for unit of replication. Recall from our discussion of conditional and marginal model in Chapter 3 that the "unit(trt)" line must be treated as "residual" and, thus, *not* appear in the linear predictor for Gaussian models, but it may and often should appear as a random effect in the linear predictor for non-Gaussian models. This residual-vs.-linear predictor effect is one aspect of overdispersion, which we cover in greater depth in Chapter 11.

The blocked design has one additional feature:

Topographical		Treatment		Combined	
Source	**d.f.**	**Source**	**d.f.**	**Source**	**d.f.**
Block	$b-1$			Block	$b-1$
		trt	$t-1$	trt	$t-1$
Unit(block)	$b(n-1)$	"Parallels"	$bn-t$	Block × trt or "residual"	$bn-b-t+1$
Total	$bn-1$	Total	$bn-1$	Total	$bn-1$

Now the topographical aspect has two elements. In this limited sense, we have a multilevel model, but we still only have a single factor and only one possible unit of replication: the block–treatment combinations. Hence, from a GLMM perspective, we have a single-level model. Again, the block–treatment effect must be treated as residual and hence not appear in the linear predictor for Gaussian models, but it may appear in certain non-Gaussian models.

If we also include subsampling, for example, the students in our school/classroom, we have

Topographical		Treatment		Combined	
Source	**d.f.**	**Source**	**d.f.**	**Source**	**d.f.**
Block	$b-1$			Block	$b-1$
		trt	$t-1$	trt	$t-1$
Unit(block)	$b(n-1)$			Block × trt	$bn-b-t+1$
Subunit(unit)	$bn(s-1)$	"Parallels"	$bns-t$	Subunit(block × trt) Residual	$bn(s-1)$
Total	$bns-1$	Total	$bns-1$	Total	$bns-1$

The solid line below the "unit(block)" row divides the skeleton ANOVA into the upper part, where treatment level randomization occurs, and the lower part—sampling units that cannot be randomized independently of the block–treatment combinations. However, they do have their own variation, over and above block and unit(block) variation. The topographical aspect has three levels: block, unit, and subunit. The model must in some way account for all three.

A comment particularly important to hierarchical linear model users: In HLM language, subunit(block × trt) is called *level 1*, that is, the smallest unit in the study. In the previous skeleton ANOVA, the blocked design without subsampling, the block–treatment combination is the *level 1* term. This is where HLM language can be somewhat unhelpful, even confusing. HLM *level* does not distinguish between unit of observation and unit of replication. With subsampling, the unit of replication is *level 2*, without subsampling the unit of replication is *level 1*. This does not have to be a problem, but it can be confusing. HLM users do need to make an extra effort here, because HLM language does not emphasize design concepts.

For the environmental hazard observational study, a possible skeleton ANOVA might look like this

Topographical		**Treatment**		**Combined**	
Source	**d.f.**	**Source**	**d.f.**	**Source**	**d.f.**
		Exposure	1	Exposure	1
Site	$s - 1$			Site(exposure)	$s - 2$
Pair(site)	$s(p - 1)$			Pair(site, exposure)	$s(p - 1)$
Individual(pair)	$sp(n - 1)$	"Parallels"	$spn - 2$	Individual(pair…) Residual	$sp(n - 1)$
Total	$spn - 1$	Total	$spn - 1$	Total	$spn - 1$

We still have a single-level design with respect to the treatment factor under investigation, but we now have three layers of nesting. Structures like these are notoriously prone to *pseudo-replication*: in this case, mistaking pair or individual as the unit of replication. They clearly are not in this setup. Note that in HLM language, the true unit of replication, site, is the *level 3* component of the model. Again, HLM level and unit of replication have no transparent connection to one another.

The university type–major example finally gives us a multilevel model.

Topographical		**Treatment**		**Combined**	
Source	**d.f.**	**Source**	**d.f.**	**Source**	**d.f.**
		Type	$t - 1$	Type	$t - 1$
Institution	$c - 1$			Institution(type)	$c - t$
		Major	$m - 1$	Major	$m - 1$
		Type × major	$(t - 1)(m - 1)$	Type × major	$(t - 1)(m - 1)$
Program(institution)	$c(p - 1)$			Major × institution(type)	$c(p - 2) - t(m - 1)$
Student(program)	$cp(s - 1)$	"Parallels"		Student(program) Residual	$cp(s - 1)$
Total	$cps - 1$	Total	$cps - 1$	Total	$cps - 1$

Here we have two distinct units of replication: types to institutions and majors to programs. Notice that we partition the skeleton ANOVA into three tiers: two for the levels relevant to the treatment design, type/institution and major/program, and one for subsampling, individual students. As usual, for Gaussian models, we must treat student effects as residual but for some non-Gaussian models, student effects should be represented in the linear predictor. Also, we have some flexibility with the treatment component. If different types of institutions have different majors, we can substitute a single nested effect—major(institution)—in place of separate main effect (major) and interaction (major × institution) effects.

The disconnected incomplete block design from Figure 2.5—10 blocks each size 3, 6 treatments, with one set (treatments 0, 1, 2) always together in five blocks and the other set (treatments 3, 4, and 5) always in the other five blocks—gives us another view of a multilevel structure. The skeleton ANOVA

Topographical		Treatment		Combined	
Source	d.f.	Source	d.f.	Source	d.f.
		Set	1	Set	1
Block	9			Block(set)	8
		trt(set)	4	trt(set)	4
Unit(block)	20			Unit(block) or "residual"	16
Total	29	Total	29	Total	29

reveals the multilevel aspect. The units of replication are blocks for set and units within block for treatments within a set. If the treatment design has additional structure—for example, factorial with "set" corresponding to factor A and treatments within each set corresponding to levels a factor B—then we can replace "set" by A in the block level and "trt(set)" by B and A × B in the unit level.

We can vary these structures in endless ways depending on the requirements and constraints of particular studies. Federer and King (2007) fill an entire book with variations on the split-plot. The key feature of any multilevel study—design, observational study, or survey—is the existence of at least two different-sized units of replication. This means that all multilevel models must be mixed models. Variation associated with the smallest unit may be embedded in the distribution of $\mathbf{y}|\mathbf{b}$. Indeed, it *must be* for observations with a Gaussian distribution. On the other hand, variation associated with the larger unit(s) *must be* modeled in the linear predictor—that is, at a minimum, $\mathbf{b}$ must include the larger unit of replication as a random model effect. In other words, if we have a multilevel structure, we must have a $\mathbf{Zb}$ component of the linear predictor.

8.3 Role of Blocking in Multilevel Models

Are block effects fixed or random? This has been an enduring controversy ever since Yates (1940) developed recovery of inter-block information, the original random-block-effect methodology. This issue has particular relevance for multilevel models because of the effect the fixed-or-random decision has on estimability.

In modeling, we have two types of nonestimability—*true* and *spurious*. Examples of true nonestimability include functions defined on the intercept alone or on isolated treatment effects in an ANOVA-type model: $\eta + \tau_i$ is estimable, η and τ_i separately are not. Interaction effects in main-effects-only designs and certain interaction contrasts in factorial designs with missing factorial combinations provide other common examples of true nonestimability. The most egregious examples of *spurious* nonestimability occur in multilevel designs. Most of these can be attributed directly to modeling blocks as fixed either intentionally or, more commonly, using software that does not recognize the difference between fixed and random effects and assesses estimability inappropriately as a result.

In Section 8.3.1, we discuss the various aspects of the block-effects-fixed or block-effects-random controversy in more detail. In Section 8.3.2, we specifically consider the impact on estimability in multilevel models.

8.3.1 "Block Effects Fixed vs. Block Effects Random" Revisited

Are block effects fixed or random? Framing the question in linear predictor terms, do we write the model as $\eta + \tau_i + \rho_j$ where ρ_j denotes the block effects and is part of $\mathbf{X\beta}$, or do we write the model $\eta + \tau_i + r_j$ where r_j denotes the block effects, has an assumed distribution and is part of **Zb**? The controversy has two primary aspects, one definitional and one practical; the latter concerns the small sample behavior of estimation and inference with each model.

Let us begin by reviewing definitions. By definition, a fixed model effect represents a factor with a finite number of levels, and our design provides for observing *every* level of that factor. A random model effect derives from a factor whose levels constitute a population. We cannot observe the entire population; instead, we observe representatives of the population via some form of sampling.

With random effects we *represent*, and with fixed effects we *observe exhaustively*. Consequently, inference for fixed effects does not extend beyond the levels actually observed (with the exception of interpolation between quantitative levels using a regression model), whereas inference for random effects extends to the population the observed levels represent. From a definitional viewpoint, the crux of the controversy amounts to this: Do we take the blocks observed to be the entire population, or do we understand them to represent a target population? Are we content with confining inference to only the blocks we actually observe—that is, narrow or block-specific inference—or does our objective call for broad or population-averaged inference? Advocates of the block-effects-random perspective contend that, by purely definitional criteria, assuming fixed-block effects appears to be an incoherent modeling decision.

Fixed-block advocates typically sidestep the definitional issues and focus on practical issues. What are these?

First, if we limit the discussion to balanced data—that is, randomized *complete* block designs with no missing data—and estimable functions defined on treatment *differences*, the controversy is moot. Fixed- and random-block-effect analyses yield identical results.

What about unbalanced data? In real life, missing data happen. Furthermore, in many contexts, we gain efficiency using *incomplete* block designs instead of complete blocks. We will discuss the latter issue in much more detail in Chapter 16, which concerns power, sample size, and planning designs.

This raises the second major practical point. For unbalanced block designs, Yates (1940) showed that recovering inter-block information and combining it with intra-block information improves efficiency. Combined inter–intra-block analysis was shown later to be equivalent to block-effects-random mixed model analysis (see, e.g., Bhattacharya, 1998). Random-block analysis is more efficient—*provided we have a known block variance*. What happens when we do not? An important question, because in real life we never do. Here we have the *real* controversy, which can be stated as follows: How good does the block variance estimate need to be in order to benefit from recovery of inter-block information? What does it take to get a good estimate?

The primary fixed-block argument turns on the supposition that in studies with a small number of blocks (e.g., in agriculture, studies with 3–5 blocks are the rule), there are too few block degrees of freedom. Ergo this precludes a good estimate of block variance, so recovering inter-block information is counterproductive. If true, fixed-block advocates have a case. If not, they do not.

So how bad is it? Let us consider a case that fixed-block advocates might call the random-block model's nightmare scenario. Here is the plot plan:

	Treatments	
Block 1	1	2
Block 2	1	3
Block 3	2	3

We have three treatments and three blocks, each of size 2. The design only allows two degrees of freedom to estimate the block variance. How do the fixed-block and random-block analyses compare? To answer this question we turn to simulation. Consider two scenarios: (1) all three treatment means equal (e.g., $\mu_1 = \mu_2 = \mu_3 = 0$) and (2) treatments differ by 15 units (e.g., $\mu_1 = 0$, $\mu_2 = 15$, and $\mu_3 = 30$). Scenario 1 allows us to assess type I error control; scenario 2 allows us to evaluate power. Assume Gaussian data. Without loss of generality, set $\sigma^2 = 1$. In theory, for $\alpha = 0.05$, the rejection rate for H_0: $\mu_1 = \mu_2 = \mu_3$ should be 5% under scenario 1. Under scenario 2, the rejection rate for random-block analysis depends on σ_B^2. For $\sigma_B^2 = 0.25$, the block variance used in this simulation, the resulting rejection rates, in theory, should be 0.806 and 0.849, respectively, under fixed- and random-block analysis. Readers can easily verify that changing σ_B^2 does not change the bottom line findings. What are the findings? The following table provides a summary for 1000 simulated experiments:

Scenario 1: all means equal

	Fixed-Block	Random-Block
Rejection rate (rejections/1000)	0.052	0.052
Mean estimate (here, e.g., average $\hat{\mu}_1$)	0.032	0.030
Std deviation, observed sampling dist of $\hat{\mu}_1$	0.829	0.808
Average std error ($\hat{\mu}_1$)	0.628	0.718
Diff estimate (here, e.g., $\hat{\mu}_2 - \hat{\mu}_1$)	0.042	0.048
Std dev, observed sampling dist of $\hat{\mu}_2 - \hat{\mu}_1$	1.124	1.090
Average std error $\hat{\mu}_2 - \hat{\mu}_1$	0.928	0.809
Block variance estimate $\hat{\sigma}_B^2$	na	0.551
Error variance estimate $\hat{\sigma}^2$	1.014	0.674

Scenario 2: all means different

	Fixed-Block	Random-Block
Rejection rate (rejections/1000)	0.819	0.910
Mean estimate (here, e.g., average $\hat{\mu}_1$)	−0.030	−0.031
Std deviation, observed sampling dist of $\hat{\mu}_1$	0.863	0.832
Average std error ($\hat{\mu}_1$)	0.606	0.719
Diff estimate (here, e.g., $\hat{\mu}_2 - \hat{\mu}_1$)	15.017	14.993
Std dev, observed sampling dist of $\hat{\mu}_2 - \hat{\mu}_1$	1.164	1.125
Average std error $\hat{\mu}_2 - \hat{\mu}_1$	0.896	0.801
Block variance estimate $\hat{\sigma}_B^2$	na	0.594
Error variance estimate $\hat{\sigma}^2$	0.957	0.640

Given the design, with two replicates per treatment and only one degree of freedom for residual, we should not expect much out of either analysis. Certainly, we would not want to use this design in an actual study. That said, if too few blocks does compromise the random-block-effect analysis, we should see the impact maximized for this design. In fact, we do not. To be sure, the estimates of σ_B^2 and σ^2 *per se* are not accurate. However, the type I error and power characteristics compare very favorably to the fixed-block analysis.

This occurs because the test statistic for treatments depends on the aggregate estimate of $Var[\mathbf{K}'(\mathbf{X}'\mathbf{V}^{-1}\mathbf{X})^{-}\mathbf{K}]$, where $\mathbf{K}$ defines $H_0{:}\,\mu_1 = \mu_2 = \mu_3$ in the estimable function $\mathbf{K}'\boldsymbol{\beta}$. The quality of this estimate, not σ_B^2 and σ^2 *individually,* is what matters. The bottom line: This is hardly a slam dunk for the fixed-block analysis—if anything, the random-block analysis seems to outperform the fixed-block model slightly overall—especially with regard to the test of treatment effect.

So far, we have considered the single-level case. If we were keeping score, on the definitional side, we have a win by forfeit for the random-block model; on the practical side, we have essentially a tie for designs with few blocks and a slight edge to the random-block model for designs with more blocks. If we limit our attention to single-level designs, there is less than meets the eye to this controversy. We now turn our attention to the multilevel model, where the real consequences lie.

8.3.2 Fixed Blocks, Multilevel Designs, and Spurious Nonestimability

Let us begin with the disconnected design from Figure 2.5. Recall that the plot plan was as follows:

Block	Treatments		
1	0	1	2
2	0	1	2
3	0	1	2
4	0	1	2
5	0	1	2
6	3	4	5
7	3	4	5
8	3	4	5
9	3	4	5
10	3	4	5

We saw in Chapter 2 that a naive linear predictor for this design, $\eta + \tau_i + \rho_j$, assuming the block effects, ρ_j, fixed, yields estimability issues for marginal treatment means and differences involving treatments not together in common blocks. For example, consider the marginal mean, defined as $\eta+\tau_i+1/10\sum_{j=1}^{10}\rho_j$. For the marginal mean to be estimable, there must exist coefficients a_{ij} such that $\sum_{i,j} a_{ij}(\eta+\tau_i+\rho_j)=\eta+\tau_i+1/10\sum_j \rho_j$.

This means $\sum_{i,j} a_{ij}=1$ (to get η), $\sum_j a_{ij}=1$ and $\sum_j a_{i'j}=0$, for all $i'\neq i$ (to get τ_i) *and* $\sum_i a_{ij}=1/10$ for all $j=1,2,\ldots,10$. It is easy to satisfy the first two conditions simultaneously, but the first two conditions preclude the third. Alternatively, we can show that **K** does not satisfy the $\mathbf{K}'(\mathbf{X}'\mathbf{X})^{-}\mathbf{X}'\mathbf{X} = \mathbf{K}'$ criterion and hence $\mu + \tau_i + 1/10\sum_j \rho_j$ is not estimable.

Given the definitional perspective on block effects, this example affords us our first look at spurious nonestimability. If we treat block effects as random, recovery of inter-block information gives us an accurate analysis for what this design actually is: a nested factorial design, where the factorial structure is defined by *set* and *treatment(set)*. This is the skeleton ANOVA we saw at the end of Section 8.2.2.

With a fixed-block model, we can repair the estimability problem by using a less naive version of the linear predictor: $\eta + \alpha_i + \beta(\alpha)_{ij} + \rho(\alpha)_{ik}$, where α denotes the set effect, $\beta(\alpha)$ denotes treatments within sets, and $\rho(\alpha)$ denotes block effects, now seen to be nested within sets. In practice, most fixed-block advocates would concede that $\rho(\alpha)$, in the context of the design viewed this way, is a whole-plot effect—that is, unit of replication with respect to set—and therefore random. However, we should note that fixed-effect advocates tend to prefer ordinary least squares software, which *does* treat $\rho(\alpha)$ as fixed when solving estimating equations, determining standard errors, and, most importantly for this discussion, assessing estimability.

The aforementioned repair "works" (up to a point) for balanced data. Let us now consider an unbalanced multilevel design. Littell et al. (2006) illustrated the analysis of a split-plot experiment for which the whole-plot was an incomplete block design whose structure is similar to that shown in the previous section. The following shows the essential features of the plot plan:

Block 1			Block 2			Block 3		
A1	B1	B2	A1	B1	B2	A2	B1	B2
A2	B1	B2	A3	B1	B2	A3	B1	B2

The levels of factor A are randomly assigned to rows that run all the way across each block. The levels of factor B are randomly assigned to the two cells within each row in a block, that is, within each unit of replication with respect to factor A. In skeleton ANOVA terms, we have

Topographical		Treatment		Combined	
Source	d.f.	Source	d.f.	Source	d.f.
Block	$r-1$			Block	$r-1$
		A	2	A	2
Row(block)	r			Block × A Row(block) | A	$r-2$
		B	1	B	1
		A × B	2	A × B	2
Cell(row, block)	$2r$	"Parallels"	$4r-6$	Block × A × B cell(row, block) | A, B or "residual"	$2r-3$
Total	$4r-1$	Total	$4r-1$	Total	$4r-1$

where r denotes the number of blocks: r could be 3 or it could be 6, 9, etc., if we use more than one block of each type. The standard design of experiments textbook gives the linear predictor for this design as

$$\eta_{ijk} = \eta + \alpha_i + \beta_j + (\alpha\beta)_{ij} + \rho_k + (\alpha\rho)_{ik}$$

If we treat blocks—meaning ρ_j and $(\alpha\rho)_{ij}$—as fixed effects, we render all A and A × B marginal means nonestimable, all A differences nonestimable, and all A × B differences involving different levels of A—which includes simple effects of A given B—nonestimable. This provides a blatant example of spurious nonestimability. No one would argue that $(\alpha\rho)_{ik}$ should be treated as a fixed effect as it is the unit-of-replication effect with respect to factor A.

The problem usually arises when software designed for ordinary least squares (e.g., SAS® PROC GLM) is used to compute the analysis from a design like this. PROC GLM cannot distinguish fixed from random effects and, as a result, inappropriately assesses marginal means, differences, and all other estimable functions as if $(\alpha\rho)_{ik}$ is fixed. This is why computations for mixed models—including *all* multilevel models—should be done only with software expressly designed for mixed model analysis.

8.4 Working with Multilevel Designs

In previous sections, we defined the multilevel structure, reviewed the use of the skeleton ANOVA (WWFD) process to identify multilevel structure, and took a close look at the impact proper specification of block effects has on multilevel analysis. Having dealt with all these preliminaries, we now look at practical aspects of working with multilevel models. We start by working through examples of multilevel analysis.

8.4.1 Examples of Multilevel Structures

In this section, we consider three examples of multilevel analysis—specifying the model, translating the model into the required GLIMMIX statements, and interpreting relevant output. In this section, we confine our attention to Gaussian data. In future chapters, we will consider these same designs but with non-Gaussian data.

Example 8.1 Nested factorial structure

The data for this example appears in Data Set 8.1. The design follows the Figure 2.5 structure we considered in the previous section—naively called a disconnected design but more accurately called a nested factorial. The full specification of the model is

1. Linear predictor: $\eta_{ijk} = \eta + \alpha_i + \beta(\alpha)_{ij} + r(\alpha)_{ik}$, where $r(\alpha)$ denotes block within set
2. Distributions:
   a. $y_{ijk} | r(\alpha)_{ik} \sim NI(\mu_{ijk}, \sigma^2)$
   b. $r(\alpha)_{ik}$ i.i.d. $N\left(0, \sigma_R^2\right)$
3. Link: identity, $\eta_{ijk} = \mu_{ijk}$

The GLIMMIX statements are

```
proc glimmix data=ten_blk_six_trt;
  class block set trt;
  model y = set trt(set);
  random block(set);
  lsmeans set trt(set)/slice=set slicediff=set;
```

Relevant output:

Covariance Parameter Estimates		
Cov Parm	**Estimate**	**Standard Error**
block(set)	60.5495	34.1719
Residual	22.7507	8.0436

The variance component estimates are $\hat{\sigma}_R^2 = 60.55$ for blocks and $\hat{\sigma}^2 = 22.75$ for observations conditional on the block effects.

Type III Tests of Fixed Effects				
Effect	**Num DF**	**Den DF**	***F* Value**	**Pr > *F***
set	1	8	0.04	0.8509
trt(set)	4	16	4.91	0.0089

The F-value for "trt(set)" tests H_0: all $\beta(\alpha)_{ij} = 0$, which we can more meaningfully express as $H_0: \eta_{00} = \eta_{01} = \eta_{02}$ and $\eta_{13} = \eta_{14} = \eta_{15}$. Its p-value is 0.0089, which tells us (1) testing the main effect of "set" constitutes an "uninteresting hypothesis" and (2) since "trt(set)" is an aggregated hypothesis, we need to split it into its components, $H_0: \eta_{00} = \eta_{01} = \eta_{02}$ and $H_0: \eta_{13} = \eta_{14} = \eta_{15}$, using the SLICE option to obtain an informative interpretation.

Tests of Effect Slices for trt(set) Sliced by set				
set	**Num DF**	**Den DF**	***F* Value**	**Pr > *F***
0	2	16	0.03	0.9709
1	2	16	9.80	0.0017

This tells us that we have evidence of a treatment effect within set 1 ($p = 0.0017$) but not within set 0 ($p = 0.9709$). A look at the treatment means makes this result intelligible:

trt(set) Least Squares Means			
set	**trt**	**Estimate**	**Standard Error**
0	0	8.2000	4.0817
0	1	8.9000	4.0817
0	2	8.7400	4.0817
1	3	4.4400	4.0817
1	4	7.2800	4.0817
1	5	17.1600	4.0817

Finally, the SLICEDIFF listing show the pair-wise simple effects of treatment within set.

Simple Effect Comparisons of trt(set) Least Squares Means by set

Simple Effect Level	trt	_trt	Estimate	Standard Error	DF	t Value	Pr > |t|
set 0	0	1	−0.7000	3.0167	16	−0.23	0.8194
set 0	0	2	−0.5400	3.0167	16	−0.18	0.8602
set 0	1	2	0.1600	3.0167	16	0.05	0.9584
set 1	3	4	−2.8400	3.0167	16	−0.94	0.3605
set 1	3	5	−12.7200	3.0167	16	−4.22	0.0007
set 1	4	5	−9.8800	3.0167	16	−3.28	0.0048

We see that the treatment within set 1 effect results from statistically significant differences $\eta_{13} - \eta_{15}$ and $\eta_{14} - \eta_{15}$ ($p = 0.0007$ and $p = 0.0048$, respectively). In other words, the mean of treatment 5 is significantly different from the means of treatments 3 and 4.

Example 8.2 Incomplete strip-plot

Data for this example appears in the SAS Data and Program Library as Data Set 8.2. The treatment design is a 3 × 3 factorial. The topographical design has nine blocks each consisting of two rows and two columns or four total units. Levels of factor A are assigned at random to rows in each block, with the following restriction: three blocks receive levels A1 and A2 only, three blocks receive A1 and A3 only, and the other three blocks receive A2 and A3 only. Factor B levels are randomly assigned to columns in each block, with the following restriction: one of the blocks that received a given pair of levels of A must be assigned B1 and B2, one of them must receive B1 and B3 and the other block must receive B2 and B3. Thus, our multilevel structure is an incomplete block design in two dimensions.

The skeleton ANOVA process helps us visualize the required linear predictor:

Topographical		Treatment		Combined	
Source	d.f.	Source	d.f.	Source	d.f.
Block	8			Block	8
		A	2	A	2
Row(block)	9			Row(block) | A Block × A	7
		B	2	B	2
Column(block)	9			Column(block) | B Block × B	7
		A × B	4	A × B	4
Row × column(block)	9	"Parallels"	27	Row × column(block) | A, B Block × A × B or "residual"	5
Total	35	Total	35	Total	35

Note that while there are two randomization processes, there are effectively three sizes of units of replication: row for levels of A, columns for levels of B, and row–column intersections for A × B combinations. A model for these data is given as follows:

1. Linear predictor: $\eta_{ijk} = \eta + \alpha_i + \beta_j + (\alpha\beta)_{ij} + r_k + (ar)_{ik} + (br)_{jk}$
2. Distributions:
   a. $y_{ijk} | r_k, (ar)_{ik}, (br)_{jk} \sim NI(\mu_{ijk}, \sigma^2)$
   b. r_k i.i.d. $N\left(0, \sigma_R^2\right)$
   c. $(ar)_{ik}$ i.i.d. $N\left(0, \sigma_{AR}^2\right)$
   d. $(br)_{jk}$ i.i.d. $N\left(0, \sigma_{BR}^2\right)$
3. Link: identity $\eta_{ijk} = \mu_{ijk}$

The GLIMMIX statements for this model, including activation of ODS graphics to enable producing an interaction plot using the LSMEANS statement and MEANPLOT option:

```
ods graphics on;
proc glimmix data=ds8d2;
  class block a b;
  model y=a|b/ddfm=kr;
  random intercept a b/subject=block;
  lsmeans a*b/plot=meanplot(sliceby=a join) slicediff=a;
```

Consistent with our discussion of standard error and test statistic bias with unbalanced mixed models, we use the DDFM=KR to invoke the Satterthwaite degrees of freedom approximation and the Kenward–Roger bias correction. The interaction plot appears as follows:

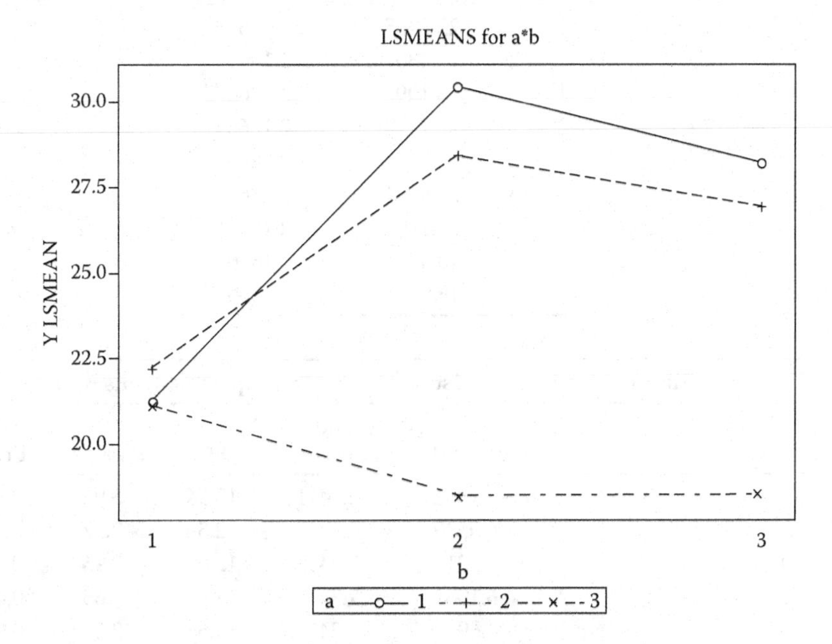

We see no visual evidence of a B effect in conjunction with A3. With A1 and A2, the mean at B1 appears to differ from the B2 and B3 means. The A–B combinations formed from A1, A2, B2, and B3 show slight visual differences. Formal statistics below can resolve whether the differences are statistically significant, but subject matter specialists need to determine whether these differences are large enough to matter, formal statistics notwithstanding.

The variance component estimates and overall effect tests are

Covariance Parameter Estimates			
Cov Parm	**Subject**	**Estimate**	**Standard Error**
Intercept	block	2.6345	7.9462
a	block	9.7784	6.0842
b	block	8.1609	5.1496
Residual		2.2451	1.4001

Type III Tests of Fixed Effects				
Effect	**Num DF**	**Den DF**	***F* Value**	**Pr > *F***
a	2	8.592	6.59	0.0184
b	2	8.405	2.31	0.1589
a*b	4	5.994	10.56	0.0070

The F-statistic for A × B provides strong evidence ($p = 0.007$) of an interaction, consistent with visual evidence from the interaction plot. At this point, inference should focus on simple effects. Accordingly, we look at the LSMEANS and simple effect results:

a*b Least Squares Means			
a	**b**	**Estimate**	**Standard Error**
1	1	21.2455	2.1866
1	2	30.4307	2.1866
1	3	28.1994	2.1866
2	1	22.2115	2.1866
2	2	28.4077	2.1866
2	3	26.9061	2.1866
3	1	21.1293	2.1866
3	2	18.4717	2.1866
3	3	18.5231	2.1866

Simple Effect Comparisons of a*b Least Squares Means by a							
Simple Effect Level	**b**	**_b**	**Estimate**	**Standard Error**	**DF**	***t* Value**	**Pr> |*t*|**
a1	1	2	−9.1852	2.3411	12.56	−3.92	0.0019
a1	1	3	−6.9539	2.3411	12.56	−2.97	0.0112
a1	2	3	2.2313	2.3411	12.56	0.95	0.3585
a2	1	2	−6.1962	2.3411	12.56	−2.65	0.0206
a2	1	3	−4.6946	2.3411	12.56	−2.01	0.0670
a2	2	3	1.5016	2.3411	12.56	0.64	0.5328
a3	1	2	2.6577	2.3411	12.56	1.14	0.2775
a3	1	3	2.6063	2.3411	12.56	1.11	0.2865
a3	2	3	−0.05138	2.3411	12.56	−0.02	0.9828

These results provide formal support for the visual characterization given earlier: at A3, none of the B differences approach statistical significance; the means range between 18.47 and 21.13 with a standard error of 2.19. Within A1 and A2, the B2 and B3 means differ from B1 but not from each other, the B1 vs. B3 difference being larger than the B2 vs. B3 difference. We could go on. For example, the means for the combinations of B1 with A1 and A2 appears to be similar to the set of all B means at A3. We leave this as an exercise. What we see here is the basic flow of the analysis.

Notice that inference on the treatment design itself does not differ from inference on any other factorial treatment design. The *only* difference involves properly defining the model's random effects to be consistent with the design structure. We do this to make sure we use the correct standard errors and test statistics.

Example 8.3 Response surface design with incomplete blocking

In Section 7.6.1 we introduced second-order polynomial regression for all factors quantitative treatment designs. Recall that this model's attractiveness stems partly from the ability to use its efficient, incomplete factorial designs. One such design, popular in response surface applications, is the Box–Behnken design (Box and Behnken, 1960). This example uses a three-factor Box–Behnken design—see response surface textbooks, for example, Myers et al. (2009) for comprehensive description of this and other response surface designs. Each factor has three levels—as is common practice with response surface design and analysis, we code the three levels as –1, 0, and 1. We construct the design by forming treatment combinations of level 0 of each factor with a 2^2 factorial using levels –1 and 1 from the other two factors. Because we have three factors here, this gives us 12 treatment combinations, 3 sets of 4 treatment combinations each. To these, we add one additional treatment combination consisting of the 0 level of all factors—called a "center point" in design terminology.

The data for this example appears in the SAS Data and Program Library as Data Set 8.3. The multilevel aspect of these data results from the incomplete blocking used. Each set of four treatment combinations associated with the 0 level of a particular factor appear to together, two blocks per set, each block size 4. The design has six such blocks, two per set. A seventh block, consisting of four units all receiving the center point, completes the design. Notice that there are two randomization stages: the first randomly assigns sets to blocks; the second randomly assigns treatment combinations within a set to units within a block. This two-stage randomization makes this a multilevel design. Also notice that the multiple levels do not correspond neatly to a single factor in the design, as they did with Examples 8.1 and 8.2. This example appears here precisely to make the point that multilevel structure can take on many forms. Modeling practitioners need to be alert. This example also underlines the weakness of depending on recipes. It is easier to learn to think these things through than learn a recipe comprehensive enough to cover all modeling contingencies—assuming you could devise such a recipe!

How would we use the WWFD skeleton ANOVA process to help construct the linear predictor? Here it is

Topographical		Treatment		Combined	
Source	d.f.	Source	d.f.	Source	d.f.
Block	6			Block	6
		f(A, B, C)	9	*f*(A, B, C)	9
Units(block)	21	"Parallels"		Unit(block) | A, B, C	12
Total	27	Total	27	Total	27

The function $f(A,B,C)=\beta_0+\sum_{i=A,B,C}\beta_i X_i+\sum_{i=A,B,C}\beta_{ii}X_i^2+\sum_{i\neq i'=A,B,C}\beta_{ii'}X_iX_{i'}$, the second-order response surface model for factors A, B, and C, where the βs denote regression equations and X_i; i = A, B, C denote quantitative levels of the three factors.

A model for these data can be written as

1. Linear predictor: $\eta_{ijkl} = \eta_{ijk} + r_{\ell k}$, where $r_{\ell k}$ denotes the block effects and $\eta_{ijk}=f(A,B,C)+(\alpha\beta\gamma)_{ijk}=\beta_0+\sum_{i=A,B,C}\beta_i X_i+\sum_{i=A,B,C}\beta_{ii}X_i^2+\sum_{i\neq i'=A,B,C}\beta_{ii'}X_iX_{i'}+(\alpha\beta\gamma)_{ijk}$ $(\alpha\beta\gamma)_{ijk}$ is a catchall lack-of-fit term
2. Distributions:
   a. $y_{ijkl}|r_l \sim NI(\mu_{ijkl},\sigma^2)$
   b. r_l i.i.d. $N(0,\sigma_R^2)$
3. Link: identity, $\eta_{ijkl}=\mu_{ijkl}$

The GLIMMIX statements appear in three steps. In step 1, we assess the model for lack of fit. Assuming no evidence of lack of fit, we fit the full second-order response surface, dropping $(\alpha\beta\gamma)_{ijk}$ from the linear predictor. If there are second-order terms that appear to be negligible, we can drop them and run a third step to fit a more parsimonious model. The GLIMMIX statements for step 1 are as follows:

```
proc glimmix;
  class block aa bb cc;
  model y=a|a|b|b|c|c@2 aa*bb*cc/ htype=1 ddfm=kr;
  random intercept/subject=block;
```

Notice we define AA, BB, and CC as CLASS variables to form the lack-of-fit term AA*BB*CC. We used this trick in Section 7.6 when we introduced quantitative factor treatment designs. Also, we use sequential testing strategy here because we want to assess lack of fit *after* fitting *f*(A, B, C). Because we have an incomplete block design, we use the Kenward–Roger adjustment. Relevant output is as follows:

Type I Tests of Fixed Effects

Effect	Num DF	Den DF	*F* Value	Pr > *F*
a	1	12	19.14	0.0009
a*a	1	3	10.33	0.0488
b	1	12	21.45	0.0006
a*b	1	12	0.76	0.4014
b*b	1	3	16.81	0.0263
c	1	12	2.01	0.1815
a*c	1	12	1.03	0.3301
b*c	1	12	5.05	0.0442
c*c	1	3	0.70	0.4647
aa*bb*cc	3	12	0.22	0.8776

In particular, because we are testing lack of fit, only the result for AA*BB*CC concerns us at this point. $F = 0.22$ and $p = 0.8776$ indicate no evidence of lack of fit. We proceed to step 2.

Step 2 requires only a change in the CLASS and MODEL statements. We drop the lack of fit term from MODEL and the corresponding effects in the CLASS statement. The new statements are as follows:

```
class block;
model y=a|a|b|b|c|c@2/htype=1 ddfm=kr;
```

Relevant output is as follows:

Type I Tests of Fixed Effects				
Effect	**Num DF**	**Den DF**	***F* Value**	**Pr > *F***
a	1	15	22.65	0.0003
a*a	1	3	10.33	0.0488
b	1	15	25.38	0.0001
a*b	1	15	0.90	0.3589
b*b	1	3	16.81	0.0263
c	1	15	2.38	0.1437
a*c	1	15	1.22	0.2869
b*c	1	15	5.98	0.0273
c*c	1	3	0.70	0.4647

Notice the linear-by-linear term A*B and the quadratic term C*C. Their *F*-values are both less than 1. Also, the *p*-value for the linear-by-linear term A*C is 0.2869. This casts doubt on whether these three terms should remain in the model. The linear main effect of C also has a *p*-value we would normally declare nonsignificant. However, the linear-by-linear B*C term *is* statistically significant: A regression model including $\beta_{BC}X_BX_C$ but not β_CX_C violates the spirit of sequential testing and would create interpretation problems for the factor B linear and quadratic main effects. Therefore, a plausible step-3-reduced model that follows from step 2 would be

$$f(A,B,C) = \beta_0 + \sum_{i=A,B,C} \beta_i X_i + \sum_{i=A,B} \beta_{ii} X_i^2 + \beta_{BC} X_B X_C$$

The GLIMMIX statements, amended for step 3, are

```
proc glimmix;
  class block;
  model y=a|a b|b b|c/ solution htype=1 ddfm=kr;
  random intercept/subject=block;
```

Relevant output is as follows:

Covariance Parameter Estimates			
Cov Parm	**Subject**	**Estimate**	**Standard Error**
Intercept	block	0.6776	0.8160
Residual		1.8206	0.6245

Solutions for Fixed Effects					
Effect	**Estimate**	**Standard Error**	**DF**	***t* Value**	**Pr > |*t*|**
Intercept	50.0850	0.8244	4	60.75	<.0001
a	1.6000	0.3373	17	4.74	0.0002
a*a	−3.3025	0.8244	4	−4.01	0.0161
b	1.6938	0.3373	17	5.02	0.0001
b*b	−3.5150	0.8244	4	−4.26	0.0130
c	0.5188	0.3373	17	1.54	0.1425
b*c	−1.1625	0.4770	17	−2.44	0.0261

We can use these regression coefficients to generate a plot of the estimated response surface. A SAS program for doing so is

```
data rs3;
  do a=-1 to 1;
    do b=-1 to 1;
      do c=-1 to 1;
      y_hat=50.085+1.60*a+1.694*b+0.519*c-3.303*a*a-3.515*b*b-1.163*b*c;
      output;
    end;
    end;
  end;
proc sort data=rs3;
  by c;
proc g3d;
  by c;
  plot a*b=y_hat/xticknum=3 yticknum=3;
```

This creates a three-dimensional plot showing the predicted response over levels of A and B for each level of C:

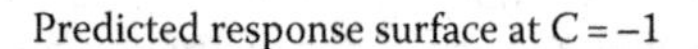

Predicted response surface at C = −1

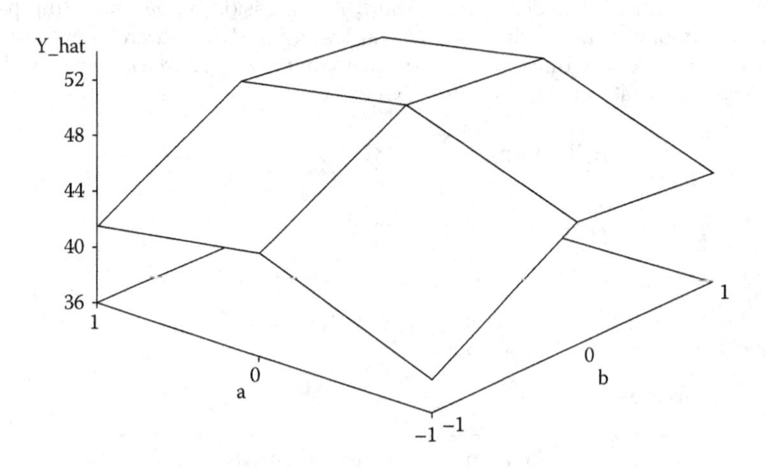

Predicted response surface at C = 0

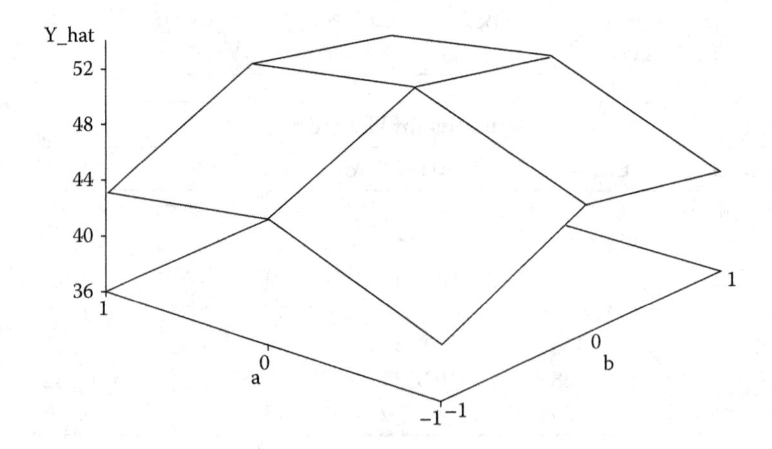

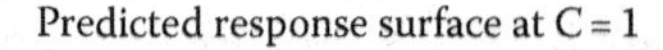
Predicted response surface at C = 1

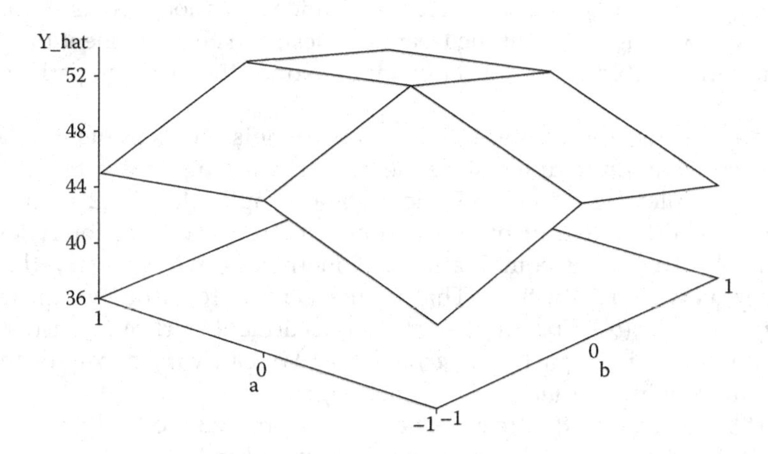

We can see that factor C has a slight impact of the shape of the response surface. The overall picture is a slightly tilted two-dimensional quadratic surface over the levels of A and B with a maximum response appearing near the center point. The main point in this example is that fitting the surface accurately requires making appropriate decisions about lack of fit and which effects to retain. We cannot do this unless we specify the multilevel error structure correctly. This example illustrates that multilevel mixed model and response surface methodology are compatible—in examples like this, it is imperative that we use both.

Another nonstandard multilevel structure not shown here but worth mentioning is a mixed-up split-plot presented by Milliken and Johnson (2008). A multifactor study was conducted by a multistate consortium. The protocol called for a split-plot design. This part of the protocol was communicated, but the part about which factor was to be assigned to the whole-plot (larger unit of replication) and which factor was to be assigned to the split-plot (smaller unit of replication) was not. As a result, the position of factors A and B was reversed for the various states' data. Viewing the structure as an incomplete block design—along the lines of the disconnected/nested factor example—reveals the analysis to be a straightforward, though unusual, application of multilevel modeling.

8.4.2 Multifactor Treatment and Multilevel Design Structures: How They Fit Together

Each of the three examples in the previous section used a specific treatment structure. In fact, we have considerable flexibility in the way we integrate the topographical aspect of multilevel models with the treatment aspect.

Consider Example 8.1 in the previous section. We used the linear predictor $\eta_{ijk} = \eta + \alpha_i + \beta(\alpha)_{ij} + r(\alpha)_{ik}$. Recalling the types of treatment structures we considered in Chapter 7, we can change the treatment component to other forms if the specifics of the treatment design call for it. For instance, if the treatment levels within each set are actually identical, so that our treatment design is in reality a 2 × 3 factorial, we can revise the linear predictor accordingly, replacing the previous with $\eta_{ijk} = \eta + \alpha_i + \beta_j + (\alpha\beta)_{ij} + r(\alpha)_{ik}$. If the treatment levels within each set are quantitative and amenable to polynomial regression, we could use $\eta_{ijk} = \beta_{0i} + \beta_{1i}X_j + \beta_{2i}X_j^2 + r(\alpha)_{ik}$. If instead we had a more nonlinear shape, we might prefer the linearized Hoerl, $\eta_{ijk} = \beta_{0i} + \beta_{1i}X_j + \beta_{2i}\log(X_j) + r(\alpha)_{ik}$. Notice that all of these are variations on $\eta_{ijk} = \eta_{ij} + r(\alpha)_{ik}$. The nested, factorial, and regression models simply represent different ways of writing η_{ij} to suit the treatment design. The component that models the multilevel design structure, $r(\alpha)_{ik}$, never changes. Notice that it is compatible with every variation on η_{ij}.

To illustrate, let us apply alternative mean models to Example 8.2. These data provide a somewhat more interesting demonstration than Data Set 8.4.1 because of the added complexity: 8.4.1 is a basic, balanced split-plot; 8.4.2 is an incomplete strip-plot with a different unit of replication with respect to factor A levels, factor B levels, and AB treatment combinations. Nonetheless, we can vary the treatment model exactly as we do with 8.4.1. That is, the generic form of the linear predictor, $\eta_{ijk} = \eta_{ij} + r_k + (ar)_{ik} + (br)_{jk}$, partitions into the treatment part, η_{ij}, and the multilevel design, or "topographical" part, $r_k + (ar)_{ik} + (br)_{jk}$. We can vary η_{ij} any of the ways we have seen while leaving $r_k + (ar)_{ik} + (br)_{jk}$ unchanged.

Let us start by expressing the treatment aspect as a nested model, that is, $\eta_{ij} = \eta + \alpha_i + \beta(\alpha)_{ij}$. The GLIMMIX statements to implement this model are

```
proc glimmix data=ds8d2;
  class block a b;
  model y=a b(a)/ddfm=kr;
  random intercept a b/subject=block;
  lsmeans b(a)/slice=a slicediff=a;
  lsmestimate b(a)'b1+b2 avg for a1 and a2' 1 1 0 1 1 0 0 0 0,
                   'b1+b2 avg for a3' 0 0 1 0 0 1 0 0 0,
                   'b3 avg for a1 and a2' 0 0 0 0 0 0 1 1 0/
                     divisor=4,2,2;
  estimate 'b1&2 - b3 diff for a1&2' b(a) 1 1 -2 1 1 -2 0 0 0,
           'b1&2 - b3 diff for a3' b(a) 0 0 0 0 0 0 1 1 -2/
                     divisor=4,2;
```

We might do this to focus inference on the simple effects of B given A. Notice the use of LSMESTIMATE and ESTIMATE statements to estimate means and differences for groupings that emerged from our earlier analysis. Using the nested form of the mean model facilitates these estimates by making them less prone to estimability issues. Notice also—and importantly—that we have not changed the random statement at all. We change the mean model to redirect focus on aspects of treatment analysis we wish to emphasize, but we leave the multilevel design model alone.

Relevant output follows (but we skip output that repeats what was already shown previously):

Type III Tests of Fixed Effects

Effect	Num DF	Den DF	*F* Value	Pr > *F*
a	2	8.592	6.59	0.0184
b(a)	6	7.771	7.34	0.0070

The B(A) tests tells us there are differences among the factor B means within each level of A, but as we have seen before with this kind of statistic, its aggregation prevents useful interpretation. Better to look at SLICE output or take an even closer look using the ESTIMATE and LSMESTIMATE output.

Tests of Effect Slices for b(a) Sliced by a				
a	Num DF	Den DF	*F* Value	Pr > *F*
1	2	12.56	8.37	0.0049
2	2	12.56	3.81	0.0508
3	2	12.56	0.84	0.4534

This is better. We see that there are significant B effects within A1 and A2 but not A3. Taking these differences apart, we have

Least Squares Means Estimates						
Effect	Label	Estimate	Standard Error	DF	*t* Value	Pr > |*t*|
b(a)	b1 + b2 avg for a1 and a2	25.5739	1.4576	13.15	17.55	<.0001
b(a)	b1 + b2 avg for a3	27.5528	1.8003	18.22	15.30	<.0001
b(a)	b3 avg for a1 and a2	19.8005	1.8469	17.49	10.72	<.0001

and

Estimates					
Label	Estimate	Standard Error	DF	*t* Value	Pr > |*t*|
b1 and 2—b3 diff for a1 and 2	−1.9789	1.8302	9.693	−1.08	0.3058
b1 and 2—b3 diff for a3	1.2774	2.0275	12.56	0.63	0.5399

The LSMESTIMATE output lists estimate $(\eta_{11} + \eta_{12} + \eta_{21} + \eta_{22})/4$, $(\eta_{31} + \eta_{32})/2$ and $\eta_{13} + \eta_{23}/2$, respectively. The ESTIMATE output lists $\eta_{11} + \eta_{12} + \eta_{21} + \eta_{22}/4 - \eta_{31} + \eta_{32}/2$ and $\frac{\eta_{13} + \eta_{23}}{2} - \eta_{33}$, respectively. We do this because the analysis from Example 8.2 suggested that the means of B1 and B2 were similar within A1 and A2, so it makes sense to average these four means. The means of B1 and B2 did not differ within A3, so it also makes sense to average them. The means for B3 differed, from B1 and B2, but B3 combined with A1 did not differ from B3 combined with A2, hence the third LSMESTIMATE. The ESTIMATE statements obtain the estimated differences between these grouping. We do this when we want to know the size of the difference, not just whether a difference exists.

Let us take this a bit further. Suppose the levels of B are quantitative. On the interaction plot, the pattern of means suggests a possible quadratic regression—certainly over the levels of B within A1 and A2. We can model this with a qualitative–quantitative model: here specifically, $\eta_{ij} = \beta_{0i} + \beta_{1i}X_j + \beta_{2i}X_j^2$, where X_j denotes the jth quantitative level of factor B. The GLIMMIX statements are

```
proc glimmix data=ds8d2;
  class block a b;
  model y=a xb(a) xb_sq(a)/noint s ddfm=kr;
  contrast 'quadratic equal, a1 and a2' xb_sq(a) 1 -1 0;
  contrast 'quadratic equal, a3 v a1&2' xb_sq(a) 1 1 -2;
  contrast 'regression equal, a1 and a2' xb(a) 1 -1 0,
                                        xb_sq(a) 1 -1 0;
  contrast 'regression equal, a3 v a1&2' xb(a) 1 1 -2,
                                        xb_sq(a) 1 1 -2;
  random intercept a b/subject=block;
```

The variables XB=B and XB_SQ=B*B. We must define them in the data step. They allow GLIMMIX to use B as a CLASS variable to specify the multilevel design model but use XB and XB_SQ as direct variables for X_j and X_j^2, respectively. Notice that the RANDOM statement remains unchanged. The contrast statements allow us to test whether the quadratic regression coefficients are equal for A1 and A2 and, if they are, whether they differ from A3. The second set of contrasts pools the linear and quadratic terms into a common test of the regression. The NOINT option gives us a full-rank model: the SOLUTION vector thus provides the actual regression estimates without any need for further manipulation.

Relevant output is as follows:

Solutions for Fixed Effects

Effect	a	Estimate	Standard Error	DF	*t* Value	Pr > |*t*|
a	1	0.6438	7.3559	14.01	0.09	0.9315
a	2	8.3175	7.3559	14.01	1.13	0.2771
a	3	26.4960	7.3559	14.01	3.60	0.0029
xb(a)	1	26.3100	8.1939	12.56	3.21	0.0071
xb(a)	2	17.7429	8.1939	12.56	2.17	0.0503
xb(a)	3	−6.7212	8.1939	12.56	−0.82	0.4274
xb_sq(a)	1	−5.7083	2.0275	12.56	−2.82	0.0150
xb_sq(a)	2	−3.8489	2.0275	12.56	−1.90	0.0809
xb_sq(a)	3	1.3545	2.0275	12.56	0.67	0.5162

For example, the estimated regression equation for A1 is $0.64 + 26.31X_j - 5.71X_j^2$. The similarity of the regression over B for A1 and A2 and the very different regression over B for A3 shows clearly in these estimates. We can also see that the linear and quadratic coefficients under A3 are not statistically significant—putting it more precisely, we fail to reject

H_0: $\beta_{13} = 0$ and H_0: $\beta_{23} = 0$ for any reasonable α level. For A3, a better fit might result from just using the intercept. We will pursue this below.

Type III Tests of Fixed Effects

Effect	Num DF	Den DF	*F* Value	Pr > *F*
a	3	9.721	6.29	0.0120
xb(a)	3	7.95	8.30	0.0078
xb_sq(a)	3	7.95	6.23	0.0175

We show the "type III tests of fixed effects" here more because they are *not* useful than because they provide "relevant" output (they do not). Each line in this listing is an overly aggregated test statistic. The three degrees of freedom come from combining each of the three parameters over all levels of factor A for the intercept, linear, and quadratic regression, respectively, and testing them simultaneously. Compare these results to those mentioned earlier for individual regression coefficients; you can see that the "type III tests" tell you nothing useful.

Finally, the contrast results are as follows:

Contrasts

Label	Num DF	Den DF	*F* Value	Pr > *F*
Quadratic equal, a1 and a2	1	5.994	1.14	0.3275
Quadratic equal, a3 v a1 and 2	1	5.994	16.48	0.0067
Regression equal, a1 and a2	2	5.994	1.20	0.3653
Regression equal, a3 v a1 and 2	2	5.994	19.92	0.0023

These confirm what we have seen from other vantage points. The regressions over B for A1 and A2 are similar, but different from the regression for A3.

Finally, we noted earlier that the tests of the individual regression coefficients suggest that it might be better to model responses for A3 using just the intercept, because B appears to have no effect within A3. A reduced model would be

$$\eta_{ij} = \beta_{0i} + I(\text{A} < 3)\beta_{1i}X_j + I(\text{A} < 3)\beta_{2i}X_j^2$$

where $I(\text{A} < 3)$ is an indicator function, equal to 1 for levels A1 and A2, equal to 0 for A3. This removes the regression over B for A3. To implement this with GLMMIX, we need to define indicator functions in the data set. In the following program, B_1 is 1 for A1, 0 otherwise; B_2 = 1 for A2, 0 otherwise; and B_3 = 1 for A3, 0 otherwise. The GLIMMIX statements are as follows:

```
proc glimmix data=ds8d2;
  class block a b;
  model y=b_1 b_2 b_3 b_1*xb b_2*xb b_1*xb_sq b_2*xb_sq/noint s ddfm=kr;
  contrast 'intercept equal, a1 and a2' b_1 1 b_2 -1;
  contrast 'linear equal, a1 and a2' b_1*xb 1 b_2*xb -1;
  contrast 'quadratic equal, a1 and a2' b_1*xb_sq 1 b_2*xb_sq -1;
  random intercept a b/subject=block;
```

Notice that the MODEL statement now includes no CLASS variables at all. All three intercepts appear and the linear and quadratic terms for A1 and A2 only. Note the form of the CONTRAST statements. Each effect only has one coefficient because each term in the MODEL is a single parameter.

> *Students*: you should be able to write the exact form of the estimable function for each contrast.

The output is as follows:

Solutions for Fixed Effects

Effect	Estimate	Standard Error	DF	*t* Value	Pr > |*t*|
b_1	−3.8940	6.1140	10.62	−0.64	0.5377
b_2	3.8518	6.1140	10.62	0.63	0.5420
b_3	19.4432	1.7083	14.23	11.38	<.0001
b_1*xb	30.5099	6.7255	8.932	4.54	0.0014
b_2*xb	21.9245	6.7255	8.932	3.26	0.0099
b_1*xb_sq	−6.5456	1.6641	8.932	−3.93	0.0035
b_2*xb_sq	−4.6890	1.6641	8.932	−2.82	0.0203

Contrasts

Label	Num DF	Den DF	*F* Value	Pr > *F*
Intercept equal, a1 and a2	1	7.767	1.42	0.2683
Linear equal, a1 and a2	1	6.217	1.51	0.2638
Quadratic equal, a1 and a2	1	6.217	1.15	0.3229

The first table gives the estimated coefficients for the reduced model. Note that $\hat{\beta}_{03} = 19.44$ estimates the expected response for A3 over all levels of B. The contrasts confirm that no statistically significant evidence exists to conclude that the regression parameters for A1 and A2 are different. We do not show it here, but we could further simplify the model by pooling β_{0i}, β_{1i}, and β_{2i} over A1 and A2.

8.5 Marginal and Conditional Multilevel Models

We introduced marginal models in Chapter 3 and provided additional background in Chapters 5 and 6. We have seen that Gaussian variance-component-only models can be restated as compound symmetry marginal models. For non-Gaussian models, we can do something analogous with working correlation matrices, but unlike the Gaussian case, the two parameterizations are not equivalent.

When does the conditional vs. marginal distinction matter for multilevel models? Primarily, the issue arises when we have a negative variance component estimate. To see how this works, we revisit the design structure used in Example 8.2 but with different data. First, we consider the Gaussian case. Then we briefly comment on how our approach would differ for the non-Gaussian case. We will return to the non-Gaussian case in more detail in future chapters.

8.5.1 Gaussian Data

Data Set 8.4 uses the same structure as Data Set 8.2, but the data are different. Using the same GLIMMIX statements we used in Example 8.2, the variance component estimates are

Covariance Parameter Estimates

Cov Parm	Subject	Estimate	Standard Error
Intercept	block	3.6346	6.9245
a	block	0	
b	block	8.2437	7.9234
Residual		10.2151	4.1255

Notice that the estimate of the variance among units of replication for factor A, $\hat{\sigma}^2_{AR}$, has been set to zero. This occurs when the REML solution vector $\hat{\boldsymbol{\sigma}}$ converges to a value such that $\tilde{\sigma}^2_{AR} < 0$. Because negative solutions lie outside the parameter space, common practice is to set these variance component estimates to zero.

Stroup and Littell (2002) found that the set-to-zero practice adversely affects the power characteristics and confidence interval coverage. The recommended fix is to allow $\hat{\sigma}^2_{AR}$ to remain negative. In GLIMMIX, we do this by including NOBOUND as an option in the PROC GLIMMIX statement. The resulting variance component estimate using NOBOUND are

Covariance Parameter Estimates

Cov Parm	Subject	Estimate	Standard Error
Intercept	block	7.8469	7.3265
a	block	−6.5950	5.4300
b	block	2.6751	8.8876
Residual		17.5683	10.1898

How does NOBOUND affect inference on the A and B treatment effects?

Without NOBOUND

Type III Tests of Fixed Effects

Effect	Num DF	Den DF	*F* Value	Pr > *F*
A	2	14.19	4.86	0.0247
B	2	9.058	17.14	0.0008
a*b	4	15.14	3.86	0.0235

With NOBOUND

Type III Tests of Fixed Effects

Effect	Num DF	Den DF	*F* Value	Pr > *F*
a	2	5.349	10.41	0.0143
b	2	8.395	20.21	0.0006
a*b	4	14.42	3.80	0.0262

We see the impact of NOBOUND for these data: main effect test statistics increase with NOBOUND; interaction test decreases slightly. The NOBOUND impact is design specific—main effect and interaction tests will not always be affected this way. What *is* globally true is that simulations consistently show that the NOBOUND results are more accurate in terms of type I error control and power characteristics. How do we assess the latter? In Chapter 16, we will discuss GLMM-based power analysis. When we say NOBOUND is more accurate in terms of power characteristics, we mean that when treatment means are unequal, using NOBOUND yields rejection rates more consistent with what theory predicts using the methods shown in Chapter 16 than rejection rates obtained without using NOBOUND.

While NOBOUND yields more accurate inference, it leaves us with the awkward problem of explaining exactly what we mean by a negative variance. Here, for example, what do we mean by $\hat{\sigma}^2_{AR} = -6.595$?

An alternative, for Gaussian data, reexpressed the affected variance component as a compound symmetry covariance. In our example, the GLIMMIX statements

```
proc glimmix data=ds8d2;
  class block a b;
  model y=a|b/ddfm=satterth;
  random intercept b/subject=block;
  random_residual_/type=cs subject=block*a;
```

yield the variance component estimates

Covariance Parameter Estimates

Cov Parm	Subject	Estimate	Standard Error
Intercept	block	7.8469	7.3265
b	block	2.6751	8.8876
CS	block*a	−6.5950	5.4300
Residual		17.5683	10.1898

Notice that this is simply the NOBOUND result reexpressed, along the lines of the variance component model and compound symmetry model equivalence shown in Chapter 3. The main difference: a *covariance* of −6.595 is well defined in statistical theory; a *variance* of −6.595 is not.

8.5.2 Non-Gaussian Models

As we saw in Chapter 3, the equivalence of the marginal CS and conditional variance component models does not extend to non-Gaussian models. The marginal model targets a different parameter—the mean of the marginal distribution—whereas the conditional model targets the expected value of $\mathbf{y}|\mathbf{b}$, usually a recognizable parameter from a member of the exponential family.

Given this nonequivalence, when we have a negative variance component estimates in a non-Gaussian GLMM, NOBOUND is the appropriate fix. The compound symmetry working correlation really cannot be regarded as a viable alternative. We will return to this theme, working through examples to illustrate, in subsequent chapters. For example, in Chapter 11 we will work through an example with count data in detail.

8.6 Summary

- Multilevel studies are defined by the presence of at least two sources of random variation.
- The simplest multilevel study is the completely randomized design with subsampling.
- For Gaussian data, multilevel studies must, at a minimum, have one random effect in the linear predictor to account for the larger unit. The smallest unit's variation is embedded in the conditional distribution of the observation given the random effects.
- For one-parameter non-Gaussian distributions, variation associated with the smallest level may also be included in the linear predictor. Failure to do so is a common cause of overdispersion, an issue discussed in more detail beginning in Chapter 10.
- Mastery of the WWFD process is an effective aid to accurate multilevel model construction.
- Design structures and treatment structures may be mixed and matched. With few exceptions, no treatment structure necessarily precludes a design structure or vice versa.
- Negative variance component estimates affect power and type I error control. In general, a negative solution set to zero biases test statistics and is therefore a potential problem.
- For Gaussian data, allowing the variance estimate to remain negative, while difficult to explain when interpreting the variance estimate *per se*, is the best way to keep test statistics unbiased and hence maintain control over type I error.
- Multilevel models can often be reexpressed as compound symmetry covariance models. Doing so eliminates the issue of reporting negative variance estimates and keeps test statistics and standard errors unbiased. Reexpression as compound symmetry covariance becomes even more important for non-Gaussian GLMMs, where the NOBOUND option is more prone to induce poor estimating equation performance.

Exercises

8.1 Parts of this problem use data contained in file `Ch _ 8 _ Problem1.sas`. The study compares four treatments' deterioration as a result of exposure to a degradant. Data are from seven batches, randomly selected representatives of a population of batches. Each batch has enough material to be divided in half—one treatment can be randomly assigned to a half-batch and another treatment assigned to the other half-batch. Material is exposed to the degradant; in the data set, **X** denotes exposure level with values of X equal to 1, 2, 4, 8, and 16 where the higher the value, the greater the exposure. The response variable is denoted **Y**; you can assume it has a Gaussian distribution. Also, it is well known than within the range of **X** observed in this study, **Y** responds linearly to increasing **X**. Lower **Y** is bad; higher **Y** is good.

The assignment of treatments to batch halves was as follows:

Batch	Treatment: Half I	Treatment: Half II
1	1	2
2	3	4
3	3	4
4	3	4
5	1	2
6	1	2
7	2	4

a. Write a complete description of the statistical model associated with this study, consistent with the description and assumptions given earlier.

b. Is this design connected? Explain.

c. The objective of the research is to identify the "best" treatment. To the researchers, "best" means the treatment with the greatest resistance to the ill effects of increasing degradant exposure (greater **Y** implies more resistant). Consistent with your model in (a), analyze the data using PROC GLIMMIX. Use *relevant* results to address the researchers' objective. Which treatment is best and why? How do the treatments rank and why?

8.2 This problem uses data contained in file `Ch _ 8 _ Problem2.sas`. A multistate study was conducted. The treatment design is a 2^2 factorial. The factors are denoted **A** and **B** in the data set, with two levels, denoted zero and one for each factor. The original protocol called for each state to use a split-plot, with factor A as the whole-plot factor and factor B as the split-plot factor. States 1, 2, and 3 got it right; states 4 and 5 got the protocol wrong: they ran factor B in the whole-plot and factor A in the split-plot. The response variable is **Y**, which you may assume has a Gaussian distribution. The researchers now want a single combined analysis on the entire data set.

a. Write a statistical model, with all required elements, to implement the desired analysis.

Hint: Use the fact that we can look at a split-plot structure as a disconnected incomplete block design. Give the treatment combinations a consistent identity over *all* of the blocks in the study. Writing the plot plan in this way and working through the WWFD process should yield a useable approach.

b. Implement the analysis consistent with (a) using Proc GLIMMIX. Write a short report characterizing the effects of the two treatment factors. Keep in mind that it is as important to report "how big" and "what kind" of effect a treatment has as it is to report whether the effect is statistically significant. Cite relevant evidence.

c. If you assumed random-block effects, what happens if you change the assumption to fixed blocks? If you assumed fixed blocks, what happens if you change your assumption to random-block effects? "What happens" includes what changes do you make in the GLIMMIX statements and what changes occur in the output listing?

8.3 The data sets **Ch8_Problem3a.sas** and **Ch8_Problem3b.sas** in the SAS Data Set and Program Library contain contain data from a strip-plot experiment with six blocks and a 3 × 3 treatment design—three levels of A, three levels of B. Visualize a block of a strip-plot as follows:

	B_1	B_2	B_3
A_1	---------------→		
A_2	---------------→		
A_3	---------------→		

In each block, levels of A are randomly assigned to a row and applied across the row; levels of B are randomly assigned to a column and applied down the entire column. Each block has a potentially unique assignment of A to rows and B to columns.

a. Show the WWFD process leading to a skeleton ANOVA for this design.

b. Write the linear predictor that follows from (a). Include assumptions about model effects considered random.

c. For which of the following distributions of the response conditional on the random model effects would it be reasonable to include a term corresponding to block*a*b in the linear predictor?

   i. Gaussian_____ reasonable to include______ not reasonable

   ii. Beta_____ reasonable to include______ not reasonable

   iii. Binomial_____ reasonable to include______ not reasonable

   iv. Poisson_____ reasonable to include______ not reasonable

   v. Negative Binomial_____ reasonable to include______ not reasonable

d. *True or false*: If you fit a negative binomial to the linear predictor you wrote in (a), you implicitly assume that the block*a*b (unit-level) effect has a gamma distribution.

9

Best Linear Unbiased Prediction

9.1 Review of Estimable and Predictable Functions

In Chapters 7 and 8, we considered strategies for inference on fixed model effects only using estimable functions. In this chapter, we turn our attention to predictable functions, inference involving both fixed and random model effects.

Recall the general form of the linear predictor, $\boldsymbol{\eta} = \mathbf{X}\boldsymbol{\beta} + \mathbf{Zb}$, of estimable functions, $\mathbf{K}'\boldsymbol{\beta}$, where $\mathbf{K}$ satisfies estimability criteria given in (5.1) and (5.2), and of predictable functions, $\mathbf{K}'\boldsymbol{\beta} + \mathbf{M}'\mathbf{b}$. For fixed-effects-only models, inference must necessarily be restricted to estimable functions. For mixed models, linear mixed models (LMMs) and generalized linear mixed models (GLMMs), inference can apply to estimable functions alone or to predictable functions.

In Chapter 3, we introduced the terms *broad* and *narrow* inference. Broad inference uses estimable functions *only*. Mechanically, this means $\mathbf{M} = \mathbf{0}$. More precisely, broad inference connotes estimating or testing $\mathbf{K}'\boldsymbol{\beta}$ *given* $\mathbf{M}'\mathbf{b} = \mathbf{0}$. In words, this means that we do not restrict attention to a particular set of levels of the random effect, $\mathbf{b}$, but intend inference to apply broadly across the entire population represented by the levels of $\mathbf{b}$. For Gaussian data, this implicit conditioning is easily overlooked, because LMMs can always be expressed as marginal linear models (LMs) with the covariance structure of $\mathbf{Zb}$ embedded in $\mathbf{V} = \text{Var}(\mathbf{y})$. For non-Gaussian data, we need to keep this implicit conditioning in mind, because of the distinction between conditional and marginal GLMMs first introduced in Chapter 3.

Narrow inference can serve one of two purposes:

1. Inference focuses on a particular level or set of levels of $\mathbf{b}$. For example, Henderson (1950, 1963) originally developed the mixed model equations specifically to estimate sire breeding values in genetic studies where sires were random model effects. Other examples include batch-specific regression (shown in Chapter 3); patient-specific effects in clinical trials; site- or location-specific effects, for example, in multi-clinic medical trials or on-farm agricultural trials; worker-specific inference in quality improvement (see McLean et al., 1991); and teacher-value-added scores in assessment of student achievement (Sanders et al., 1997). The latter is a recent—and controversial—application of mixed model inference. We provide an overview in Section 9.4.
2. Focus remains primarily on the fixed effects, but nonzero $\mathbf{M}$ is used to limit inference about $\mathbf{K}'\boldsymbol{\beta}$ to particular subsets of the population represented by $\mathbf{b}$. In a multilocation study, for example, a multi-clinic medical study or an on-farm agricultural study, we might wish to see if treatment effects are consistent across

the entire population or if there is evidence of subpopulations that interact with the fixed effects. Obviously, if these subpopulations are well known in advance of the study, they should be explicitly included in the treatment design as fixed effects. However, when subpopulations are suspected, but we do not know for sure if they exist, and if they do, we do not understand them well enough to include in the design, BLUPs defined on $\mathbf{K}'\boldsymbol{\beta}+\mathbf{M}'\mathbf{b}$ can provide important exploratory tools.

Estimation for $\boldsymbol{\beta}$ and $\mathbf{b}$ differ, as shown in Chapter 4, because $\boldsymbol{\beta}$ is a fixed parameter whereas $\mathbf{b}$ is a random vector with a multivariate normal distribution. Estimates of realized values of $\hat{\mathbf{b}}$ must take this distribution into account.

Section 9.2 focuses on the use of predictable functions for inference on levels of $\mathbf{b}$ deemed to be of specific interest. Section 9.3 focuses on uses of predictable functions to manipulate the inference space. Sections 9.2 and 9.3 introduce fundamental ideas for standard random-effects-only or mixed models. While these two sections use examples with Gaussian data to illustrate the main ideas, aside from the addition of model scale and data scale, these issues apply equally to predictable functions with non-Gaussian data. Finally, in Section 9.4, we explore an advanced application of predictable functions using a nonstandard $\mathbf{Z}$ matrix, illustrated by the value-added model that is currently of intense interest in educational improvement efforts.

9.2 BLUP in Random-Effects-Only Models

The term *random-effects-only model* refers to a mixed model whose only fixed effect is the intercept. Generically, we denote the linear predictor as $\boldsymbol{\eta} = \eta + \mathbf{Zb}$. We can illustrate the important features of best linear unbiased prediction (BLUP)—that is, predictable-function-based inference where the random model effects are of primary interest—using the two simplest random-effects-only models. They are

1. The one-way model
   a. Linear predictor:
   $$\eta_i = \eta + a_i \tag{9.1}$$
   b. a_i i.i.d. $N\left(0,\sigma_A^2\right)$
2. The two-way nested model
   a. Linear predictor:
   $$\eta_{ij} = \eta + a_i + b(a)_{ij} \tag{9.2}$$
   b. a_i i.i.d. $N\left(0,\sigma_A^2\right)$; $b(a)_{ij}$ i.i.d. $N\left(0,\sigma_B^2\right)$

Let us first consider the one-way model.

9.2.1 One-Way Random Effects Model

Data for this example appear in the SAS Data and Program Library as Data Set 9.1. Assume a Gaussian response variable, that is, $y_{ij}|a_i \sim NI(\mu_i, \sigma^2)$, and identity link, $\eta_i = \mu_i$. There are 12 levels of factor A.

We start by comparing key results for the random effects model (9.1) with results we would obtain if we consider A effects to be fixed. Denote the fixed effects version of (9.1) by

$$\eta_i = \eta_F + \alpha_i$$

For the purposes of this example, let η_R denote the intercept of the random effects model.

Our first comparison is the estimate of the overall mean, defined as $\mu = \bar{\mu}_{\bullet} = (1/12)\sum_{i=1}^{12} \mu_i$. For the random effects version of the model, we can obtain this estimate using the GLIMMIX statements:

```
proc glimmix data=c9_ex1;
  class a;
  model y=/solution;
  random intercept/subject=a solution;
  estimate 'overall mean' intercept 1;
```

We add the SOLUTION options to the MODEL and RANDOM statements because we will use these estimates to illustrate several points later. Before we do this, let us look at the GLIMMIX statements for the fixed effects version of this model and compare results for $\hat{\mu}$:

```
proc glimmix data=c9_ex1;
  class a;
  model y=a/solution;
  estimate 'overall mean' intercept 1/e;
```

The random effects version of this model produces the output

Estimates

Label	Estimate	Standard Error
overall mean	15.9000	0.7232

Compare this with the fixed effects version

Estimates

Label	Estimate	Standard Error
overall mean	15.9000	0.2526

The estimates are identical, both equal $\bar{y}_{\bullet\bullet} = 15.9$, but the standard errors are not. For the random-effects-only model, only the intercept η is fixed and, hence, the estimable function for the overall mean is $\hat{\eta} = \hat{\mu} = 15.9$ with standard error

$$\sqrt{\frac{n\hat{\sigma}_A^2 + \hat{\sigma}^2}{an}} = \sqrt{\frac{2 \times 5.5114 + 1/5308}{12 \times 2}} = 0.7232$$

On the other hand, for the fixed-effects-only model, η by itself does not satisfy the estimability criterion: $\eta + (1/12)\sum_{i=1}^{12} \alpha_i$ is the estimable function for the overall mean and its standard error is

$$\sqrt{\frac{\hat{\sigma}^2}{an}} = \sqrt{\frac{1.5308}{24}} = 0.2526$$

What happens if we define a BLUP analogously to the fixed effects estimable function, that is, using the predictable function $\eta + (1/12)\sum_{i=1}^{12} a_i$? We can do this by adding the following statement to the GLIMMIX program for the random effects model:

```
estimate 'overall mean narrow' intercept 12 | intercept 1 /
  subject 1 1 1 1 1 1 1 1 1 1 1 1 divisor=12;
```

This produces

Estimates		
Label	**Estimate**	**Standard Error**
overall mean narrow	15.9000	0.2526

Notice that this *predictable function* is identical to the similarly defined *estimable function* for the model with fixed A effects. This illustrates two important principles:

1. Defining an effect as fixed narrows the inference space to only the levels that are actually observed. Hence, for balanced data sets, averaging over the levels of a random effect (here, factor A) has the same narrowing impact on the inference space as defining that factor to be fixed. In fact, we can use the BLUP to make explicit the inference space that the fixed effect model implicitly assumes.
2. Including an effect in the predictable function removes the corresponding variance component from the standard error. Here, the standard error for the predictable function η *includes* $\hat{\sigma}_A^2$, whereas $\eta + (1/12)\sum_{i=1}^{12} a_i$ *removes* $\hat{\sigma}_A^2$ from the standard error.

Now let us examine the "means" of the levels of factor A. For the A-effects-fixed model, these are true treatment means, corresponding to the estimable function $\eta + \alpha_i$. For the

A-effects-random model, these are BLUPs, defined by the predictable function $\eta + a_i$. We obtain the former using the LSMEAN statement:

```
lsmeans a;
```

For the latter, we use ESTIMATE statements:

```
estimate 'blup a_1' intercept 1 | intercept 1/subject 1 0;
estimate 'blup a_2' intercept 1 | intercept 1/subject 0 1 0;
estimate 'blup a_3' intercept 1 | intercept 1/subject 0 0 1 0;
estimate 'blup a_4' intercept 1 | intercept 1/subject 0 0 0 1 0;
estimate 'blup a_5' intercept 1 | intercept 1/subject 0 0 0 0 1 0;
estimate 'blup a_6' intercept 1 | intercept 1/subject 0 0 0 0 0 1 0;
estimate 'blup a_7' intercept 1 | intercept 1/subject 0 0 0 0 0 0 1 0;
estimate 'blup a_8' intercept 1 | intercept 1/subject 0 0 0 0 0 0 0 1 0;
estimate 'blup a_9' intercept 1 | intercept 1/subject 0 0 0 0 0 0 0 0 1 0;
estimate 'blup a_10' intercept 1 | intercept 1/subject 0 0 0 0 0 0 0 0
0 1 0;
estimate 'blup a_11' intercept 1 | intercept 1/subject 0 0 0 0 0 0 0 0 0
0 1 0;
estimate 'blup a_12' intercept 1 | intercept 1/subject 0 0 0 0 0 0 0 0 0
0 0 1;
```

Table 9.1 shows the resulting estimates and standard errors.

TRT_MEAN and TRT_SE_MEAN are mean and standard error from the A-effects-fixed model. In estimable function and estimating-equation terms, these are estimates of $\eta + \alpha_i$ with $\hat{\eta}$ and $\hat{\alpha}_i$ obtained from solving $\mathbf{X'X\beta} = \mathbf{X'y}$, where $\boldsymbol{\beta}$ denotes the parameter vector consisting of η and $\{\alpha_i; i = 1, 2,\ldots, 12\}$. Here, we know the estimate and standard error are $\bar{y}_{i\bullet}$ and $\sqrt{\sigma^2/an}$, respectively.

We obtain TRT_BLUP and TRT_SE_BLUP from the solution to the mixed model equations for the A-effects-random model. Earlier, we saw that $\hat{\eta} = 15.9$. From (4.20), we know that the random vector BLUP $\hat{\mathbf{b}} = \mathbf{ZGV}^{-1}(\mathbf{y} - \mathbf{X\beta})$. Here, the random effect vector $\mathbf{b}$ consists

TABLE 9.1

LSMEANS from A-Effects-Fixed and BLUPs from A-Effects-Random Models.

Obs	a	Trt_Mean	Trt_SE_Mean	Trt_BLUP	Trt_SE_BLUP
1	1	16.15	0.87488	16.1195	0.82453
2	2	17.35	0.87488	17.1732	0.82453
3	3	12.15	0.87488	12.6073	0.82453
4	4	16.35	0.87488	16.2951	0.82453
5	5	14.40	0.87488	14.5829	0.82453
6	6	17.85	0.87488	17.6122	0.82453
7	7	20.05	0.87488	19.5439	0.82453
8	8	18.40	0.87488	18.0951	0.82453
9	9	17.50	0.87488	17.3049	0.82453
10	10	12.60	0.87488	13.0024	0.82453
11	11	12.90	0.87488	13.2658	0.82453
12	12	15.10	0.87488	15.1976	0.82453

entirely of the random A effects, that is, $\mathbf{b}' = \mathbf{a}' = \begin{bmatrix} a_1 & a_2 & \cdots & a_{12} \end{bmatrix}$. Also, $\mathbf{Z} = \mathbf{I}_{12} \otimes \begin{bmatrix} 1 \\ 1 \end{bmatrix}$, $\mathbf{G} = \sigma_A^2 \mathbf{I}_{12}$, and $\mathbf{X\beta} = \boldsymbol{\eta}$ where $\boldsymbol{\eta}$ is a 24 × 1 vector. From this, we can show that for the ith element of $\mathbf{a}$

$$\hat{a}_i = \left(\frac{\sigma_A^2}{\left(\sigma^2 + n\sigma_A^2\right)/n} \right) (\bar{y}_{i\bullet} - \bar{y}_{\bullet\bullet})$$

For example, for the first level of A, $\bar{y}_{1\bullet} = 16.15$, hence,

$$\hat{a}_1 = \left(\frac{2 \times 5.5114}{1.5308 + (2 \times 5.5114)} \right) (16.15 - 15.9) = 0.2195$$

Hence, the BLUP for level 1 of factor A is $\hat{\eta} + \hat{a}_1 = 15.9 + 0.2195 = 16.1125$. Note that the fixed effect $\hat{\alpha}_1 = \bar{y}_{1\bullet} - \bar{y}_{\bullet\bullet}$.

Notice that in variance-components-only models, $\mathbf{GV}^{-1} = \mathbf{G}(\mathbf{ZGZ}' + \mathbf{I}\sigma^2)^{-1}$ and $\mathbf{G}$ is block diagonal with each diagonal consisting of $\mathbf{I}$ times the variance component for the corresponding random effect. The resulting ratio will always be <1. We can think of this as the *shrinkage* term; the BLUP is sometimes called a *shrinkage estimator* because it "shrinks" the fixed effect estimate of the factor effect (in this case, $\hat{\alpha}_1 = 0.25$) toward zero, with the magnitude of shrinkage depending on the variance of the associated factor, here σ_A^2. For larger variance, we have less compelling basis to attenuate estimates of factor level effects, so we see less shrinkage. However, as the variance decreases, large $\bar{y}_{1\bullet} - \bar{y}_{\bullet\bullet}$ are treated as increasingly implausible because small σ_A^2 implies that the a_i should be tightly distributed about zero. Hence, as σ_A^2 decreases, shrinkage increases. In Table 9.1, we see that for all A means less than the overall mean of 15.9, the corresponding BLUP is greater, whereas for all A means greater than 15.9, the corresponding BLUP is less.

Why shrinkage? When does shrinkage make sense and when does it not? The answer lies in the treatment design. If we have a standard "these are the treatments, and we want to compare them" design, then we should consider treatment effects fixed. However, suppose we want to compare all of the sires in a breeding trial, all of the teachers at a school, all of the patients at a clinic, all of the workers at a plant or office, and all of the athletes on a team. These sires, teachers, patients, workers, athletes, etc., do not exist in a vacuum. They exist in a larger context, and we know something about the distribution of their effects. The sires in our trial (or teachers, patients, workers, athletes, etc.) have distribution characteristics similar enough to sires in other trials that we can assume that in this sense they belong to a larger population. We can use information about the distribution of effects to refine our estimates. In these cases, assuming treatment effects to be random and using shrinkage via predictable functions makes sense.

9.2.2 Two-Way Random Effects Nested Model

In the two-way nested model, we have levels of factor B nested within levels of factor A and one or more observations on each AB combination. The linear predictor and assumptions about the random A and B within A effects were given earlier in (9.2). In the examples in this section, we assume Gaussian data, specifically

- $y_{ijk} | a_i, b(a)_{ij} \sim NI(\mu_{ij}, \sigma^2)$, $k = 1, 2, \ldots, n_{ij}$; $n_{ij} \geq 1$,
- Link: identity

TABLE 9.2
Balanced Three-Level Nested Design

A_1				A_2				A_3				A_4			
B_{11}		B_{12}		B_{21}		B_{22}		B_{31}		B_{32}		B_{41}		B_{42}	
✓	✓	✓	✓	✓	✓	✓	✓	✓	✓	✓	✓	✓	✓	✓	✓
A_5				A_6				A_7							
B_{51}		B_{52}		B_{61}		B_{62}		B_{71}		B_{72}					
✓	✓	✓	✓	✓	✓	✓	✓	✓	✓	✓	✓				

TABLE 9.3
Unbalanced Three-Level Nested Design

A_1			A_2			A_3			A_4			A_5		
b_{11}		b_{12}	b_{21}		b_{22}	b_{31}		b_{32}	b_{41}		b_{42}	b_{51}		b_{52}
✓	✓	✓	✓	✓	✓	✓	✓	✓	✓	✓	✓	✓	✓	✓
A_6			A_7			A_8			A_9			A_{10}		
b_{61}		b_{62}	b_{71}		b_{72}	b_{81}		b_{82}	b_{91}		b_{92}	$b_{10,1}$		
✓	✓	✓	✓	✓	✓	✓	✓	✓	✓	✓	✓	✓		

Let B_i denote the number of levels of B observed in combination with the ith level of A. The B_i need not be equal. At least one B_i should be strictly >1 but not all are required to be; the same is true for n_{ij}, the number of observations on the ijth AB combination.

We will consider two data sets described by this model, one balanced, the other unbalanced. The designs appear in Tables 9.2 and 9.3; the corresponding data appear in the SAS Data and Program Library as Data Sets 9.2 and 9.3.

The balanced design (Table and Data Set 9.2) has seven levels of factor A, two levels of B nested within each level of A and two observations per AB combination. The unbalanced design has 10 levels of factor A. There are two levels of B nested within nine of the A levels and one level of B with the 10th level of A. For the nine levels of A with two levels of B, one AB combination has two observations, and the other has only one. For the 10th level of A, there is only one observation.

Why the unbalanced design? A look at the skeleton ANOVA explains why.

Source of Variation	d.f.—Balanced Design (9.2)	d.f.—Unbalanced Design (9.3)
A	6	9
B(A)	7	9
obs(A,B)	14	9
total	27	27

The balanced design disproportionately assigns degrees of freedom to obs(A,B) and hence to estimating σ^2 at the expense of estimating σ_A^2 and σ_B^2. The unbalanced design distributes degrees of freedom equally among the model effects. In fixed effects models, priority goes to maximizing the accuracy of the estimate of σ^2. Hence, the balanced design seems more appropriate. However, in a random effects model, all variance components must be estimated accurately, especially if we also desire BLUPs. For the random effects model, we can make a strong case for preferring the unbalanced design.

9.2.2.1 Analysis: Balanced Case

As with the one-way design, it is instructive to compare the analysis using the random-effects-only model with a model that treats A and B as fixed. This comparison is particularly important because many statistical software programs compute results using ordinary least squares—even though they purport to have mixed model capabilities, we quickly discover that they do not. Examples in SAS include PROC GLM and PROC GENMOD. The examples in this section add to our growing list of reasons why such software should *never* be used in conjunction with models that contain random effects.

For both the balanced and unbalanced designs, the GLIMMIX statements to implement the two-way nested model given in (9.2) are

```
proc glimmix;
  class a b;
  model y=;
  random intercept b/subject=a;
```

The GLIMMIX statements to implement the fixed effect analog of (9.2) are

```
proc glimmix;
  class a b;
  model y=a b(a);
```

Let us begin, as we did with the one-way design, by estimating the overall mean using the ESTIMATE statement

```
estimate 'overall mean' intercept 1/e;
```

We know that for the random effects model, the estimable function for the overall mean is η and, because we use the identity link, $\hat{\eta} = \hat{\mu}$. For the fixed effects model, we know that η alone is not estimable—we must average over the A and B(A) effects. This yields the estimable function $\eta + \bar{\alpha}_{\bullet} + \overline{\beta(\alpha)}_{\bullet\bullet}$ and hence the implied coefficients are 1/7 for the A effects and 1/14 for the B(A) effects. We can verify this using the coefficient output from the E option in the ESTIMATE statement. Also, we can add the following ESTIMATE statement to the random effects model to compute a BLUP using the same coefficients as the fixed effects "overall mean":

```
estimate 'overall mean narrow' intercept 14 | intercept 2 b 1 1 /
  subject 1 1 1 1 1 1 1 divisor=14 e;
```

For the random effects model, the overall mean output is

Label	Estimate	Standard Error
overall mean	19.2571	0.6456
overall mean narrow	19.2571	0.2205

The corresponding output for the fixed effects model is

Label	Estimate	Standard Error
overall mean	19.2571	0.2205

As with the one-way model, the estimates are equal but the standard errors differ. For the fixed effects estimate and the random effects narrow BLUP, the standard error involves only σ^2. For the estimate of the true overall mean, assuming A and B(A) are random effects, the actual standard error is

$$\sqrt{\frac{bn\hat{\sigma}_A^2 + n\hat{\sigma}_B^2 + \hat{\sigma}^2}{abn}}$$

For the A levels, assuming the fixed effects model, we can compute the estimable functions $\eta + \alpha_i + \overline{\beta(\alpha)}_{i\bullet}$—noting that $\eta + \alpha_i$ alone is not estimable. For the random effects model, we have a choice of two forms of predictable function:

- Broad: $\eta + a_i$
- Narrow: $\eta + a_i + \overline{b(a)}_{i\bullet}$

The latter matches the LSMEAN coefficients but, like the LSMEANS, restricts inference to *only* the levels of B actually observed. The former—the broad inference BLUP—extends inference to the entire population that the observed B(A) represent. In this sense, the broad-space BLUPs are closer to the inference space implicitly intended in most applications.

For the fixed effects model estimable functions and the random effects model broad-inference-space BLUPs, we use the same LSMEANS and ESTIMATE commands used in the one-way model. For the narrow-inference-space BLUPs, the GLIMMIX statement, for example, for level 1 of A, is

```
estimate 'blup a_1 narrow' intercept 2 | intercept 2 b 1 1/subject 1 0
divisor=2;
```

Table 9.4 shows the resulting estimates/predictors and their associated standard errors for the three alternatives.

TABLE 9.4

Fixed Effect LSMEANS and Broad and Narrow BLUPs for the Levels of Factor A

Obs	*a*	A_mean	A_se_mean	A_blup	A_se_blup	A_nrrw_blup	A_se_nrrw_blup
1	1	17.200	0.58340	17.7116	0.75556	17.4400	0.55346
2	2	19.375	0.58340	19.3457	0.75556	19.3612	0.55346
3	3	19.300	0.58340	19.2893	0.75556	19.2950	0.55346
4	4	19.150	0.58340	19.1766	0.75556	19.1625	0.55346
5	5	20.050	0.58340	19.8528	0.75556	19.9575	0.55346
6	6	22.300	0.58340	21.5432	0.75556	21.9450	0.55346
7	7	17.425	0.58340	17.8807	0.75556	17.6388	0.55346

TABLE 9.5

Fixed Effect LSMEANS and Broad and Narrow BLUPs with KR Standard Error Bias Adjustment

Obs	*a*	A_mean	A_se_mean	A_blup	A_se_blup	A_nrrw_blup	A_se_nrrw_blup
1	1	17.200	0.58340	17.7116	0.87313	17.4400	0.58201
2	2	19.375	0.58340	19.3457	0.87313	19.3612	0.58201
3	3	19.300	0.58340	19.2893	0.87313	19.2950	0.58201
4	4	19.150	0.58340	19.1766	0.87313	19.1625	0.58201
5	5	20.050	0.58340	19.8528	0.87313	19.9575	0.58201
6	6	22.300	0.58340	21.5432	0.87313	21.9450	0.58201
7	7	17.425	0.58340	17.8807	0.87313	17.6388	0.58201

None of the estimates are the same. As expected, the broad inference BLUPs have larger standard errors because they include uncertainty that results from extending inference to the population represented by B(A)—that is, the standard errors depend on σ_B^2 as well as σ^2. The broad-space BLUPs, labeled A_BLUP, also show the pattern of shrinkage described for the one-way model. The pattern of shrinkage for the narrow-space BLUPs is more complex and not as neatly characterized.

The standard errors shown in Table 9.4 are "naive"—they do not correct for bias resulting from using estimated variance components. This is not an issue for the fixed effect model LSMEANS, but it is for the BLUPs, even though the data are balanced. Strictly speaking, we should use the DDFM = KR option for the random effects model BLUPs. Table 9.5 shows the adjusted results. Note that the estimates do not change, but the standard errors of the two sets of BLUPs do.

Finally, the broad-inference-space BLUPs use the same coefficients as that of a model with A effects fixed and B(A) effects random. Such a model would yield LSMEANS equal to the LSMEANS shown here for the fixed-effects-only model, but the standard errors would be close (but not equal) to those for the broad-space BLUPs, because they, too, would include σ_B^2.

9.2.2.2 Unbalanced Case

The design follows the format shown in Table 9.3. The data appear as Data Set 9.3 in the SAS Data and Program Library. The model is identical to that used for the balanced case. Only the number of B levels for the *i*th level of A and the number of observations for the *ij*th B(A) level change. The GLIMMIX CLASS, MODEL, and RANDOM statements are identical to those used in the balanced case.

As before, we start with the overall mean, using the ESTIMATE statement

```
estimate 'overall mean' intercept 1/e;
```

For the random-effects-only model, we get

Label	Estimate	Standard Error
overall mean	18.5812	0.6113

For the fixed-effects-only model, we get

Label	Estimate	Standard Error
overall mean	Nonest	

The nonestimability results from the coefficients the fixed effects model's default algorithm selects, shown in the output for the E option:

Coefficients for Estimate Overall Mean			
Effect	***a***	***b***	**Row1**
Intercept			1
a	1		0.1
a	2		0.1
a	3		0.1
a	4		0.1
a	5		0.1
a	6		0.1
a	7		0.1
a	8		0.1
a	9		0.1
a	10		0.1
b(*a*)	1	1	0.0526
b(*a*)	1	2	0.0526
b(*a*)	2	1	0.0526
b(*a*)	2	2	0.0526
b(*a*)	3	1	0.0526
b(*a*)	3	2	0.0526
b(*a*)	4	1	0.0526
b(*a*)	4	2	0.0526
b(*a*)	5	1	0.0526
b(*a*)	5	2	0.0526
b(*a*)	6	1	0.0526
b(*a*)	6	2	0.0526
b(*a*)	7	1	0.0526
b(*a*)	7	2	0.0526
b(*a*)	8	1	0.0526
b(*a*)	8	2	0.0526
b(*a*)	9	1	0.0526
b(*a*)	9	2	0.0526
b(*a*)	10	1	0.0526

Notice that the estimable function weights each level of A and each level of B(A) equally—seemingly a sensible strategy. However, there is no linear combination of observations that corresponds to this weighting, and the function, therefore, fails the estimability criterion.

The actual overall sample mean, $\bar{y}_{\bullet\bullet\bullet} = (1/28)\sum_{i,j,k} y_{ijk} = 18.45$. In model terms, this is

$$\left(\frac{1}{28}\right)\left\{\sum_{i=1}^{9}\left[(\eta+\alpha_i+b(a)_{i1})+2(\eta+\alpha_i+b(a)_{i2})\right]+\eta+\alpha_{10}+b(a)_{10,1}\right\}$$

and we need the following estimate statement

```
estimate 'overall sample mean' intercept 28 a 3 3 3 3 3 3 3 3 3 1
  b(a) 1 2 1 2 1 2 1 2 1 2 1 2 1 2 1 2 1 2 1/e divisor=28;
```

to capture the needed estimate. For the random-effects-only model, we need to rewrite the RANDOM statement in order to define the corresponding BLUP. The statements are

```
random a b(a);
estimate 'overall sample mean' intercept 28 | a 3 3 3 3 3 3 3 3 3 1
  b(a) 1 2 1 2 1 2 1 2 1 2 1 2 1 2 1 2 1 2 1/e divisor=28;
```

Notice that all coefficients for A and B(A) must appear to the right of the vertical slash in the random-effects-only model. Both the fixed-effects-only and random-effects-only models give estimate/BLUP = 18.45 and standard error 0.126.

Because the data are unbalanced intentionally by design to achieve precision balance among A, B(A), and observation(AB), it would seem more reasonable with the fixed effects design to weight levels of A equally, rather than using the uncorrected sample mean. After all, this is the fundamental concept underlying adjusted marginal mean estimation. We can do this with the fixed-effects-only model using the following ESTIMATE statement:

```
estimate 'overall lsmean' intercept 10 a 1 1 1 1 1 1 1 1 1 1 1/e
divisor=10;
```

The resulting coefficients are

Coefficients for Estimate Overall Lsmean

Effect	*a*	*b*	Row1
Intercept			1
a	1		0.1
a	2		0.1
a	3		0.1
a	4		0.1
a	5		0.1
a	6		0.1
a	7		0.1
a	8		0.1
a	9		0.1
a	10		0.1

Coefficients for Estimate Overall Lsmean

Effect	*a*	*b*	Row1
b(*a*)	1	1	0.05
b(*a*)	1	2	0.05
b(*a*)	2	1	0.05
b(*a*)	2	2	0.05
b(*a*)	3	1	0.05
b(*a*)	3	2	0.05
b(*a*)	4	1	0.05
b(*a*)	4	2	0.05
b(*a*)	5	1	0.05
b(*a*)	5	2	0.05
b(*a*)	6	1	0.05
b(*a*)	6	2	0.05
b(*a*)	7	1	0.05
b(*a*)	7	2	0.05
b(*a*)	8	1	0.05
b(*a*)	8	2	0.05
b(*a*)	9	1	0.05
b(*a*)	9	2	0.05
b(*a*)	10	1	0.1

The estimate and standard error for the fixed effects model are

Label	Estimate	Standard Error
overall lsmean	18.7200	0.1389

We can use the same coefficients to define the BLUP analog for the random-effects-only model. Also, because the coefficients of predictable functions do not need to satisfy estimability criteria, we can use the coefficient the fixed-effects-only model was unable to use in the default "overall mean" to define a random-effects-only model BLUP. The two ESTIMATE statements are

```
estimate 'overall lsmean narrow' intercept 20 | a 2 2 2 2 2 2 2 2 2 2
  b(a) 1 1 1 1 1 1 1 1 1 1 1 1 1 1 1 1 1 1 2/e divisor=20;
estimate 'overall mean fixed nonest' intercept 190 | a 19 19 19 19 19 19
19 19 19 19
  b(a) 10 10 10 10 10 10 10 10 10 10 10 10 10 10 10 10 10 10 10/e
  divisor=190;
```

The results are as follows:

Label	Estimate	Standard Error
overall lsmean narrow	18.7011	0.1383
overall mean fixed nonest	18.5812	0.1436

Interestingly, the estimate of the BLUP using the coefficients the fixed effects model selected but could not use happens to yield the same point estimate, 18.58, as the true overall mean estimate, $\hat{\eta} = \hat{\mu}$, in the random-effects-only model. The fixed-effects-only model selected the correct coefficients—this is another manifestation of spurious nonestimability that results when fixed effects computing tactics are used for random model effects.

> The primary reason we spend time on this issue here is that when you use fixed-effects-only software, such as SAS PROC GLM, you are doing exactly what we have shown here for the fixed-effects-only model. The take-home message is that if the effects are random, computing as if they are fixed gives you the wrong answers. It is inappropriate. Do not do it!

9.3 Gaussian Data with Fixed and Random Effects

In the previous section, we considered BLUP, also known as shrinkage estimation, for random-effects-only models. These models only permit covariance component estimation and BLUP, so they must be the focus of inference. For mixed models this is not necessarily the case: often, we focus on fixed effects means and differences, with covariance estimation merely an essential step to get the needed standard errors and test statistics. However, BLUP can be useful with mixed models when we want to confine inference to a particular subset of the random effect. For example, we might want to know the treatment mean or difference just for *this* patient, not for *all* patients, or for this *group* of patients. For this, we use subject-specific BLUP. We saw a glimpse of this technique in Chapter 3 when we introduced random coefficient regression. In this section, we expand our repertoire.

Looking at BLUP in the mixed model case also serves a more academic purpose. As we have already seen in the previous section, in the days before mixed model software was available, linear model software with fixed-effects-only algorithms (usually ordinary least squares) and a few "mixed model band-aids"—for example, the RANDOM statement in SAS PROC GLM—was the only available choice for mixed model analysis. As with the random-effects-only case, seeing what such software produces compared with *true* mixed model analysis is instructive.

To illustrate these ideas, we use Data Set 9.4 in the SAS Data and Program Library. The data consist of two factors, A and B. Think of factor A as a treatment with two levels. Think of factor B as a sample of locations—or patients, teachers, or workers—with eight levels. Thus, we can regard A effects as fixed and B effects as random. There are two observations per AB combination—think of them as two plots per treatment per location or the equivalent for patients, teachers, or workers. We will consider two aspects of this data set. In the first subsection, we work through the correct mixed model analysis of these data. In the second subsection, we look at what "fixed effects software with 'band-aids' for mixed-models" produces and how it compares to true mixed model analysis.

9.3.1 Mixed-Model Analysis with BLUP to Modify the Inference Space

For these data, we assume the following model:

1. Linear predictor: $\eta_{ij} = \eta + \alpha_i + b_j + (ab)_{ij}$
2. Distributions:

$$\text{a. } y_{ijk} \mid b_j, (ab)_{ij} \sim NI\left(\mu_{ij}, \sigma^2\right) \tag{9.3}$$

   b. b_j i.i.d. $N\left(0, \sigma_B^2\right)$
   c. $(ab)_{ij}$ i.i.d. $N\left(0, \sigma_{AB}^2\right)$
3. Link: identity, $\eta_{ij} = \mu_{ij}$

The items we would logically be interested in are the estimates of the variances (σ^2, σ_B^2, and σ_{AB}^2); the treatment means, that is, estimable functions $\eta + \alpha_1$ and $\eta + \alpha_2$; the treatment difference; and treatment BLUPs specific to a particular level or levels of B.
The GLIMMIX statement needed to obtain these are

```
proc glimmix data=c9_ex4;
  class a b;
  model y=a/solution;
  random intercept a/ subject=b solution;
  lsmeans a/diff;
  estimate 'blup a_1 @ b_1' intercept 1 a 1 0 | intercept 1 a 1 0/
  subject 1 0;
  estimate 'blup a_2 @ b_1' intercept 1 a 0 1 | intercept 1 a 0 1/
  subject 1 0;
  estimate 'blup diff @ b_1' a 1 -1 | a 1 -1/subject 1 0;
  estimate 'blup diff @ b_2' a 1 -1 | a 1 -1/subject 0 1 0;
  estimate 'blup diff @ b_3' a 1 -1 | a 1 -1/subject 0 0 1 0;
  estimate 'blup diff @ b_4' a 1 -1 | a 1 -1/subject 0 0 0 1 0;
  estimate 'blup diff @ b_5' a 1 -1 | a 1 -1/subject 0 0 0 0 1 0;
  estimate 'blup diff @ b_6' a 1 -1 | a 1 -1/subject 0 0 0 0 0 1 0;
  estimate 'blup diff @ b_7' a 1 -1 | a 1 -1/subject 0 0 0 0 0 0 1 0;
  estimate 'blup diff @ b_8' a 1 -1 | a 1 -1/subject 0 0 0 0 0 0 0 1;
```

We use the LSMEANS statement to obtain treatment means and differences. The first two ESTIMATE statements define the two treatment BLUPs specific to the first level of factor B. The next eight statements define the treatment differences specific to each level of factor B. Output of interest is as follows:

Covariance Parameter Estimates

Cov Parm	Subject	Estimate	Standard Error
Intercept	B	4.0237	4.8209
a	B	6.2562	4.3667
Residual		3.6472	1.2895

These are the estimates $\hat{\sigma}_B^2$ (4.02), $\hat{\sigma}_{AB}^2$ (6.26), and $\hat{\sigma}^2$ (3.65), respectively.

Solutions for Fixed Effects

Effect	*a*	Estimate	Standard Error	*DF*	*t* Value	*Pr* > |*t*|
Intercept		12.2750	1.2300	7	9.98	<0.0001
a	1	2.5437	1.4212	7	1.79	0.1166
a	2	0	.	.	.	.

These are the solutions for $\hat{\eta}$, $\hat{\alpha}_1$, and $\hat{\alpha}_2$ subject to the "last effect zero" convention used by the SAS generalized inverse SWEEP operator.

Type III Tests of Fixed Effects

Effect	Num DF	Den DF	F Value	*Pr* > *F*
a	1	7	3.20	0.1166

Solution for Random Effects

Effect	*a*	Subject	Estimate	Std Err Pred	*DF*	*t* Value	*Pr* > |*t*|
Intercept		b 1	−0.9840	1.5056	16	−0.65	0.5227
a	1	b 1	−0.4528	1.7536	16	−0.26	0.7995
a	2	b 1	−1.0771	1.7536	16	−0.61	0.5477
Intercept		b 2	0.3509	1.5056	16	0.23	0.8187
a	1	b 2	−0.2862	1.7536	16	−0.16	0.8724
a	2	b 2	0.8317	1.7536	16	0.47	0.6417
Intercept		b 3	−0.1856	1.5056	16	−0.12	0.9034
a	1	b 3	−0.9936	1.7536	16	−0.57	0.5789
a	2	b 3	0.7051	1.7536	16	0.40	0.6930
Intercept		b 4	0.8998	1.5056	16	0.60	0.5585
a	1	b 4	1.0310	1.7536	16	0.59	0.5648
a	2	b 4	0.3680	1.7536	16	0.21	0.8364
Intercept		b 5	−0.4850	1.5056	16	−0.32	0.7515
a	1	b 5	1.8709	1.7536	16	1.07	0.3019
a	2	b 5	−2.6249	1.7536	16	−1.50	0.1539
Intercept		b 6	−1.8198	1.5056	16	−1.21	0.2444
a	1	b 6	−2.0124	1.7536	16	−1.15	0.2680
a	2	b 6	−0.8171	1.7536	16	−0.47	0.6475
Intercept		b 7	2.8583	1.5056	16	1.90	0.0758
a	1	b 7	−0.4468	1.7536	16	−0.25	0.8021
a	2	b 7	4.8910	1.7536	16	2.79	0.0131
Intercept		b 8	−0.6347	1.5056	16	−0.42	0.6790
a	1	b 8	1.2899	1.7536	16	0.74	0.4726
a	2	b 8	−2.2767	1.7536	16	−1.30	0.2126

The solutions labeled *"Intercept"* are the $\hat{b}_j$ BLUPs; those labeled *"a 1"* and *"a 2"* are the $\widehat{(ab)}_{ij}$ BLUPs.

a Least Squares Means		
a	**Estimate**	**Standard Error**
1	14.8188	1.2300
2	12.2750	1.2300

Note that the A LEAST SQUARES MEANS are the estimable functions $\hat{\eta} + \hat{\alpha}_1$ and $\hat{\eta} + \hat{\alpha}_2$, respectively. The difference, shown in the following, is $\hat{\alpha}_1 - \hat{\alpha}_2$:

Differences of a Least Squares Means						
a	*_a*	**Estimate**	**Standard Error**	***DF***	***t* Value**	***Pr* > |*t*|**
1	2	2.5437	1.4212	7	1.79	0.1166

Estimates					
Label	**Estimate**	**Standard Error**	***DF***	***t* Value**	***Pr* > |*t*|**
blup a_1 @ b_1	13.3820	1.2463	7	10.74	<0.0001
blup a_2 @ b_1	10.2140	1.2463	7	8.20	<0.0001
blup diff @ b_1	3.1680	1.7108	7	1.85	0.1065
blup diff @ b_2	1.4259	1.7108	7	0.83	0.4321
blup diff @ b_3	0.8451	1.7108	7	0.49	0.6364
blup diff @ b_4	3.2067	1.7108	7	1.87	0.1030
blup diff @ b_5	7.0395	1.7108	7	4.11	0.0045
blup diff @ b_6	1.3484	1.7108	7	0.79	0.4565
blup diff @ b_7	−2.7941	1.7108	7	−1.63	0.1464
blup diff @ b_8	6.1104	1.7108	7	3.57	0.0091

The B_1 specific BLUPs of A levels A_1 and A_2 are $\hat{\eta}+\hat{\alpha}_1+\hat{b}_1+\widehat{(ab)}_{11}$ and $\hat{\eta}+\hat{\alpha}_2+\hat{b}_1+\widehat{(ab)}_{21}$, respectively. Thus, the B_1 specific BLUP difference is the estimate $\hat{\alpha}_1-\hat{\alpha}_2+\widehat{(ab)}_{11}-\widehat{(ab)}_{21}$. In general, the B_j specific BLUP differences are estimates of the predictable function $\alpha_1-\alpha_2+(ab)_{1j}-(ab)_{2j}$.

We can also group levels of B together if we know, or suspect, that certain levels have common characteristics and we want an aggregate estimate. For example, B levels 1, 4, 5, and 8 appear to have relatively large treatment differences, whereas B levels 2, 3, 6, and 7 do not. We could define predictable function

$$\mu+\alpha_i+\left(\frac{b_1+b_4+b_5+b_8}{4}\right)+\left(\frac{(ab)_{i1}+(ab)_{i4}+(ab)_{i5}+(ab)_{i8}}{4}\right)$$

for the individual treatment BLUPs at locations 1, 4, 5, and 8 and, for the difference,

$$\alpha_1 - \alpha_2 + \left(\frac{(ab)_{11} + (ab)_{14} + (ab)_{15} + (ab)_{18}}{4}\right) - \left(\frac{(ab)_{21} + (ab)_{24} + (ab)_{25} + (ab)_{28}}{4}\right)$$

The following ESTIMATE statements, added to the aforementioned GLIMMIX program, obtain the BLUP estimates:

```
estimate 'blup a1 @ b_1,4,5,8' intercept 4 a 4 0 | intercept 1 a 1 0 /
  subject 1 0 0 1 1 0 0 1 divisor=4 e;
estimate 'blup a2 @ b_1,4,5,8' intercept 4 a 0 4 | intercept 1 a 0 1 /
  subject 1 0 0 1 1 0 0 1 divisor=4 e;
estimate 'blup diff @ b_1,4,5,8' a 4 -4 | a 1 -1 /
  subject 1 0 0 1 1 0 0 1 divisor=4 e;
estimate 'blup a1 @ b_2,3,6,7' intercept 4 a 4 0 | intercept 1 a 1 0 /
  subject 0 1 1 0 0 1 1 0 divisor=4 e;
estimate 'blup a2 @ b_2,3,6,7' intercept 4 a 0 4 | intercept 1 a 0 1 /
  subject 0 1 1 0 0 1 1 0 divisor=4 e;
estimate 'blup diff @ b_2,3,6,7' a 4 -4 | a 1 -1 /
  subject 0 1 1 0 0 1 1 0 divisor=4 e;
```

Notice that the coefficients before the vertical bar (|) define $\mathbf{k}'\boldsymbol{\beta} = 4\eta + 4\alpha_i$ and after the vertical bar the basic form $b_j + (ab)_{ij}$. SUBJECT selects $b_j + (ab)_{ij}$ for the levels of B with 1 in the coefficient list and does not select those with a 0. DIVISOR divides all of these terms by 4. The resulting listing is

Estimates		
Label	**Estimate**	**Standard Error**
blup a1 @ b_1,4,5,8	15.4525	0.6460
blup a2 @ b_1,4,5,8	10.5714	0.6460
blup diff @ b_1,4,5,8	4.8812	0.8994
blup a1 @ b_2,3,6,7	14.1850	0.6460
blup a2 @ b_2,3,6,7	13.9786	0.6460
blup diff @ b_2,3,6,7	0.2063	0.8994

Let us now see what happens when we attempt to do this analysis with ordinary least squares.

9.3.2 Relationship between BLUP and Fixed Effect Estimators

True mixed model software did not appear until the 1990s. In a relatively short time, less than a decade, mixed model methodology went from being an indifferently understood curiosity to mainstream. There was awareness of the need to use the right error term to construct test statistics—for example, for a split-plot. Before computers became widely available, this was done with hand-calculated analysis of variance tables. Once computers appeared, ordinary least squares software, of which SAS PROC GLM provides a prototypic example, predominated. Introduced in 1976, PROC GLM was *the* flagship of SAS linear models procedures until PROC MIXED and GENMOD appeared in the early 1990s (GLIMMIX appeared in 2005).

Use of GLM for mixed model applications is still prevalent, so it is instructive to see what happens when we use it in an attempt to do the mixed model analysis described in the previous section. The statements are

```
proc glm data=c9_ex4;
  class a b;
  model y=a|b;
  random b a*b/test;
  lsmeans a/stderr e;
  lsmeans a/stderr e=a*b e;
  estimate 'a diff' a 1 -1/e;
  estimate 'a1 @ b_1' intercept 1 a 1 0 b 1 0 0 0 0 0 0 0 a*b 1 0;
  estimate 'a2 @ b_1' intercept 8 a 0 8 b 1 0 0 0 0 0 0 0 a*b 0 0 0 0 0 0
  0 0 1;
  estimate 'diff @ b_1' a 1 -1 a*b 1 0 0 0 0 0 0 0 -1 0 0 0 0 0 0 0;
  estimate 'diff @ b_2' a 1 -1 a*b 0 1 0 0 0 0 0 0 0 -1 0 0 0 0 0 0;
  estimate 'diff @ b_3' a 1 -1 a*b 0 0 1 0 0 0 0 0 0 0 -1 0 0 0 0 0;
  estimate 'diff @ b_4' a 1 -1 a*b 0 0 0 1 0 0 0 0 0 0 0 -1 0 0 0 0;
  estimate 'diff @ b_5' a 1 -1 a*b 0 0 0 0 1 0 0 0 0 0 0 0 -1 0 0 0;
  estimate 'diff @ b_6' a 1 -1 a*b 0 0 0 0 0 1 0 0 0 0 0 0 0 -1 0 0;
  estimate 'diff @ b_7' a 1 -1 a*b 0 0 0 0 0 0 1 0 0 0 0 0 0 0 -1 0;
  estimate 'diff @ b_8' a 1 -1 a*b 0 0 0 0 0 0 0 1 0 0 0 0 0 0 0 -1;
  estimate 'a1 @ b_1,4,5,8' intercept 4 a 0 4 b 1 0 0 1 1 0 0 1
   a*b 1 0 0 1 1 0 0 1 0 0 0 0 0 0 0 0/divisor=4;
  estimate 'a2 @ b_1,4,5,8' intercept 4 a 0 4 b 1 0 0 1 1 0 0 1
   a*b 0 0 0 0 0 0 0 0 1 0 0 1 1 0 0 1/divisor=4;
  estimate 'diff @ b_1,4,5,8' a 4 -4 a*b 1 0 0 1 1 0 0 1 -1 0 0 -1 -1
  0 0 -1 /
   divisor=4;
  estimate 'a1 @ b_2,3,6,7' intercept 4 a 0 4 b 0 1 1 0 0 1 1 0
   a*b 0 1 1 0 0 1 1 0 0 0 0 0 0 0 0 0/divisor=4;
  estimate 'a2 @ b_2,3,6,7' intercept 4 a 0 4 b 0 1 1 0 0 1 1 0
   a*b 0 0 0 0 0 0 0 0 0 1 1 0 0 1 1 0/divisor=4;
  estimate 'diff @ b_2,3,6,7' a 4 -4 a*b 0 1 1 0 0 1 1 0 0 -1 -1 0 0 -1 -1 0 /
   divisor=4;
```

The RANDOM statement directs the procedure to determine expected mean squares under the assumption that B and A × B effects are random. Test statistics for the A, B, and A × B effects are determined and listed in a table *in addition to* the default analysis of variance output—which (incorrectly) uses the estimate of σ^2 for all test statistics. The first LSMEANS statement shows what happens on default. PROC GLM uses $\hat{\sigma}^2$ to determine all standard errors—a reasonable strategy when all effects are fixed but inappropriate when some effects are random. The second statement is as close as PROC GLM can get to the correct standard error for the LSMEANS. However, as we shall see, it is still not right. The LSMEANS statement in PROC GLM does not have a DIFF option, so we have to use an ESTIMATE statement to compute estimates of treatment differences. Unfortunately, PROC GLM's ESTIMATE statement does not have an option to override the use of $\hat{\sigma}^2$ as the basis for standard errors. We shall see the result as we go through the listing. Finally, all of the B-specific estimates are given in their GLM-syntax forms.

Let us now work through the results. First, the PROC GLM version of the "Type III tests of Fixed Effects" appears as follows:

Source	*DF*	Type III *SS*	Mean Square	*F* Value	*Pr* > *F*
a	1	51.765312	51.765312	3.20	0.1166
b	7	225.782188	32.254598	2.00	0.1910
*Error: MS(a*b)*	7	113.117187	16.159598		

These tests derive from the expected mean squares

Source	Type III Expected Mean Square
a	Var(Error) + 2 Var(a*b) + Q(a)
b	Var(Error) + 2 Var(a*b) + 4 Var(b)
*a*b*	Var(Error) + 2 Var(a*b)

We see that if we want to test H_0: $\alpha_1 - \alpha_2 = 0$, the *F*-ratio $MS(A)/MS(A^*B)$ isolates "*Q*(*A*)," the quadratic form $(\alpha_1 - \alpha_2)^2$ contained in the non-centrality parameter of *F*. We see that $F = 3.20$ and $p = 0.1166$ agree with the mixed model analysis.

The least squares means are another story. The default output is

a	*y* LSMEAN	Standard Error
1	14.8187500	0.4774403
2	12.2750000	0.4774403

The point estimates agree with the mixed model analysis, but the standard error here is 0.477, compared with 1.23 with PROC GLIMMIX (or, equivalently for these data, MIXED). This is because $0.477 = \sqrt{\hat{\sigma}^2/nb}$, which we know is not right. The second LSMEANS statement in the GLM program uses the option E = A*B, directing the procedure to use the expected mean square for A × B—$\sigma^2 + n\sigma^2_{AB}$—to compute the standard error, that is, $s.e.(\text{LSMean}) = \sqrt{\left(\hat{\sigma}^2 + n\hat{\sigma}^2_{AB}\right)/nb}$. This yields

a	*y* LSMEAN	Standard Error
1	14.8187500	1.0049751
2	12.2750000	1.0049751

The standard error, 1.005, is still not 1.23. Working from the model (9.3), the standard error should be $\sqrt{\left(\hat{\sigma}^2 + n\hat{\sigma}^2_{AB} + n\hat{\sigma}^2_B\right)/nb}$. There is no option available in PROC GLM to compute the correct standard error. We can, however, use predictable functions in the mixed model analysis to help understand what GLM is *implicitly* computing. If we add the following statements to the GLIMMIX program in Section 9.3.1,

```
estimate 'blup a_1 narrow' intercept 8 a 8 0 | intercept 1 a 1 0 /
  subject 1 1 1 1 1 1 1 1 divisor=8;
estimate 'blup a_2 narrow' intercept 8 a 0 8 | intercept 1 a 0 1 /
  subject 1 1 1 1 1 1 1 1 divisor=8;
estimate 'blup a_1 intermediate' intercept 8 a 8 0 | intercept 1 /
  subject 1 1 1 1 1 1 1 1 divisor=8;
estimate 'blup a_2 intermediate' intercept 8 a 0 8 | intercept 1 /
  subject 1 1 1 1 1 1 1 1 divisor=8;
```

we obtain the results

Estimates

Label	Estimate	Standard Error
blup a_1 narrow	14.8188	0.4774
blup a_2 narrow	12.2750	0.4774
blup a_1 intermediate	14.8188	1.0050
blup a_2 intermediate	12.2750	1.0050

Remember that ordinary least squares programs such as PROC GLM do not distinguish between fixed and random effects—there is no distinction between $\mathbf{X\beta}$ and $\mathbf{Zb}$ in implementing the estimating equations. Thus, from an ordinary least squares perspective, $\eta + \alpha_i$—the actual least squares mean consistent with model (9.3)—is not estimable. To make it estimable, ordinary least squares must add the average over the levels of B and the levels of AB within *i*th level of A. The resulting coefficients are those of the "narrow BLUP" shown in the estimate statements earlier. Notice that the resulting estimate and standard error of "BLUP A_1 NARROW" are identical to the default LSMEAN in PROC GLM. What GLM computes is a mean whose standard error is applicable *only* to those levels of B actually observed in the data. Inference, at least the confidence interval width, may *not* be applied to *any* other part of population the levels of B represent.

Now consider the predictable functions called "INTERMEDIATE." Notice that they are specific to the B main effect levels but not A × B. This is equivalent to inference on a model with A and B fixed main effects only. The A × B effects become part of distribution $y_{ijk} \sim NI\left(\mu_{ij}, \sigma_P^2\right)$, where "*P*" signifies that the variance components for A × B and observation have been pooled. This broadens inference a bit, but it is still restricted to the levels of B actually observed; we still cannot legitimately use confidence intervals based to this standard error to make statements about the entire population.

Finally, Table 9.6 shows the B-specific ordinary least squares estimates and standard errors GLM computes compared with the BLUPs we computed earlier.

Observe the impact of shrinkage on the various estimators. The group specific estimates for B = {2, 3, 6, 7}, by the way, are correct: the "average" over the two levels of A really does reverse. If model (9.3) accurately described the data—if the B effects do indeed have a probability distribution and are not fixed—then the BLUPs more accurately characterize what is going on. They have used information about the distribution of B to refine the estimates, whereas the ordinary least squares analysis has not.

TABLE 9.6

B-Specific OLS Estimates and BLUP for Data Set 9.4

blup_label	blup	se_blup	ols_label	ols_est	se_ols_est
blup a_1 @ b_1	13.3820	1.24632	a1 @ b_1	13.2500	1.35041
blup a_2 @ b_1	10.2140	1.24632	a2 @ b_1	9.9000	1.35041
blup diff @ b_1	3.1680	1.71082	diff @ b_1	3.3500	1.90976
blup diff @ b_2	1.4259	1.71082	diff @ b_2	1.1000	1.90976
blup diff @ b_3	0.8451	1.71082	diff @ b_3	0.3500	1.90976
blup diff @ b_4	3.2067	1.71082	diff @ b_4	3.4000	1.90976
blup diff @ b_5	7.0395	1.71082	diff @ b_5	8.3500	1.90976
blup diff @ b_6	1.3484	1.71082	diff @ b_6	1.0000	1.90976
blup diff @ b_7	−2.7941	1.71082	diff @ b_7	−4.3500	1.90976
blup diff @ b_8	6.1104	1.71082	diff @ b_8	7.1500	1.90976
blup a1 @ b_1,4,5,8	15.4525	0.64598	a1 @ b_1,4,5,8	15.7250	0.67520
blup a2 @ b_1,4,5,8	10.5714	0.64598	a2 @ b_1,4,5,8	10.1625	0.67520
blup diff @ b_1,4,5,8	4.8812	0.89939	diff @ b_1,4,5,8	5.5625	0.95488
blup a1 @ b_2,3,6,7	14.1850	0.64598	a1 @ b_2,3,6,7	13.9125	0.67520
blup a2 @ b_2,3,6,7	13.9786	0.64598	a2 @ b_2,3,6,7	14.3875	0.67520
blup diff @ b_2,3,6,7	0.2063	0.89939	diff @ b_2,3,6,7	−0.4750	0.95488

9.4 Advanced Applications with Complex Z Matrices

Sections 9.2 and 9.3 covered inference on predictable functions for standard mixed models. The ideas covered in these sections apply equally to Gaussian and non-Gaussian data. While Sections 9.2 and 9.3 used simple designs (completely randomized and randomized block), the techniques shown extend in a straightforward manner to more complex designs (as long as the **Zb**, random effects part of the linear predictor can be formed using conventions shown in Chapters 1 and 2).

For some mixed models, the conventions for constructing **Zb** we have used to this point do not adequately capture the process we are trying to model. One important class of examples comes from quantitative genetics, where the **Z** matrix must include information about relationships among offspring, not merely 0–1 indicators or regression variables. See Wright (1922) for the seminal publication on relationship coefficients, Henderson (1984, 1985) for definitive treatment of the use of these coefficients in mixed models, and Gilmour (2007) for a contemporary overview.

Another newer and less extensively discussed issue in linear model literature is value-added modeling, which is gaining increasing attention particularly in educational achievement research. These were introduced by Sanders et al. (1997) and further discussed by Raudenbush (2004). McCaffery et al. (2004) compare value-added models and discuss issues concerning their application in depth.

The essential idea of value-added modeling is to estimate the teacher contribution to growth in student achievement. Teacher effects are defined as random for reasons laid out earlier in this chapter. No matter what group of teachers is being observed in any given data set, they always constitute a subset of a larger population of teachers about whom the general characteristics of their effects are known. Their effects can be crudely estimated by fixed effects methods or refined by BLUP—a.k.a. shrinkage estimation—techniques.

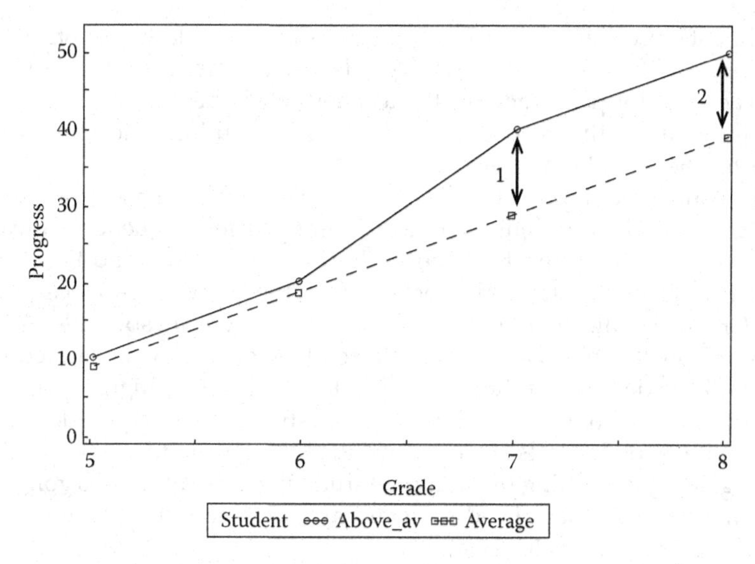

FIGURE 9.1
Student progress by grade—average and above-average student progress.

What sets value-added modeling apart from standard mixed models is the construction of the **Z** matrix. To visualize the issue, consider the plot in Figure 9.1.

The dotted line shows the mean progress of average students from the end of the 5th grade (5 on the horizontal axis) to the end of 8th grade. The solid line shows the progress of a hypothetical student who has made average progress through 5th grade, continues to progress at the average rate through 6th grade, then makes above-average progress in the 7th grade, and, finally, remains above average through the end of 8th grade.

Clearly, the progress from 6 to 7 on the grade axis occurred on the student's 7th grade teacher's watch. So the first increment (the double arrow labeled 1) would clearly be attributed to the 7th grade teacher. The question arises with the 8th grade score—does the 8th grade teacher get all the credit, even though after the 7th grade jump the student's 8th grade rate of progress was average or does the 7th grade teacher get some, or all, of the credit? Now consider how this narrative translates to a **Z** matrix. During the time span shown in Figure 9.1 the student has been under the watch of three teachers. Denote the three teacher effects as t_6, t_7, and t_8. In a conventional mixed model, the **Zb** matrix associated with the teacher effects would be

$$\begin{bmatrix} 1 & 0 & 0 \\ 0 & 1 & 0 \\ 0 & 0 & 1 \end{bmatrix} \begin{bmatrix} t_6 \\ t_7 \\ t_8 \end{bmatrix}$$

Under this model, the student's 8th grade teacher would receive all the credit for the student's above-average score at the conclusion of the 8th grade. The alternative, called a *layered* model, defines **Zb** as

$$\begin{bmatrix} 1 & 0 & 0 \\ 1 & 1 & 0 \\ 1 & 1 & 1 \end{bmatrix} \begin{bmatrix} t_6 \\ t_7 \\ t_8 \end{bmatrix}$$

Thus, the student's progress through 6th grade is attributable to the 6th grade teacher, through 7th grade to the sum of 6th and 7th grade teacher impact, and so forth. This allows the model to account for persistence of the teacher's effect beyond the year when the student was actually under that teacher's watch. We all can think of teachers whose impact lasted long after the school year was over.

Data Set 9.5 contains a hypothetical student scores from 3 years—grades 6, 7 and 8, 2 teachers per grade, and 48 students. There are eight possible sequences of teachers a student could have. If we let A and B be the 2 6th grade teachers, C and D be the 7th grade teachers, and E and F be the 8th grade teachers, then the possible sequences are ACE, ACF, ADE, ADF, BCE, BCF, BDE, and BDF. For each sequence, there are six students. Admittedly, this is an impossibly balanced, idealized data set. Live data for value-added models tend to be messy and the data sets quite large. Usually, specialized, high capacity computing equipment not available to linear models graduate students are required—as one specialist puts it, these kinds of data sets "bring conventional software to its knees." This example is intended to give the beginning linear model student a sense of what is going on in value-added modeling without specialized software and high capacity equipment distracting from the main purpose of this textbook.

This said, a model for these data is as follows:

1. Linear predictor: $\eta_{ijk} = \eta + s_i + \gamma_j + t_{k(6)} + I(j \geq 7)t_{k(7)} + I(j = 8)t_{k(8)}$, where s_i denotes the effect of the ith student, γ_j denotes the effect of the jth grade, and $t_{k(j)}$ denotes the effect of the kth teacher of grade j. Thus, $k(6) = A, B$, $k(7) = C, D$, and $k(8) = E, F$. Notice that the layered model means there will be a $t_{k(6)}$ for every grade level linear predictor, $t_{k(7)}$ will appear in 7th and 8th grade linear predictors only and $t_{k(8)}$ only appears in 8th grade linear predictors.
2. Distributions:
   a. s_i i.i.d. $N\left(0, \sigma_S^2\right)$.
   b. $t_{k(j)}$ i.i.d. $N\left(0, \sigma_T^2\right)$.
   c. $\mathbf{y}_i | s_i, t_{k(j)} \sim N(\boldsymbol{\mu}_i, \boldsymbol{\Sigma})$, where $\mathbf{y}_i' = \begin{bmatrix} y_{i5} & y_{i6} & y_{i7} & y_{i8} \end{bmatrix}$ the vector observations over the 4 exams on the ith student and $\boldsymbol{\mu}_i$ denotes the corresponding mean vector. The matrix $\boldsymbol{\Sigma}$ denotes the covariance among the scores for each students. Because these are repeated measures on each student, in theory they may be correlated. In this example, because we will not cover repeated measures until Chapter 14, we assume they are not. Interested readers could pursue the issue of correlated observations on each student using concepts and methods presented in Chapter 14.
3. Link: identity.

The GLIMMIX statements required for this model are

```
proc glimmix data=bin_data;
  class year;
  effect tchr=collection(z1-z6);
  model score= year/ solution;
  random tchr/solution;
  random intercept/subject=stu_id;
  estimate 'tchr=B' intercept 1 | tchr 1 0;
  estimate 'tchr=B' intercept 1 | tchr 0 1 0;
```

The variables Z1 through Z6 are the columns of the layered **Z** matrix corresponding to the effects of teachers A through F. The EFFECT statement assembles these columns into a single **Z** matrix with the name TCHR. In the RANDOM statement involving TCHR, GLIMMIX "knows" to use the **Z** matrix defined in the EFFECT statement.
Selected results are as follows:

Type III Tests of Fixed Effects

Effect	Num DF	Den DF	*F* Value	*Pr* > *F*
year	3	3.002	1.35	0.4057

This result is not surprising: Progress scores are typically normalized so that values that do not change from year to year indicate average progress. With such scoring systems, a significant year effect would raise suspicion.

Covariance Parameter Estimates

Cov Parm	Subject	Estimate	Standard Error
Intercept	stu_id	10.0215	2.0679
tchr		0.7377	0.6028
Residual		0.009161	0.001103

The variance component for student (STU_ID) can be interpreted as the measure of variation among student baseline scores at the start of the study (end of 5th grade/start of 6th grade). These estimates suggest that *in this data set,* after accounting for variability among students, teachers, and years, little variability remains among student progress scores. Remember, this is a hypothetical, simulated data set—any resemblance or lack thereof to actual achievement data is purely coincidental.

Solution for Random Effects

Effect	tchr	Estimate	Std Err Pred	*DF*	*t* Value	*Pr* > |*t*|
tchr	z1	−0.5983	0.6075	2.999	−0.98	0.3974
tchr	z2	0.5983	0.6075	2.999	0.98	0.3974
tchr	z3	−0.02154	0.6075	2.998	−0.04	0.9739
tchr	z4	0.02154	0.6075	2.998	0.04	0.9739
tchr	z5	−0.8645	0.6075	2.999	−1.42	0.2499
tchr	z6	0.8645	0.6075	2.999	1.42	0.2499

These are the estimated BLUPs of the teacher effects: the teacher scores. These represent the primary objective of a value-added model. Notice that they sum to zero within each grade pairing. This is purely an artifact of the balance of this data set. In a live data set, we would be exceedingly unlikely to see balance like this because our assignment of students to three-teacher sequences would never occur in the real world. However, we do interpret these teacher BLUPs the same way we would with live data. The ESTIMATES indicate how average progress over the 3 year period for students who had that teacher during the grade taught by that teacher differs from the overall average. Thus, teacher Z1 (which corresponds to teacher A) students averaged 0.60 units lower on the progress scale, whereas teacher Z2 (teacher B) students averaged 0.60 units higher.

The *t*-value gives the number of standard deviations above or below average for each teacher score. While there is no evidence that any of these teacher scores is out of line, we do notice that teachers A and E have lower scores. As this is a textbook on linear model concepts and methodology, we will not get into a discussion of how these scores may or may not be used. We only note that their use would be highly context dependent. The role of a linear models expert should be to sort out legitimate and illegitimate interpretations of these scores in view of mixed model theory.

Estimates

Label	Estimate	Standard Error	*DF*	*t* Value	*Pr* > |*t*|
tchr = A	23.7238	0.5894	16.5	40.25	<0.0001
tchr = B	24.9203	0.5894	16.5	42.28	<0.0001

The estimates are of predictable functions of the form $\hat{\eta}+\hat{\bar{\gamma}}_{\bullet}+\hat{t}_{k(j)}$. Shown here are BLUPs for the two 6th grade teachers. These are simply the teacher scores expressed on the overall progress scale. For these data, the sample mean student overall progress is 23.32. Teacher A is roughly 0.6 units below that; teacher B is 0.6 units above.

9.5 Summary

This chapter focuses on the uses of predictable functions $\mathbf{K}'\boldsymbol{\beta} + \mathbf{M}'\mathbf{b}$ for intentionally narrow inference. Such estimation goes variously under the terms BLUP, subject-specific inference, and shrinkage estimations. The main idea is that estimation and inferential statistics involving **b** must take into account the fact that **b** is a random vector, not a fixed parameter.

This chapter covered the following:

- Random effects models with inference on specific levels of **b**.
- Mixed models, highlighting those from multilocation or multi-occasion studies with interest in location- or occasion-specific effects, for example, location-specific treatment means and differences.
- More complex mixed models where the **Z** matrix must include nonstandard coefficients, for example, resulting from genetic relationships or persistence of teacher impact on student achievement.
- For the first two applications listed previously, GLIMMIX CONTRAST, ESTIMATE, and LSMESTIMATE functions can be used to compute most BLUPs of interest.
- More complex **Z** matrices must be defined explicitly, for example, by using the EFFECT option in GLIMMIX.

Like fixed effects inference in mixed models with estimated covariance parameters, predictable functions must take into account issues of approximate degrees of freedom and standard error bias that motivated the Satterthwaite approximation and Kenward–Roger adjustment. In the predictable function case, bias correction builds on the work of Prasad and Rao (1990) and Harville and Jeske (1992).

- The Kenward–Roger option DDFM=KR implements bias correction based on Prasad–Rao and Harville–Jeske.
- The chapter presented Gaussian examples. In principle, non-Gaussian GLMMs work the same way.
- Only conditional models can be used for predictable function estimation and inference. Marginal models have no **b** in the linear predictor—ergo no possibility of $\mathbf{K}'\boldsymbol{\beta} + \mathbf{M}'\mathbf{b}$.
- Older textbooks that distinguish between fixed and random effects by saying that only variance component estimation is of interest for the latter are obviously dated and misleading.

10

Rates and Proportions

10.1 Types of Rate and Proportion Data

In the previous three chapters, we have considered overarching issues of how treatment design structure and "topographical" or study design structure drive linear predictor construction and subsequent inference. While the strategies presented in these chapters apply to all linear models—Gaussian and non-Gaussian—all of our examples in Chapters 7 through 9 featured Gaussian response variables. As a result, these chapters focused mostly on *mixed model issues*. In this chapter, we expand our repertoire. This is the first chapter in the application section that explicitly considers *generalized* linear model (GLM) issues in conjunction with mixed models. We begin with data for rates and proportions.

By rate and proportion data, we mean data whose expected value lies between 0 and 1. We can divide rates and proportions into two types: *discrete* and *continuous*. Discrete proportions arise when the unit of observation consists of N distinct entities, Y of which have some attribute of interest. N must be a positive integer and Y must be a nonnegative integer $\leq N$. Thus, the observed proportion must be a discrete fraction: $0/N, 1/N, \ldots, N/N$. Typically we use the binary or binomial distribution if $N = 1$ and the binomial distribution when $N \geq 2$. For these distributions, $0 \leq E(Y/N) = \pi \leq 1$; inference targets model effects on π. Continuous rates arise when we measure responses such as percent of leaf area infected or percent of land area flooded. Like the binomial π, continuous rates must be between 0 and 1, but, unlike the binomial, rates clearly do not result from a set of distinct Bernoulli trials. Instead, the beta distribution is most often used.

In Sections 10.2 and 10.3, we consider various aspects of modeling with binary and binomial data. In Chapter 3, we made extensive use of examples with binomial data to introduce GLM concepts. We do not need to repeat these. As a result, we present more advanced binomial generalized linear mixed models (GLMMs). The examples in Section 10.2 use log or logit links. In Section 10.3, we explore alternative link functions for binomial data. Section 10.4 focuses on continuous rates and the beta distribution.

10.2 Discrete Proportions: Binary and Binomial Data

All of the examples in Chapters 7 and 8 used Gaussian data. We used REML estimates of all variance components and the likelihood-based mixed model equations to estimate model parameters and random effects. With Gaussian data, we have no decisions to make regarding estimation: The residual likelihood for covariance components and the likelihood for the linear predictor are both well defined and tractable. We saw in Chapter 4 that with non-Gaussian mixed models, we have a different situation. We have either a well-defined but intractable likelihood, or we have an equally intractable quasi-likelihood. In either case, we need to decide whether to use a linearization method (e.g., pseudo-likelihood [PL]) or integral approximation (e.g., Laplace or quadrature).

We begin this section with a comparison of PL, Laplace, and quadrature small sample performance using a data set that illustrates an extreme boundary condition. Recall that PL has the advantage of allowing us to use the full range of mixed model techniques, but the downside of being much more vulnerable to boundary condition data sets. Integral approximation is much less prone to poor performance, but using it limits our flexibility in terms of models we can consider and inference techniques we can employ. The first example illustrates what we mean by a "boundary condition" for binary and binomial data and gives us a sense of when PL begins to struggle.

The remaining section surveys three common scenarios involving binomial response variables. The first example is a nested factorial, similar in structure to the design we introduced via Figure 2.5. The second a response surface design with a binomial response. The third is a three-way contingency table typical of categorical data. Finally, the last section explores what "residual" means—and does *not* mean—in the context of binomial data.

10.2.1 Pseudo-Likelihood or Integral Approximation

An extreme form of binary data occurs when researchers collect a single binary observation on each unit of replication. For this kind of data, all linear model procedures struggle. We can do the calculations with ordinary least squares, but the results at best are meaningless and at worst do harm if important decisions rest on interpretation drawn from such data. If we make a genuine effort to analyze such data appropriately, from a linear model perspective, we want to fit a binomial GLM—or GLMM if the study design calls for random model effects.

This type of design represents what we call a *boundary condition* for discrete proportion data. All models have a smallest unit—the unit of observation. For discrete proportions, the data we collect on this unit involves Y out of N occurrences of an outcome of interest. In this context, N is called the *cluster size*. Usually, we assume Y has a binomial distribution, meaning N denotes the number of independent Bernoulli trials comprising the observations.

For relatively large N, PL performs well. As N approaches 1, the linearization underlying PL loses accuracy; PL performance deteriorates but the integral approximation methods continue to provide accurate estimates. As N gets even closer to 1, the Laplace integral approximation loses accuracy. Finally, even quadrature may struggle. This gives us a general picture, but what exactly do we mean by "relatively large N?" How do we know when to stop using PL? At what point does Laplace become unreliable as well?

We answer these questions using simulation. To illustrate, consider the random intercept model:

- Linear predictor: $\eta_i = \beta_0 + c_i; \quad i = 1, 2, \ldots, C$
- c_i i.i.d. $N(0, \sigma^2)$
- $y_{ij} \mid c_i \sim \text{binary}(\pi_i); \quad j = 1, \ldots, N$
- Link: logit = $\text{logit}(\pi_i) = \log[\pi_i/(1 - \pi_i)]$

C denotes the number of random subjects or clusters, and β_0 indirectly determines the average probability, that is, $\pi = 1/(1+e^{-\beta_0})$ Suppose we set $\beta_0 = -1$ (and therefore $\pi \cong 0.27$), $\sigma^2 = 1$, and $C = 200$, then see what happens when we set *N* to 1, 2, 4, 10, 20, and 50. With $N = 50$, all estimation methods should be well behaved. For lower values of *N*, we should see PL yield progressively less accurate results. As *N* approaches 1, we can see when—or if—each estimation method begins to struggle.

We evaluate performance by three criteria. First we look at the average variance component estimate $\hat{\sigma}^2$. We set $\sigma^2 = 1$, so the average $\hat{\sigma}^2$ should be close to 1 if the estimation procedure behaves well. Next we look at the average $\hat{\beta}_0$; we set $\beta_0 = -1$ so the average $\hat{\beta}_0$ should be close to −1. Finally, we look at the standard error of $\hat{\beta}_0$. By definition, the standard error is an estimate of the standard deviation of the sampling distribution of $\hat{\beta}_0$. We should see close agreement between the observed standard deviation of the $\hat{\beta}_0$ and the average standard error computed by the estimation procedure. Since PROC GLIMMIX uses adaptive quadrature, we can also track the number of quadrature points the GLIMMIX algorithm selects. Here are results for one set of 1000 simulated data sets.

Case 1: $C = 200$ clusters; $N = 1$ binary observations per cluster

Method	**Average $\hat{\sigma}^2$**	**Average $\hat{\beta}_0$**	**Std dev of $\hat{\beta}_0$ Sampling Distribution**	**Average std err($\hat{\beta}_0$)**	**Q-Points**	
RSPL	0.024	−0.839	0.159	0.155		
MSPL	0*	−0.839	0.159	0.155		
Laplace	438.2 (median = 0)	−3.225	4.402	0.435		
					9	0.917
Quadrature	4.120	−1.394	0.651	4.846	11	0.083

* MSPL consistently produced negative solutions for $\hat{\sigma}^2$ that were set to 0.

Comment: Consider Case 1 cautionary note to researchers about collecting binary data at the unit level. In this example, when $N = 1$, all estimation procedures perform atrociously. Sometime, however, binary data are unavoidable. Later in this section we consider an example.

Case 2: $C = 200$ clusters; $N = 2$ binary observations per cluster

Method	**Average $\hat{\sigma}^2$**	**Average $\hat{\beta}_0$**	**Std dev of $\hat{\beta}_0$ Sampling Distribution**	**Average std err(β_0)**	**Q-Points**	
RSPL	0.411	−0.848	0.117	0.119		
MSPL	0.396	−0.848	0.117	0.118		
Laplace	0.673	−0.966	0.158	0.155		
					7	0.999
Quadrature	1.103	−1.013	0.167	0.168	9	0.001

Case 3: $C = 200$ clusters; $N = 4$ binary observations per cluster

Method	**Average $\hat{\sigma}^2$**	**Average $\hat{\beta}_0$**	**Std dev of $\hat{\beta}_0$ Sampling Distribution**	**Average std err($\hat{\beta}_0$)**	**Q-Points**	
RSPL	0.613	−0.870	0.0938	0.0965		
MSPL	0.602	−0.870	0.0937	0.0961		
Laplace	0.842	−0.988	0.116	0.114		
					5	0.368
Quadrature	1.001	−1.000	0.117	0.118	7	0.632

Case 4: $C = 200$ clusters; $N = 10$ binary observations per cluster

Method	**Average $\hat{\sigma}^2$**	**Average $\hat{\beta}_0$**	**Std dev of $\hat{\beta}_0$ Sampling Distribution**	**Average std err($\hat{\beta}_0$)**	**Q-Points**	
RSPL	0.788	−0.915	0.0833	0.0815		
MSPL	0.780	−0.914	0.833	0.0812		
Laplace	0.950	−1.000	0.0931	0.0902		
Quadrature	0.998	−1.002	0.0933	0.0916	5	1.000

Case 5: $ZC = 200$ clusters; $N = 20$ binary observations per cluster

Method	**Average $\hat{\sigma}^2$**	**Average $\hat{\beta}_0$**	**Std dev of $\hat{\beta}_0$ Sampling Distribution**	**Average std err($\hat{\beta}_0$)**	**Q-Points**	
RSPL	0.865	−0.941	0.0737	0.0759		
MSPL	0.859	−0.941	0.0737	0.0756		
Laplace	0.974	−0.997	0.0794	0.0809		
Quadrature	0.994	−0.998	0.0795	0.0815	5	1.000

Case 6: $C = 200$ clusters; $N = 50$ binary observations per cluster

Method	**Average $\hat{\sigma}^2$**	**Average $\hat{\beta}_0$**	**Std dev of $\hat{\beta}_0$ Sampling Distribution**	**Average std err($\hat{\beta}_0$)**
RSPL	0.934	−0.970	0.0712	0.0727
MSPL	0.928	−0.970	0.0712	0.0725
Laplace	0.991	−0.999	0.0737	0.0750

Case 6 omits quadrature because the previous cases persuasively show that for $N \geq 3$, quadrature sets the gold standard for accuracy. No point belaboring the obvious for $N = 50$. Laplace achieves reasonable accuracy for $N \geq 10$. For PL, N must be considerably greater.

Important caveat: These results give an accurate sense of the relative accuracy of quadrature, Laplace, and PL for too small, very small, small, and moderate cluster sizes in the sense that none of them do well when the cluster size is "too small," quadrature does well with "very small" or larger, Laplace does well with "small" or larger, and PL requires at least "moderate" cluster sizes to yield useable results. However, what constitutes "too small," "very small," "small," and "moderate" is context specific. In general, PL does not need cluster sizes of 50 or more to perform acceptably. *If you find yourself in uncharted territory*, you should do exactly what we tried to role-model here: simulate data according the process you are trying to model and see for yourself how the estimation procedures perform.

Students are encouraged to try other values of β_0, σ^2, and *C*. Obviously, these results should not be interpreted to mean "always use quadrature." We encounter many situations in which the model is too complex for quadrature or the required computing time is prohibitive. In most cases, quadrature and Laplace results are virtually indistinguishable, but Laplace takes a couple of seconds and quadrature may take a couple of hours.

Finally, and perhaps most importantly, extreme boundary conditions are usually symptomatic of badly designed studies. It is important that our knowledge of how estimation procedures behave and what their limitations are be factored into study designs.

10.2.2 Example of Explanatory-Response Models

Three examples are presented in this section. Simpler binomial models were discussed in Chapter 3. Our discussion here presumes that readers are familiar with the issues covered in those earlier examples.

Example 10.1 A mixed model with binary data

While basing a GLMM on a binary response at the unit level is not ideal, in many studies there is no alternative. The following example, courtesy of Bello et al. (2006), illustrates one such case. The data pertinent to this example appear in the SAS Data and Program Library as Data Set 10.1. The data used in this example are only a representative subset of the data Bello et al. used in their research. Also, the analyses in this example are not the analyses Bello et al. used. The models and analysis shown here were chosen for demonstration purposes.

The study involved a comparison of four synchronization treatments whose purpose was to regulate ovulation and assess its impact of reproductive efficiency. Cows were blocked by 8 weekly groups as they entered the point in their reproductive cycle at which treatment was appropriate. Cows in each group were randomly assigned to one of four treatment groups, denoted by the variable TRT in the data set. The identities of the treatments are blinded in Data Set 10.1. Cows were classified according to whether they were in their first, second, etc., lactation. The variable LACN denotes this classification. In these types of studies, the distinction between first, second and later lactations is of interest, but distinguishing among third, fourth, etc., lactations adds little if any useful information. Accordingly, the analyses in this example use the variable LACT, coded as 1 (first lactation), 2 (second lactation), or 3 (later than second). Bello et al. only distinguished between first lactation and later lactations. Measurements were taken on three covariates of potential interest: these are referred to in Data Set 10.1 as XM1, XP2, and XP3 rather than by the actual names of the covariates. The two response variables shown in this example: SynchOv (synchronized ovulation in response to the treatment protocol, 1 = Yes, 0 = No) and CR35 (pregnancy outcome at 35 da, 1 = pregnant, 0 = not).

The full GLMM considered was as follows:

1. Linear predictor: $\eta_{ij} = \eta + \tau_i + \alpha_j + (\tau\alpha)_{ij} + \sum_c \beta_c X_c + g_k$, where τ denotes the treatment effect, α denotes the lactation code, X_c denotes the cth covariate, β_c denotes the regression coefficient associated with the cth covariate, and g_k denotes the kth group effect.
2. Distributions:
   a. g_k i.i.d. $N(0, \sigma_G^2)$ initially. Modifications of the random model effects and covariance assumptions were required for the response SynchOv and are discussed in the following.
   b. $y_{ijk} \mid g_k \sim \text{Binary}(\pi_{ijk})$, where y_{ijk} can denote either SynchOv or CR35.
3. Link: $\eta_{ijk} = \log[(\pi_{ijk}/(1 - \pi_{ijk})]$.

Let us first consider the analysis of CR35. We implement the model using the following SAS statements:

```
ods graphics on;
proc glimmix data=binary_cow_data method=quadrature;
class Group Trt Lact;
model cr35=Trt|Lact XM1 XP2 XP3/dist=binary;
random intercept/subject=group;
lsmeans Trt*Lact/slicediff=(Lact Trt) ilink plot=meanplot(sliceby=Trt
join ilink);
run;
```

Note the use of METHOD=QUADRATURE in view of our discussion earlier regarding estimation methods with binary data.

The selected results are as follows:

Fit Statistics for Conditional Distribution	
−2 Log L(cr35 | r. effects)	93.85
Pearson chi-square	80.10
Pearson chi-square/DF	0.93

Covariance Parameter Estimates

Cov Parm	Subject	Estimate	Standard Error
Intercept	Group	0.07708	0.3380

Type III Tests of Fixed Effects

Effect	Num DF	Den DF	*F* Value	Pr > *F*
Trt	3	64	0.97	0.4133
Lact	2	64	0.08	0.9276
Trt*Lact	6	64	1.23	0.3040
XM1	1	64	0.15	0.7039
XP2	1	64	3.29	0.0742
XP3	1	64	4.00	0.0496

In addition to accuracy considerations, analysis with Laplace or quadrature allows us to obtain the "conditional distribution fit statistics," specifically Pearson χ^2/df. Recall that this

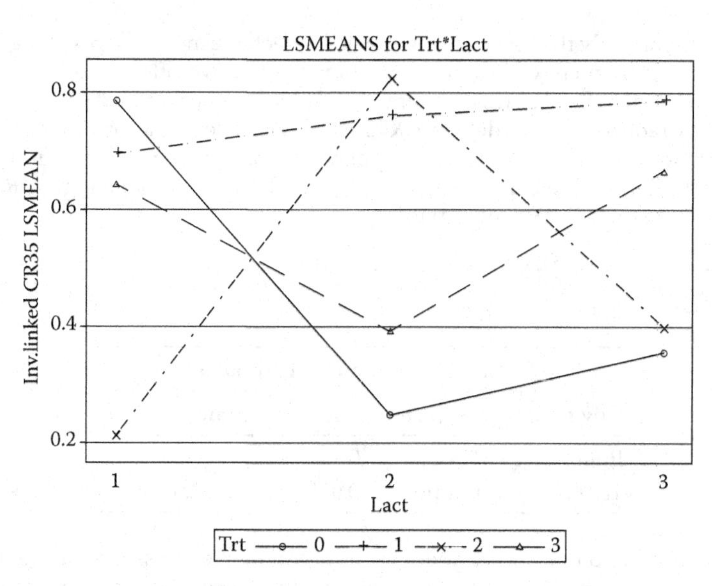

FIGURE 10.1
Interaction plot for CR35 binary data.

statistic helps assess the goodness-of-fit of the model. We interpret $\chi^2/df \gg 1$ as a symptom of overdispersion, possibly because the linear predictor is incomplete or the distribution has in some way been misspecified. Conditional distribution Pearson $\chi^2/df = 0.93$ fit statistic indicates that we have no evidence of overdispersion. The estimate of the group variance $\hat{\sigma}_G^2 = 0.077$ means we do not have negative variance component issues. The approximate *F*-values provide evidence of XP2 and XP3 covariate effects on the 35 da pregnancy rates, but little evidence of TRT or LACT effects. Figure 10.1 shows the interaction plot.

Despite the lack of significant TRT effects, we might view Figure 10.1 as suggestive, pending confirmatory evidence, that TRT = 1 yields relatively high expected pregnancy rates regardless of lactation number, whereas the other treatments produce high rates only conditionally on certain lactation number. The lack of statistical power for TRT and LACT may be an artifact of this being a sample of a larger data set, but we should note that lack of power is commonly associated with binary data in small data sets. In Chapter 16 we discuss power and precision analysis. We can use methods presented in Chapter 16 to address the following question. How many cows in how many groups do we would need to show differences that appear in Figure 10.1 to be statistically significant if in fact they are real differences.

For the response variable SynchOv, we can proceed along the same lines, except we encounter additional estimation issues. For SynchOv, when we attempt to fit the model with all three covariates, we get nonsense results such as the following table of type III tests of fixed effects:

Type III Tests of Fixed Effects

Effect	Num DF	Den DF	*F* Value	Pr > *F*
Trt	3	65	Infty	<.0001
Lact	2	65	Infty	<.0001
Trt*Lact	6	65	Infty	<.0001
XM1	1	65	0.00	0.9969
XP2	1	65	Infty	<.0001
XP3	1	65	Infty	<.0001

It turns out that only the covariate XM1 yields a tractable model for response SynchOv. Again, this is a frequent occurrence with binary data: we often must scale back the model we attempt to fit until we find something that produces at least something of use.

Fitting the reduced model deleting XP2 and XP3 yields a negative estimate of the group variance σ_G^2. We know this can happen if there is also unit-level—in this case a cow-level—variation. We can add a cow-level random effect by changing RANDOM statement in the GLIMMIX program to

```
random intercept Trt*Lact/subject=group;
```

This yields the following result:

Covariance Parameter Estimates			
Cov Parm	**Subject**	**Estimate**	**Standard Error**
Intercept	Group	0	
Trt*Lact	Group	0.9387	2.3260

We saw in Chapter 3 that these random effects can be reexpressed as a cow-level random model effect with compound symmetry covariance. The group variance $\hat{\sigma}_G^2 = 0$ occurs because there is a negative solution that may indicate negative compound symmetry covariance. The modified linear predictor is

$$\eta_{ij} = \eta + \tau_i + \alpha_j + (\tau\alpha)_{ij} + \beta_{XM1}X_{XM1} + u_{ijk}$$

where $\text{Cov}(\mathbf{u}_k) = \sigma_U^2 \begin{bmatrix} 1 & \rho & \cdots & \rho \\ & 1 & \cdots & \rho \\ & & \cdots & \cdots \\ & & & 1 \end{bmatrix}$ and the $\mathbf{u}_k$ is nominally a $n_k \times 1$ vector consisting of the cow-level effects for the n_k cows in the kth group.

The GLIMMIX statements are

```
proc glimmix data=Binary_Cow_Data method=laplace /*method=quadrature*/;
class group Trt Lact;
model SynchOv=Trt|Lact XM1/dist=binary;
random Trt*Lact/subject=group type=CS;
parms (0.930.1);
lsmeans Trt*Lact/slicediff=(Lact Trt) ilink plot=meanplot(sliceby=Trt
join ilink);
run;
```

Note the use of METHOD=LAPLACE in place of METHOD=QUADRATURE. The quadrature method is not meant for covariance structures and typically does not perform well, if at all. The selected results are as follows:

Fit Statistics for Conditional Distribution	
Pearson chi-square/DF	0.49

Covariance Parameter Estimates		
Cov Parm	**Subject**	**Estimate**
Variance	Group	0.9051
CS	Group	−0.07543

Type III Tests of Fixed Effects				
Effect	**Num DF**	**Den DF**	***F* Value**	**Pr > *F***
Trt	3	51	0.56	0.6429
Lact	2	51	0.70	0.4991
Trt*Lact	6	51	0.61	0.7184
XM1	1	37	4.36	0.0438

As with CR35, the main evidence in the type III tests of fixed effects is the covariate effect, in this case XM1. The fits statistics gives no evidence of overdispersion. The interaction plot (Figure 10.2) is suggestive of possible TRT simple effects at given lactation numbers, but we lack statistical power to make statements with strong *p*-value support.

Example 10.2 Nested-factorial design with binomial data

The data for this example appear in the SAS Data and Program Library as Data Set 10.2. The design uses the same format shown in Figure 2.5: 10 blocks, each with 3 units; 6 treatments divided into 2 sets. Set A_0 consists of treatments labeled B_0, B_1, and B_2. Set A_1 contains treatments B_3, B_4, and B_5. Each set is randomly assigned to five blocks. Each treatment is randomly assigned to one unit within each block. The response variable for each unit is binomial. A model describing this process is as follows:

1. Linear predictor: $\eta_{ijk} = \eta + \alpha_i + \beta(\alpha)_{ij} + r_k$ where α, β, and r denote set, treatment, and block effects, respectively
2. Distributions:
   a. $y_{ijk} \mid r_i \sim \text{Binomial}(n_{ijk}, \pi_{ijk})$
   b. r_k i.i.d. $N\left(0, \sigma_R^2\right)$
3. Link: $\eta_{ijk} = \text{logit}(\pi_{ijk}) = \log[\pi_{ijk}/(1 - \pi_{ijk})]$

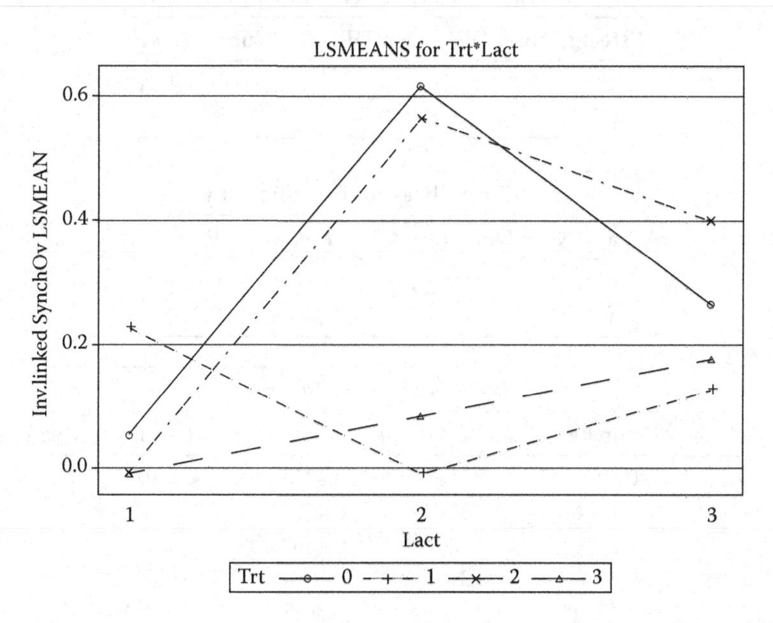

FIGURE 10.2
Interaction plot for SynchOv binary data.

The GLIMMIX statements to implement this analysis are

```
proc glimmix data=ch10_exB1 method=quad;
class block a b;
model f/n=a b(a);
random a/subject=block;
lsmeans b(a)/slice=a slicediff=a ilink;
```

The program shown here uses estimation by quadrature (METHOD=QUAD). Readers can verify that the Laplace method yields virtually identical results and the default PL results are similar. For this data set, you can take your choice. Following guidelines from Chapter 7, this GLIMMIX program uses the LSMEANS statement to obtain standard statistics for a factorial structure.

The listing begins with

Fit Statistics for Conditional Distribution	
−2 Log L(f | r effects)	140.87
Pearson chi-square	22.41
Pearson chi-square/DF	0.75

Here $\chi^2/df = 0.75$; the Pearson chi-square statistics gives no evidence calling our distribution and linear predictor assumptions into question.

The following shows the standard listing we use for the nested-factorial design:

Covariance Parameter Estimates

Cov Parm	Subject	Estimate	Standard Error
a	block	0.3770	0.1932

Type III Tests of Fixed Effects

Effect	Num DF	Den DF	*F* Value	Pr > *F*
a	1	8	0.58	0.4670
b(a)	4	16	2.45	0.0887

Tests of Effect Slices for b(a) Sliced by a

a	Num DF	Den DF	*F* Value	Pr > *F*
0	2	16	0.43	0.6569
1	2	16	4.46	0.0289

b(a) Least Squares Means

a	b	Estimate	Standard Error	Mean	Standard Error Mean
0	0	−0.3070	0.3141	0.4238	0.07671
0	1	−0.1091	0.3195	0.4728	0.07963
0	2	−0.1657	0.3207	0.4587	0.07962
1	3	−0.8548	0.3141	0.2984	0.06576
1	4	−0.4359	0.3105	0.3927	0.07406
1	5	−0.2275	0.3126	0.4434	0.07715

Simple Effect Comparisons of b(a) Least Squares Means by a							
Simple Effect Level	b	_b	Estimate	Standard Error	DF	*t* Value	Pr > |*t*|
a0	0	1	−0.1980	0.2213	16	−0.89	0.3843
a0	0	2	−0.1413	0.2233	16	−0.63	0.5359
a0	1	2	0.05667	0.2313	16	0.24	0.8096
a1	3	4	−0.4189	0.2105	16	−1.99	0.0639
a1	3	5	−0.6273	0.2136	16	−2.94	0.0097
a1	4	5	−0.2084	0.2078	16	−1.00	0.3308

The "covariance parameter estimates" show $\hat{\sigma}_R^2 = 0.377$. For the logit link, we interpret σ_R^2 as the variance of log odds among blocks—here, log odds averaged over the units within a block. Interpreting the "type III tests for fixed effects" is problematic: the p-value for B(A) suggests something may be going on, but, as we have seen in similar examples from Chapters 7 and 8, the B(A) test is overly aggregated. Better to look at the individual slices. Here we see that for the test of equal treatment means within set A_0, $p > 0.65$, whereas within set A_1, $p = 0.0289$. The SLICEDIFF output and the "LSMeans" column labeled "means"—the $\hat{\pi}_i$ on the data scale—help tell the story. The estimated probabilities for treatments 4 and 5 are 0.39 and 0.44, respectively, and not significantly different ($p > 0.33$). For treatment 3, $\hat{\pi}_3 = 0.298$ and there is moderate evidence ($p = 0.0634$) that it differs from treatment 4 and stronger evidence ($p = 0.0097$) that it differs from treatment 5.

All of the treatment comparisons stated earlier, use model-based statistics. Because we have used integral approximation in this example, we might reasonably ask how these results compare to empirical, sandwich estimators. We begin with the SLICE listing. Using sandwich estimators does not change the fact that the B(A) statistic from the "type III tests of fixed effects" is overly aggregated. Also, we want to avoid interpreting the main effect of A until we are sure we do not have a significant B(A) effect. Therefore, we bypass the overall tests of A and B(A). The SLICE results using sandwich estimators are

Tests of Effect Slices for b(a) Sliced by a				
a	Num DF	Den DF	*F*-Value	Pr > *F*
0	2	16	0.32	0.7340
1	2	16	7.30	0.0056

The sandwich estimator produces noticeably inflated F-values relative to the model-based estimates. In Chapter 5, sandwich estimators were introduced with the caveat that they can generate upwardly biased test statistics when used with small data sets. We see this here. Better to use the Morel et al. (2003) bias-corrected sandwich estimators, the EMPIRICAL=MBN option in the GLIMMIX PROC statement. The adjusted SLICE output is as follows:

Tests of Effect Slices for b(a) Sliced by a				
a	Num DF	Den DF	*F* Value	Pr > *F*
0	2	16	0.17	0.8473
1	2	16	2.65	0.1012

Morel et al. developed their correction for GEE-type models. Our model in this example is not a GEE-type model (we could not use integral approximation if it was). With certain applications, the MBN correction over-adjusts—this appears to be an example of the resulting overkill. The F-values now show noticeable *de*flation.

Recall from Chapter 5, the MBN correction, as Morel et al. presented it, is $\delta_m \phi \mathbf{V}(\hat{\boldsymbol{\sigma}})$, where $\delta_m = \min(0.5, k/m-k)$ and $\phi = \max\left[1, trace\left\{\left(\mathbf{X}'\mathbf{V}^{-1}\mathbf{X}\right)^{-}\sum_i \left[\mathbf{V}_i^{-1}(\mathbf{y}_i - \mathbf{X}_{i\,i})(\mathbf{y}_i - \mathbf{X}_{i\,i})'\mathbf{V}_i^{-1}\right]\Big/k\right\}\right]$. See Chapter 5 for details.

PROC GLIMMIX allows us to adjust these corrections, by defining $\delta_m = \min[1/D, k/(m-k)]$ and $\phi = \max[R, trace\{\bullet\}]$. On default, $D = 2$ and $R = 1$, but we can replace either or both of these with user-specified values, with the restriction that $D \geq 1$ and $0 \leq R \leq 1$. With these data, the trace in ϕ is already 1, so there is no point in changing R, but increasing D reduces the magnitude of the correction. For example, the statement

```
proc glimmix data=ch10_exB1 method=quad empirical=mbn(d=12);
```

produces the output

Tests of Effect Slices for b(a) Sliced by a

a	Num DF	Den DF	*F*-Value	Pr > *F*
0	2	16	0.22	0.8053
1	2	16	4.52	0.0278

Simple Effect Comparisons of b(a) Least Squares Means by a

Simple Effect Level	b	_b	Estimate	Standard Error	DF	*t* Value	Pr > |*t*|
a0	0	1	−0.1980	0.3006	16	−0.66	0.5196
a0	0	2	−0.1413	0.3479	16	−0.41	0.6901
a0	1	2	0.05667	0.2567	16	0.22	0.8281
a1	3	4	−0.4189	0.2676	16	−1.57	0.1370
a1	3	5	−0.6273	0.2286	16	−2.74	0.0144
a1	4	5	−0.2084	0.1517	16	−1.37	0.1884

b(a) Least Squares Means

a	b	Estimate	Standard Error	Mean	Standard Error Mean
0	0	−0.3070	0.4891	0.4238	0.1194
0	1	−0.1091	0.5162	0.4728	0.1287
0	2	−0.1657	0.4815	0.4587	0.1196
1	3	−0.8548	0.2361	0.2984	0.04944
1	4	−0.4359	0.2774	0.3927	0.06615
1	5	−0.2275	0.1673	0.4434	0.04129

Notice the impact relative to the model-based statistics. The F-values for SLICE are similar. The standard errors are larger for the estimated logits (ESTIMATE) and probabilities (MEAN) within set A_0 but smaller for the estimates within A_1. The simple-effect tests result in somewhat more conservative p-values.

Bear in mind that this comparison applies to this data set only. We should not jump to conclusions about the behavior of the MBN correction relative to model-based statistics even among nested-factorial GLMMs. Our use of $D = 12$ here is distinctly *ad hoc*, for demonstration purposes only, with the only criterion being similarity of the SLICE F-values for the model-based and adjusted-MBN statistics.

If we were doing this in real life, we would run a simulation to assess the type I error and power characteristics of the model-based statistics. If they seem reasonable, we probably want to leave well enough alone. If they show bias, for example, the inflation of type I error rate that motivated the Kenward–Roger correction, then a reasonable next step would be to assess the behavior of the MBN correction, trying different R and D values to see what combination appears to be most accurate.

Why is all this necessary? Keep in mind that GLMM theory is relatively new, and useable software such as PROC GLIMMIX is even newer. In Chapters 1 though 3, we established why, in principle, we prefer GLMMs over pre-GLMM practices, but we also have to recognize that we are in a transitional period in which investigation of the small-sample behavior of GLMM methodology is a work in progress. This should not be seen as a flaw in the GLMM; it is statistical science at work.

Example 10.3 Response surface with binomial data

These data appear in the SAS Data and Program Library as Data Set 10.3. The study uses a Box–Behnken response-surface design with incomplete blocking, the same structure as Data Set 8.3 except here the data are binomial. There are three quantitative factors, called A, B, and C, each having levels –1, 0, and 1. Our objective is to estimate a parsimonious second-order polynomial regression model for these data. To do this, we use the same linear predictor that we used for Data Set 8.3 but modify the distribution and link functions. Specifically,

1. Linear predictor: $\eta_{ijkl} = \eta_{ijk} + r_k$, where r_k denotes the block effects and $\eta_{ijkl} = f(\mathrm{A,B,C}) + (\alpha\beta\gamma)_{ijk} = \beta_0 + \sum_{i=\mathrm{A,B,C}} \beta_i X_i + \sum_{i=\mathrm{A,B,C}} \beta_{ii} X_i^2 + \sum_{i \neq i'=\mathrm{A,B,C}} \beta_{ii'} X_i X_{i'} + (\alpha\beta\gamma)_{ijk}$
   $(\alpha\beta\gamma)_{ijk}$ is a catchall lack-of-fit term
2. Distributions:
   a. $y_{ijk} \mid r_i \sim \text{Binomial}(n_{ijkl}, \pi_{ijkl})$
   b. r_i i.i.d. $N\left(0, \sigma_R^2\right)$
3. Link function: $\text{logit}(\pi_{ijkl}) = \log[\pi_{ijkl}/(1 - \pi_{ijkl})]$

As with any response-surface analysis, fitting the model involves a three-step process:

1. Assess lack of fit from the intended regression model (here a second-order polynomial).
2. If warranted by step 1, fit the regression model and see if it can be reduced.
3. Contingent on step 2, fit the final model.

The following GLIMMIX statements allow us to implement step 1:

```
proc glimmix method=quad;
class block a b c;
model y/n=xa|xa|xb|xb|xc|xc@2 a*b*c/htype=1;
random intercept/subject=block;
```

In this case, the adaptive quadrature procedure selects 1 quadrature point. This means that you would get exactly the same results if you use METHOD=LAPLACE. Key results of interest here are the Pearson χ^2/df and the test of A*B*C. The former measures possible lack of fit of the assumed conditional distribution of the overall linear predictor. The latter specifically addresses possible lack of fit of the second-order polynomial regression relative to the mean model. Note that there are *two* lack-of-fit assessments: one for the entire model (Pearson χ^2) and one for the treatment model only (does the second-order polynomial adequately describe η_{ijk}). In Gaussian regression analysis, lack of fit only applies to the latter. For response surface designs like this one, we usually think of lack of fit in the Gaussian, mean model sense. The non-Gaussian model requires an adjustment in mindset. The relevant output is as follows:

Fit Statistics for Conditional Distribution	
Pearson chi-square/DF	0.40

Type I Tests of Fixed Effects

Effect	Num DF	Den DF	*F* Value	Pr > *F*
xa	1	12	11.49	0.0054
xa*xa	1	12	8.54	0.0128
xb	1	12	2.02	0.1806
xa*xb	1	12	0.05	0.8253
xb*xb	1	12	31.20	0.0001
xc	1	12	3.52	0.0851
xa*xc	1	12	12.13	0.0045
xb*xc	1	12	4.87	0.0476
xc*xc	1	12	0.04	0.8474
a*b*c	**3**	**12**	**0.54**	**0.6660**

The statistics in **bold** are the ones of interest in step 1. We see that we have no evidence of lack of fit. We proceed to step 2 using the GLIMMIX statements

```
proc glimmix method=quad;
class block;
model y/n=xa|xa|xb|xb|xc|xc@2/htype=1;
random intercept/subject=block;
```

The relevant output is as follows:

Type I Tests of Fixed Effects

Effect	Num DF	Den DF	*F* Value	Pr > *F*
xa	1	15	12.65	0.0029
xa*xa	1	15	8.78	0.0097
xb	1	15	2.41	0.1412
xa*xb	1	15	0.02	0.8966
xb*xb	1	15	31.70	<.0001
xc	1	15	3.39	0.0853
xa*xc	1	15	12.05	0.0034
xb*xc	1	15	4.66	0.0475
xc*xc	1	15	0.01	0.9070

We search the "Type I tests" for candidates to eliminate from the model. Two such candidates are XC*XC, the quadratic effect of factor C, and XA*XB, the linear A by linear B interaction. Their *p*-values are both roughly 0.90. XB and XC, the linear effects of B and C, are arguably nonsignificant, but they are both contained in higher-order effects that are statistically significant—XB*XB ($p < 0.0001$) and XB*XC ($p = 0.0475$). The reduced model for η_{ij} is thus $\beta_0 \sum_{i=A,B,C} \beta_i X_i + \sum_{i=A,B} \beta_{ii} X_i^2 + \beta_{AC} X_A X_C + \beta_{BC} X_B X_C$. We can now fit this using the GLIMMIX statements

```
proc glimmix method=quad;
class block;
model y/n=xa|xa xb|xb xa|xc xb*xc/solution;
random intercept/subject=block;
```

"Solutions for fixed effects" provide the relevant output:

Solutions for Fixed Effects

Effect	Estimate	Standard Error	DF	*t* Value	Pr > |*t*|
Intercept	0.7014	0.2301	4	3.05	0.0381
xa	0.3610	0.1011	16	3.57	0.0026
xa*xa	−0.8837	0.2307	16	−3.83	0.0015
xb	0.1551	0.09879	16	1.57	0.1360
xb*xb	−1.3075	0.2321	16	−5.63	<.0001
xc	0.1700	0.09225	16	1.84	0.0840
xa*xc	−0.4630	0.1334	16	−3.47	0.0032
xb*xc	0.2741	0.1269	16	2.16	0.0463

The values in ESTIMATE column are the estimated regression coefficients for the final model. We can use them to determine predicted values *on the model scale*, that is, predicted logits. If we want predicted probabilities, we have to apply the inverse link. At this point, we could proceed by plotting the estimated surface, as we did in Example 8.3, or determine an optimal combination of A, B, and C levels, or whatever else the objectives require.

10.2.3 Models for Contingency Tables

Contingency tables are standard tools for categorical data analysis. Log-linear models provide a connection between GLMs and contingency tables. These models are especially valuable when working with contingency tables that have three or more factors. In this section, we briefly introduce the logic of the log-linear model using a two-way contingency table and then work through an example with three factors.

A two-way contingency table has the following structure:

	B	
A	1	2
1	f_{11}	f_{12}
2	f_{21}	f_{22}

We classify every observation in the data set by its A characteristic (1 or 2) and its B characteristic (1 or 2). The A and B categories must be exhaustive and mutually exclusive—every member of the population must belong to either A1 or A2 and to either B1 or B2. No one can belong to A1 and A2 simultaneously. The number of observations classified as A1 × B1, we denote by f_{11}. In general, f_{ij} denotes the number of $A_i \times B_j$ observations.

To construct a log-linear model, we assume that observations "arrive" in $A_i \times B_j$ cells according to a Poisson process. Also, we denote the probability that a given member of the population belongs to the $A_i \times B_j$ cell by π_{ij}. If we take a total of N observations, the expected number of observations in the $A_i \times B_j$ cell is $N\pi_{ij}$. Thus, $f_{ij} \sim \text{Poisson}(N\pi_{ij})$. If the probability of belonging to A_i and B_j are independent, then $\pi_{ij} = \pi_i\pi_j$, where π_i denotes the marginal $P\{\text{obs} \in A_i\}$ and π_j denotes the marginal $P\{\text{obs} \in B_j\}$. Under independence, $f_{ij} \sim \text{Poisson}(N\pi_i\pi_j)$.

Using the canonical link, we fit a GLM to f_{ij}. The link function is $\log(N\pi_{ij})$, which, under independence, can be written as $\log(N\pi_i\pi_j) = \log(N) + \log(\pi_i) + \log(\pi_j)$. We rewrite the latter as $\eta_{ij} = \eta + \alpha_i + \beta_j$. We can add the term $(\alpha\beta)_{ij}$. This gives us a saturated model; we interpret $(\alpha\beta)_{ij}$ as the departure from independence. Also, in categorical data, we interpret $(\alpha\beta)_{ij}$ as *association* not interaction.

If the table has three or more factors, the log-linear model expands to include main effects, two-way association terms, three-way association terms, etc. Herein lies the appeal of log-linear models. Three- and higher-way contingency tables are clumsy to work with. Log-linear models allow much more streamlined methodology.

Example 10.4

Let us look at how this works for a three-way contingency table. Here are the raw data in contingency table form. These data also appear in the SAS Data and Program Library as Data Set 10.4.

		C	
A	B	1	2
1	1	94	171
	2	50	18
2	1	11	101
	2	12	18

The shaded portion shows the levels of factors A, B, and C, respectively. The observed frequencies, which we can denote by f_{ijk} for category $A_i \times B_j \times C_k$, are shown in the unshaded area. For example, $f_{111} = 94$.

We write the log-linear model for these data as

$$\log(\pi_{ijk}) = \eta + \alpha_i + \beta_j + \gamma_k + (\alpha\beta)_{ij} + (\alpha\gamma)_{ik} + (\beta\gamma)_{jk} + (\alpha\beta\gamma)_{ijk}$$

Use the following GLIMMIX statements to fit this model:

```
proc glimmix data=ct;
class a b c;
model freq=a|b|c/d=poisson ddfm=none chisq;
```

The options DDFM=NONE and CHISQ reflect the fact that we do not have replication as we would in most of our other examples. These options direct GLIMMIX to compute χ^2 test statistics. The output is as follows:

Type III Tests of Fixed Effects

Effect	Num DF	Den DF	Chi-Square	*F* Value	Pr > ChiSq	Pr > *F*
a	1	Infty	50.65	50.65	<.0001	<.0001
b	1	Infty	61.59	61.59	<.0001	<.0001
a*b	1	Infty	4.67	4.67	0.0307	0.0307
c	1	Infty	14.58	14.58	0.0001	0.0001
a*c	1	Infty	27.97	27.97	<.0001	<.0001
b*c	1	Infty	35.50	35.50	<.0001	<.0001
a*b*c	1	Infty	0.11	0.11	0.7392	0.7392

The denominator degrees of freedom *DenDF=Infty* result from the DDFM=NONE option. We ignore the results associated with *F* and focus on those for *ChiSq*. Mimicking the usual strategy for factorial treatment designs, we start with the three-way association, A*B*C. Three-way association occurs when a two-way association depends on the third factor. That is, the AB association assumes one form in conjunction with C_1 but assumes an altogether different character with C_2. For example, if A represents an environmental stress (1 = not present, 2 = present), B represents a medical condition (1 = not present, 2 = present), and C represents gender, a three-way association could occur if A and B were associated, say, for men but not women, or if A and B were associated for both genders, but the *strength* of association was gender dependent.

For these data, we have no evidence of a three-way association by any rational standard of evidence ($p > 0.70$), so we move on to the two-way associations. Here we have evidence of an AB association ($p = 0.0307$) and very strong evidence of AC and BC associations ($p < 0.0001$). If we continue with our hypothetical factor identities, this would tell us we have evidence that the incidence of the medical condition was associated with exposure to environmental stress, (AB association) and strongly associated with gender (BC association). In addition, we have evidence that exposure to environmental stress and gender are associated (AC association).

We can reexpress the model and use ESTIMATE statements to shed further light on these associations. For example, suppose we want to break down the association between exposure to environmental stress and incidence of the medical condition. We can use the following GLIMMIX statements:

```
proc glimmix data=ct;
class a b c;
model freq=c(a*b)/d=poisson chisq ddfm=none s;
estimate 'a1 odds @ c1' c(a*b) 1 0 -1 0 0 0 0 0/exp;
estimate 'a1 odds @ c2' c(a*b) 0 1 0 -1 0 0 0 0/exp;
estimate 'a2 odds @ c1' c(a*b) 0 0 0 0 1 0 -1 0/exp;
estimate 'a2 odds @ c2' c(a*b) 0 0 0 0 0 1 0 -1/exp;
estimate 'odds ratio @ c1' c(a*b) 1 0 -1 0 -1 0 1 0/exp cl;
estimate 'odds ratio @ c2' c(a*b) 0 1 0 -1 0 -1 0 1/exp cl;
estimate 'comb odds ratio' c(a*b) 1 1 -1 -1 -1 -1 1 1/exp cl;
```

The first four ESTIMATE statements calculate the odds of B1 relative to B2 for each level of A and C. For example, the first statement defines $\log(\pi_{111}/\pi_{121})$ on the model scale and π_{111}/π_{121} on the data scale. If B1 is "medical condition not present" and B2 denotes "medical condition present," then we have an estimate of the likelihood of not having the condition relative to having the condition given A1 and C1—we could suppose A1 denotes "not exposed to environmental stress" and C1 denotes females. We can compute the odds ratio for the AB association condition on each level of C—these are the ESTIMATE statements entitled "`odds ratio @ c1`" and "`odds ratio @ c2`." Finally, we can compute the combined odds ratio over both levels of C. The output is as follows:

Estimates

Label	Estimate	Standard Error	t Value	Pr > $\|t\|$	Exponentiated Estimate	Exponentiated Lower	Exponentiated Upper
a1 odds @ c1	0.6313	0.1750	3.61	0.0003	1.8800	1.3340	2.6494
a1 odds @ c2	2.2513	0.2478	9.09	<.0001	9.5000	5.8452	15.4401
a2 odds @ c1	−0.08701	0.4174	−0.21	0.8349	0.9167	0.4045	2.0774
a2 odds @ c2	1.7247	0.2558	6.74	<.0001	5.6111	3.3984	9.2646
odds ratio @ c1	0.7183	0.4526	1.59	0.1125	2.0509	0.8446	4.9800
odds ratio @ c2	0.5265	0.3562	1.48	0.1393	1.6931	0.8424	3.4029
comb odds ratio	1.2448	0.5760	2.16	0.0307	3.4723	1.1229	10.7371

The column labeled "estimate" is on the model scale—log (odds) and log(odds ratio)—and the column "Exponentiated estimate" is on the data scale. Because of the way we assigned coefficients in the ESTIMATE statements, we interpret the ODDS as the greater the estimate, the better, that is, the greater the odds of *not* having the medical condition. If we want to be able to read these as the odds of *having* the condition, we can reverse the signs on the coefficients in the ESTIMATE statement. Notice that the odds ratios individually do not have impressive p-values, but the combined odds ratio's p-value is more convincing. This appears to be more an artifact of sample size than anything else. Finally, notice that the p-value for the combined odds ratio is identical to the p-value for the A*B effect in the "type III tests of fixed effects": $p = 0.0307$.

As a final note, in categorical data analysis, multidimensional analyses are sometimes "collapsed"—that is, the frequencies are summed over a factor, yielding a simpler table. For example, here, if we sum over factor C, we get the collapsed table:

	B	
A	1	2
1	265	68
2	112	30

We can compute the log-linear model analysis for this table using the GLIMMIX statements

```
proc glimmix data=ct;
class a b;
model freq=a|b/d=poisson ddfm=none chisq;
```

We do this without any need to manipulate the data set. The result is as follows:

Type III Tests of Fixed Effects						
Effect	Num DF	Den DF	Chi-Square	*F* Value	Pr > ChiSq	Pr > *F*
a	1	Infty	46.44	46.44	<.0001	<.0001
b	1	Infty	118.03	118.03	<.0001	<.0001
a*b	1	Infty	0.03	0.03	0.8617	0.8617

Notice that the A*B association now shows no evidence of statistical significance ($p > 0.86$). This is an example of *Simpson's paradox* (Simpson, 1951). If we collapse a table over a factor that is associated with other factors in the study (factor C clearly is), we risk distorting the conclusions regarding the association of the two remaining factors. In this case, the distortion is severe. By collapsing the table over factor C, not only do we reach an erroneous conclusion regarding the AB association, but we completely miss the possibly very important associations that also exist between C and the other two factors.

Simpson's paradox is a variation on one of the cardinal rules in statistical science: *never throw away data.*

10.3 Alternative Link Functions for Binomial Data

Although we regard the logit link as the default link for binomial models, nothing says we *must* use it. The logit link's primary rationale stems from its being the canonical parameter in the binomial log-likelihood. In addition, it has a natural interpretation as the log odds in categorical data analysis. However, in certain applications with binomial data, other link functions are preferred, either because they facilitate more transparent interpretation or because, for certain binomial data sets, logit link models cannot fit the data accurately and, as a result, give spurious and misleading results.

In the section, we consider examples of two most common alternative link functions—the *probit* link and the *complimentary-log-log* link.

We encountered the probit link in Chapter 1 when we first began to develop the thought process leading to the GLM. The probit assumes an underlying, unobservable Gaussian process that is only visible to us as a "success" or "failure." The probit has advantages in disciplines that have well-developed theory based on the Gaussian distribution. Quantitative genetics is an important example—Gaussian variance component models have genetic meaning, and the probit link allows these meanings to remain intact when the response variable of interest is binomial.

The complementary-log-log, defined as $\eta = \log(-\log(1-\pi))$, is useful for data where most of the probabilities lie close to zero or close to one. This is one of binomial data's important "boundary conditions." Whereas the probit and logit are symmetric over the parameter space, the complementary-log-log does not require symmetry and is thus, in theory, better able to fit binomial models for events that either occur very infrequently or occur almost—but not quite—all the time.

Example 10.5 Probit link example

This example revisits the nested-factorial Data Set 10.1. In the previous section, we fit the model using the logit link. Let us now fit it with a probit link and compare and contrast the results. The model is identical to the model we used in Example 10.1 except for the link function. We replace $\eta_{ijk} = \text{logit}(\pi_{ijk})$ by $\eta_{ijk} = \Phi^{-1}(\pi_{ijk})$, where $\Phi^{-1}(\bullet)$ denotes the inverse Gaussian, or normal, cumulative distribution function. It is more common to specify a probit model by the inverse link. For this example, we write it as $\pi_{ijk} = \Phi(\eta + \alpha_i + \beta(\alpha)_{ij} + r_k)$, where the terms of the linear predictor are all defined as previously in Example 10.1. Use the following GLIMMIX statements to implement the analysis:

```
proc glimmix data=ch10_exB1 method=quad;
class block a b;
model f/n=a b(a)/link=probit;
random a/subject=block;
lsmeans b(a)/slice=a slicediff=a ilink;
```

The only change in the GLIMMIX program is the addition of the LINK=PROBIT option in the MODEL statement. The relevant results are as follows:

Covariance Parameter Estimates

Cov Parm	Subject	Estimate	Standard Error
a	block	0.1397	0.07062

Tests of Effect Slices for b(a) Sliced by a

a	Num DF	Den DF	*F* Value	Pr > *F*
0	2	16	0.46	0.6382
1	2	16	4.51	0.0280

b(a) Least Squares Means

a	b	Estimate	Standard Error	Mean	Standard Error Mean
0	0	−0.1879	0.1912	0.4255	0.07495
0	1	−0.06462	0.1948	0.4742	0.07753
0	2	−0.09283	0.1952	0.4630	0.07754
1	0	−0.5246	0.1906	0.2999	0.06625
1	1	−0.2705	0.1896	0.3934	0.07292
1	2	−0.1406	0.1910	0.4441	0.07546

Simple Effect Comparisons of b(a) Least Squares Means by a

Simple Effect Level	b	_b	Estimate	Standard Error	DF	*t* Value	Pr > |*t*|
a0	0	1	−0.1233	0.1357	16	−0.91	0.3772
a0	0	2	−0.09508	0.1367	16	−0.70	0.4966
a0	1	2	0.02822	0.1418	16	0.20	0.8448
a1	0	1	−0.2541	0.1281	16	−1.98	0.0647
a1	0	2	−0.3840	0.1302	16	−2.95	0.0094
a1	1	2	−0.1299	0.1284	16	−1.01	0.3268

These results differ only slightly from the analysis using the logit link. The biggest difference is the variance component estimate: $\hat{\sigma}_R^2 = 0.1397$ using the probit link; $\hat{\sigma}_R^2 = 0.377$ using the logit. In genetics, where the variance has an interpretation in terms of the heritability of a trait, this distinction matters. In other cases, where the variance estimate is just a stepping stone to ensure accurate standard errors and test statistics, the difference simply accounts for the fact that the link functions are not the same. The slice *F* values are 0.46 and 4.51 for the probit, 0.43 and 4.46 with the logit link—a negligible difference.

Standard errors for the data scale means—that is, the $\hat{\pi}_{ij}$ obtained using the inverse link—are determined using the delta rule. Denoting the LSMEAN estimates on the model scale—those that appear in the "estimate" column—by $\hat{\eta}_{ij}$, we obtain the data scale estimates from $\hat{\pi}_{ij} = \Phi(\hat{\eta}_{ij}) = \int_{-\infty}^{\hat{\eta}_{ij}} 1/\sqrt{2\pi}\, e^{-z^2/2} dz$. Using the delta rule,

$$s.e.\,(\hat{\pi}_{ij}) = \left.\frac{\partial\Phi(\eta_{ij})}{\partial\eta_{ij}}\right|_{\hat{\eta}_{ij}} s.e.\,(\hat{\eta}_{ij}) = \left(\frac{1}{\sqrt{2\pi}} e^{-\frac{\hat{\eta}_{ij}^2}{2}}\right) \times s.e.\,(\hat{\eta}_{ij}).$$

What we see here is typical. The probit and logit links describe similar functional relationships between η and π. As a result, they tend to produce similar results. From a purely statistical standpoint, there is no reason to prefer one over the other; we can generally expect them to produce essentially interchangeable conclusions regarding the treatment effects. The main reason to choose one over the other has more to do with the subject matter discipline that gives rise to the data and its conventions and intellectual history.

Example 10.6 Complementary log-log example

These data use the same incomplete strip-plot structure given in Example 8.2. However, the data for this example are binomial. They are given in SAS Data and Program Library, Data Set 10.5. The model uses the same linear predictor as Example 8.2. Only the distribution and link function change. The model is as follows:

1. Linear predictor: $\eta_{ijk} = \eta + \alpha_i + \beta_j + (\alpha\beta)_{ij} + r_k + (ar)_{ik} + (br)_{jk}$
2. Distributions:
   a. $y_{ijk} \mid r_k, (ar)_{ik}, (br)_{jk} \sim \text{Binomial}(n_{ijk}, \pi_{ijk})$
   b. r_k i.i.d. $N(0, \sigma_R^2)$
   c. $(ar)_{ik}$ i.i.d. $N(0, \sigma_{AR}^2)$
   d. $(br)_{jk}$ i.i.d. $N(0, \sigma_{BR}^2)$
3. Link: ? we will compare $\eta_{ijk} = \text{logit}(\pi_{ijk})$ and $\eta_{ijk} = \log(-\log(1 - \pi_{ijk}))$

The latter link is the complementary log-log. The inverse link is $\pi_{ijk} = 1 - \exp[\exp(\eta_{ijk})]$. Readers familiar with nonlinear models will recognize this as a Gompertz model with a fixed intercept at 0 and a fixed asymptote at 1. The reason we may need to consider this link function is that the sample proportions in Data Set 10.5 are all large, most of them greater than 0.9 and quite a few of them equal to 1.

We first fit the logit link to see how we do. The GLIMMIX statements are

```
proc glimmix data=ds8d2 method=laplace;
class block a b;
model y/n=a|b; /* note: default binomial link is the logit */
random intercept a b/subject=block;
output out=check_fit_l pred(ilink)=p_hat_logit;
```

The OUTPUT statement creates a new data set containing the predicted $\hat{\pi}_{ijk}$ allowing us to compare them to the observed sample proportions, $p_{ijk} = y_{ijk}/n_{ijk}$. Here are the selected results:

Fit Statistics for Conditional Distribution	
−2 Log L(y | r effects)	98.85
Pearson chi-square	19.04
Pearson chi-square/DF	0.53

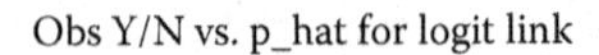

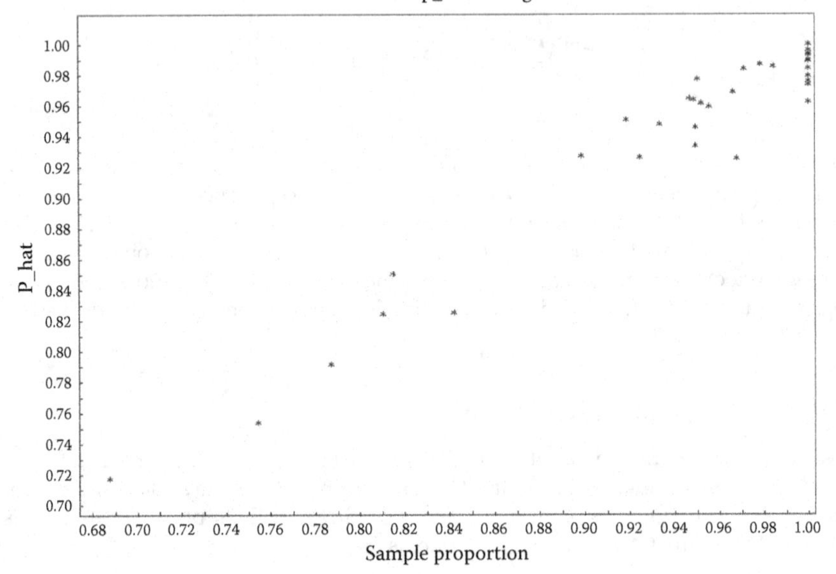

The correlation between the p_{ijk} and $\hat{\pi}_{ijk}$ is 0.97. The Pearson χ^2/df does not indicate any problems. From these indications, the logit-based model seems to fit pretty well.

How does the complementary log-log link compare? The GLIMMIX statements are identical to those we use for the logit, except for the MODEL statement:

```
proc glimmix data=ds8d2 method=laplace;
class block a b;
model y/n=a|b/ link=ccll;
random intercept a b/subject=block;
output out=check_fit_c pred(ilink)=p_hat_ccll;
```

LINK=CCLL overrides the default link and substitutes the complementary log-log. The results are as follows:

Fit Statistics for Conditional Distribution	
−2 Log L(y | r effects)	90.67
Pearson chi-square	13.24
Pearson chi-square/DF	0.37

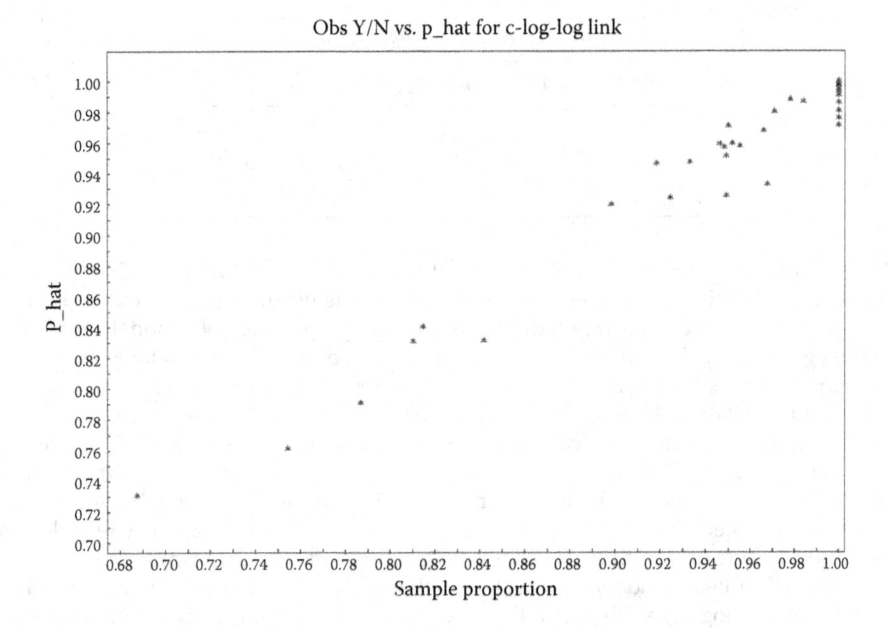

The correlation between the p_{ijk} and $\hat{\pi}_{ijk}$ is 0.98. The plot of p_{ijk} vs. $\hat{\pi}_{ijk}$ looks similar to the plot for the logit predicted values.

Let us now look at the variance component estimates and the tests of the A and B effects.

For the logit model,

Covariance Parameter Estimates			
Cov Parm	**Subject**	**Estimate**	**Standard Error**
Intercept	block	0.03972	0.5650
a	block	1.55E-18	.
b	block	0.5504	.

Type III Tests of Fixed Effects				
Effect	**Num DF**	**Den DF**	***F* Value**	**Pr > *F***
a	2	7	40.33	0.0001
b	2	7	148.42	<.0001
a*b	2	5	61.23	0.0003

For the complementary log-log model,

Covariance Parameter Estimates			
Cov Parm	**Subject**	**Estimate**	**Standard Error**
Intercept	block	0.006363	0.05019
a	block	0.008462	0.02296
b	block	0.07005	0.05069

Type III Tests of Fixed Effects				
Effect	Num DF	Den DF	F Value	Pr > F
a	2	7	1.92	0.2161
b	2	7	4.83	0.0481
a*b	4	5	0.83	0.5583

We immediately see a problem with the logit output. Although we know that the standard errors of variance component estimates have little useful meaning, they do speak loudly when they are not computed. This is usually symptomatic of a model estimation process gone badly. The "type III tests of fixed effects output" does not look any better. The numerator degrees of freedom for A*B are incorrect, and all of the *F*-values appear to be wildly inflated. Although the logit model has determined the predicted values accurately, it has not done so for other needed information. In the SASLOG, we find a warning "`NOTE: At least one element of the gradient is greater than 1e-3.`" While this is not *always* a problem, *it often is* an indication that, while the procedure has converged, the solution to which it has converged is not necessarily reliable and the resulting output should be taken with a grain of salt. The problem here is that the logit link is ill suited to model data so close to the boundary of π_{ijk}'s parameter space. The complementary log-log analysis, on the other hand, is less prone to such pathologies.

Using the complementary log-log, we can now proceed with the analysis. For these data, our focus would likely be on the main effects of B, as B appears to be the only effect approaching statistical significance.

10.3.1 Role of "Residual" in Binomial Models

Early in Chapter 1, we established that "residual" as it has been commonly understood in modeling—as "**e**" in $\mathbf{y} = \mathbf{X}\boldsymbol{\beta} + \mathbf{e}$—has no place in generalized linear models. Not GLMs nor GLMMs. As we introduced the skeleton analysis of variance (ANOVA) processes in Chapter 2 and applied it to increasingly more complex designs in Chapter 8, we began to realize that the ANOVA source of variation we habitually call "error" or "residual" has a specific meaning for models with Gaussian data. It enables estimation of σ^2, the variance of $\mathbf{y}|\mathbf{b}$.

For non-Gaussian models—in particular those with a distribution of $\mathbf{y}|\mathbf{b}$ belonging to the one-parameter exponential family—there is no scale parameter to estimate. As a result, the final source of variation must either play a different role or play no role at all. We saw in Chapter 3 a hint of what this role might be. In certain data sets, we see evidence that the mean parameter—π in the binomial case—is perturbed on the unit of observation level, over and above variation accounted for by the "usual" linear predictor. By "usual" linear predictor, we mean the linear predictor that accounts for all sources of variation up to but not including the unit-of-observation term—"error" or "residual" in conventional ANOVA.

The usual linear predictor requires that we assume no unit-to-unit perturbation occurs. If we think about it, we should realize that we do not make this assumption with Gaussian data. Unit-to-unit perturbation is precisely what σ^2 models. If unit-to-unit variation occurs with Gaussian data, it is a good bet that it occurs with one-parameter exponential family data as well. When it does, and our model does not account for it, we often see evidence of it in the Pearson χ^2/df fit statistic.

This is one aspect of *overdispersion*, an important issue in statistical modeling and one that is unique to non-Gaussian models, particularly those with one-parameter exponential family distributions such as the binomial and Poisson. Overdispersion occurs when the observed variance in the data exceeds what it should be under the assumed distribution. An important cause of overdispersion is an under-specified model. In this case, we under-specify by leaving out the last term in the skeleton ANOVA when it is in fact affecting $E(\mathbf{y} \mid \mathbf{b})$. We "fix" the problem not by including an **e** term in the linear predictor—that is not allowed—but by including a term corresponding to the named source of variation on the last line of the skeleton ANOVA.

Let us look at an example to see how this works.

Example 10.7 Binomial with evidence of under-specified linear predictor

We use the same design format as Data Set 10.3—a Box–Behnken design run in incomplete blocks. These data appear in the SAS Data and Program Library as Data Set 10.6. We start with the same linear predictor and the same GLIMMIX statements used in Example 10.3. This time, the results are:

Fit Statistics for Conditional Distribution	
−2 Log L(y | r effects)	186.84
Pearson chi-square	99.49
Pearson chi-square/DF	3.55

Type III Tests of Fixed Effects				
Effect	**Num DF**	**Den DF**	***F* Value**	**Pr > *F***
—	—		—	—
a*b*c	3	12	9.96	0.0014

Notice two problems. First, the Pearson $\chi^2/df = 3.55 \gg 1$. This indicates that we have misspecified either the distribution of $\mathbf{y}|\mathbf{b}$ or the linear predictor. In this case, we are seeing evidence of linear predictor underspecification. The second problem appears to be mean model lack of fit, signaled by the A*B*C statistic's p-value of 0.0014. The mean model lack of fit is actually an artifact of underspecification of the model's random component; type I error rates tend to be wildly inflated with underspecified **Zb**.

We address the problem by returning to the skeleton ANOVA we derived for this design in Example 8.3. The final term in the ANOVA was "Unit(block)|A,B,C," the effect of unit within block adjusted for the factor A, B, and C effects. We make this understandable to GLIMMIX using the following statements:

```
proc glimmix method=quad;
class block a b c;
model y/n=xa|xa|xb|xb|xc|xc@2 a*b*c/htype=1;
random intercept a*b*c/subject=block;
covtest 'overD?'. 0;
```

Two changes appear in this program. One occurs in the RANDOM statement: we add A*B*C. In conjunction with SUBJECT=BLOCK, GLIMMIX reads this as the block × treatment source of variation. We know this to be mathematically identical to "error" in conventional ANOVA tables for blocked designs. The second change: we add a COVTEST statement to formally test the hypothesis that the variance component we have added is equal to zero.

If we run this with METHOD=QUAD, we get the following error message in the SASLOG: "ERROR: Insufficient resources to perform adaptive quadrature with 3 quadrature points. METHOD=LAPLACE, corresponding to a single point, may provide a computationally less intensive possibility." We have just encountered our first example of a variance structure beyond quadrature's capacity. We use METHOD=LAPLACE instead and obtain the following:

Fit Statistics for Conditional Distribution	
−2 Log L(y | r effects)	141.52
Pearson chi-square	49.92
Pearson chi-square/DF	1.78

Covariance Parameter Estimates			
Cov Parm	**Subject**	**Estimate**	**Standard Error**
Intercept	block	0.2211	0.3608
a*b*c	block	0.6320	0.4105

Type I Tests of Fixed Effects				
Effect	**Num DF**	**Den DF**	***F* Value**	**Pr > *F***
xa	1	3	6.71	0.0811
xa*xa	1	3	3.53	0.1568
xb	1	3	0.00	0.9606
xa*xb	1	3	0.03	0.8746
xb*xb	1	3	8.44	0.0622
xc	1	3	1.22	0.3504
xa*xc	1	3	0.24	0.6577
xb*xc	1	3	0.04	0.8462
xc*xc	1	3	0.42	0.5619
a*b*c	3	9	1.95	0.1918

Tests of Covariance Parameters					
Based on the Likelihood					
Label	**DF**	**−2 Log Like**	**ChiSq**	**Pr > ChiSq**	**Note**
overD?	1	210.73	13.20	0.0001	MI

We see that the Pearson $\chi^2/df = 1.78$. This is still greater than 1 but much better than 3.55. The COVTEST $\chi^2 = 13.2$, $p < 0.0001$, allowing us to reject $H_0{:}\sigma^2_{RT} = 0$, where σ^2_{RT} denotes the block × treatment variance. Together these tell us that adding the term accounting for unit-to-unit perturbation was a good thing. Having done this, we also see that the spurious indication of mean model lack of fit is now gone: the *p*-value for A*B*C is 0.1918.

The one disturbing aspect of this output is that adding the BLOCK*A*B*C variance component has confused GLIMMIX's default denominator degrees of freedom algorithm. According to the skeleton ANOVA we derived in Example 8.3, we should have 12 denominator degrees of freedom for the tests of the second-order polynomial regression effects. We need to use the DDF option in the GLIMMIX MODEL statement to override the default. Our choice is whether to use 12 degrees of freedom or to accept some penalty. Recall that the skeleton ANOVA degrees of freedom can be "naive" degrees of freedom when we use estimated variance components. If we do accept a penalty, what should it be? If these were Gaussian data, we would use the Satterthwaite approximate degrees of freedom as a matter of course. The LAPLACE method does not allow the Satterthwaite option, but we could obtain the GLMM *ad hoc* Satterthwaite analog using PL. We do not want to use the rest of the PL output, but running the aforementioned program without METHOD=LAPLACE yields

Type I Tests of Fixed Effects

Effect	Num DF	Den DF	*F* Value	Pr > *F*
xa	1	7.309	2.09	0.1894
xa*xa	1	3.558	1.42	0.3062
xb	1	7.433	0.04	0.8404
xa*xb	1	9.625	0.00	0.9726
xb*xb	1	4.188	3.32	0.1395
xc	1	5.451	0.72	0.4332
xa*xc	1	5.405	0.18	0.6903
xb*xc	1	5.482	0.08	0.7832
xc*xc	1	4.893	0.13	0.7335
a*b*c	3	6.521	0.49	0.7036

The Satterthwaite degrees of freedom vary from 3.6 to 9.6. Using 3.6 is probably exacting too harsh a penalty; 9.6 is probably too optimistic. The mean is 5.5; the median is 5.4. If we decide to use the mean d.f., the GLIMMIX statements are

```
proc glimmix method=laplace;
class block a b c;
 model y/n=xa|xa|xb|xb|xc|xc@2 a*b*c/htype=1 ddf=5.5,5.5,5.5,5.5,5.5,
 5.5,5.5,5.5,5.5;
random intercept a*b*c/subject=block;
```

The output is as follows:

Type I Tests of Fixed Effects

Effect	Num DF	Den DF	*F* Value	Pr > *F*
xa	1	5.5	6.71	0.0447
xa*xa	1	5.5	3.53	0.1137
xb	1	5.5	0.00	0.9591
xa*xb	1	5.5	0.03	0.8698
xb*xb	1	5.5	8.44	0.0300
xc	1	5.5	1.22	0.3158
xa*xc	1	5.5	0.24	0.6431
xb*xc	1	5.5	0.04	0.8403
xc*xc	1	5.5	0.42	0.5417
a*b*c	3	9	1.95	0.1918

In fairness, if this way of determining denominator degrees of freedom strikes you as too *ad hoc*, you will get no argument from the author. For these data, whether you use 5.5 or 12 or something in between, the results remain the same: The regression depends on quadratic B and linear A main effects only—no interaction and no effect of C in any form. For other data sets we will not be so lucky. This remains an open area of research in GLMM methodology.

The important point to take from this section concerns the role of the last term in the skeleton ANOVA. The convention in generalized linear modeling has been to use the same linear predictor for a non-Gaussian model that we use for its Gaussian analog. For two-parameter exponential families, including the beta distribution we encounter later, this is probably appropriate. However, it is not always the right thing to do for one-parameter exponential families, including the binomial.

One common strategy not shown here, but often used, is to add an additional scale parameter. In GLIMMIX, you do this by adding a `random _residual_` statement. We will look at this approach in more detail in Chapter 11—it is a common (and not always effective) "fix" for overdispersion in count data. It has two problems. First, adding a scale parameter creates a quasi-likelihood. We can no longer use integral approximation, and quasi-likelihood ties inference to a marginal model that may not necessarily focus inference on our intended target. Second, adding a scale parameter is a form of design-model disconnect. The point of working through the WWFD skeleton ANOVA routine is to identify sources of variation implicit in the process of conducting the study design. As a first approximation, the model should attempt to respect the design from which the data came.

10.4 Continuous Proportions

While the binomial distribution is generally applicable to discrete proportions, it cannot be applied to continuous proportion data. The binomial requires that we be able to specify the response variable in terms of Y and N. When the response is "proportion of leaf area affected" or "proportion of land area flooded," we have Y bounded between 0 and 1 but no obvious N.

The distribution of choice for continuous proportions is the beta distribution. The beta random variable is bounded between 0 and 1, and the shape of the p.d.f. is extremely flexible. In Section 10.4.1, we give a very brief presentation of parameterization of the beta distribution used for GLMs, which differs from the way we usually see it in mathematical statistics texts. In Section 10.4.2, we work through a model with continuous proportion data using the beta distribution.

10.4.1 Beta Distribution

In a standard mathematical statistics presentation, the beta p.d.f. is as follows:

$$f(y;\alpha,\beta)=\frac{1}{B(\alpha,\beta)}y^{\alpha-1}(1-y)^{\beta-1}$$

where $0 < y < 1$; $\alpha, \beta > 0$; $B(\alpha, \beta)=\Gamma(\alpha)\Gamma(\beta)/\Gamma(\alpha + \beta)$. The mean and variance of the beta distribution are $E(y)=\alpha/(\alpha+\beta)$ and $Var(y)=\alpha\beta/(\alpha+\beta+1)(\alpha+\beta)^2$.

Ferrari and Cribari-Neto (2004) give an alternative parameterization more amenable to generalized linear modeling. They parameterize the beta distribution in terms of its mean, μ, and a scale parameter, ϕ. Relating their parameterization to the standard form, $\mu=\alpha/(\alpha+\beta)$, $\phi=\alpha+\beta$ and $Var(y)=[\mu(1-\mu)/(1+\phi)]$. Also, the canonical parameter is $\theta=\log[(\mu/(1-\mu))]$, that is, the logit link.

10.4.2 Continuous Proportion Example Using the Beta Distribution

The data for this example appear in the SAS Data and Program Library as Data Set 10.7. Two treatments, called "trt 0" and "trt 1," are randomly assigned to 12 runs each. Each run can be divided into six subunits. Each subunit is randomly assigned a dose level. The dose levels are 0, 1, 2, 3, 4, and 5. The doses are not doses of the treatments, so trt_0 × dose_0 is not the same treatment as trt_1 × dose_0. Describing this study in skeleton ANOVA form, we have

Topo		Treatment		Combined	
Source	**d.f**	**Source**	**d.f**	**Source**	**d.f**
		trt	1	trt	1
run	23			run(trt)	23 – 1 = 22
		dose	5	dose	5
		trt × dose	5	trt × dose	5
subunit(run)	120	"parallels"	132	trt × run(dose) residual	120 – 10 = 110
Total	143	Total	143	Total	143

Suppose we also have reason to believe that we have a linear response over dose levels. The resulting model is

1. Linear predictor: $\eta_{ijk} = \eta + \tau_i + \beta_{1i}D_j + (\tau\delta)_{ij} + r(\tau)_{ik}$, where τ denote treatment effects, β_{1i} denotes the linear regression coefficient for the ith treatment, D_j denotes the jth dose level, $(\tau\delta)_{ij}$ denotes the dose × treatment effects over and above linear, interpreted as lack of fit, and $r(\tau)$ denotes the run effects
2. Distributions:
   a. $y_{ijk}|r(\tau)_{ik} \sim \text{Beta}(\mu_{ijk}, \phi)$
   b. $r(\tau)_{ik}$ i.i.d. $N\left(0, \sigma_R^2\right)$
3. Link: $\eta_{ijk} = \log[\mu_{ijk}/(1 - \mu_{ijk})]$

Use the following GLIMMIX statements to begin the analysis. Our initial focus is on lack of fit, both the Pearson χ^2/df, the overall distribution and linear predictor diagnostic, and $(\tau\delta)_{ij}$, the departure from linear in the mean model.

```
proc glimmix data=rates method=quad;
class run trt dose;
model proportion=trt d(trt) trt*dose/d=beta htype=1;
random intercept/subject=run(trt);
```

The variable D in the model statement is the dose level in direct variable form. D(TRT), thus, models separate linear regressions over dose levels for each treatment. TRT*DOSE is defined on the CLASS variable DOSE—it represents lack of fit. We show METHOD=QUAD here. The adaptive procedure selected 5 quadrature points. You can verify that the analysis using METHOD=LAPLACE yields virtually identical results.

In general, PL appears to perform less well for GLMMs with beta response variables. Relevant results at this point are

Fit Statistics for Conditional Distribution	
−2 Log L(proportion | r effects)	−416.45
Pearson chi-square	125.08
Pearson chi-square/DF	0.87

Type I Tests of Fixed Effects

Effect	Num DF	Den DF	*F* Value	Pr > *F*
trt	1	22	16.79	0.0005
d(trt)	2	110	49.90	<.0001
trt*dose	8	110	1.49	0.1705

We see that the Pearson χ^2/df shows no evidence of overall distribution or linear predictor lack of fit and TRT*DOSE shows no evidence of lack of fit from linear regression. We proceed to step 2:

```
proc glimmix data=rates;
class run trt dose;
model proportion=trt d(trt) /noint s htype=1 d=beta;
random intercept/subject=run(trt);
contrast 'equal intercepts' trt 1 -1;
contrast 'equal slopes' d(trt) 1 -1;
```

The MODEL statement defines the fixed component of the linear predictor as $\eta_{ij} = \beta_{0i} + \beta_{1i}D_j$, where β_{0i} corresponds to TRT in the MODEL statement. This is the standard qualitative-quantitative unequal slopes regression model described in Chapter 7. Our primary focus at this point is on the two CONTRAST statements. The first tests H_0: $\beta_{0,0} = \beta_{0,1}$, the equality of the intercepts. The second tests whether the regression lines over dose for each treatment are parallel—formally, H_0: $\beta_{1,0} = \beta_{1,1}$. The results are as follows:

Solutions for Fixed Effects

Effect	trt	Estimate	Standard Error	DF	*t* Value	Pr > |*t*|
trt	0	0.6965	0.1841	22	3.78	0.0010
trt	1	0.8054	0.1896	22	4.25	0.0003
d(trt)	0	0.2846	0.05654	118	5.03	<.0001
d(trt)	1	0.5541	0.06604	118	8.39	<.0001

Contrasts

Label	Num DF	Den DF	*F* Value	Pr > *F*
Equal intercepts	1	22	0.17	0.6837
Equal slopes	1	118	10.07	0.0019

The "solutions for fixed effects" give the estimated regression coefficients. We can use these regression equations to obtain predicted responses *on the model scale* for a given treatment and dose level. Keep in mind that we need to apply to inverse link to the value we calculate to convert it to the data scale, that is, to a proportion between 0 and 1.

We see that the estimated intercepts are very similar, but the slopes are not. This is confirmed by the test of equal intercept and slope coefficients. Based on these results, we would expect the two treatments to respond similarly at dose=0 but diverge as the dose level increases.

One output listing we specifically *do not* give under relevant output is the "type I test of fixed effects." There is a case to be made for calling this "the SAS output linear models students cannot resist misinterpreting." Here is the output:

Type I Tests of Fixed Effects

Effect	Num DF	Den DF	F Value	Pr > F
trt	2	22	139.43	<.0001
d(trt)	2	118	46.00	<.0001

What this table most assuredly does *not* tell us is anything about the effect of treatment. Recall that in our linear predictor for this run, TRT denotes the intercept coefficients, β_{0i}. The F-test for TRT tests H_0: $\beta_{0,0} = 0$ and $\beta_{0,1} = 0$. It does *not* test treatment effect; it simply tests whether the intercepts are nonzero, that is, on the model scale, is the expected response nonzero at dose=0? Similarly, D(TRT) is a two degree of freedom term that jointly tests H_0: $\beta_{1,0} = 0$ and $\beta_{1,1} = 0$.

> *So students*, do yourself a favor and make your linear models instructor proud of you. Keep track of what the model parameters are and resist the urge to write "F = 139.43 and $p < 0.0001$, therefore we have a significant treatment effect." Your grade will be better, and your instructor will not have to pull out so much hair.

We could proceed by obtaining predicted values and treatment differences at specified dose levels. For example, we can add the following statements to our GLIMMIX program to obtain predicted values and treatment differences at each DOSE:

```
lsmeans trt/diff at dose=0 ilink oddsratio;
lsmeans trt/diff at dose=1 ilink oddsratio;
lsmeans trt/diff at dose=2 ilink oddsratio;
lsmeans trt/diff at dose=3 ilink oddsratio;
lsmeans trt/diff at dose=4 ilink oddsratio;
lsmeans trt/diff at dose=5/ilink oddsratio;
```

The output is shown here for doses 0 and 5 only:

trt Least Squares Means

trt	dose	Estimate	Standard Error	DF	t Value	Pr> |t|	Mean	Standard Error Mean
0	0.00	0.6961	0.1835	22	3.79	0.0010	0.6673	0.04075
1	0.00	0.8056	0.1891	22	4.26	0.0003	0.6912	0.04036

Differences of trt Least Squares Means

trt	_trt	dose	Estimate	Standard Error	DF	*t* Value	Pr> |*t*|	Odds Ratio
0	1	0.00	−0.1095	0.2629	22	−0.42	0.6812	0.896

trt Least Squares Means

trt	dose	Estimate	Standard Error	DF	*t* Value	Pr> |*t*|	Mean	Standard Error Mean
0	5.00	2.1188	0.2088	22	10.15	<.0001	0.8927	0.02000
1	5.00	3.5738	0.2563	22	13.94	<.0001	0.9727	0.006803

Differences of trt Least Squares Means

trt	_trt	dose	Estimate	Standard Error	DF	*t* Value	Pr> |*t*|	Odds Ratio
0	1	5.00	−1.4550	0.3102	22	−4.69	0.0001	0.233

We see that at DOSE=0, the expected proportions, given on the data scale under MEAN, are quite close together—0.67 and 0.69 for treatment 0 and 1, respectively. These values arise by computing the ESTIMATE on the model scale from the estimated regression equation and then applying the inverse link. For example, for treatment 0, the intercept is 0.696. Since D = 0, this is the estimate on the model scale. Applying the inverse link, we get $1/(1 + e^{-0.696}) = 0.667$. At D = 5, the predicted value on the model scale for treatment 0 is $0.696 + 0.2845 \times 5 = 2.119$. The estimated proportion on the data scale is thus $1/(1 + e^{-2.119}) = 0.893$. Finally, we see that for treatment 0, the expected proportion does not increase as D increases as much it does under treatment 1. At D = 0, the odds ratio is close to 0.9. By dose 5 it has fallen to 0.23.

10.5 Summary

1. There are two types of proportion data: discrete and continuous.
2. Discrete proportion data tends to imply the binomial distribution.
3. Discrete proportions can apply to the full range of models you would fit with Gaussian data—split-plot, response surface, etc.
4. Discrete proportion models can also be used for contingency table analysis via log-linear models—valuable tool for categorical data analysis.
5. Continuous proportion data tends to imply the beta distribution.
6. Logit and probit links are standard for Binomial GLM and GLMM:
   a. Statistically there is little to choose
   b. Choice often driven by subject-matter concerns

7. Logit links are the norm for beta GLM and GLMM.
8. Complementary log-log is useful if the proportion is close to 0 or 1.
9. Integral approximation is generally the method of choice:
   a. Not a big deal with binomial data and moderate cluster size: PL, Laplace, and quadrature results interchangeable
   b. Small cluster size: use Laplace or quad
   c. Cluster size = 1: binary data; problematic, avoid if possible, but useful analyses possible with quad or Laplace if one approaches modeling carefully
   d. Beta: experience to date suggests use Laplace or quad when possible
10. Last term in the skeleton ANOVA:
   a. For one-parameter distribution (read binomial), not accounting for it may manifest itself as overdispersion.
   b. For two-parameter distribution (read beta), leaving it out of linear predictor (as you would with Gaussian models) appears to be the right thing to do.

Exercises

10.1 This problem uses data from SAS Data and Program Library/Exercises **chapter10_prob1.sas**. These are binary data from a study conducted as follows. Twenty-five locations were selected at random. At each location, there were four matched pairs selected at random. Each pair was assigned one level of factor A (factor A had four levels). There were two levels of factor B. Each member of the pair was randomly assigned to receive one of the levels of B. Thus, the treatment design was a 4×12 factorial and the topographical design was a split-plot: The whole plot experimental unit was the pair; the split-plot experimental unit was the member of the pair; and whole-plot units were blocked by location. The response was binary: Each member either showed the response of interest (denoted $Y = 1$ in the data set) or not (denoted $Y = 0$).

a. Write a skeleton ANOVA for this study design (e.g., following the WWFD process).
b. Write a generalized linear mixed model following from (a).
c. Implement the analysis of these data, for example, using the GLIMMIX procedure. Summarize relevant results.

10.2 This problem uses data from SAS Data and Program Library/Exercises **chapter10_prob2.sas**. The design is similar to Problem 1, except that at each location, there are 80 matched pairs. Twenty are assigned at random to each level of Factor A and then levels of B are randomly assigned to members of each pair as before.

a. Write a skeleton ANOVA for this study design (e.g., following the WWFD process).
b. Write a generalized linear mixed model following from (a) assuming a binary distribution.
c. Implement the analysis of these data consistent with your model in (b), for example, using the GLIMMIX procedure. Summarize relevant results.

d. Alternatively, you can write a generalized linear mixed model following from (a) assuming a binomial distribution with $N=10$ (the number of matched pairs per A×B combination). Write the model in this form.
e. Implement the analysis of these data consistent with your model in (d), for example, using the GLIMMIX procedure. Summarize relevant results.
f. Compare and contrast your results in (c) and (e).

10.3 Consider a completely randomized design with three treatments. Suppose the probabilities of a favorable outcome are $\pi_1=0.60$, $\pi_2=0.75$, and $\pi_3=0.90$, respectively, for the three treatments.

Write a program to simulate data generated by these three treatments under the following assumptions:

a. There are 100 units of replication per treatment and one observation per unit of replication (hence the observations are binary).
b. There are 10 units of replication per treatment and 10 observations per unit of replication (hence the observations are binary).
c. Like (a) and (b) except at the unit of replication level, there is a random unit effect that is i.i.d. $N(0,\sigma_U^2)$, where $\sigma_U^2=0.5$. This means that for part (a), $\eta_{ij}=\text{logit}(\pi_i)+u_{ij}$ where $u_{ij}\sim NI\,(0, 0.5)$ for each of the 100 observations per treatment, and the actual observation results from a binary $\left[1/\left(1+e^{-\eta_{ij}}\right)\right]$ random variable. For part (b), $\eta_{ij}=\text{logit}\,(\pi_i)+u_{ij}$ where $u_{ij}\sim NI\,(0, 0.5)$ for each of the 10 units of replication per treatment and the actual observations result from a Binomial $\left(10,\left[1/\left(1+e^{-\eta_{ij}}\right)\right]\right)$ random variable.
d. For each data-generating scenario in the preceding text, write the appropriate GLM or GLMM that describes the processes generating the observation.
e. For each scenario, generate 100 simulated data sets and obtain the average performance of the models you gave in (d) for relevant estimates of interest (e.g., the estimated probability of each treatment on the data scale, the estimated variance components for the scenarios in part (c), etc.)

   Note: Once you get these simulations working, you can vary the probabilities, the number of units of replication, the number of Bernoulli trials (Binomial N) per unit of observation, the variance σ_U^2, the method of estimation (PL, Laplace, quadrature), etc., to explore the small sample behavior of binomial and binary GLMMs.

10.4 An alternative GLMM for binomial data is the beta-binomial. Refer to Problem 3, part (c), the logit-normal process, where binomial observations result from generating probabilities of $\tilde{\pi}_{ij}=1/\left(1+e^{-\eta_{ij}}\right)$ where $\eta_{ij}=NI\,(\text{logit}\,(\pi_i)+\sigma_U^2)$. This is the standard assumption of logistic GLMMs, the main focus of Chapter 10. Alternatively, the generating probabilities could arise from a beta distribution, that is, $\tilde{\pi}_{ij}\sim\text{indep Beta}\left(\alpha_i,\beta_i\right)$ where the Beta random variable's expected value, $\alpha_i/\left(\alpha_i+\beta_i\right)$, equals π_i as defined at the beginning of Problem 3.

a. Let $\beta=5$ for each treatment. Determine α_i for the three treatments so that their probabilities π_i equal those given in Problem 3.
b. Generate a set of data from the beta-binomial process assuming 10 units of replication and $N=10$ for each unit of replication for each treatment. Analyze the data in three ways:

   i. Binomial GLM, no random unit effect (pay attention to the Pearson χ^2/df)
   ii. Binomial GLMM assuming a normally distributed random unit effect
   iii. Beta GLMM (define the response variable as PCT=Y/N and the distribution as beta)

   Compare and contrast results.

c. Generate 100 simulated data sets with the structure from (b). Compare the average behavior of the three models in (b) with respect to estimates of interest.

11

Counts

11.1 Introduction

The term "counts" refers to data with a nonnegative integer response variable. As the name implies, counts arise from studies tracking the number of occurrences, for example, number of defects in a quality improvement study, number of disease events in a medical study, number of insects or birds or weeds in ecological or agricultural studies, number of website hits on the Internet, etc. The Poisson distribution figures prominently in the modeling of count data. However, not all counting processes give rise to the Poisson distribution.

In Section 11.1.1, we briefly review the Poisson distribution's assumptions and present an overview of the issues these assumptions create in working with generalized linear models (GLMs) and generalized linear mixed models (GLMMs). In Section 11.1.2, we work through implementation of a basic GLM with count data. Section 11.2 focuses on overdispersion, the paramount modeling concern with count data. In Section 11.3, we consider alternatives to the Poisson for counts. Section 11.4 concerns aspects of conditional and marginal modeling either unique to or of special concern with count data. Finally, Section 11.5 deals with modeling strategies for data with more zero counts than any distribution in the exponential family can accommodate.

11.1.1 Count Data and the Poisson Distribution

We write the Poisson p.d.f. and log-likelihood, respectively, as

- $$f(y) = \frac{e^{-\lambda}\lambda^{y}}{\Gamma(y+1)} \tag{11.1}$$
- $$\ell(\lambda; y) = y\log(\lambda) - \lambda - \log\left[\Gamma(y+1)\right] \tag{11.2}$$

The expected value and variance are $E(y) = Var(y) = \lambda$. The canonical parameter is $\theta = \log(\lambda)$. As with the binomial distribution, for GLM purposes, the canonical link is the usual link function for Poisson data. Unlike binomial GLMs, for which commonly used alternative links such as the probit and complementary log-log are common, GLMs for count data rarely use link functions other than the log.

Like the binomial, the Poisson belongs to the one-parameter exponential family. The rigidity of the Poisson mean–variance relationship makes Poisson-based GLMs and

GLMMs especially vulnerable to overdispersion. Overdispersion is arguably *the* single biggest modeling issue for count data. If not accounted for, overdispersed count data can produce severely inflated type I error rates and inadequate confidence interval coverage.

Pre-GLM modeling of count data used Gaussian linear model or linear mixed model methods with either the observed counts or transformed counts, typically either the square root or the natural log. The log transformation requires some adjustment for zero counts.

To illustrate pre-GLM analysis and compare and contrast it to a basic analysis using a Poisson GLM, let us work through an example with count data from a completely randomized study design.

11.1.2 Example Comparing Pre-GLM ANOVA-Based Analysis to Poisson GLM

The data from this example are from a two-treatment completely randomized study design. We have five observations on each of two treatments. We compare analyses based on two models, the normal approximation and the Poisson GLM. First, the normal approximation is as follows:

- Linear predictor: $\eta_i = \eta + \tau_i$, where τ denotes treatment effects (11.3)
- Distribution: $y_{ij} \sim NI(\mu_{ij}, \sigma^2)$, where y_{ij} denotes the observed count for the jth unit of replication on the ith treatment ($i=1,2; j=1,2,3,4,5$)
- Link: identity, $\eta_{ij} = \mu_{ij}$

We recognize this as the standard analysis of variance model. Justification rests on the central limit theorem: For sufficiently large counts, we can assume Poisson random variables have an approximate normal distribution. One weakness with the analysis of variance (ANOVA) model stems from the Poisson's mean–variance relationship: ANOVA assumes homogeneity of variance; Poisson data necessarily violate this assumption except when the treatment means are equal.

This gives rise to an alternative form of the ANOVA model. The linear predictor and link remain the same, but we replace the response variable by $t(y_{ij})$, where $t(\bullet)$ denotes a transformation. The transformation is meant to stabilize variance, improve the normal approximation or both. The ANOVA procedure then proceeds using $t(y_{ij})$. Resulting estimates are on the model scale, but we can "back-transform" (a pre-GLM ANOVA expression) these estimates using the inverse transformation function $t^{-1}(\bullet)$ and the standard errors using the delta rule. For count data, standard recommendations include $t(y_{ij}) = \sqrt{y_{ij}}$ and $t(y_{ij}) = \log(y_{ij})$, or $\log(y_{ij} + \varepsilon)$; $\varepsilon > 0$ when $y_{ij} = 0$. McCullagh and Nelder (1989) suggest $t(y_{ij}) = y_{ij}^{2/3}$ for a better approximation to normality. Schabenberger and Pierce (2002) suggest $t(y_{ij}) = \sqrt{y_{ij} + 0.375}$ for small y_{ij}, for example, when the data consist of a substantial proportion of single digit counts.

The second model is the Poisson GLM:

- Linear predictor: $\eta_i = \eta + \tau_i$ (11.4)
- Distribution: $y_{ij} \sim$ independent Poisson (λ_i)
- Link function: $\eta_i = \log(\lambda_i)$

How do these analyses compare? First, let us review the implementation. The data for this example appear in the SAS Data and Program Library as Data Set 11.1. Using GLIMMIX, the following statements specify analysis of y_{ij} using the normal approximation:

```
proc glimmix data=CRD_Counts;
  class trt;
  model count=trt;
  lsmeans trt/diff;
```

The selected output is as follows:

Type III Tests of Fixed Effects

Effect	Num DF	Den DF	*F* Value	Pr>F
trt	1	8	4.47	0.0674

trt Least Squares Means

trt	Estimate	Standard Error	DF	t Value	Pr > |t|
1	4.2000	1.3379	8	3.14	0.0138
2	8.2000	1.3379	8	6.13	0.0003

Differences of trt Least Squares Means

trt	_trt	Estimate	Standard Error	DF	t Value	Pr > |t|
1	2	−4.0000	1.8921	8	−2.11	0.0674

The column labeled "Standard Error" in the "trt Least Squares Means" should trouble us. For Poisson data, the mean should equal the variance. Therefore, the standard errors should be $\sqrt{4.2/5} = 0.92$ and $\sqrt{8.2/5} = 1.28$ for treatments 1 and 2, respectively (because the estimated means are 4.2 and 8.2). Instead, both standard errors are 1.34. This follows from standard ANOVA assumptions but violates Poisson assumptions.

What about the pre-GLM fix, the transformation? Let us try the log transformation first. The GLIMMIX statements are

```
proc glimmix data=CRD_Counts;
  class trt;
  if count=0 then logc=log(count+0.1); else logc=log(count);
  model logc=trt;
  lsmeans trt/diff;
  ods output lsmeans=lsm;
run;
data data_scale_means;
  set lsm;
  lamda_hat=exp(estimate);
  StdErrLamda=exp(estimate)*stderr;
proc print data=data_scale_means;
  var trt estimate stderr lamda_hat stderrlamda;
```

The IF-THEN-ELSE statement defines the transformation. The data set contains one COUNT = 0 observation. We add a small positive increment when COUNT = 0 to enable log calculation. If you just use the statement LOGC = LOG(COUNT), then GLIMMIX treats all COUNT = 0 observations as missing. The ODS OUTPUT statement creates a new data set containing the "trt Least Squares Means" table. The additional data step computes the "back-transformation"—in this case e^{LSMean}, where "estimate" is the data set name for the treatment LSMeans—and specifies the delta rule for the "back-transformed standard error."

The selected output is as follows:

Type III Tests of Fixed Effects

Effect	Num DF	Den DF	*F* Value	Pr > F
trt	1	8	2.24	0.1727

trt Least Squares Means

trt	Estimate	Standard Error	DF	t Value	Pr > |t|
1	0.8148	0.5785	8	1.41	0.1967
2	2.0398	0.5785	8	3.53	0.0078

Obs	trt	Estimate	StdErr	lamda_hat	StdErrLamda
1	1	0.8148	0.5785	2.25879	1.30682
2	2	2.0398	0.5785	7.68928	4.44862

Notice that the p-value for the test of equal treatment means is 0.1727, compared to 0.0674 for the test using the untransformed data. Either the transformation adversely affected the power of the test, or the untransformed data is giving us a type I error. The estimated difference is on the model scale—that is, it is the difference between the means of $\log(y_{ij})$. We cannot back-transform this estimate and obtain anything meaningful because the transformation is nonlinear. The estimated λ_i are 2.26 and 7.69 compared to 4.2 and 8.2 with the untransformed data. The problem with the back-transformation is that $\exp\left[(1/5)\sum_{j=1}^{5}\log(y_{ij})\right]$ is not quite the same thing as the quantity we seek, $\exp(\hat{\eta}_{ij})$. With a transformation, there is no way to estimate the latter; the back-transformation is as close as we can get, but it is not the same thing.

We could use similar programs to compute the analyses for the other transformations listed earlier. If we do this, we must adjust the back-transformations accordingly. For example, for the square root transformation, the statements would be

```
lamda_hat=estimate**2;
StdErrLamda=2*estimate*stderr;
```

In the STDERRLAMDA statement, we use 2*ESTIMATE*STDERR because $2 \times \text{LSMean} = \partial(\text{LSmean}^2)/\partial\text{LSMean}$, the required term for the delta rule.

For the GLM, we use the GLIMMIX statements

```
proc glimmix data=CRD_Counts;
  class trt;
  model count=trt/d=poisson;
  lsmeans trt/diff ilink;
```

This is the original program for the untransformed data but with D=POISSON added to the MODEL statement and ILINK added to the LSMEANS statement. The selected output is as follows:

Type III Tests of Fixed Effects

Effect	Num DF	Den DF	*F* Value	Pr>F
trt	1	8	6.22	0.0373

trt Least Squares Means

trt	Estimate	Standard Error	DF	t Value	Pr> |t|	Mean	Standard Error Mean
1	1.4351	0.2182	8	6.58	0.0002	4.2000	0.9165
2	2.1041	0.1562	8	13.47	<.0001	8.2000	1.2806

Differences of trt Least Squares Means

trt	_trt	Estimate	Standard Error	DF	t Value	Pr> |t|
1	2	−0.6690	0.2683	8	−2.49	0.0373

While not shown in this output, we should check the "Conditional Distribution Fit Statistics" for the Pearson χ^2/df statistic. For these data, $\chi^2/df = 1.60$—indicating lack of fit in the form of overdispersion does not appear to be an issue. The p-value is 0.0373 versus 0.0674 and 0.1727 we have seen in previous analyses. Notice the "trt Least Squares Means" estimates in the "mean" column are identical to those obtained using the untransformed data, but the standard errors here agree with Poisson theory. As with the analysis of the log transformed data, the estimated difference is on the model, log scale and is not amenable to inverse linking.

Which analysis should we report? As Walt Federer was fond of saying, "Don't argue, take data." The reasonable way to answer this question is to run a statistical experiment, a simulation, and find out. Here is a comparison of the methods based on 1000 simulated experiments using the same design as this example—two treatments, five mutually independent observations each—with $y_{1j} \sim$ Poisson (5) and $y_{2j} \sim$ Poisson (10). The criteria are

- Observed number of rejections/1000 of H_0: $\tau_1 = \tau_2$ for $\alpha = 0.05$
- The estimated difference, $\hat{\tau}_1 - \hat{\tau}_2$, the standard deviation of the observed sampling distribution of $\hat{\tau}_1 - \hat{\tau}_2$ and the average $s.e.(\hat{\tau}_1 - \hat{\tau}_2)$
- The estimates of each treatment mean, $\hat{\lambda}_i$, the standard deviation of its sampling distribution and its average standard error

TABLE 11.1

Small Sample Performance of Competing Methods to Model λ_i

	Normal Approximation					
Criteria	y	**log(y)**	$\sqrt{y}$	$y^{2/3}$	$\sqrt{y+0.375}$	**Poisson GLM**
Rejection rate H_0:$\tau_1=\tau_2$	0.721	0.676	0.720	0.724	0.722	0.711
Average $\hat{\tau}_1-\hat{\tau}_2$	5.041	0.750	0.958	1.751	0.926	0.708
std dev $\hat{\tau}_1-\hat{\tau}_2$	1.807	0.286	0.345	0.622	0.330	0.263
Average s.e.($\hat{\tau}_1-\hat{\tau}_2$)	1.685	0.262	0.321	0.581	0.308	0.249
Average $\hat{\lambda}_1$	4.997	4.520	4.762	4.845	4.792	4.997
std dev $\hat{\lambda}_1$	1.004	1.104	1.038	1.018	1.017	1.004
Average s.e.($\hat{\lambda}_1$)	1.192	0.848	0.977	1.036	1.119	0.995
Average $\hat{\lambda}_2$	10.038	9.618	9.831	9.901	9.840	10.038
std dev $\hat{\lambda}_2$	1.429	1.439	1.426	1.435	1.424	1.429
Average s.e.($\hat{\lambda}_2$)	1.192	1.902	1.417	1.320	2.221	1.413

The standard error of the estimator is supposed to estimate the standard deviation of the estimator's sampling distribution, so we should see close agreement between these two. Disparity indicates poor performance of the analysis method. The estimators should be close to the parameters they are intended to estimate. While not shown here, you can verify that all of these methods yield observed rejection rates of approximately 5% with data simulated under the null hypothesis with y_{ij} ~ Poisson (5), so type I error control is not an issue for any of these methods. The simulation includes alternative transformations given earlier in this section. Table 11.1 shows the simulation results.

The untransformed normal approximation yields accurate estimates of λ_i and $\lambda_1-\lambda_2$, but its standard deviation-to-standard error agreement is unimpressive, to say the least. None of the transformations compete well for estimation of λ_i. The square root transformations' standard deviation-standard error performance appears to be reasonable except for $\sqrt{y+0.375}$ with larger λ_i. The log transformation not only does a poor job estimating λ_i, but its standard error performance is atrocious, and it is the only method showing a noticeably adverse impact on power. The Poisson GLM does exactly what it is supposed to do.

11.2 Overdispersion in Count Data

Not all processes that give rise to count data can be modeled as Poisson. If we assume that a response variable, Y, has a Poisson distribution, we implicitly assume that $E(Y)=\lambda=Var(Y)$. This is not always what we see. In some disciplines, ecology for example, count data are common, but count data that satisfy Poisson assumptions are not. More often, we see data whose variance exceeds its mean, often substantially. We call this *overdispersion*.

11.2.1 Overdispersion Defined

Formally, overdispersion is said to occur whenever the observed variance exceeds the theoretical variance under the assumed distribution. We usually understand overdispersion to apply to the distribution of $\mathbf{y}|\mathbf{b}$. Overdispersion cannot occur with Gaussian data, because the mean μ and variance σ^2 are unrelated; a given μ does not, in principle, preclude any value of σ^2 in the parameter space. In GLM terms, the Gaussian variance function $V(\mu) = 1$. Overdispersion is not an intelligible concept for Gaussian data. On the other hand, for any distribution or quasi-likelihood with a nontrivial variance function, overdispersion is theoretically possible. It is especially a possibility for distributions belonging to the one-parameter exponential family because they lack a scale parameter to mitigate the mean–variance relationship. For the Poisson, if the mean equals λ, so must the variance. No wiggle room. As a result, Poisson models are exceptionally vulnerable to overdispersion.

To illustrate what overdispersion might look like, consider a multisite study, where at each site we take count data at several positions. Examples where we might see something like this include environmental monitoring, ecology, epidemiology, and various agricultural disciplines. In an agricultural study, for example, we might sample several fields and, within each field, use multiple transects to obtain counts of insects, weeds or whatever we are investigating.

In our example, suppose we have 20 sites, and in each site we sample 10 transects, or positions. We then calculate the sample variance among the counts on transects within each site and plot them. Figure 11.1 shows one possible plot; Figure 11.2 shows another.

Figure 11.1 shows what we expect to see if the Poisson accurately characterizes the probability distribution of $\mathbf{y}|\mathbf{site}$. The mean–variance relationship is approximately linear with a slope of 1. Figure 11.2 is what we often see: the variance–mean relationship is closer to quadratic than linear. Figure 11.2 shows the signature plot of overdispersed count data. Overdispersion can arise in count data because of the following:

1. The mean model is misspecified, for example, we fit a linear regression model to data that show a curvilinear relationship to the explanatory variable on the model scale.
2. The linear predictor is underspecified. We saw an example of this in Chapter 10, Section 10.3.1. Other ways this can happen include omitting a unit-of-replication

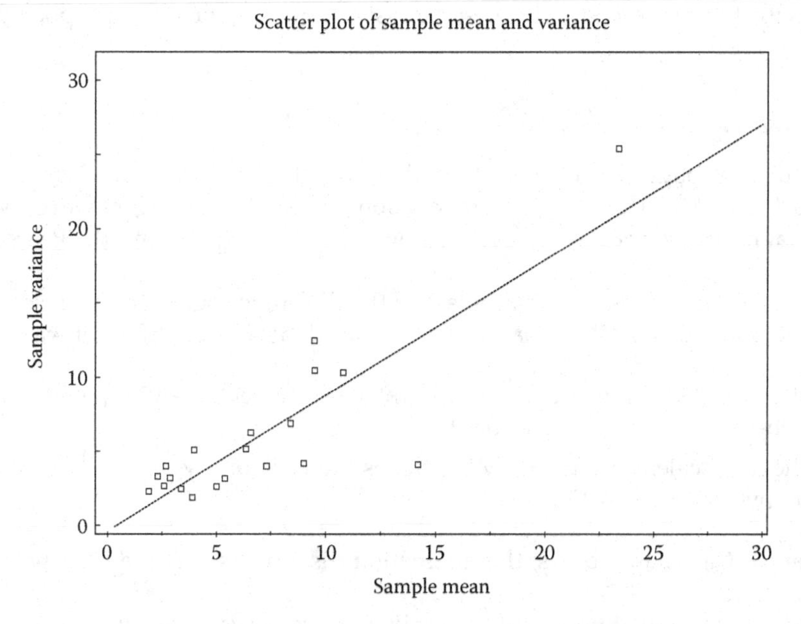

FIGURE 11.1
Plot of mean vs. variance, multisite example.

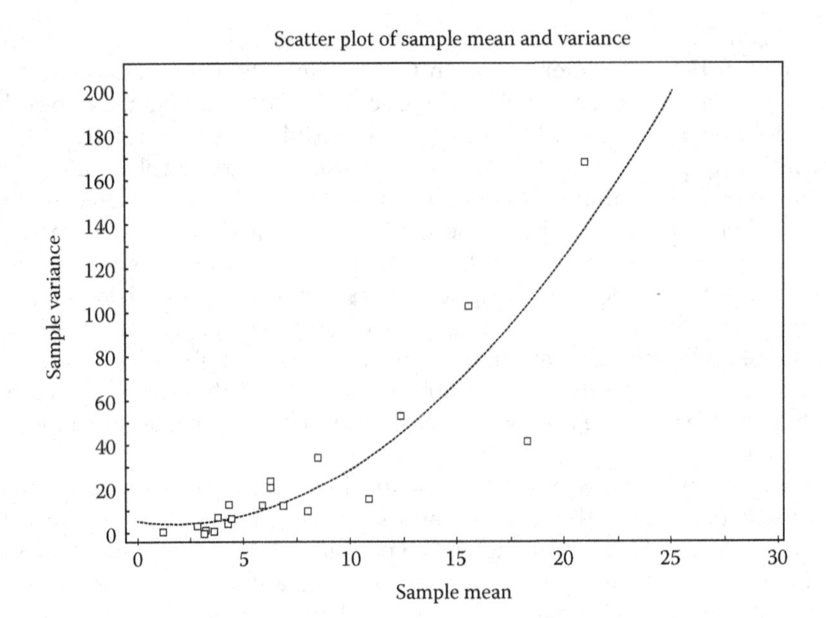

FIGURE 11.2
Plot of mean vs. variance, multisite data.

term in a multilevel model, for example, leaving the whole plot error term out of the model for split-plot data.

3. We assume the wrong distribution. We will consider alternative count distributions as we work through this chapter.

If we fail to account for overdispersion, we underestimate standard errors and inflate test statistics with the usual consequences: excessive type I error rate and poor confidence interval coverage.

11.2.2 Detecting Overdispersion

If we are lucky enough to have data that allow constructing a mean–variance plot such as Figures 11.1 or 11.2, detecting overdispersion is easy. In general, however, we do not. The two diagnostics we *do* have at our disposal are residual plots and the Pearson χ^2/df fit statistic.

In ordinary least squares, residual refers to the difference between the observation and its predicted value. In GLMMs, we can think of residuals in two different ways:

- On the model scale: for a true GLMM, the residual vector is $\mathbf{y}^* - \hat{\boldsymbol{\eta}}'$ where $\mathbf{y}^*$ is the pseudo-variate as defined in Chapter 4.
- On the data scale: for a true GLMM, the residual vector is $\mathbf{y} - \hat{\boldsymbol{\mu}} = \mathbf{y} - h(\hat{\boldsymbol{\eta}})$, where $h(\bullet)$ is the inverse link function.

Notice that for Gaussian models, the distinction disappears because LM and LMMs use the identity link.

The residuals as defined previously are called *raw residuals*. Often, we can learn more by adjusting the residuals to the scale of measurement. We do this by dividing the residual by some measure of standard deviation or standard error. The Pearson residual uses the estimated variance of the conditional distribution:

- On the *model* scale, the *Pearson residual* is $\frac{\mathbf{y}^* - \hat{\eta}}{\sqrt{\widehat{Var}(\mathbf{y}^* \mid \mathbf{b})}}$ (11.5)
- On the *data* scale, the *Pearson residual* is $\frac{\mathbf{y} - h(\hat{\eta})}{\sqrt{\widehat{Var}(\mathbf{y} \mid \mathbf{b})}}$ (11.6)

The studentized residual uses the estimated variance of the residual:

- On the *model* scale, the *studentized residual* is $\frac{\mathbf{y}^* - \hat{\eta}}{\sqrt{\widehat{Var}(\mathbf{y}^* - \hat{\eta})}}$ (11.7)
- On the *data* scale, the *studentized residual* is $\frac{\mathbf{y} - h(\hat{\eta})}{\sqrt{\widehat{Var}(\mathbf{y} - \hat{\mu})}}$ (11.8)

For diagnosing overdispersion, we have six potential plots: raw, Pearson, or studentized residuals versus the estimated linear predictor, $\hat{\eta}$, on the model scale, and raw, Pearson, or studentized residuals versus the estimated mean, $\hat{\mu} = h(\hat{\eta})$, on the data scale. Evidence of overdispersion may show up on some, but not all, plots. We obtain residual plots via PROC GLMIMMIX using the PLOT option.

For our multisite example, using the data from Figure 11.2, assuming site effects random $\mathbf{y} \mid \mathbf{site} \sim$ Poisson, we have the GLMM as follows:

1. Linear predictor: $\eta_i = \eta + f_i$ where f denotes site (a.k.a. "field") effects
2. Distributions:
   a. $y_{ij} \mid f_i \sim$ indep Poisson (λ_i)
   b. f_i i.i.d. $N(0, \sigma_F^2)$
3. Link: $\eta_i = \log(\lambda_i)$

The GLIMMIX statements are as follows:

```
ods graphics on;
procglimmix data=p plot=residualpanel(unpack ilink)
  plot=residualpanel(unpack noilink) plot=studentpanel(unpack noilink);
  class f;
  model count_pw= /d=poisson;
  random intercept/subject=f;
```

Notice that we can have multiple PLOT statements. RESIDUALPANEL calls for the raw residuals. STUDENTPANEL calls for Student residuals. While not shown here, PEARSONPANEL calls for Pearson residuals. ILINK calls for data-scale residuals; NOILINK calls for model-scale residuals. UNPACK allows the plots to presented one-at-a-time; otherwise they appear in sets of four plots, including a histogram, Q-Q plot, and box-plot. Here, we only want to look at residual vs. predicted. The three example plots are

a. Data-scale raw residuals

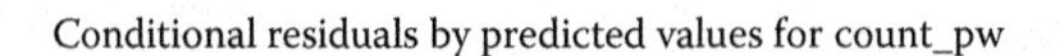

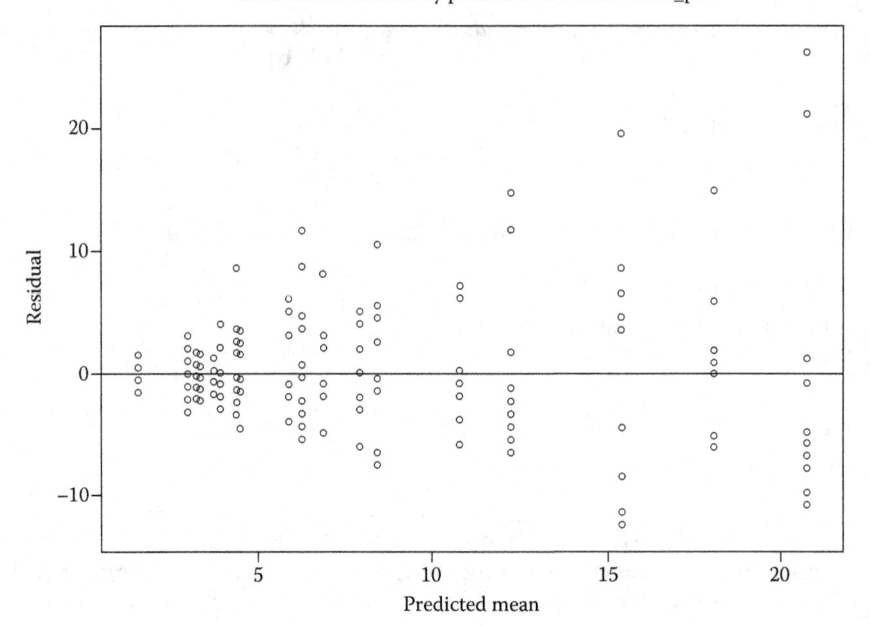

b. Model-scale raw residuals

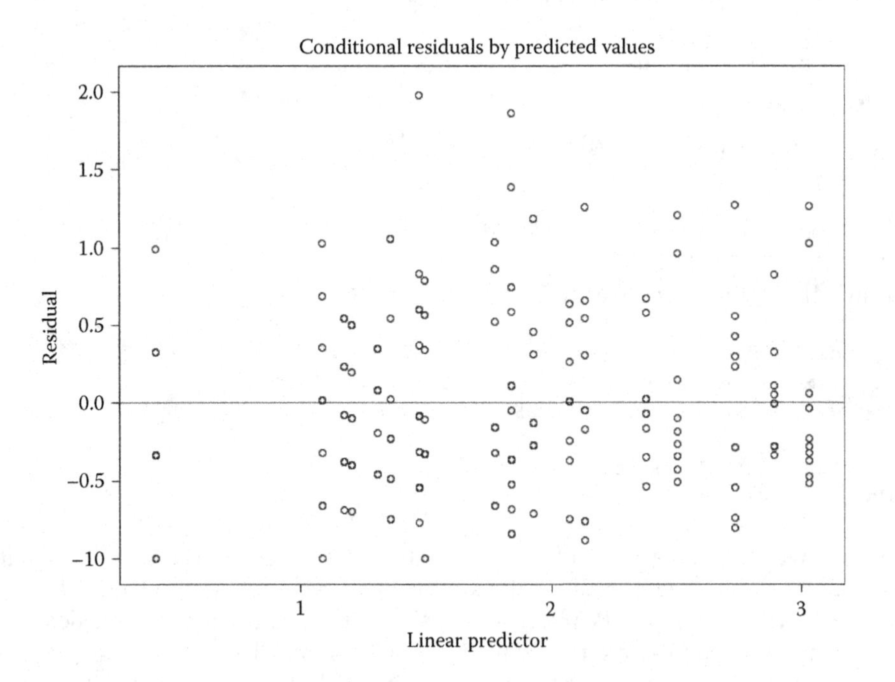

c. Model-scale studentized residuals

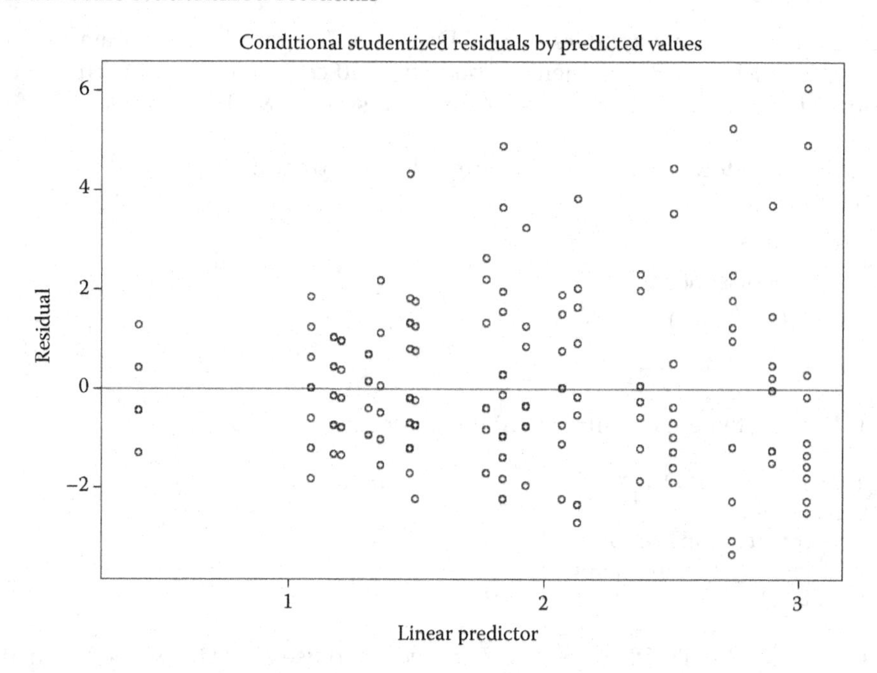

Plot (a) shows obvious visual evidence of overdispersion. As the "predicted mean" $\hat{\mu}$ increases, the associated residuals become more widely dispersed. The variance should increase as a function of the mean, but not as rapidly as we see in this plot. Plot (c) also shows evidence of overdispersion. On the model scale—the log—we should not see the variance-adjusted residuals become more variable as $\hat{\eta}$ increases. Plot (b) is not very informative.

The other primary diagnostic is the Pearson χ^2/df. For these data, the fit statistics output appears as

Fit Statistics	
−2Reslogpseudo-likelihood	547.11
Generalized chi-square	470.85
Gener. chi-square/DF	2.37

Notice that we have the *generalized* χ^2/df, not the *Pearson* statistic. We see here the default output using PL estimation with a GLMM. The generalized chi-square statistic measures the combined goodness of fit of the conditional distribution and the random model effects. As such, it is an overly aggregated statistic: it is the *conditional* distribution of $\mathbf{y}|\mathbf{b}$ that we assume is Poisson, not the *joint* distribution of $\mathbf{y}$ and $\mathbf{b}$. To assess the former, we need the conditional distribution Pearson χ^2/df. We have to use METHOD=LAPLACE or METHOD=QUAD to obtain it. Under the Laplace method, $\chi^2/df=2.26$.

These are the usual statistics that would allow us to diagnose overdispersion in the data from Figure 11.2 during the course of standard analysis of a Poisson GLMM.

We now work through an example to illustrate the various strategies for coping with overdispersed count data.

11.2.3 Strategies

The data for this example appears in SAS Data and Program Library, Data Set 11.3. The study design consists of 3 treatments observed in 10 complete blocks with count as the response variable. We start the analysis using a Poisson GLMM. The model is as follows:

1. Linear predictor: $\eta_{ij}=\eta+\tau_i+r_j$, where τ denotes treatment effects and r denotes block effects
2. Distributions:
   a. $y_{ij}|r_j \sim \text{Poisson}(\lambda_{ij})$
   b. r_j i.i.d. $N\left(0,\sigma_R^2\right)$
3. Link function: $\eta_{ij}=\log(\lambda_{ij})$

The GLIMMIX statements to implement the model are as follows:

```
proc glimmix method=laplace data=RCBD_counts;
  class blk trt;
  model y=trt/dist=poisson;
  random intercept/subject=blk;
lsmeans trt/ilink;
```

Notice that we use METHOD=LAPLACE (or we could use METHOD=QUAD) in order to get the Pearson χ^2/df fit statistic. The "fit statistics" output is as follows:

Fit Statistics for Conditional Distribution	
−2 log L(y | r. effects)	296.08
Pearson chi-square	206.53
Pearson chi-square/DF	6.88

This is very strong evidence of overdispersion. We note the "type III tests of fixed effects" and the LSMeans so we can compare them with subsequent analyses that account for overdispersion.

Type III Tests of Fixed Effects

Effect	Num DF	Den DF	*F* Value	Pr>F
trt	2	18	48.49	<.0001

trt Least Squares Means

trt	Estimate	Standard Error	Mean	Standard Error Mean
1	1.5481	0.3540	4.7023	1.6647
2	2.0564	0.3469	7.8176	2.7120
3	2.7151	0.3416	15.1061	5.1609

The striking feature of this output is the *F*-value of 48.49. Given the strong evidence of overdispersion, we should consider the *F*-statistic to be highly suspect.
There are three ways we can deal with overdispersion:

1. Add a scale parameter; replace $Var(y_{ij}|r_j)=\lambda_{ij}$ by $Var(y_{ij}|r_j)=\phi\lambda_{ij}$. Said another way, we replace the Poisson conditional log-likelihood $y\log(\lambda)-\lambda-\log[\Gamma(y+1)]$ by the Poisson-like quasi-likelihood $y\log(\lambda)-\lambda/\phi$. The premise for this "fix" rests on there being a $\phi>1$ that accurately models the observed variance. This is the "classical" fix in GLM literature—and in many cases the least likely to be effective, as we shall see in the following.
2. Revisit the skeleton ANOVA for the complete-block design. Recall that we have a block × treatment source of variation that we omitted from the linear predictor simply because we followed the Gaussian model analogy. In a Gaussian model, we assume this source of variation is "residual" to enable the estimation of σ^2. The Poisson belongs to the one-parameter exponential family. Here, the block × treatment source of variation may have a legitimate place in the linear predictor. The underlying concept is similar to Example 10.7 in the previous chapter.
3. Use a different distribution. There are other ways to develop counting processes that do not lead to the Poisson. Prominent among them is the negative binomial distribution. As we will see in Section 11.3.1, we can motivate the negative binomial via a process that may make more sense than the Poisson, particularly in many biological scenarios.

Let us now look at each of these alternatives.

11.2.3.1 Scale Parameter

The GLIIMIX statements to invoke the scale parameter are

```
proc glimmix /*method=laplace*/ data=RCBD_counts;
  class blk trt;
  model y=trt/dist=poisson;
  random intercept/subject=blk;
  random _residual_;
  lsmeans trt/ilink;
```

We retain the DIST = POISSON specification but add a statement RANDOM _ RESIDUAL_. Notice that we cannot implement this program with integral approximation; we suppress METHOD = LAPLACE, so estimation occurs using pseudo-likelihood. This causes a scale parameter to be estimated and subsequently used as an adjustment for all standard errors and test statistics. GLIMMIX uses the generalized $\chi^2/df=\hat{\phi}$. McCullach and Nelder (1989) use the deviance statistic for $\hat{\phi}$. All standard errors from the default Poisson analyses are multiplied by $\sqrt{\hat{\phi}}$, and all *F*-values are divided by $\hat{\phi}$. Notice that because we have created a quasi-likelihood, we cannot use Laplace or quadrature: we must use PL or its equivalent in other software packages. The relevant output is as follows:

Covariance Parameter Estimates			
Cov Parm	**Subject**	**Estimate**	**Standard Error**
Intercept	blk	0.8201	0.5173
Residual (VC)		10.7695	3.4792

Type III Tests of Fixed Effects				
Effect	**Num DF**	**Den DF**	***F* Value**	**Pr>F**
trt	2	18	4.50	0.0260

trt Least Squares Means							
trt	**Estimate**	**Standard Error**	**DF**	**t Value**	**Pr > |t|**	**Mean**	**Standard Error Mean**
1	1.7867	0.4775	18	3.74	0.0015	5.9698	2.8507
2	2.2950	0.4176	18	5.50	<.0001	9.9248	4.1444
3	2.9538	0.3678	18	8.03	<.0001	19.1780	7.0544

The "Residual (VC)" in the "Covariance Parameter Estimates" output is the scale parameter estimate, $\hat{\phi}=10.77$. The *F*-value was 48.49 in the default Poisson analysis. Now it is 4.50 (=48.49/$\hat{\phi}$). For these data, we would still conclude a statistically significant treatment effect, assuming $\alpha-0.05$, but we can easily see that accounting for overdispersion could be the difference between concluding a highly significant treatment effect and a failure to reject the hypothesis of no treatment effect. Including the scale parameter affects the block variance estimate $\hat{\sigma}_R^2$. This in turn slightly affects the treatment mean *estimates*, but the main impact is on the standard errors. This is not necessarily a good thing. If overdispersion results from underspecifying the random component of the model (see "strategy b" immediately below), then the means we see here target the marginal distribution, raising all the ambiguity of interpretation issues we discussed in Chapter 3.

At one time, the scale parameter was considered the standard fix for overdispersion. As we gain experience with GLMMs, this is increasingly less true. First, the scale parameter implies quasi-likelihood. Recall the first paragraph of this textbook: a model should describe a plausible process giving rise to the data. With the scale parameter, we do not have a true likelihood and, hence, fail to satisfy one criterion for a statistical model—there is no *definable* probability process that gives us an expected value of λ and a variance of $\phi\lambda$. Second, studies, for example, Young et al. (1998), suggest that the scale parameter does not effectively control type I error. We can see why from Figure 11.2. The scale parameter implies that the variance should increase linearly with the mean. Under the Poisson assumption (Figure 11.1), the variance increases linearly with a slope of 1. With a scale parameter, the relationship should still be linear but with a slope >1. Instead, Figure 11.2 depicts a *quadratic* increase.

Our task, then, is to find a mechanism that accounts for a quadratic increase in the variance as a function of the mean *and* defines a legitimate probability process.

11.2.3.2 "What Would Fisher Do?" Revisited

One such method involves the same strategy we employed in Example 10.7. If λ_{ij} is being randomly perturbed from unit-to-unit within a block, over and above just treatment effect, our linear predictor does not account for it. The skeleton ANOVA derived by applying the WWFD process to this design has a block × treatment source of variation that can be included in the linear predictor for a GLMM whose distribution belongs to the one-parameter exponential family. This gives us the modified linear predictor

$$\eta_{ij} = \eta + \tau_i + r_j + (rt)_{ij}, \text{ where } (rt)_{ij} \text{ i.i.d. } N\left(0, \sigma_{RT}^2\right).$$

We can use the following GLIMMIX statements:

```
proc glimmix method=laplace data=RCBD_counts;
  class blk trt;
  model y=trt/dist=poisson;
  random intercept trt/subject=blk;
  lsmeans trt/ilink;
```

The only difference between this and the original GLIMMIX program for these data is the inclusion of the TRT term in the RANDOM statement; in conjunction with SUBJECT = BLK, this defines the block × treatment random effect. The relevant results are as follows:

Fit Statistics for Conditional Distribution	
−2 log L(y | r. effects)	117.20
Pearson chi-square	4.95
Pearson chi-square/DF	0.17

This provides strong evidence that we have effectively eliminated overdispersion.

Covariance Parameter Estimates			
Cov Parm	**Subject**	**Estimate**	**Standard Error**
Intercept	blk	0.6081	0.4596
trt	blk	0.9523	0.3977

The estimated block × treatment variance component is $\hat{\sigma}_{RT}^2 = 0.95$.

Type III Tests of Fixed Effects				
Effect	**Num DF**	**Den DF**	***F* Value**	**Pr > F**
trt	2	18	1.43	0.2656

This *F*-value compares to 4.50 with the scale parameter and 48.49 for the "naive" Poisson GLM. We get a sense here of why the scale parameter does not effectively control type I error in situations like the one portrayed in this example: accounting for the quadratic relationship between mean and variance has a much more drastic impact on the *F*-statistic.

trt Least Squares Means

trt	Estimate	Standard Error	DF	t Value	Pr > |t|	Mean	Standard Error Mean
1	1.5558	0.4330	18	3.59	0.0021	4.7387	2.0520
2	1.8484	0.4243	18	4.36	0.0004	6.3494	2.6939
3	2.3601	0.4160	18	5.67	<.0001	10.5924	4.4066

The values we see here in the "Mean" column, 4.7, 6.3, and 10.6, respectively, for treatments 1, 2, and 3, unlike the estimates in the previous two analyses, are *bona fide* broad inference space estimates of Poisson rate parameters.

While not shown here, we could also use COVTEST for a likelihood ratio test of $H_0{:}\sigma^2_{RT} = 0$. This would formally test the legitimacy of including $(rt)_{ij}$ in the linear predictor.

11.2.3.3 Alternative Distributions

Another way to account for random perturbation of the Poisson rate parameter is to change our assumed distribution. As we will formally demonstrate in Section 11.3.1, one way to motivate the negative binomial is to express it as a Poisson(λ) random variable where λ is itself a random variable with a Gamma distribution. The negative binomial has expected value $E(y)=\lambda$ and variance $Var(y)=\lambda+\phi\lambda^2=\lambda(1+\phi\lambda)$. Notice that this also accounts for the quadratic mean–variance relationship. Like the beta distribution, the negative binomial has a scale parameter, meaning that we should not include $(rt)_{ij}$ in the linear predictor. The model we want to fit is thus

1. Linear predictor: $\eta_{ij}=\eta+\tau_i+r_j$
2. Distributions:
   a. $y_{ij}|r_j \sim$ Negative Binomial(λ_{ij},ϕ)
   b. r_k i.i.d. $N\left(0,\sigma^2_R\right)$
3. Link function: $\eta_{ij}=\log(\lambda_{ij})$

Like the beta distribution, for GLM purposes, we use a different parameterization for the negative binomial distribution than we typically see in mathematical statistics textbooks. We also think of the negative binomial differently. In probability theory, we motivate the negative binomial as the number of times it takes to have a given number of Bernoulli successes. Here, we use it to model counting processes with more clustering or clumping together and, hence, more variance than we see in a Poisson process.

The GLIMMIX statements to implement the analysis are identical to the "naive" Poisson, except we replace DIST = POISSON by DIST = NEGBIN:

```
proc glimmix method=laplace data=RCBD_counts;
  class blk trt;
  model y=trt/dist=negbin;
  random intercept/subject=blk;
  lsmeans trt/ilink;
```

The relevant output is as follows:

Fit Statistics for Conditional Distribution	
−2 log L(y | r. effects)	197.61
Pearson chi-square	20.49
Pearson chi-square/DF	0.68

Pearson $\chi^2/df = 0.68$ shows no evidence of overdispersion. The conditional fit statistics provide no clear guidance as to which model is better: the negative binomial model or the Poisson model with the additional block × treatment random effect in the model.

Covariance Parameter Estimates

Cov Parm	Subject	Estimate	Standard Error
Intercept	bik	0.6364	0.4570
Scale		0.8182	0.2928

The SCALE estimate is $\hat{\phi} = 0.82$. This is not the same as the scale parameter for the Poisson model with the RANDOM _RESIDUAL_ statement. They affect the mean–variance relationship differently. For example, if $\lambda = 10$, the Poisson quasi-likelihood scale parameter would imply that the variance of $y_{ij}|r_k$ is $\phi\lambda = 10.77 \times 10 = 107.7$. The negative binomial implies that the variance is $\lambda + \phi\lambda^2 = 10 + 0.82 \times 10^2 = 92$. For $\lambda = 15$, the variances would be $\phi\lambda = 10.77 \times 15 = 161.55$ and $\lambda + \phi\lambda^2 = 15 + 0.82 \times 15^2 = 199.5$. You can see how this changes the mean–variance relationship.

The test statistic for treatment effect

Type III Tests of Fixed Effects

Effect	Num DF	Den DF	*F* Value	Pr>F
trt	2	18	1.33	0.2893

is similar to what we obtained with the Poisson plus block × treatment model. The LSMeans:

trt Least Squares Means

trt	Estimate	Standard Error	DF	t Value	Pr > |t|	Mean	Standard Error Mean
1	2.0182	0.4259	18	4.74	0.0002	7.5248	3.2051
2	2.2404	0.4196	18	5.34	<.0001	9.3970	3.9429
3	2.7713	0.4265	18	6.50	<.0001	15.9795	6.8156

These differ from the two sets of means seen previously. The estimates of the negative binomial rate parameter appear in the column labeled "mean." They have a similar interpretation to the estimates obtained from the Poisson + block × treatment model but they stem from a different process. Again, there is nothing in the statistics that says decisively whether to use the negative binomial or the Poisson + block × treatment model. Both of them, however, are clearly desirable relative to the "naive" Poisson model and the Poisson + scale parameter model.

At this point, we may want to test $H_0{:}\,\phi=0$, noting that if $\phi=0$, then the variance of the negative binomial equals the variance of the Poisson. We defer this until the next section, in which we present the formal aspects of the negative binomial. There is theory to the effect that as $\phi \to 0$, the negative binomial $\to$ Poisson, but basing a likelihood ratio test for $\phi=0$ is not as simple as it appears.

11.3 More on Alternative Distributions

In the previous example we looked at two ways to perturb λ_{ij} at the unit-of-observation level, one by adding a unit-of-observation random effect to the linear predictor, the other by assuming a negative binomial rather than a Poisson distribution. We motivated the former by literal application of WWFD to a response variable belonging to the one-parameter exponential family. For the latter, we change the perturbation from a Gaussian random model effect to a process with a gamma distribution. In Section 11.3.1, we develop the reasoning behind this.

At the end of the previous section we said testing $H_0{:}\,\phi=0$, the negative binomial scale parameter, *seems* like an intuitive way to assess "Poisson-ness", but it presents a problem. In Section 11.3.2, we will see why, and present a second alternative distribution, the generalized Poisson.

11.3.1 Negative Binomial

In standard probability theory, we motivate the *negative* binomial using an alternative take on the *binomial*. Instead of asking the binomial question—what is the probability of exactly Y successes out of N independent Bernoulli trials—we ask the question, "what is the probability that it will take exactly N independent Bernoulli trials for us to have Y successes?" Following this line of development, we write the negative binomial p.d.f. as

$$f(y;N,\pi)=\binom{N-1}{y-1}\pi^{y-1}(1-\pi)^{N-y} \tag{11.9}$$

This motivates the negative binomial as a rate-and-proportion problem. This is not how we generally use the negative binomial in GLMs.

An alternative approach to the negative binomial motivates it as a counting distribution. Following an approach presented by Hilbe (2007), if we assume that the Poisson rate parameter is itself a random variable with a Gamma distribution, that is, $Y|\lambda u \sim \text{Poisson}(\lambda u)$ and $u \sim \text{Gamma}\left(1/\phi, \phi\right)$, we can show that the marginal distribution of Y is negative binomial. Recalling Example 10.7, $u \sim \text{Gamma}\left(1/\phi, \phi\right)$ replaces $(rt)_{ij}$ i.i.d. $N\left(0, \sigma^2_{RT}\right)$ as the process perturbing λ at the unit-of-observation level. Here is the derivation.

Suppose we have a count with a Poisson distribution with rate parameter λu, where λ is a constant and $u \sim \text{Gamma}\left(\frac{1}{\phi},\phi\right)$. In other words, we have a Poisson rate parameter centered at λ but being randomly perturbed by u. The conditional distribution of the observed count is thus $f(y|u) = e^{-\lambda u}(\lambda u)^y/y!$. The marginal distribution of y is thus

$$\int_0^\infty \frac{e^{-\lambda u}(\lambda u)^y}{y!}\frac{1}{\Gamma\left(\frac{1}{\phi}\right)(\phi)^{\frac{1}{\phi}}}u^{\left(\frac{1}{\phi}\right)-1}e^{-\frac{u}{\phi}}\,du = \frac{\lambda^y\phi^{-\frac{1}{\phi}}}{y!\Gamma\left(\frac{1}{\phi}\right)}\int_0^\infty e^{-\lambda u}u^y u^{\left(\frac{1}{\phi}\right)-1}e^{-\left(\frac{1}{\phi}\right)u}\,du$$

Now, noting $y! = \Gamma(y+1)$, collecting terms in the integral, and multiplying and dividing by key constants, we have

$$\frac{\lambda^y\phi^{-\frac{1}{\phi}}\Gamma\left(y+\left(\frac{1}{\phi}\right)\right)}{\Gamma(y+1)\Gamma\left(\frac{1}{\phi}\right)\left(\lambda+\left(\frac{1}{\phi}\right)\right)^{y+\left(\frac{1}{\phi}\right)}}\int_0^\infty \frac{1}{\Gamma\left(y+\left(\frac{1}{\phi}\right)\right)\left(\frac{1}{\left(\lambda+\left(\frac{1}{\phi}\right)\right)}\right)^{y+\left(\frac{1}{\phi}\right)}}u^{y+\left(\frac{1}{\phi}\right)-1}e^{-\left(\lambda+\left(\frac{1}{\phi}\right)\right)u}\,du$$

Now, note that the integral is over a $\text{Gamma}\left(y+\left(\frac{1}{\phi}\right),\lambda+\left(\frac{1}{\phi}\right)\right)$ distribution and is, hence, equal to one, leaving us with

$$\binom{y+\left(\frac{1}{\phi}\right)-1}{y}\left(\frac{\lambda}{\lambda+\frac{1}{\phi}}\right)^y\left(\frac{\left(\frac{1}{\phi}\right)}{\lambda+\frac{1}{\phi}}\right)^{\frac{1}{\phi}} = \binom{y+\left(\frac{1}{\phi}\right)-1}{y}\left(\frac{\lambda\phi}{1+\lambda\phi}\right)^y\left(\frac{1}{1+\lambda\phi}\right)^{\frac{1}{\phi}} \tag{11.10}$$

Hence, the marginal distribution of y is negative binomial with mean λ and variance $\lambda(1+\lambda\phi)$. While not the parameterization we see in a mathematical statistics textbook, (11.10) facilitates fitting negative binomial GLMs and GLMMs.

Another way to envision the difference between the Poisson and negative binomial counting process is to return to the development of the Poisson distribution in probability theory. Imagine a grid with N cells. Each cell can be unoccupied or occupied by exactly one object. Let $\pi = P\{\text{cell occupied}\}$. Assume all cells are mutually independent. Then the probability that Y cells are occupied is $\binom{N}{y}\pi^y(1-\pi)^{N-y}$. Now, if we keep the total size of the grid constant but divide it into more smaller cells, so the cell size $\rightarrow 0$ and $N \rightarrow \infty$ while holding $N\pi$ constant, then the limiting distribution of Y is Poisson(λ), where $\lambda = N\pi$.

Young and Young (1998) provide an alternative motivation for the negative binomial as a biological process. Say we have plants and insects. An insect can either be on a given plant or not. Let $\pi = P\{\text{insect on plant}\}$. Suppose the insects arrive one at a time, each one going to our target plant with probability π and independently of other insects' behavior. Then the probability our plant has exactly Y insects is $\binom{N}{y}\pi^y(1-\pi)^{N-y}$. Now, if we let $N \rightarrow \infty$, such that $N\pi$ remains constant, then we have the same limiting distribution, $Y \sim \text{Poisson}(\lambda)$ where $\lambda = N\pi$.

On the other hand, suppose that once an insect enters the system, the next insect's behavior is NOT independent. As a simple example, imagine that the first insect goes to our plant with $\pi=0.5$. Now for the second insect, instead of $\pi=0.5$ irrespective of the first insect, the insect goes to our plant with $\pi=2/3$ if the first insect is on our plant, but $\pi=1/3$ otherwise. This means that for $N=2$, the probabilities of $Y=0$, 1, or 2 insects on our plant are each 1/3.

Instead of the standard development of a binomial process, we now have the beginnings of Bose–Einstein statistics. If we continue in this vein, with each new insect *more* likely to "select" our plant when more insects are already there and *less* likely to select it when fewer insects are there, we have a process who limiting distribution is geometric, a special case of the negative binomial with scale parameter $\phi=1$. With some tweaking, this process generalizes to the negative binomial. Thinking about it this way, the negative binomial seems to be a natural model for many biological processes, whereas the Poisson seems distinctly unrealistic. Insects cluster in favorable locations and avoid unfavorable locations. People tend to like to live in cities, or at least towns with jobs and amenities. Biological entities rarely pick a spot at random on the map without regard to local conditions.

Returning to (11.10) before concluding this section, recall at the conclusion of Section 11.2, we wanted to test the negative binomial scale parameter—specifically $H_0\colon\phi=0$. For the negative binomial model, $Var(y|b)=\lambda(1+\phi\lambda)$; hence, if $\phi=0$, then $Var(y|b)$ reduces to λ, the Poisson variance. In this sense, we could regard $H_0\colon\phi=0$ as a test of whether the Poisson or negative binomial distribution provides the better fit. If we try to do this with the COVTEST statement in PROC GLIMMIX, we get the following:

Tests of Covariance Parameters					
		Based on the Likelihood			
Label	**DF**	**–2 Log Like**	**ChiSq**	**Pr>ChiSq**	**Note**
Scale=0	1	6E21	6E21	<.0001	MI

Notice that the "–2LogLike" and "ChiSq" are both 6E21—a nonsense number indicating that the computation could not be done. Inspecting (11.10) we see why. The likelihood ratio test requires determining the log-likelihood under the null hypothesis, but one of the terms in the log-likelihood is $\left(1/\phi\right)\left(1/1+\lambda\phi\right)$—undefined at $\phi=0$. In other words, the likelihood ratio test of $H_0\colon\phi=0$ for a negative binomial model is undefined. There are a couple of ways around this. One way would be to compare the –2 log-likelihood Fit Statistics from the Poisson and negative binomial models and treat the difference as a $\chi^2_{(1)}$ statistic. A better way is to use the generalized Poisson distribution, shown in the following.

11.3.2 Generalized Poisson

Joe and Zhu (2005) present the following generalization of the Poisson p.d.f.:

$$f(y;\lambda,\xi)=\frac{\lambda}{y!}(\lambda+\xi y)^{y-1}e^{-(\lambda+\xi y)} \tag{11.11}$$

Note that the expected value and variance are $E(y)=\mu=\lambda/(1-\xi)$ and $Var(y)=\mu/(1-\xi)^2$, respectively. We interpret ξ as an overdispersion parameter.

Notice that for $\xi = 0$, we have

$$f(y;\lambda,0) = \frac{\lambda}{y!}(\lambda)^{y-1} e^{-(\lambda)} = \frac{e^{-\lambda}\lambda^{y}}{y!}.$$

That is, at $\xi = 0$, we have the standard Poisson distribution. However, unlike the negative binomial, we have a log-likelihood at $\xi = 0$ that we can evaluate, meaning that we can compute a likelihood ratio statistic to test H_0: $\xi = 0$.

The generalized Poisson is not among the distributions available with the DISTRIBUTION = options in PROC GLIMMIX, but Example 38.14 in the SAS/STAT® (2008) documentation shows program statements to implement the generalized Poisson. For Data Set 11.3, the statements are

```
proc glimmix method=laplace data=RCBD_counts;
class blk trt;
  model y=trt/ link=log s;
    random int / subject=blk;
    xi = (1 - 1/exp(_phi_));
    _variance_ = _mu_ / (1-xi)/(1-xi);
    if (_mu_=.) or (_linp_ =.) then _logl_ =.;
    else do;
      mustar= _mu_ - xi*(_mu_-y);
      if (mustar < 1E-12) or (_mu_*(1-xi) < 1e-12) then
        _logl_ = -1E20;
      else do;
        _logl_ = log(_mu_*(1-xi)) + (y-1)*log(mustar) -
         mustar - lgamma(y+1);
      end;
    end;
  lsmeans trt/ilink;
  covtest 'scale=0'. 0;
title 'Generalized Poisson model';
```

The program statements define user-specified log-likelihood and variance functions. You *must* use METHOD = LAPLACE or METHOD = QUAD for programs with user-defined distributions. Notice that the variable XI is the overdispersion parameter ξ. All variables that begin and end with an underscore, for example, _PHI_, are reserved variable names that are part of GLIMMIX's internal architecture. GLIMMIX defines the scale parameter _PHI_ as nonnegative only, and hence the reparameterization in the program to define the appropriate bounds: $0 < \xi < 1$. Consult the SAS/STAT documentation for more information. The relevant output is as follows:

Fit Statistics for Conditional Distribution	
−2 log L(y | r. effects)	198.42
Pearson chi-square	15.68
Pearson chi-square/DF	0.52

This compares with χ^2/df of 0.68 for the negative binomial. The AICC statistics are 227.4 for the negative binomial and 228.81 for the generalized Poisson. Neither the χ^2/df nor the AICC statistic provide compelling evidence favoring one distribution over the other, although the negative binomial appears to be *slightly* better using both criteria. The generalized Poisson is a heavier-tailed distribution, meaning that we would expect zero counts and very large counts to be somewhat more prevalent under the generalized Poisson than under the negative binomial. Examining the data may help.

Covariance Parameter Estimates			
Cov Parm	**Subject**	**Estimate**	**Standard Error**
Intercept	blk	0.3968	0.3524
Scale		1.4338	0.3394

The "scale" estimate here refers to SAS internal variable _PHI_. Using the definition mentioned earlier, we calculate the generalized Poisson $\hat{\xi}$ as $1-\left(1/e^{-PHI\_}\right)=1-\left(1/e^{1.4338}\right)=0.762$.

Type III Tests of Fixed Effects				
Effect	**Num DF**	**Den DF**	***F* Value**	**Pr>F**
trt	2	18	1.68	0.2148

This compares with $F=1.33$ for the negative binomial model. We could go ahead and compare the estimated treatment means, etc. These are not shown here. The differences between the generalized Poisson and negative binomial models for these data are neither striking nor consequential. This, however, may not necessarily be true for other data sets.

Tests of Covariance Parameters					
		Based on the Likelihood			
Label	**DF**	**−2 Log Like**	**ChiSq**	**Pr>ChiSq**	**Note**
Scale=0	1	340.11	123.80	<0.0001	MI

This output is *the* compelling reason to run the model. Here we have a legitimate test of the "Poisson-ness" of the data. $\chi^2=123.8$ provides overwhelming evidence that $\xi>0$ and, hence, the standard Poisson does *not* fit the data. Legitimate inference must come from one of the models that account for overdispersion: generalized Poisson, negative binomial, or Poisson with block × treatment included in the linear predictor.

11.4 Conditional and Marginal

In Chapter 3, we first encountered the distinction between the conditional and marginal model. We saw that for non-Gaussian GLMMs the two not only produce different results, but they target different parameters. Consequently, we must not be vague about model construction or interpretation. We need to be sure that the way we are thinking about the problem agrees with the parameters our model targets. Non-Gaussian marginal and conditional models constitute an open invitation for sloppy interpretation.

In Chapters 4 and 5, we encountered GEE-type models. These are non-Gaussian GLMMs that use working correlation structures embedded in $Var(\mathbf{y}|\mathbf{b})$, the "R-side" variance. A true GEE model has a fixed-effects-only linear predictor, $\boldsymbol{\eta}=\mathbf{X}\boldsymbol{\beta}$, and all variation embedded on the "R-side." GEE models are necessarily quasi-likelihood models, because the presence of a working correlation structure guarantees the nonexistence of any feasible probability mechanism that corresponds to the model as written.

What is the relevance here? In certain circles, GEEs are a popular way to deal with overdispersion in multilevel models with counts. This is partly attributable to the fact that GEE methodology occurred and GEE-capable software became available somewhat earlier than GLMM software. For example, in SAS®, PROC GENMOD was available with a GEE option in the early 1990s, but PROC GLIMMIX was not released until 2005. Because it was the only game in town for over a decade, the GEE approach became entrenched, and old habits die hard.

To fully appreciate the issues involved, let us consider a multilevel example. The data appear in the Chapter 11 Appendix as Data Set 11.4. The treatment design used a 7×4 factorial structure—seven levels of factor A and four levels of factor B. Four blocks were divided into seven units to which levels of factor A were randomly assigned. Each unit was divided into four subunits to which levels of factor B were randomly assigned. The skeleton ANOVA is thus

Topographical		**Treatment**		**Combined**	
Source	**d.f.**	**Source**	**d.f.**	**Source**	**d.f.**
Block	3			Block	3
		A	6	A	6
Unit(block)	24			Block × A	18
		B	3	B	3
		A × B	18	A × B	18
Subunit(unit)	84	"Parallels"	18	B × block(A) Residual	63
Total	111	Total	111	Total	111

The data were counts. A "first approximation" model following the standard process is thus

1. Linear predictor: $\eta_{ijk}=\eta+\alpha_i+\beta_j+(\alpha\beta)_{ij}+r_k+(ra)_{ik}$, where α, β, and r denote A, B, and block effects, respectively
2. Distributions
   a. $y_{ijk}|r_k, (ra)_{ik} \sim Poisson(\lambda_{ijk})$
   b. r_k i.i.d. $N\left(0,\sigma_R^2\right)$
   c. $(ra)_{ik}$ i.i.d. $N\left(0,\sigma_{RA}^2\right)$
3. Link function: $\eta_{ijk}=\log(\lambda_{ijk})$

As usual, we start with a preliminary run using METHOD=LAPLACE. With these data, the Pearson $\chi^2/df=4.50$, meaning we need to identify a suitable amendment to the model to account for overdispersion. Essentially, this retraces the steps shown in Section 11.2. We will not repeat them here. Instead, let us ask how the GEE approach differs. More importantly, let us explore what implicitly goes with the territory.
The GEE-type model is

1. Linear predictor: $\eta_{ijk}=\eta+\alpha_i+\beta_j+(\alpha\beta)_{ij}+r_k$
2. Distributions
   a. $y_{ijk}|r_k \sim Poisson(\lambda_{ijk})$
   b. r_k i.i.d. $N\left(0,\sigma_R^2\right)$
   c. Working covariance: $Var\left(\mathbf{y}_{ik}\mid r_k\right)=\mathbf{I}\boldsymbol{\lambda}_{ik}\left(\phi\begin{bmatrix}1 & \rho & \rho & \rho\\ & 1 & \rho & \rho\\ & & 1 & \rho\\ & & & 1\end{bmatrix}\right)$, where $\mathbf{y}'_{ik}=\begin{bmatrix}y_{i1k} & y_{i2k} & y_{i3k} & y_{14k}\end{bmatrix}$, the observations of the 4 subunits within each unit, and $\boldsymbol{\lambda}'_{ik}=\begin{bmatrix}\lambda_{i1k} & \lambda_{i2k} & \lambda_{i3k} & \lambda_{i4k}\end{bmatrix}$, the variance functions associated with the y_{ijk}
3. Link function: $\eta_{ijk}=\log(\lambda_{ijk})$

$Var(\mathbf{y}_{ik}|r_k)$ follows from the general structure we developed in Chapters 4 and 5, $Var(\mathbf{y}|\mathbf{b})=\mathbf{V}_\mu^{1/2}\mathbf{A}_W\mathbf{V}_\mu^{1/2}$. Notice that the working covariance matrix here, $\mathbf{A}_W=\left(\phi\begin{bmatrix}1 & \rho & \rho & \rho\\ & 1 & \rho & \rho\\ & & 1 & \rho\\ & & & 1\end{bmatrix}\right)$, has a scale parameter. This means that a GEE-type model has a built-in overdispersion parameter. Like it or not, with the GEE-type model, we get the following:

1. Estimation and inference that targets the mean of the marginal distribution. As we have seen in previous examples in other chapters, the marginal mean is *not* the Poisson rate parameter λ_{ijk}. If you interpret the analysis from a GEE-type model as if you have estimates of Poisson rate parameters, that is an example of vague and sloppy interpretation. As Deming used to say in his famed 3 day courses, "Shame!"
2. A linear overdispersion adjustment. That is, the working correlation implicitly defines a quasi-likelihood such that the conditional variance is $\phi\lambda_{ijk}$.

The latter consequence is especially a problem for counting processes for where the negative binomial scenario is more plausible. This is because using a GEE-type model defined on a negative binomial yields a particularly ill-conceived model. We would have a working covariance with an embedded scale parameter *and* a distribution defined in part by a scale parameter. The two scale parameters are not the same thing, but they would have overlapping interpretations and only be partially identifiable. At best, the resulting estimates would be next to impossible to interpret. At worst, the estimating equations would be so ill conditioned that we might not get a solution, and even if we do, it might be utter nonsense.

This said, for curiosity's sake, if nothing else, let us see what we get with the GEE-type Poisson model here. The GLIMMIX statements are

```
proc glimmix data=sp_counts;
  class a b blk;
  model count=a b a*b/dist=Poisson;
  random intercept/subject=blk;
  random _residual_/type=cs subject=blk*a;
  lsmeans a*b/ilink;
```

The selected results are as follows:

Covariance Parameter Estimates

Cov Parm	Subject	Estimate	Standard Error
Intercept	blk	0.06116	0.08064
CS	a*blk	3.4556	1.8846
Residual		9.2290	1.6631

Type III Tests of Fixed Effects

Effect	Num DF	Den DF	*F* Value	Pr>F
a	6	81	1.85	0.2893
b	3	81	2.44	0.0707
*a***b*	18	81	1.05	0.4193

ab* Least Squares Means**

a	*b*	Estimate	Standard Error	DF	t Value	Pr > |t|	Mean	Standard Error Mean
1	1	2.3130	0.5699	81	4.06	0.0001	10.1043	5.7588
1	2	2.0959	0.6323	81	3.31	0.0014	8.1327	5.1425
1	3	2.9561	0.4220	81	7.01	<0.0001	19.2228	8.1117
1	4	1.8575	0.7094	81	2.62	0.0105	6.4076	4.5459
				...				
7	1	3.9417	0.2759	81	14.29	<0.0001	51.5073	14.2125
7	2	3.7699	0.2958	81	12.74	<0.0001	43.3746	12.8317
7	3	3.1945	0.3789	81	8.43	<0.0001	24.3982	9.2447
7	4	2.9938	0.4148	81	7.22	<0.0001	19.9622	8.2800

In the interest of space, only the LSMeans for factor A levels 1 and 7 appear. How do these compare with the GLMM? Recall that the Poisson GLMM showed strong evidence of over-dispersion. We omit the steps here, but the negative binomial does turn out to be the model of choice for the GLMM. Its model is identical to what we saw earlier, except we replace the

Poisson by the negative binomial as the conditional distribution of the observations. The results are as follows:

Covariance Parameter Estimates

Cov Parm	Subject	Estimate	Standard Error
Intercept	blk	0.002400	0.02765
a	blk	0.1222	0.07103
Scale		0.3459	0.06876

Type III Tests of Fixed Effects

Effect	Num DF	Den DF	*F* Value	Pr>F
a	6	18	2.70	0.0473
b	3	63	2.13	0.1054
*a*b*	18	63	1.18	0.3018

***a*b* Least Squares Means**

a	*b*	Estimate	Standard Error	DF	t Value	Pr > |t|	Mean	Standard Error Mean
1	1	2.3399	0.3801	63	6.16	<0.0001	10.3800	3.9459
1	2	2.1273	0.3872	63	5.49	<0.0001	8.3924	3.2497
1	3	2.9782	0.3736	63	7.97	<0.0001	19.6534	7.3421
1	4	1.8231	0.4110	63	4.44	<0.0001	6.1910	2.5446
				...				
7	1	3.8096	0.3576	63	10.65	<0.0001	45.1342	16.1390
7	2	3.7238	0.3540	63	10.52	<0.0001	41.4203	14.6607
7	3	3.1810	0.3596	63	8.85	<0.0001	24.0715	8.6558
7	4	2.9132	0.3654	63	7.97	<0.0001	18.4164	6.7285

Except for the block variance, the covariance parameter estimates for the two analyses are not comparable. While the magnitude of $\hat{\sigma}_R^2$ appears small in both cases, the GEE estimate is roughly 30 times the GLMM estimate (0.06 vs. 0.002). If we are merely using BLK to remove variation, then this discrepancy has little impact. If σ_R^2 has an intrinsic interpretation or if we need to use it to plan subsequent studies, then we will need to think carefully about whether the GEE or GLMM is best targeting our intended inference. The *magnitudes* of the *F*-values do not differ greatly (*in this case*) but the *conclusions* may differ considerably. Both tell us that the A*B interaction appears to be negligible. However, the GEE main effects for both A and B appear to be marginally significant, whereas the GLMM main effect of A shows a stronger level of significance, and the *p*-value for the main effect of B is less convincing. The estimates of the expected counts appear in the MEAN column of the LSMeans table. They do not appear to be all that different except that relatively large means are consistently greater for the GEE than the GLMM. This is a reflection of the characteristic skewness of the marginal distribution—GEEs will always tend to overestimate larger counts.

11.5 Too Many Zeroes

Overdispersion in count data can result from a number of causes. So far, we have considered two: under-specified linear predictors and assuming the wrong distribution. Usually, the former occurs with Poisson models that omit the last term in the skeleton ANOVA. The latter commonly results from assuming a Poisson process when the negative binomial more realistically describes how the counts arise. These both assume that given the right linear predictor or the right distribution, a model with a single distribution is possible. For some forms of overdispersion, we have more zeroes than we would expect under *any* of the discrete counting distributions. We call this the *too many zeroes* problem.

Two types of models—*zero-inflated* and *hurdle*—suggested by Lambert (1992) and Mullahy (1986), respectively, allow use of linear models in the presence of too many zeroes. Both use mixtures of a binary, on-off process and a discrete counting distribution, the two most common being the Poisson and the negative binomial. Both models define an "off" phase, in which only zero counts are possible, and an "on" phase, during which nonzero counts may occur. They differ as follows:

- In zero-inflated models, zero *or* nonzero counts may occur in the "on" part of the process, with the probability of 0, 1, 2, etc., determined by the counting distribution in the "on" part of the process.
- In hurdle models, only nonzero counts may be observed in the "on" part of the process.

Thus, the distinction is what happens in the "on" part of the process—in zero-inflated models, zero counts are still possible; in hurdle models, zero counts only happen when the process is "off."

Example to illustrate zero-inflated vs. hurdle models
We can think of the distinction between zero-inflated and hurdle models using the following illustrations.

Zero-inflated. In one scenario, suppose we count cars going through an intersection with a traffic light. Assuming that people obey the law, when the light is red, cars must stop—they cannot proceed through the intersection. When we observe the system when the light is red, we always observe a zero count. When the light is green, cars are free to move through the intersection. However, even though the light is green, it is still possible that traffic is light and no cars go through the intersection. When the light is green, any count—including zero—is possible.

Another scenario: imagine that our study involves counting eagles. If one of our sampling sites is, say, in Antarctica, we will certainly not see any eagles. The system is "off"—zero counts are certain. For sampling sights, say, in one of the Great Plains river flyways, there are certainly eagles to be seen. But on any given sampling occasion, we may or may not see an eagle.

Hurdle. Suppose we count the number of times individuals visit the doctor in a given year. We have two types of individuals: those who visited the doctor during the past year and those who did not. For those who did not, the number of visits they made to the doctor must be zero. For those who did, the number of visits must be at least one. Here, the "off" state means "did not visit doctor" which necessarily means a count of zero visits. The "on" state means "did visit doctor" which necessarily means the count is one or greater. No zero counts in the "on" state.

Another scenario: we want to model the number of insurance claims people make, perhaps to compare their behavior under two different policies. We have those who did not file a claim and those who did. If they did file a claim, then the count must be at least one. The hurdle model's value rests with the fact that it allows the probability of not filing a claim to be arbitrary and not restricted by the distribution of the nonzero counts.

11.5.1 Formal Description of Zero-Inflated and Hurdle Models

Both the zero-inflated and hurdle models must be described in two parts. We call these process 1, the on-off part, and process 2, the counting part.

Process 1: This is a binary process. The count process is "off" with probability π and "on" with probability $1-\pi$. When the process is "off," only zero counts are observed. We call π the *inflation probability*. This is because zero counts are "inflated"—more numerous that they would be with any single discrete counting distribution—when the system is "off."

Process 1 is the same for zero-inflated and hurdle models. The models differ in process 2.

Process 2 for zero-inflated models: When the system is "on," counts may take on values 0, 1, 2,... according to a discrete probability distribution. Thus, zero-inflated models can be described generically as follows. Let Y denote the random variable corresponding to observations from a zero-inflated model:

$$\Pr\{Y=y\}=\begin{cases}\pi+(1-\pi)f(0) & \text{for } y=0\\ (1-\pi)f(y) & \text{for } y=1,2,\ldots\end{cases} \tag{11.12}$$

where $f(y)$ is the discrete probability distribution function, for example, Poisson or negative binomial.

Process 2 for hurdle models: When the system is "on," counts must be nonzero. Their distribution is expressed by a zero-truncated p.d.f, whose form is $\frac{f(y)}{1-f(0)}$, $y=1,2,\ldots$. In words, this is the distribution for the nonzero values of y scaled by $1-f(0)$ so that $\sum_{y=1}^{\infty}\frac{f(y)}{1-f(0)}=1$.

The resulting probability distribution for the hurdle model is thus

$$\Pr\{Y=y\}=\begin{cases}\pi & \text{for } y=0\\ (1-\pi)\dfrac{f(y)}{1-f(0)} & \text{for } y=1,2,\ldots\end{cases} \tag{11.13}$$

11.5.2 GLMM for Poisson and Negative Binomial Zero-Inflated and Hurdle Models

We discuss four types of models for too many zeros:

1. Zero-inflated Poisson (or *ZIP*) model. That is, the p.d.f. $f(y)$ as denoted earlier follows a Poisson distribution.
2. Zero-inflated negative binomial (or *ZINB*) model. The p.d.f. follows a negative binomial.

3. Poisson hurdle model.
4. Negative binomial hurdle model.

The four forms define the conditional distribution of the GLMM. Thus, the generic description of the "too many zeroes" GLMM is

1. Linear predictor: $\boldsymbol{\eta} = \mathbf{X}\boldsymbol{\beta} + \mathbf{Zb}$
2. Distributions:
   a. $\mathbf{y}|\mathbf{b}$: p.d.f. as shown earlier. ZIP, ZINB, Poisson, or NB hurdle
   b. $\mathbf{b} \sim N(\mathbf{0}, \mathbf{G})$
3. Link: $\boldsymbol{\eta} = \log(\boldsymbol{\mu}|\mathbf{b})$

Unlike all of the GLMMs we have seen to this point, none of the SAS linear model software can implement these models. PROC GENMOD has limited ZIP and ZINB capability, but GENMOD is a fixed-effects-only program so it cannot be considered an option. To implement "too many zero" GLMMs, we have to use nonlinear mixed model software that will accept a user-defined log-likelihood for $\mathbf{y}|\mathbf{b}$. In SAS this means PROC NLMIXED. The next section contains example programs. First, we need to describe the process and the required log-likelihoods.

The NLMIXED programs require that we specify the following:

- An inflation parameter on the logit or probit link scale. Why logit or probit? Because the inflation probability is a parameter of the binary distribution. We usually use the logit because it is the canonical parameter. Letting α_0 denote the inflation parameter on the logit scale implies that we estimate the inflation probability by $\hat{\pi} = 1/(1+e^{-\hat{\alpha}_0})$.
- Linear predictor: $\boldsymbol{\eta} = \boldsymbol{X\beta} + \boldsymbol{Zb}$.
- Link: for both Poisson and negative binomial we use the log link, $\eta = \log(\lambda)$.
- Hence, $\hat{\boldsymbol{\lambda}} = e^{X\hat{\beta}+Z\hat{b}}$.
- Log-likelihood—must be defined—the four are written as shown in the following.

Zero-inflated log-likelihood

Applying the log to the generic form given in Equation 11.12 yields

$$\log L = \begin{cases} \log\{\pi + (1-\pi) f(0)\} & \text{for } y = 0 \\ \log(1-\pi) + \log[f(y)] & \text{for } y=1,2,\ldots \end{cases}$$

The resulting log-likelihoods are

ZIP log-likelihood

$$\log L = \begin{cases} \log\{\pi + (1-\pi) e^{-\lambda}\} & \text{for } y = 0 \\ \log(1-\pi) + \log\left[\dfrac{e^{-\lambda}\lambda^{y}}{y!}\right] = \log(1-\pi) + y\log(\lambda) - \lambda - \log(y!) & \text{for } y=1,2,\ldots \end{cases} \tag{11.14}$$

ZINB log-likelihood

$$\log L = \begin{cases} \log\left\{\pi + (1-\pi)\left(\dfrac{1}{1+\lambda\phi}\right)^{1/\phi}\right\} & \text{for } y=0 \\ \log(1-\pi) + \log\left[\dbinom{y+(1/\phi)-1}{y}\left(\dfrac{\lambda\phi}{1+\lambda\phi}\right)^{y}\left(\dfrac{1}{1+\lambda\phi}\right)^{1/\phi}\right] = & \\ \log(1-\pi) + y\log\left(\dfrac{\lambda\phi}{1+\lambda\phi}\right) - \dfrac{1}{\phi}\log(1+\lambda\phi) + \log\left\{\dbinom{y+(1/\phi)-1}{y}\right\} & \text{for } y=1,2,\ldots \end{cases} \tag{11.15}$$

Hurdle log-likelihood
The first step in obtaining the hurdle log-likelihood is to determine the **zero-truncated** form of the p.d.f., that is, $\dfrac{f(y)}{1-f(0)}$, $y=1,2,\ldots$

Zero-truncated Poisson:

$$\frac{\left(e^{-\lambda}\lambda^{y}\right)/y!}{1-e^{-\lambda}} \tag{11.16}$$

Zero-truncated negative binomial:

$$\frac{\dbinom{y+(1/\phi)-1}{y}\left(\dfrac{\lambda\phi}{1+\lambda\phi}\right)^{y}\left(\dfrac{1}{1+\lambda\phi}\right)^{1/\phi}}{1-\left(\dfrac{1}{1+\lambda\phi}\right)^{1/\phi}} \tag{11.17}$$

Now, using (11.12), it follows that the log-likelihoods for the Poisson and negative binomial hurdle models are, respectively,

Poisson hurdle log-likelihood

$$\log L = \begin{cases} \log(\pi) & \text{for } y=0 \\ \log(1-\pi) + y\log(\lambda) - \lambda - \log(y!) - \log\left[1-e^{-\lambda}\right] & \text{for } y=1,2,\ldots \end{cases} \tag{11.18}$$

Negative binomial hurdle log-likelihood

$$\log L = \begin{cases} \log(\pi) & \text{for } y=0 \\ \log(1-\pi) + y\log\left(\dfrac{\lambda\phi}{1+\lambda\phi}\right) - \left(\dfrac{1}{\phi}\right)\log(1+\lambda\phi) + \\ \quad \log\left\{\begin{pmatrix} y + 1/\phi - 1 \\ y \end{pmatrix}\right\} - \log\left[1 - \left(\dfrac{1}{1+\lambda\phi}\right)^{1/\phi}\right] & \text{for } y = 1,2,\ldots \end{cases} \tag{11.19}$$

One last item before we look at examples. We will need to use the gamma function in defining the log-likelihoods. We use the result $\Gamma(y+1)=y!$. SAS has two functions we will need: `gamma(y)` defines $\Gamma(y)$; `lgamma(y)` defines $\log[\Gamma(y)]$.

Example 11.1

The data for this example appear in SAS Data and Porgram Library, Data Set 11.5. These are simulated data from a scenario that might go something like this. Data were collected at 18 sites. There were n_i samples taken at the ith site—as few as two samples and as many as 14, depending on the site. For each sample, two observations, a predictor variable (EXPOSURE_LEVEL) and a response (TOXIN_COUNT), were taken. The objective: estimate a linear regression relating the predictor and response variable. The GLMM is

1. Linear Predictor: $\eta_{ij}=\beta_0+\beta_1X_{ij}+s_i$, where X_{ij} denotes EXPOSURE_LEVEL of the jth sample at the ith site and s_i denotes the site effect
2. Distributions:
   a. $y_{ij}|s_i \sim$ according to one of the distributions
   b. s_i i.i.d. $N(0,\sigma_S^2)$
3. Link: $\eta_{ij}=\log(\lambda_{ij})$

The NLMIXED programs for each of the four possible "too many zero" distributions are shown in the program file **four_zero_inflated_models.sas** in the SAS Data and Program Library. As an example, we look at the ZINB program:

```
proc nlmixed data=biohazard_study;
parms a0=0 scale=0.5 beta0=0 beta1=0 sd2_0=1;
     x=exposure_level;
     y=toxin_count;
/* lin predictor for zero-inflation probability        */
LinPr_Infl = a0;
/* InfProb=prob(zero) due to zero-inflation process        */
/* note: it is the logistic i-link of lp_pi       */
InfProb = 1/(1+exp(-LinPr_Infl));
/* neg bin mean (lamda)       */
eta=beta0+b0+beta1*x;
lamda=exp(eta);
/* build "y+n+1 chose y" part of neg bin logL        */
/* note SAS has no combinatoric "choose" function,
     so you use the gamma function        */
/* LGamma is the log(gamma) function        */
nn=1/scale;
```

```
/* note "pp=nn/(l_hat+nn)" is the prob in
    neg bin pdf you see in Math Stat class          */
log_choose=lgamma(y+nn)-lgamma(y+1)-lgamma(nn);
/* define logL for ZINB          */
if y=0 then ll=log(InfProb +(1-InfProb))-log(1+lamda*scale)/scale;
else ll=log(1-InfProb)+y*log(lamda*scale/(1+lamda*scale))-
log(1+lamda*scale)/scale+log_choose;
/* model & random statement              */
model y~general(ll);
random b0 ~ normal([0],[sd2_0*sd2_0]) subject=site;
/* estimate various things of interest      */
estimate 'var_site' sd2_0*sd2_0;
estimate 'yhat at x=0' exp(beta0);
estimate 'yhat at x=10' exp(beta0+beta1*10);
estimate 'yhat at x=20' exp(beta0+beta1*20);
estimate 'inflation prob' InfProb;
```

The statements in *bold* are active program statements. The other statements are comments to annotate the program. The NLMIXED program begins by setting starting values of all of the terms that will need to be estimated. The variable **LinPr _ Infl** is the inflation parameter on the logit scale. The variable **lamda** is the rate parameter, the mean of the negative binomial, obtained from **eta** the linear predictor via the inverse link function. The variables **nn** and **log _ choose** are conveniences, partly to avoid writing the expressions out repetitively and partly to help students relate terms in the log-likelihood to the parameterization of the negative binomial used in mathematical statistics texts. Notice that the MODEL statement defines the conditional distribution of the observations given the random effects. In NLMIXED you spell it out: in this case, **general(ll)** means we are supplying our own log-likelihood. The random effect distribution also has to be written out. The variance parameter is given in terms of the standard deviation—this is not required, but it does avoid the possibility of a negative solution for the variance, which often causes the NLMIXED program to crash. The SUBJECT statement is required. NLMIXED uses Gauss–Hermite quadrature only. Notice that the ESTIMATE statement in NLMIXED is defined in terms of the actual function of the parameters we desire, not a list of coefficients.
The output for this run appears as

Fit Statistics	
−2 Log likelihood	368.7
AIC (smaller is better)	378.7
AICC (smaller is better)	379.2
BIC (smaller is better)	383.2

Parameter Estimates

Parameter	Estimate	Standard Error	DF	t Value	Pr > |t|	Alpha	Lower	Upper	Gradient
a0	−18.9254	1719.92	17	−0.01	0.9913	0.05	−3647.64	3609.79	3.38E-7
scale	1.2535	0.3973	17	3.15	0.0058	0.05	0.4152	2.0917	0.000115
beta0	−1.2578	0.4022	17	−3.13	0.0061	0.05	−2.1063	−0.4092	−0.00125
beta1	0.01683	0.005292	17	3.18	0.0055	0.05	0.005666	0.02799	−0.07783
sd2_0	0.9086	0.2639	17	3.44	0.0031	0.05	0.3519	1.4654	−0.00069

Additional Estimates								
Label	Estimate	Standard Error	DF	t Value	Pr> |t|	Alpha	Lower	Upper
var_site	0.8256	0.4795	17	1.72	0.1033	0.05	−0.1861	1.8374
yhat at x=0	0.2843	0.1143	17	2.49	0.0236	0.05	0.04305	0.5255
yhat at x=10	0.3364	0.1233	17	2.73	0.0143	0.05	0.07623	0.5966
yhat at x=20	0.3981	0.1336	17	2.98	0.0084	0.05	0.1163	0.6799
Inflation prob	6.037E-9	0.000010	17	0.00	0.9995	0.05	−0.00002	0.000022

Here, the AICC fit statistic is 379.2. We can use this to compare it with the other three "too many zero" distributions. In this example, the ZINB model we are looking at did yield the smallest (best) AICC. The SCALE, BETA_0, and BETA_1 estimates are directly interpretable, respectively, as the negative binomial scale parameter (ϕ) and the regression intercept and slope on the model scale. The "additional estimates" are items of potential interest. Notice that the inflation probability here is essentially 0: the negative binomial model without zero-inflation adjustment actually provides the best fit here.

Example 11.2

The data for this example appear in SAS Data and Program Library, Data Set 11.6. A company is trying to assess the impact of their program to improve the quality of a product and reduce the number of warranty claims. "Trt 0" denotes the "preprogram" data on number of warranty claims. "Trt 1" denotes post-program data.

The company conducts a study over a one-year period, following the number of warranty claims by randomly sampled customers. The data set has 274 individuals sampled under trt "0" and 201 individuals sampled under trt "1." The object of this analysis is to see if the two treatments (trt "0" and trt "1") differ in the number of warranty claims and—if so—to characterize the difference.

In this model, the inflation probability itself may differ for the two treatments. In fact, if the program is working, it should. We can modify the likelihoods to make the inflation probability treatment specific. This is a hurdle model, as we noted earlier in describing examples we might consider ZI or hurdle models. The modified form of (11.13) to reflect treatment effect on the hurdle inflation probability is

$$\Pr\{Y = y \mid trt = i\} = \begin{cases} \pi_i & \text{for } y = 0 \\ (1-\pi_i)\dfrac{f_i(y)}{1-f_i(0)} & \text{for } y = 1,2,\ldots \end{cases} \tag{11.20}$$

where

$i=0, 1$, π_i denotes the inflation probability for treatment i

$f_i(\bullet)$ denotes the distribution given the ith treatment

The linear predictor for this model is $\eta_i = \eta + \tau_i$, where τ denotes the treatment effect. There are no random effects in this scenario.

The AICC for the Poisson hurdle is 1734.8; for the negative binomial 1549.6. The negative binomial form is the model of choice. The NLMIXED statements for the negative binomial hurdle model:

```
proc nlmixed data=claims;
   parameters scale=0.5 eta=0 tau_0=0
        a0=0 a1=0;
   y=count;
/* linear predictor for the inflation probability       */
LinPr_Infl= (trt=0)*a0+(trt=1)*a1;
/* infprob=inflation probability for zeros        */
/* assume Logit Hrdle zero-inflation       */
InfProb=1/(1+exp(-LinPr_Infl));
/* NegBin mean        */
lamda=exp(eta+(trt=0)*tau_0);
/* Build the Hurdle log likelihood */
nn=1/scale;
log_choose=lgamma(y+nn)-lgamma(y+1)-lgamma(nn);
   if y=0 then
     ll=log(infprob);
   else ll=log((1-infprob)) +
       +y*log(lamda*scale/(1+lamda*scale))-log(1+lamda*scale)/
scale+log_choose
       -(log(1-(1/(1+lamda*scale))**(1/scale)));
model y ~ general(ll);
estimate "inflation probability: trt=0" 1/(1+exp(-a0));
estimate "inflation probability: trt=1" 1/(1+exp(-a1));
estimate "NB rate: trt=0" exp(eta+tau_0);
estimate "NB rate: trt=1" exp(eta);
title 'Neg Binomial Hurdle';
```

As before, the program statements appear in *bold* and the comment statements in regular font. The program differs from the previous program in two instances. First, there are two inflation parameters, one for each treatment. This is reflected in the PARAMETERS statement that defines the starting values and in the **LinPr _ Infl** statement that defines the treatment-specific inflation parameter. Second, the log-likelihood has the hurdle form, as in (11.19) rather than the ZINB form (11.17) we saw earlier. Otherwise, this and the previous program are essentially identical.

In principle, we can make other parameters treatment specific as well. In some applications, the scale parameter may be different for two treatments. PROC GLIMMIX does not allow heterogeneous scale parameters (although that could change in future releases), but NLMIXED allows considerably greater flexibility. To vary the scale parameter, define two scale parameters in the PARAMETERS statement and then use a treatment-specific variable definition similar to the inflation parameter statement here.

The primary output of interest comes from the ADDITIONAL ESTIMATES:

Additional Estimates

Label	Estimate	Standard Error	DF	t Value	Pr > |t|	Alpha	Lower	Upper
Inflation probability: trt=0	0.4015	0.02961	475	13.56	<.0001	0.05	0.3433	0.4596
Inflation probability: trt=1	0.7612	0.03007	475	25.31	<.0001	0.05	0.7021	0.8203
NB rate: trt=0	3.3635	0.4002	475	8.40	<.0001	0.05	2.5771	4.1499
NB rate: trt=1	1.7870	0.3670	475	4.87	<.0001	0.05	1.0659	2.5080

We see that the inflation probability estimate for treatment 0 is $\hat{\pi}_0=0.40$, and for treatment 1, $\hat{\pi}_1=0.76$. The rate parameter estimates for the two treatments are $\hat{\lambda}_0=3.36$ and $\hat{\lambda}_1=1.79$. We could define additional statements to estimate the difference between the rate parameters and inflation probabilities. We can also compute the overall number of claims average. For example, for treatment 0, there is a 40% chance an individual does not make a claim, and among the 60% who do, the average number of claims is 3.36. Therefore, the *overall* average number of claims is $0.6 \times 3.36=2.02$. For treatment 1 it is 0.43. The new treatment appears to have been successful in reducing claims.

11.6 Summary

1. Rationale for GLM, GLMM with count data: normal approximation and standard transformations historically used with ANOVA and OLS linear models do not perform well.
2. Poisson is the standard choice for GLM, GLMM with counts:
   a. As a one-parameter exponential, the Poisson is very vulnerable to overdispersion.
   b. Evidence from many disciplines bears this out.
   c. *Always* begin analysis with overdispersion diagnostics
      i. Conditional Pearson χ^2/df
      ii. Residual plots
      iii. Mean–variance plots when possible
3. Fixes for overdispersion:
   a. Linear scale parameter. The classic "fix" but experience casts doubt on its effectiveness.
   b. Poisson + unit-of-observation effect from skeleton ANOVA.
   c. Replace Poisson with negative binomial or generalized Poisson.
   d. Zero-inflated and hurdle models.
4. In SAS, ZI and hurdle models require NLMIXED, other GLMMs with counts can be done with GLIMMIX.
5. In general, use integral approximation, not PL.
6. GEEs:
   a. Restricted to linear overdispersion (scale) parameter
   b. Restricted to marginal inference
   c. *Caveat emptor*

Exercises

11.1 This problem uses data in file `Ch_11_Problem1.sas`, in the SAS Data and Program Library.

These are count data from a balanced incomplete block design with seven blocks of size four and seven treatments. Assume that treatment 5 is a "control" treatment and the focus of the experiment is on identifying which treatments have expected counts significantly less than the control.

a. Analyze the data using the GLMM $\log(\lambda_{ij}) = \eta + \tau_i + b_j$, where τ denotes the treatment effect; b denotes the block effect, assumed i.i.d. $N(0, \sigma_B^2)$; and the data are assumed to have a Poisson distribution.
b. Reanalyze the data using the model $\log(\lambda_{ij}) = \eta + \tau_i + b_j + (bt)_{ij}$, where $(bt)_{ij}$ is the block*treatment interaction, assumed i.i.d. $N(0, \sigma_{BT}^2)$, and the other terms are as shown earlier.
c. Reanalyze the data using the model $\log(\lambda_{ij}) = \eta + \tau_i + b_j$, where all terms are as shown earlier except assume the data have a negative binomial distribution.
d. Compare and contrast your results from these three analyses with regard to the objective stated earlier.
e. Which analysis would be most appropriate to report and why? (Explaining "why" means citing relevant output from the analyses that indicate its appropriateness or lack thereof.)

11.2 This problem refers to data and data step statements given in file `ch_11_problem2.sas`.

The data are from a randomized complete block design with three treatments and six blocks. The response variable (Y) is a count.

a. Run a standard block-effects-random linear mixed model analysis. Make note of the variance component estimates, *F*-value for treatment, and estimates (and standard errors) of the treatment means.
b. Rerun the analysis using the log count, a standard, pre-GLMM variance stabilizing transformation. Note that "log" means *natural log*. For zero counts, you add a small increment, $\varepsilon > 0$, to the count before taking the log—for example, in SAS, `if y=0 then log_y=log(y+0.1)`. Obtain the variance component estimates, *F*-value for treatment, and the estimates (and standard errors) of the treatment means *on the data scale.*
c. Rerun the analysis as a GLMM assuming counts given blocks have a Poisson distribution. Use the same linear predictor that you did in parts (a) and (b). Obtain the variance components estimates, *F*-value for treatment, and the estimates (and standard errors) of the treatment means *on the data scale.*
d. Obtain the *Pearson chi-square/DF* fit statistic for the GLMM from part (c).
e. Rerun the GLMM from part (c) but add a random block*trt effect to the linear predictor and obtain the likelihood ratio test of $H_0: \sigma_{BT}^2 = 0$. The statistic from (d) and the test from (e) have essentially the same interpretation. Explain, briefly.
f. Characterize the impact of the change in the linear predictor between parts (c) and (e) on the F-test and estimates (and std errors) of treatment means *on the data scale.*
g. Pick the GLMM (part c or part e) most appropriate to report and compare and contrast the variance component, *F*- and data scale estimate results for the GLMM, log transform LMM, and untransformed LMM.

11.3 This problem refers to the SAS program `Ch_11_Problem3.sas`. The program generates 10,000 simulated data sets, each of which have the following structure:

- There are two treatments (denoted trt = 0 and trt = 1).
- Each treatment is observed on 10 units per treatment.

- The observations, *conditional on the unit*, have a Poisson distribution with rate parameter centered at $\lambda_0=2$ for trt=0 and $\lambda_1=5$ for trt=1.
- Units have a probability distribution: i.i.d. $N\left(0,\sigma_U^2\right)$. In this simulation, $\sigma_U^2=1$.
- The canonical parameter of the *observation* | *unit* distribution is perturbed at the unit level according to the unit probability distribution. What you see is the resulting observation on the random variable after the perturbation has occurred.

Two variables are generated by the simulation: "Observed_Count" and "True_Poisson." The former are the data one would actually see in such an experiment. The latter follow a Poisson distribution without any random disturbance attributable to the different units.

There are two PROC UNIVARIATE programs. The first gives complete descriptive statistics for the both variables—"Observed_Count" and "True_Poisson"—for the simulated data on trt=0 and for trt=1. The second provides a cleaned up look at the histograms for each variable on each treatment after deleting a handful of extreme values. A normal distribution is superimposed over the histogram so you can visualize how well the normal approximation fits.

Based on the information provided by the program, answer the following:

a. Give the GLMM that most appropriately describes the variable "Observed_Count":
   i. Linear predictor:
   ii. Distribution(s):
   iii. Link:
b. Compare and contrast the distributions of "Observed_Count" versus "True_Poisson" for each treatment. Pay attention to the mean, median, variance, and shape of each distribution.
c. Why are the distributions so different? Formally, show how the distribution of "observed count" arises—that is, give a formal description of its probability density function.
d. Suppose you fit the following:

```
proc glimmix;
class trt;
model observed_count=trt;
lsmeans trt;
```

   i. What is the default distribution assumed in this program?
   ii. What would the expected value of the LSMeans for each treatment be—approximately: base your speculation on the simulation results you described in part (b).
e. Write the GLIMMIX statements for your model given in part (a). For *your program*/model what would the expected value of the LSMeans for each treatment be—on the data scale—approximately: as with part (d), base your speculation on the simulation results you described in part (b).

11.4 This problem refers to the SAS program `Ch_11_Problem4.sas`. The program generates 10,000 simulated data sets, each of which have the following structure:

- There are two treatments (denoted trt=0 and trt=1).
- There are 10 blocks, each size 2. Each treatment is observed once in each block.

- The observations, *conditional on the unit*, have a negative binomial distribution with rate parameter centered at $\lambda_0 = 10$ for trt=0 and $\lambda_1 = 5$ for trt=1. The negative binomial scale parameter is $\phi = 0.4$.
- Block effects have a probability distribution: i.i.d. $N\left(0, \sigma_B^2\right)$. In this simulation, $\sigma_B^2 = 1$.

Two variables are generated by the simulation: "Observed_Count" and "True_Negative Binomial." The former are the data one would actually see in such an experiment. The latter follow a negative binomial distribution without any random disturbance attributable to block.

The PROC UNIVARIATE program gives descriptive statistics for the both variables—"Observed_Count" and "True_NegativeBinomial"—for the simulated data on trt=0 and trt=1.

Based on the information provided by the program, answer the following:

a. Give the GLMM that most appropriately describes the variable "Observed_Count":
   i. Linear predictor:
   ii. Distribution(s):
   iii. Link:
   iv. Anything else?
b. Compare and contrast the distributions of "Observed_Count" versus "True_NegativeBinomial" for each treatment. Pay attention to the mean, median, variance, and shape of each distribution.
c. What *should* the mean and variance of the "True_NegativeBinomial" be for each treatment? Do they agree with your simulated data in (b)?
d. Why are the distributions so different? Formally, show how the distribution of "observed count" arises—that is, give a formal description of its probability density function.
e. Suppose you fit the following:

```
proc glimmix;
class trt block;
model observed_count=trt;
random intercept/subject=block;
lsmeans trt;
```

What would the expected value of the LSMeans for each treatment be? Approximately base your speculation on the simulation results you described in part (b).

f. Suppose you add `/dist=Poisson` to the model statement in (e).
   Would you expect the resulting GLMM to provide a good fit for data arising from this process? Explain, briefly.
g. Write the GLIMMIX statements for *your* model given in part (a). For *your program*/model what should the expected value of the LSMeans for each treatment be—on the data scale? As with part (e), base your speculation on the simulation results you described in part (b).

11.5 This problem uses SAS file `ch _ 11 _ Problem5.sas`.
The data are from a strip-plot experiment with six blocks and a 3 × 3 treatment design—three levels of A, three levels of B. Visualize a block of a strip-plot as follows:

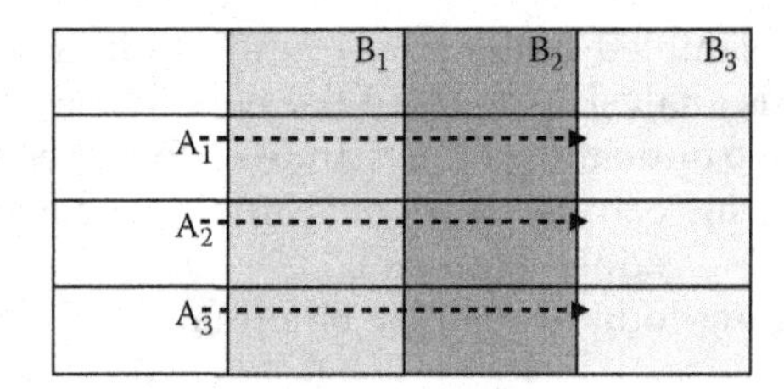

In each block, levels of A are randomly assigned to a row and applied across the row; levels of B are randomly assigned to a column and applied down the entire column. Each block has a potentially unique assignment of A to rows and B to columns.

The following parts were also in Chapter 8, Exercise 3. They are repeated here for review and because they are relevant to the rest of the problem:

a. Show the WWFD process leading to a skeleton ANOVA for this design.
b. Write the linear predictor that follows from (a). Include assumptions about model effects considered random.
c. For which of the following distributions of the response conditional on the random model effects would it be reasonable to include a term corresponding to block*a*b in the linear predictor? Check one choice per distribution.
   i. Poisson ______reasonable to include ______not reasonable
   ii. Negative binomial ______reasonable to include ______not reasonable
d. *True or false* (choose one): if you fit a negative binomial to the linear predictor you wrote in (a), you implicitly assume that the block*a*b (unit-level) effect has a gamma distribution.
   The data for these runs are counts.
e. Consider Run 1.
   i. What is the method of variance component estimation used in this run?
   ii. Would the results of this run be appropriate to report? Why or why not?
f. Consider Runs 2 and 3. Notice that they give equivalent results.
   i. Give the form of $\mathbf{V}(\mathbf{y}^*) = \mathbf{ZGZ}' + \mathbf{R}^*$ for run 2
   ii. Give the form of $\mathbf{V}(\mathbf{y}^*) = \mathbf{ZGZ}' + \mathbf{R}^*$ for run 3
   iii. Verify that they are equivalent
g. Consider Runs 4, 5, and 6.
   i. Which runs agree and which disagree?
   ii. Give the form of $\mathbf{V}(\mathbf{y}^*) = \mathbf{ZGZ}' + \mathbf{R}^*$ for each of these runs.
   iii. Use the $\mathbf{V}(\mathbf{y}^*) = \mathbf{ZGZ}' + \mathbf{R}^*$ for runs 4, 5, and 6 to explain why these runs agree or disagree.
h. Finally, consider run 7. Does this model, as specified, make sense? Would it be reasonable to base an analysis on this model? Explain.

11.6 This problem uses the data and program statements from the file `ch_11_problem6.sas`.

A policy analysis group is trying to assess the impact of two different ways of handling a particular out-patient medical issue. One policy (labeled "trt" 0) is the standard way of handling the issue. The other policy (labeled "trt" 1) that has been initiated in some randomly selected districts to address some long-standing concerns about trt "0."

The policy analysis group conducts a study over a 1 year period, following the number of visits to out-patient clinics by randomly sampled individuals. The data set has 231 individuals sampled under trt "0" and 144 individuals sampled under trt "1." The object of this analysis is to see if the two policies (trt "0" and trt "1") affect the number of visits to out-patient clinics and—if so—to characterize the difference.

Because these data are counts, six models are under consideration:

1. GLM assuming count ~ Poisson
2. GLM assuming count ~ negative binomial
3. ZIP GLM
4. ZINB GLM
5. Poisson hurdle GLM
6. Negative binomial hurdle GLM

The SAS file contains the programs you need to run all 6 analyses. Use them to answer the following:

a. Describe the link and the linear predictor common to all six models.
b. Complete the following table:

Model	AICC
Poisson GLM	
Negative binomial GLM	
ZIP	
ZINB	
Poisson hurdle	
Negative binomial hurdle	

c. Based on (b), which model should be used to address the objective of this study?
d. Using the analysis from the model you selected in (d) write a paragraph summarizing the findings provided by this analysis relative to the goal of the study. The summary should provide a clear statement of the finding and the relevant evidence supporting that statement (e.g., if you find that "there is a treatment effect"—treatment effect on what, what parameter and test statistic allows you to say that, and what is the level of significance [or confidence, if you are basing your statement on a confidence interval]?)

12

Time-to-Event Data

12.1 Introduction: Probability Concepts for Time-to-Event Data

As the name implies, time-to-event data involve response variables that concern how long it takes for an event of interest to occur. Time-to-event data occur in a wide-variety of disciplines. Examples include recovery time or survival time in medical studies, how long until a drug takes effect or its effects wear off, time till seed germination or until a plant flowers, how long a component lasts, etc.

Like the rate and proportion data covered in Chapter 10 and the count data discussed in Chapter 11, modeling time-to-event data begins by considering the characteristics the generalized linear mixed model (GLMM) probability components must address. Here, the response variable must be continuous. Typically, it is strictly positive or, at least, non-negative. Often, the distribution is skewed; it may even be U-shaped. For example, many household appliances either fail almost immediately—if they are defective—or have years of trouble-free operation before their reliability decreases as they wear out. Because the Gaussian distribution lacks most of these characteristics, it is often a poor choice for modeling time-to-event. The exponential, log-normal, gamma, Weibull, and other more advanced distributions tend to be more common—and better—choices.

For many applications, we can approach modeling much as we have in Chapters 10 and 11—that is, by fitting GLMMs whose conditional distribution of the response given the random model effects, $\mathbf{y}|\mathbf{b}$, has the desired properties. In other applications, issues unique to time-to-event modeling have given rise to specialized methods and terminology, collectively known as *survival analysis*. One important issue is censoring. In this context, units in a study are said to be censored if they are not observed for their entire lifetime. For example, in a reliability trial, we may conclude the study after a specified period of time even though not every component has failed. In medical studies, we often cannot begin observing all patients at the time their treatment begins; some join the clinical trial weeks, months, or even years after they start the treatment under study. Units for which the event of interest has not occurred by the time the study concludes are said to be *right censored*; units that initiated treatment before entering the study are said to be *left censored*. Units can have observations that are both left and right censored.

In this chapter we will consider elementary time-to-event GLMMs only. There are entire courses—indeed series of courses—on survival analysis. The purpose of this chapter is simply to provide linear model students with the essential language and GLMM–survival analysis connection. In Section 12.2, we consider GLMMs that are not formally survival analysis models but use time-to-event-appropriate distributions of $\mathbf{y}|\mathbf{b}$. Section 12.3 consists of a very basic introduction to survival analysis from a GLMM perspective. We will introduce the basic framework and consider one example with uncensored data and one example with censoring.

Before beginning Sections 12.2 and 12.3, it will help to review some probability basics in the context of time-to-event data so that the motivation and rationale for the distributions used for $\mathbf{y}|\mathbf{b}$ in the GLMMs in this chapter make sense.

The simplest time-to-event model assumes that events of interest (the component fails, the patient gets well, the plant flowers, etc.) occur according to a Poisson process. That is, the probability that w events occur during t units of time—as defined in a particular study—is $\Pr\{W=w\}=e^{-\lambda t}(\lambda t)^w/w!$, where λ denotes the Poisson rate parameter per unit of time. Thus, the time till an event occurs can be expressed as $\Pr\{W=0|T \geq t\}=e^{-\lambda t}$, the probability that *no* event occurs until at least time t. Alternatively, this can be denoted in terms of the waiting time random variable T, that is, $\Pr\{\text{time till event } T \geq t\}=1-F(t)=e^{-\lambda t}$. Thus, the cumulative distribution function of T, $F(t)=\Pr\{T \leq t\}=1-e^{-\lambda t}$. The p.d.f. is thus

$$f(t;\lambda)=\frac{\partial\left[F(t)\right]}{\partial t}=\lambda e^{-\lambda t}; \quad 0 \leq t < \infty; \lambda > 0 \tag{12.1}$$

which we recognize as the exponential distribution. This is the simplest time-to-event distribution.

The gamma can be viewed as a generalization of the exponential distribution. In mathematical statistics textbooks, the gamma p.d.f. usually appears as $f(y;\alpha,\beta)=1/(\Gamma(\alpha)\beta^\alpha)$ $y^{\alpha-1}e^{-y/\beta}$ (see, e.g., Casella and Berger [2002]). For our purposes, it is better to reexpress the p.d.f. using the random variable T, rate parameter $\lambda=1/\beta$, and shape parameter $\phi=1/\alpha$. The reexpressed gamma p.d.f. is

$$f(t;\phi,\lambda)=\frac{\lambda^{(1/\phi)}}{\Gamma(1/\phi)}t^{(1/\phi)-1}e^{-\lambda t} \quad \text{where } 0 \leq t < \infty; \phi,\lambda > 0 \tag{12.2}$$

The gamma expected value is $E(T)=\mu=\alpha\beta=(\lambda\phi)^{-1}$ and the variance is $\text{Var}(T)=\alpha\beta^2=\mu^2\phi$. The motivation for expression (12.2) is that the mean, μ, variance function, μ^2, and scale parameter, ϕ, fit the needed framework for generalized linear modeling.

We can see that if $\phi=1$, then the gamma p.d.f. (12.2) reduces to (12.1), the exponential p.d.f. Also, just as the exponential distribution can be motivated from the upper tail of the Poisson c.d.f., the gamma can be motivated as the sum of independent Poisson c.d.f.

Now we turn to two time-to-event GLMMs.

12.2 Gamma GLMMs

In this section we consider two examples. The first is a hierarchical model arising from a split-plot design. These are common in biological sciences (where response variables include time to seed germination in plant science, time to estrus in animal science, and time to first instar in entomology, to name just a few) and reliability engineering (where response variables include burn-in time or time to component failure, etc.). Examples abound in other areas of study as well. The second example is from a response surface design.

12.2.1 Hierarchical (Split-Plot) Gamma GLMM

The data for this example appear as Data Set 12.1 in the SAS Data and Program Library. The design uses four complete blocks with three units each. Treatment factor A has three levels, which are randomly assigned to units within blocks so that each level of A appears once per block. Each unit consists of two subunits. Treatment factor B has two levels. They are randomly assigned to subunits so that each level of B appears once in each unit within a block. Thus, the treatment design is a 3×2 factorial and the study design is a split-plot. The response variable, denoted DAYS in Data Set 12.1, is short for the number of days until the event of interest happens. These are simulated data, but they are based on a horticulture experiment in which the response was number of days from the time the seedling "breaks soil" (i.e., its first appearance above the soil where seeds were planted) until flowering.

In skeleton ANOVA form, we can set up the essential architecture of the model:

Topographical		Treatment		Combined	
Source	**d.f.**	**Source**	**d.f.**	**Source**	**d.f.**
Block	3			block	3
		A	2	A	2
unit(block)	8			block × A	6
		B	1	B	1
		A × B	2	A × B	2
subunit(unit)	12			B × Block(A)	9
		"parallels"	18	or "residual"	
Total	23	Total	23	Total	29

The elements of the model are as follows:

1. Linear predictor: $\eta_{ijk} = \eta + \alpha_i + \beta_j + (\alpha\beta)_{ij} + r_k + (ra)_{ik}$, where α and β refer to the effects of A and B, assumed fixed, and r and ra refer to the block and block × A effects, respectively, both assumed random
2. Distributions:

   r_k i.i.d. $N\left(0, \sigma_R^2\right)$

   $(ra)_{ik}$ i.i.d. $N\left(0, \sigma_{RA}^2\right)$

   $y_{ijk} \mid r_k, (ra)_{ik} \sim$ ind $Gamma\left(\mu_{ijk}, \phi\right)$
3. Link: typically, $\eta_{ijk} = \log(\mu_{ijk})$; canonical link: $\eta_{ijk} = 1/\mu_{ijk}$

Note that although both the exponential and gamma distributions have canonical link equal to the inverse of the mean, exponential and gamma generalized linear models more often use the computationally more stable log link. Also note that the last term in the skeleton ANOVA should not appear in the linear predictor because we need the degrees of freedom to estimate the scale parameter, ϕ. However, because the gamma variance involves both the scale parameter *and* the variance function μ^2, the last term in the skeleton ANOVA is *not* "residual" as we commonly understand it for Gaussian linear models.

12.2.1.1 What Happens If We Fit This Model Using a Gaussian LMM?

As with other examples in this textbook, it is instructive to see what happens if we "do the wrong thing"—in this case, assume $y_{ijk}|r_k, (ra)_{ik} \sim$ Gaussian and, hence, fit a linear mixed model (LMM) with an identity link.

The GLIMMIX statements for the Gaussian LMM are

```
ods graphics on;
proc glimmix data=ch12_ex1 plot=residualpanel;
  class block a b;
  model days=a|b;
  random intercept a/subject=block;
  lsmeans a*b/cl;
```

We use the ODS GRAPHICS to obtain a plot of the residuals—the PLOT=RESIDUALPANEL option in the PROC statement. The CL option in the LSMEANS statement will also be instructive for this example.

The residual panel appears as

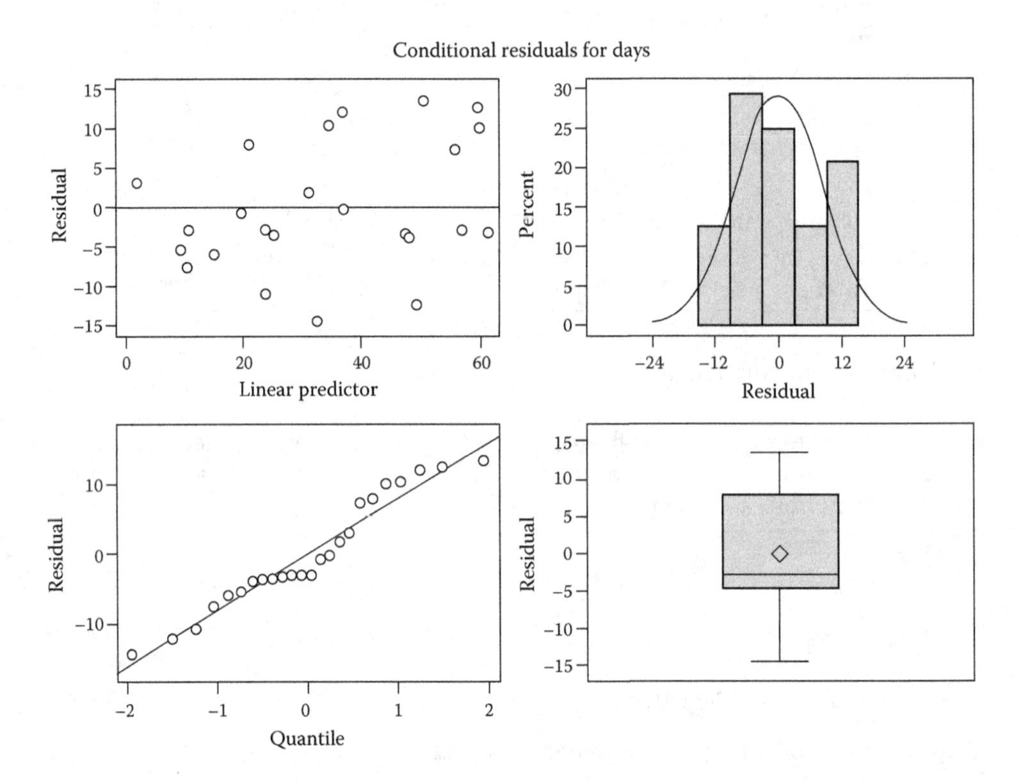

The histogram (upper right) shows an asymmetric, vaguely skewed pattern. The normal probability plot (lower left) shows marked departure from normality. The scatter plot (upper left) of residuals by predicted value suggests that the variance is not homogeneous but increases with increasing mean. In other words, these residual plots are more suggestive of an exponential or gamma distribution than they are of a Gaussian distribution.

The estimated treatment means and their 95% confidence limits appear in the LSMEANS listing as

a*b Least Squares Means

a	b	Estimate	Standard Error	DF	*t* Value	Pr > |*t*|	Alpha	Lower	Upper
1	1	30.7500	11.2412	9	2.74	0.0230	0.05	5.3206	56.1794
1	2	17.7500	11.2412	9	1.58	0.1488	0.05	–7.6794	43.1794
2	1	44.7500	11.2412	9	3.98	0.0032	0.05	19.3206	70.1794
2	2	43.5000	11.2412	9	3.87	0.0038	0.05	18.0706	68.9294
3	1	28.5000	11.2412	9	2.54	0.0320	0.05	3.0706	53.9294
3	2	40.0000	11.2412	9	3.56	0.0061	0.05	14.5706	65.4294

Notice treatment AB_{12}. The lower confidence bound for its mean, μ_{12}, is less than –7. Obviously, this is nonsense, because days to flowering cannot be negative: The plant cannot flower before it germinates!

Finally, the tests of overall effects appear as

Type III Tests of Fixed Effects

Effect	Num DF	Den DF	*F* Value	Pr>*F*
a	2	6	1.21	0.3630
b	1	9	0.04	0.8546
a*b	2	9	2.12	0.1762

This suggests that no statistically significant treatment effect can be inferred from these data—at least not based on the Gaussian LMM analysis. Let us see how this compares with a gamma GLMM.

12.2.1.2 Gamma Generalized Linear Model

The GLIMMIX statements modify those stated earlier as follows. First, we should use integral approximation, either METHOD=LAPLACE or QUADRATURE. Here, we use LAPLACE. Second, we add the DISTRIBUTION option to the MODEL statement. Finally, we add the ILINK option to the LSMEANS statement because the model-scale/data-scale distinction now matters. The affected GLIMMIX statements are

```
Proc glimmix method=laplace plot=residualpanel(ilink);
model days=a|b/d=gamma;
lsmeans a*b/ilink cl;
```

The CLASS and RANDOM statements do not change. GLIMMIX uses the log link on default when we specify D=GAMMA. While not shown here, we could specify the canonical link by adding the option LINK=INVERSE (or RECIPROCAL) to the MODEL statement.

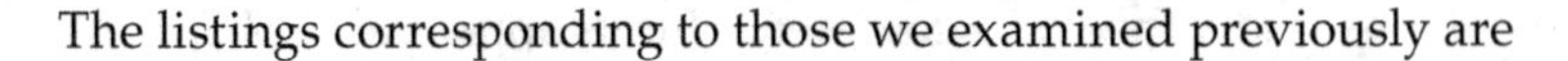
The listings corresponding to those we examined previously are

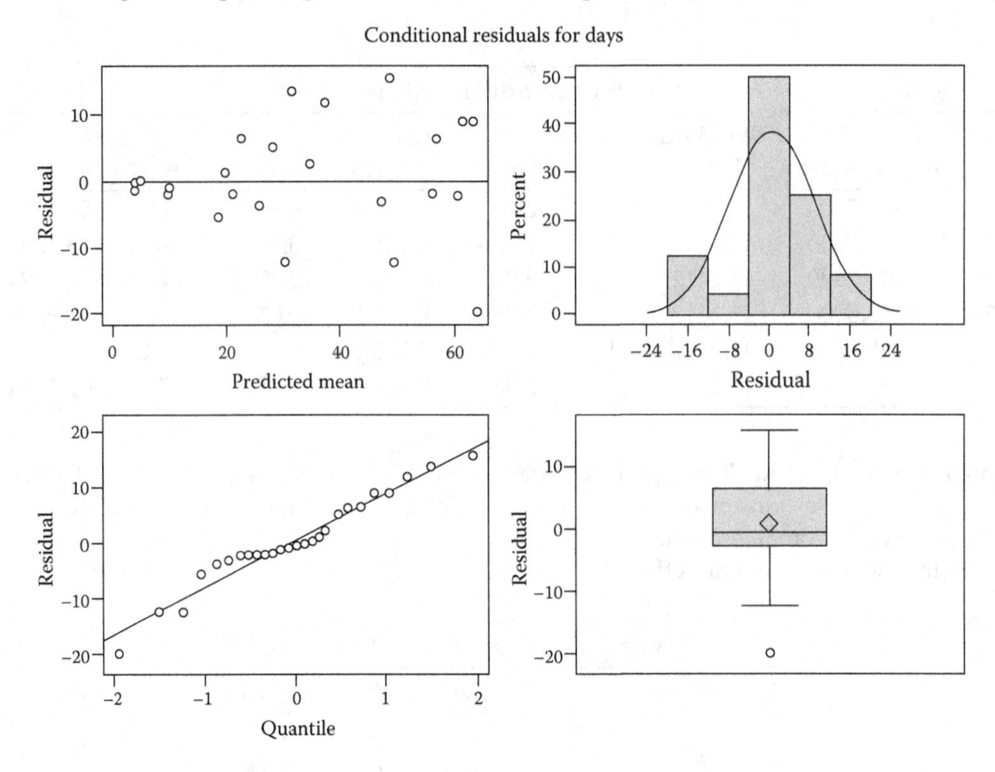

a*b Least Squares Means													
a	b	Estimate	Standard Error	DF	t Value	Pr > $\|t\|$	Alpha	Lower	Upper	Mean	Standard Error Mean	Lower Mean	Upper Mean
1	1	3.2565	0.5002	9	6.51	0.0001	0.05	2.1250	4.3881	25.9587	12.9848	8.3725	80.4840
1	2	2.5512	0.5002	9	5.10	0.0006	0.05	1.4196	3.6827	12.8219	6.4136	4.1355	39.7539
2	1	3.3475	0.5002	9	6.69	<0.0001	0.05	2.2159	4.4790	28.4306	14.2213	9.1698	88.1483
2	2	3.3603	0.5002	9	6.72	<0.0001	0.05	2.2287	4.4918	28.7973	14.4047	9.2880	89.2850
3	1	3.3100	0.5002	9	6.62	<0.0001	0.05	2.1785	4.4416	27.3861	13.6988	8.8329	84.9096
3	2	3.6033	0.5002	9	7.20	<0.0001	0.05	2.4718	4.7349	36.7204	18.3679	11.8435	113.85

Type III Tests of Fixed Effects				
Effect	Num DF	Den DF	F Value	Pr > F
a	2	6	0.59	0.5852
b	1	9	0.92	0.3621
a*b	2	9	4.60	0.0420

We now see the following:

- The residuals now appear to be more symmetric (somewhat) and much less heavy-tailed. The art/science of residual analysis for GLMMs is very much a work in progress and, generally, beyond the scope of this textbook. However, comparisons such as the one we see here *are* instructive.

- The confidence bounds *on the data scale* (the last four columns, labeled MEAN, of the LSMEANS table), while still wide, at least appear plausible. Also, all $\hat{\mu}_{ij}$ are significantly greater than 0, as you would expect for days to flowering.
- The Type III tests provide evidence of an $A \times B$ interaction. This makes sense looking at the $\hat{\mu}_{ij}$: B_2 appears to reduce days to flowering in conjunction with A_1, has no effect in conjunction with A_2, and increases days to flowering when applied with A_3. While not shown here, we could pursue this using the SLICEDIFF option.

Finally, we could ask if the gamma distribution was necessarily a good choice, or if fitting a GLMM using the exponential as the conditional distribution of $y_{ijk}|r_k$, $(ra)_{ik}$ would have sufficed. The first place to look is the covariance parameter estimates:

Covariance Parameter Estimates			
Cov Parm	**Subject**	**Estimate**	**Standard Error**
Intercept	block	0.3536	0.4633
a	block	0.5320	0.3415
Residual		0.1153	0.05434

RESIDUAL gives the estimate of the scale parameter, that is, $\hat{\phi} = 0.1153$. For the exponential to be a good candidate distribution, the scale parameter should be $\phi \approx 1$. Here, $\hat{\phi}$ seems a bit low for the exponential to be a plausible choice. Using the exponential might result in the theoretical variance being considerably greater than what we actually observe. This in turn might cause a ripple effect distorting parameter estimates, confidence bounds, and tests of significance. We could pursue this more formally using the COVTEST statement to obtain a confidence interval, as shown in Chapter 6. We leave writing the statement and running the program to the reader. The result, however, is that a 95% profile likelihood confidence interval for ϕ is [0.05, 0.34]—substantial evidence that the gamma, not the exponential, should be used.

Another implication of the scale parameter is that with the gamma, the estimated conditional variance is $(\hat{\mu}_{ij})^2 \hat{\phi}$ and, hence, ranges from approximately $(12.8)^2 \times (0.12) \cong 19.7$ for AB_{12} to approximately $(36.7)^2 \times (0.12) \cong 161.6$, whereas the Gaussian model assumes the conditional variance is constant, estimated by $\hat{\sigma}^2 = 141.8$. The gamma model's change in variance as a function of the mean seems more consistent with what we saw in the Gaussian residual plots, where the Gaussian model failed to account for the mean–variance relationship.

12.2.2 Response Surface for Time-to-Event: An Example Using the Box–Behnken Design

The data for this example appear as Data Set 12.2 in the SAS Data and Program Library. There are three factors, denoted A, B, and C. Each factor has three levels, 0, 1, and 2. Assume the levels are quantitative and the objective involves fitting a second-order polynomial regression. The study used a Box–Behnken design divided into eight runs each with four observations per run. The design is similar to the blocked Box–Behnken design used in Example 8.3. Here there are four sets of runs. One set includes the center treatment for factor A (i.e., $A = 1$) in conjunction with a 2^2 factorial defined on the outer levels (0 and 2) for factors B and C. The second set uses the center level for B, B = 1, with a 2^2 factorial defined on the outer levels of factors A and C. The third set uses center level for C, C = 1, with a 2^2 factorial for the 0 and 2 levels of factors A and B. The fourth set contains four observations on the center level of all three factors, that is, four replications of treatment combination ABC_{111}. The design has two runs of each set. Thus, the data set has a total of 32 observations.

The response variable is the processing time for a computer component, denoted R_TIME in Data Set 12.2. Lower is better. The factors represent levels of materials to be varied in the manufacturing process. The purpose of the study is to find out (1) which (if any) factor affects processing time and (2) what is the best factor level combination to use?

To address the objectives, the GLMM we fit is as follows:

1. Linear predictor: $\eta_{ijkl} = \eta_{ijk} + r_l$, where r_l denotes the block effects and

$$\eta_{ijk} = f(A,B,C) + (\alpha\beta\gamma)_{ijk} = \beta_0 + \sum_{i=A,B,C} \beta_i X_i + \sum_{i=A,B,C} \beta_{ii} X_i^2 + \sum_{i \neq i'=A,B,C} \beta_{ii'} X_i X_{i'} + (\alpha\beta\gamma)_{ijk}$$

   where

   $(\alpha\beta\gamma)_{ijk}$ is a catchall lack-of-fit term

   $X_i; i = A, B, C$ denote the quantitative levels of factors A, B, and C

   β's denote the regression coefficients

2. Distributions:

   $y_{ijkl} \mid r_l \sim$ ind $Gamma(\mu_{ijkl}, \phi)$

   r_l i.i.d. $N(0, \sigma_R^2)$

3. Link: log, $\eta_{ijkl} = \log(\mu_{ijkl})$

Notice that this model is motivated by the same skeleton ANOVA that appeared with Example 8.3: The linear predictor and distribution of the block effects are unchanged; only the conditional distribution of the observations and the link differ, to accommodate the fact that these are time-to-event data.

12.2.2.1 Gaussian LMM

As with the split-plot example in Section 12.2.1, let us first see what happens if we use a Gaussian LMM. The GLIMMIX statements are

```
proc glimmix data=classes plot=residualpanel;
  class run ca cb cc;
  model r_time=a|a|b|b|c|c@2 ca*cb*cc/solution htype=1;
  random intercept/subject=run;
```

As with previous examples, you must precede this program by activating the ODS graphics (i.e., the statement ODS GRAPHICS ON). A, B, and C denote the factor levels as direct variables. CA, CB, and CC are equal to A, B, and C but are defined as CLASS variables in order to compute the departure from second-order polynomial lack-of-fit term. That is, CA*CB*CC in the MODEL statement estimates $(\alpha\beta\gamma)_{ijk}$. HTYPE = 1 causes hypotheses to be computed in terms of sequential estimable functions. As we discussed in Chapter 7, the default HTYPE = 3 estimable functions yield nonsense when fitting regression models, especially those with lack-of-fit terms.

Results of interest are as follows:

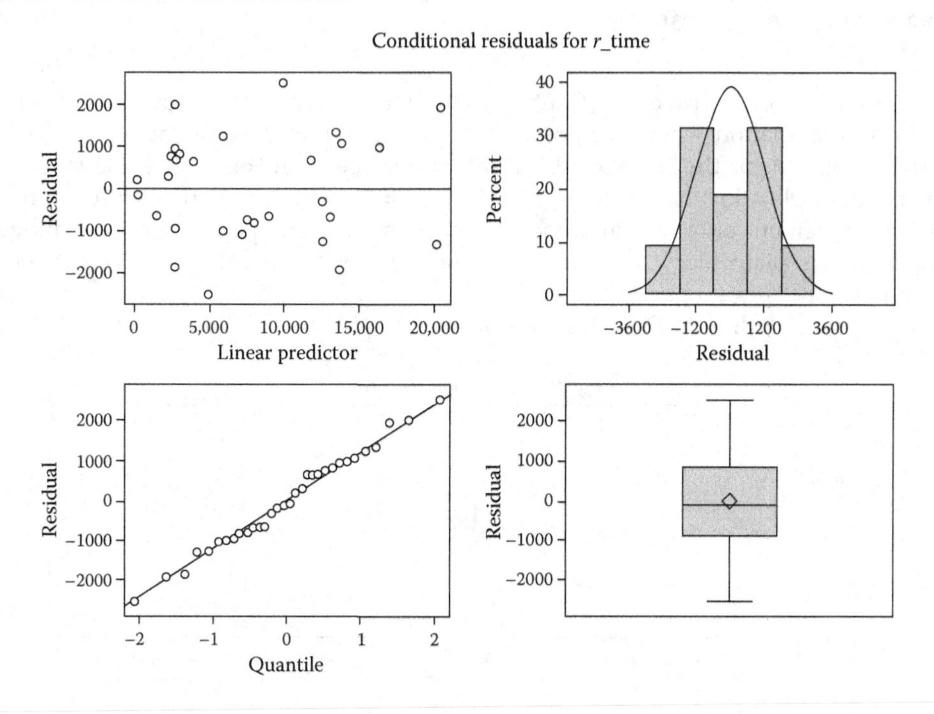

Type I Tests of Fixed Effects				
Effect	**Num DF**	**Den DF**	***F* Value**	**Pr > *F***
a	1	15	5.91	0.0281
a*a	1	15	0.00	0.9471
b	1	15	26.87	0.0001
a*b	1	15	0.98	0.3369
b*b	1	15	4.11	0.0608
c	1	15	0.24	0.6315
a*c	1	15	0.30	0.5948
b*c	1	15	0.40	0.5367
c*c	1	15	1.95	0.1828
ca*cb*cc	3	15	2.82	0.0743

The residual panel shows residuals that are symmetric but heavy-tailed. We do not see any overt gamma signature (e.g., right-skewed histogram) as we did in the split-plot example, but neither do we see any reassurance that the Gaussian model is a good choice. The Type I tests show evidence—not strong but not ignorable either—of lack of fit. That is, for the lack-of-fit term CA*CB*CC, $F = 2.82$ and $p \approx 0.074$.

12.2.2.2 Gamma GLMM

The GLIMMIX statements for the gamma GLMM are

```
proc glimmix data=classes method=laplace plot=residualpanel(ilink);
  class run ca cb cc;
  model r_time=a|a|b|b|c|c@2 ca*cb*cc/d=gamma solution htype=1;
  random intercept/subject=run;
  covtest/cl(type=prl);
```

The differences are the use of METHOD=LAPLACE (or quadrature—but we need to use integral approximation whenever possible for two-parameter exponential distributions) and ILINK option for the RESIDUALPANEL so we see it on the same scale (data scale) as the residual plots for the Gaussian LMM. The D=GAMMA option is self-evident. The COVTEST statement obtains confidence bounds for the scale parameter. This allows us to see if $\phi=1$ is plausible. If it is, we could fit the model with the simpler exponential distribution.

Results corresponding to those for the Gaussian LMM are as follows:

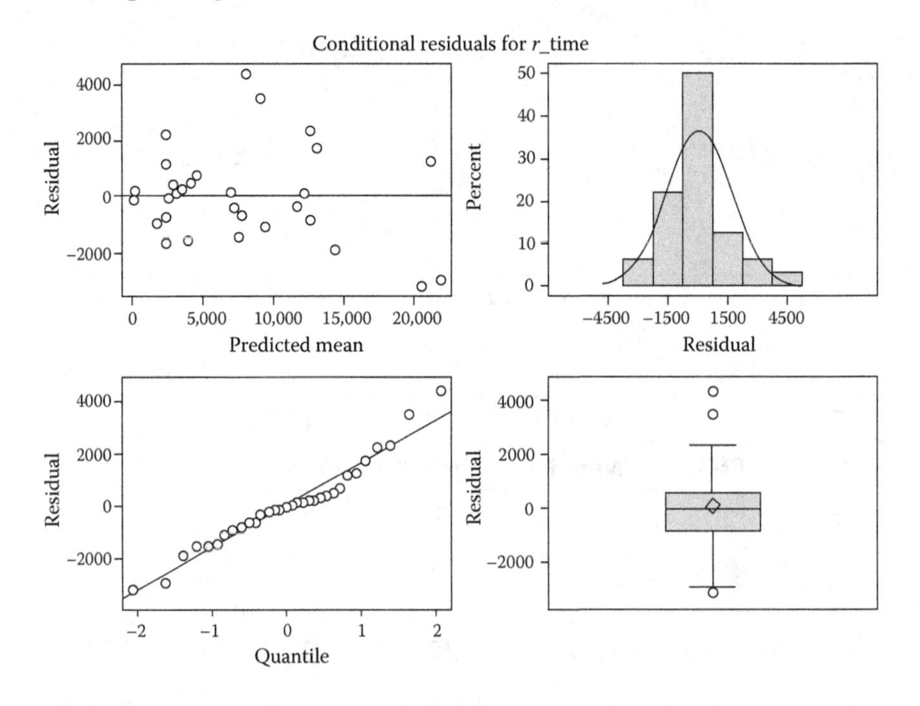

Type I Tests of Fixed Effects				
Effect	**Num DF**	**Den DF**	***F* Value**	**Pr>*F***
a	1	15	2.25	0.1548
a*a	1	15	1.30	0.2715
b	1	15	3.64	0.0758
a*b	1	15	0.01	0.9431
b*b	1	15	6.09	0.0261

(continued)

Type I Tests of Fixed Effects				
Effect	Num DF	Den DF	F Value	Pr>F
c	1	15	0.00	0.9946
a*c	1	15	0.00	0.9650
b*c	1	15	0.00	0.9511
c*c	1	15	5.40	0.0346
ca*cb*cc	3	15	0.22	0.8844

Covariance Parameter Estimates							
				Profile Likelihood 95% Confidence Bounds			
Cov Parm	Subject	Estimate	Standard Error	Lower Bound	Pr>Chisq	Upper Bound	Pr>Chisq
Intercept	run	0.7053	0.3853	0.2632	0.0500	2.4189	0.0500
Residual		0.2480	0.06815	0.1509	0.0500	0.4458	0.0500

The residual plots show lighter tails. This is about all we should read into these plots. Part of the story here is how much information plots of residuals *do not* contain in this sort of analysis, at least for the purposes of deciding which distribution fits best. McCullagh and Nelder (1989) and Hardin and Hilbe (2007) have discussions of model fitting statistics and plots that are better suited to model checking in this example.

The covariance parameter estimates show that the 95% confidence interval for the scale parameter, ϕ, is [0.15, 0.45], strongly suggesting that the exponential is not a suitable candidate distribution for this model.

The Type I tests show none of the evidence of lack of fit we saw in the Gaussian LMM. The Gaussian LMM's lack of fit is mainly an artifact of the inappropriateness of the Gaussian distribution, not a clear statement of the adequacy of the second-order polynomial regression. On the other hand, the Type I tests show strong evidence of quadratic regression over levels of B and C, no evidence of an A effect and no evidence of any interaction among the factors.

Compare these conclusions with the Gaussian LMM, which did detect the quadratic B effect, but completely missed any evidence of a C effect and picked up spurious evidence of an A effect. This is important because standard response surface analysis packages all assume normality as part of their "black box"—an unwary user would obtain the same results we see here for the Gaussian LMM. The probable results are as follows: a completely misdiagnosed assessment of what is going on *and* the resulting inappropriate decisions about levels of A, B, and C to use in manufacturing. Quite possibly, this could lead to quality and reliability problems. In the extreme, it might even put a company out of business.

We could pursue the gamma GLMM by fitting a reduced polynomial model involving B and C only and obtain predicted responses for various BC combinations. Example of GLIMMIX statements are as follows:

```
proc glimmix data=classes method=laplace;
  class run ca cb cc;
  model r_time=b|b c|c /d=gamma solution;
  random intercept / subject=run;
  estimate 'bc 0 0' intercept 1 / ilink cl;
  estimate 'bc 0 1' intercept 1 b 0 c 1 b*b 0 c*c 1 / ilink cl;
```

```
estimate 'bc 0 2' intercept 1 b 0 c 2 b*b 0 c*c 4 / ilink cl;
estimate 'bc 1 0' intercept 1 b 1 c 0 b*b 1 c*c 0 / ilink cl;
estimate 'bc 1 1' intercept 1 b 1 c 1 b*b 1 c*c 1 / ilink cl;
estimate 'bc 1 2' intercept 1 b 1 c 2 b*b 1 c*c 4 / ilink cl;
estimate 'bc 2 0' intercept 1 b 2 c 0 b*b 4 c*c 0 / ilink cl;
estimate 'bc 2 1' intercept 1 b 2 c 1 b*b 4 c*c 1 / ilink cl;
estimate 'bc 2 2' intercept 1 b 2 c 2 b*b 4 c*c 4 / ilink cl;
```

Selected output is as follows:

Solutions for Fixed Effects

Effect	Estimate	Standard Error	DF	*t* Value	Pr> |*t*|
Intercept	10.0561	0.6035	5	16.66	<.0001
b	−3.3378	1.3332	22	−2.50	0.0202
b*b	1.5524	0.6634	22	2.34	0.0288
c	−2.8481	1.3334	22	−2.14	0.0441
c*c	1.4252	0.6634	22	2.15	0.0430

Estimates

Label	Estimate	Standard Error	DF	*t* Value	Pr> |*t*|	Alpha	Lower	Upper	Mean	Standard Error Mean	Lower Mean	Upper Mean
bc 0 0	10.0561	0.6035	5	16.66	<0.0001	0.05	8.5047	11.6075	23,296	14,060	4937.72	109,914
bc 0 1	8.6331	0.5897	22	14.64	<0.0001	0.05	7.4102	9.8560	5,614.31	3,310.57	1652.73	19,072
bc 0 2	10.0604	0.6040	22	16.66	<0.0001	0.05	8.8077	11.3131	23,398	14,133	6685.70	81,885
bc 1 0	8.2706	0.5896	22	14.03	<0.0001	0.05	7.0478	9.4935	3,907.47	2,304.00	1150.33	13,273
bc 1 1	6.8477	0.5752	22	11.90	<0.0001	0.05	5.6547	8.0406	941.68	541.69	285.63	3,104.60
bc 1 2	8.2750	0.5886	22	14.06	<0.0001	0.05	7.0543	9.4957	3,924.49	2,309.99	1157.81	13,302
bc 2 0	9.5900	0.6049	22	15.85	<0.0001	0.05	8.3355	10.8445	14,618	8,842.58	4169.38	51,252
bc 2 1	8.1670	0.5886	22	13.87	<0.0001	0.05	6.9463	9.3878	3,522.89	2,073.65	1039.31	11,941
bc 2 2	9.5944	0.6060	22	15.83	<0.0001	0.05	8.3376	10.8511	14,682	8,897.26	4177.99	51,593

We see that both B and C have positive quadratic regression coefficients, which tells us that the regression equation has a minimum. We could determine the BC combination that minimizes the expected response—a standard response surface exercise. Among the BC combinations we actually observed, the center point, BC_{11} yields an expected response of 941, with 95% confidence limits [286, 3105]—noticeably faster than any other observed treatment combination.

12.3 GLMMs and Survival Analysis

Survival analysis, as the name implies, concerns statistical modeling of expected life span. Important applications include, but are certainly not restricted to, medicine and engineering. In medicine, common survival analysis examples involve studies that focus on how long patients with a serious disease live and whether a proposed new treatment extends expected lifetime. In engineering, similar analysis would apply to the time a component can be expected to function properly and the likelihood of failure after it has been in use for a given period. There are parametric, semi-parametric, and nonparametric forms of survival analysis. The former two forms, parametric and semi-parametric, are in essence generalized linear models with specialized functions and interpretations associated with the parameter estimates.

12.3.1 Basic Concepts and Terminology

Recall that we started this chapter developing distributions that make sense for time-to-event data, beginning with the probability that the first event does not occur until at least time $T \geq t$, that is, in c.d.f. terms, $1-F(t)$. In survival analysis terminology, this is called the *survivor function*, denoted $S(t)$. Formally,

$$\textit{Survivor function}: S(t) = \Pr\{\text{survival till at least time } t\} = \Pr\{T \geq t\} = 1 - F(t) \tag{12.3}$$

The *hazard function*, denoted $h(t)$, is defined as the instantaneous likelihood of failure (e.g., death in a medical setting, component ceases to function in an engineering setting) in the interval $[t, t+\Delta t]$ given survival up until time t. Formally,

$$h(t) = \lim_{\Delta t \to 0} \Pr\{t \leq T < t + \Delta t \mid T \geq t\} = \lim_{\Delta t \to 0} \frac{\Pr\{(t \leq T < t + \Delta t) \cap (T \geq t)\}}{\Pr\{T \geq t\}}$$

$$= \lim_{\Delta t \to 0} \frac{\Pr\{t \leq T < t + \Delta t\}}{\Pr\{T \geq t\}} = \lim_{\Delta t \to 0} \left[F(t + \Delta t) - F(t)\right] \frac{1}{S(t)}$$

Therefore,
Hazard function:

$$h(t) = \frac{f(t)}{S(t)} \tag{12.4}$$

The cumulative hazard function is defined as
Cumulative hazard function:

$$H(t) = \int_0^t h(x)\, dx \tag{12.5}$$

Note that (12.4) can alternatively be expressed as

$$h(t) = -\frac{\partial\{\log[S(t)]\}}{\partial t} \tag{12.6}$$

and hence $S(t) = e^{-H(t)}$ and $H(t) = -\log[S(t)]$.

Building on our development of the simplest time-to-event probability distribution in Section 12.1, let us consider the survival model for the exponential p.d.f. $f(t)=\lambda e^{-\lambda t}$. We have the following:

- Survivor function: $S(t)=1-F(t)=e^{-\lambda t}$
- Hazard function: $h(t)=\frac{f(t)}{S(t)}=\frac{\lambda e^{-\lambda t}}{e^{-\lambda t}}=\lambda$
- Cumulative hazard function: $H(t)=\int_0^t h(x)dx=\int_0^t \lambda dx=\lambda t$

Notice that for the survival model based on the exponential distribution, the hazard function $h(t)=$ the constant λ regardless of the time. This follows from the exponential distribution's basis in the Poisson process, with its well known lack of memory property. More complex survival models, which we will not consider in this textbook, assume time-dependent hazard functions. McCullagh and Nelder (1989) and Dobson and Barnett (2008) briefly discuss such extensions in the context of fixed-effects generalized linear models.

Our purpose here is to introduce the survival/GLMM connection. We do this now using the exponential p.d.f.

12.3.2 Exponential Survival GLMM for Uncensored Data

In survival analysis, we want to estimate the survivor and hazard functions—and other inference of interest that can be obtained from these functions. When we have treatment and experiment designs as described in Chapters 7 and 8, the GLMM becomes a useful tool for obtaining these estimates. If the data, conditional on the random model effects, has an exponential distribution, we know that the conditional expectation $E(y|b)=\mu=(\lambda)^{-1}$. Using the GLMM, we can estimate $\boldsymbol{\eta}=g(\boldsymbol{\mu})=\mathbf{X}\boldsymbol{\beta}+\mathbf{Zb}$. Once we have parameter estimates, we can obtain estimable functions of the model scale. Applying the inverse link gives us the $\hat{\mu}$ for the estimable functions of interest. Hence, taking $(\hat{\mu})^{-1}$ gives us $\hat{\lambda}$ for the estimable functions of interest, and these in turn give us the needed survivor and hazard functions.

To illustrate, consider the following example. The data appear as Data Set 12.3 in the SAS Data and Program Library. The treatment design is a 2×2 factorial. The study was conducted as a split-plot. Six clinics are randomly assigned to each level of factor A. Within each clinic, there are two groups of patients—each level of factor B is assigned to one of the groups within each clinic. Within each clinic–group combination, n_{ij} patients, denoted PT_ID in the data set, are observed. The response variable, survival time, is denoted Y.

Except for the lack of blocking, the study design for these data is similar to the design in Data Set 12.1. Clinics are whole-plot units of replication for factor A. Groups within clinic are the split-plot units of replication for factor B. A model for these data is thus:

1. Linear predictor: $\eta_{ijk}=\eta+\alpha_i+\beta_j+(\alpha\beta)_{ij}+c(a)_{ik}+bc(a)_{ijk}$, where α, β, and c denote A, B, and clinic effects, respectively; alternatively, we could use the "cell means" form of the linear predictor $\eta_{ijk}=\eta_{ij}+c(a)_{ik}+bc(a)_{ijk}$ since our interest will focus on $\hat{\mu}_{ij}=e^{\hat{\eta}_{ij}}$ rather than explicitly partitioned α, β, and $(\alpha\beta)$ effects

2. Distributions:

   $c(a)_{ik}$ i.i.d. $N\left(0,\sigma_C^2\right)$

   $bc(a)_{ijk}$ i.i.d. $N\left(0,\sigma_{BC}^2\right)$

   $y_{ijkl} \mid c(a)_{ij}, bc(a)_{ijk} \sim$ independent $Exponential\left(\mu_{ijk}\right)$ $\left[\text{or } Gamma\left(\mu_{ijk},\phi\right)\right]$

3. Link: $\eta_{ijk}=\log(\mu_{ijk})$ [or canonical link (reciprocal): $\eta_{ijk}=1/\mu_{ijk}$]

The primary goals of the analysis are as follows:

1. Verify that the exponential distribution provides an adequate fit for $y_{ijkl}|c(a)_{ij}, bc(a)_{ijk}$.
2. Estimate μ_{ij}, the AB_{ij} treatment combination means.
3. Use the $\hat{\mu}_{ij}$ to determine the hazard and survivor functions for each treatment.

We can approach the objective either by starting with the gamma distribution and seeing if $\phi=1$ is a plausible value of the scale parameter—that is, does the confidence interval for ϕ contain 1—or by fitting the exponential and seeing if there is evidence of over- or under-dispersion. To be thorough, let us do both.
The GLIMMIX statements for the gamma model are

```
proc glimmix method=laplace data=chapter12_example3;
  class clinic a b;
  model y=a|b/d=gamma;
  random intercept b/subject=a*clinic;
  covtest/cl(type=plr);
```

The only output of interest here concerns the scale parameter. The estimate: $\hat{\phi}=0.89$; 95% confidence interval: [0.72, 1.10]. This suggests $\phi=1$ is quite plausible and gives us justification for assuming an exponential distribution.

The GLIMMIX statements for the exponential model are

```
proc glimmix method=laplace data=chapter12_example3;
  class clinic a b;
  model y=a|b/d=exponential solution;
  random intercept b/subject=a*clinic;
  lsmeans a*b/ilink;
```

Here, we add the LSMEANS statement. The ILINK option provides the $\hat{\mu}_{ij}$ we need to compute the hazard and survivor functions. Below, we will pursue these in greater detail. First things first, the conditional fit statistics listing is

Fit statistics for conditional distribution	
−2 log L(y|r effects)	1382.48
Pearson chi-square	124.63
Pearson chi-square/DF	0.83

The Pearson $\chi^2/df=0.83$ is suitably close to 1. This is consistent with what we found with the gamma model: These data provide no evidence to suggest that the exponential model is inappropriate. The Type III tests and LSMEANS appear as follows:

Type III Tests of Fixed Effects				
Effect	**Num DF**	**Den DF**	***F* Value**	**Pr>*F***
a	1	10	6.33	0.0306
b	1	10	0.15	0.7047
a*b	1	10	0.26	0.6199

a*b Least Squares Means

a	b	Estimate	Standard Error	DF	*t* Value	Pr > |*t*|	Mean	Standard Error Mean
1	1	3.8816	0.2016	10	19.25	<0.0001	48.4994	9.7782
1	2	3.9022	0.1896	10	20.58	<0.0001	49.5094	9.3859
2	1	3.3821	0.2161	10	15.65	<0.0001	29.4333	6.3597
2	2	3.2206	0.2286	10	14.09	<0.0001	25.0439	5.7241

The Type III tests tell us that the main effect of factor A is statistically significant (assuming the criterion $\alpha = 0.05$), but there is no evidence of a B effect. The key information in the LSMEANS table comes from the column labeled MEAN: these are the $\hat{\mu}_{ij}$, the estimated mean times of survival for each treatment combination. Though not shown here, we could—and probably should—replace this table with LSMEANS A/ILINK. That is, we should focus on main effect means for the two levels of A in view of the absence of a detectable B effect.

The estimated hazard function for each treatment combination is $\hat{\lambda}_{ij} = 1/\hat{\mu}_{ij}$, for example, for AB_{11} the estimated hazard function is $\hat{\lambda} \approx 1/48.5 \approx 0.021$. We can hand-calculate these from the MEAN column or we could automate the process by adding and ODS OUTPUT statement to the GLIMMIX program and then using a DATA step to compute the hazard functions. The needed statements are

```
ods output lsmeans=mu;
run;
data hazard;
  set mu;
  hazard=1/mu;
proc print data=hazard;
run;
```

The results are as follows:

Obs	Effect	a	b	Estimate	Std Err	DF	*t* Value	Probt	Mu	Std Err Mu	Hazard
1	a*b	1	1	3.8816	0.2016	10	19.25	<0.0001	48.4994	9.7782	0.020619
2	a*b	1	2	3.9022	0.1896	10	20.58	<0.0001	49.5094	9.3859	0.020198
3	a*b	2	1	3.3821	0.2161	10	15.65	<0.0001	29.4333	6.3597	0.033975
4	a*b	2	2	3.2206	0.2286	10	14.09	<0.0001	25.0439	5.7241	0.039930

The key column is the last one, labeled HAZARD. These are the estimated hazard functions, $h_{ij}(t) = \hat{\lambda}_{ij}$ for each AB_{ij} combination. *Proportional hazard models* take this idea a step further by taking ratios of hazards for treatment comparisons of interest. For example,

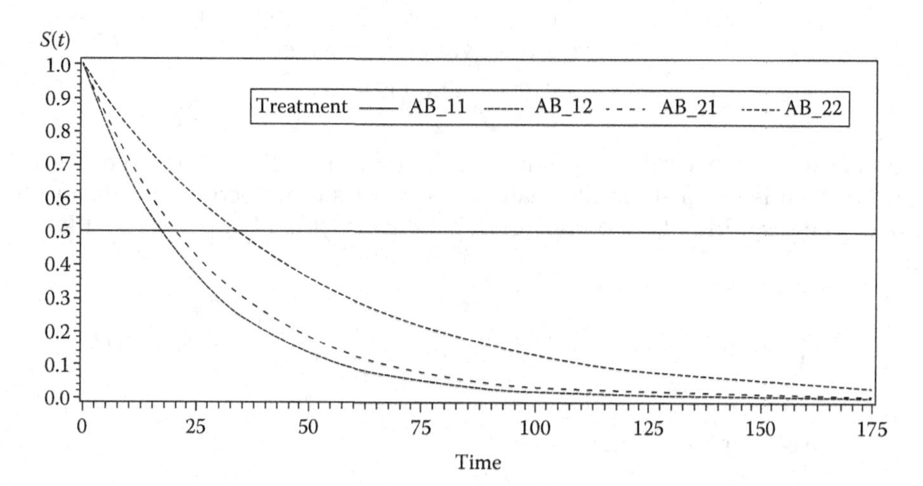

FIGURE 12.1
Plot of survivor functions for each treatment combination in Data Set 12.3 based on exponential GLMM for uncensored data.

the ratio $\hat{\lambda}_{11}/\hat{\lambda}_{21}=0.021/0.034$ gives the estimated proportional hazard for the simple effect of A given B_1.

We could compute upper and lower confidence limits for the hazard functions by adding a CL option to the LSMEANS statement and then taking the inverse of the upper and lower confidence limits of the ILINKed $\hat{\mu}_{ij}$. We leave this as an exercise for readers.

We can also use the $\hat{\lambda}_{ij}$ to compute the estimated survivor function, $S_{ij}(t)=e^{-t\hat{\lambda}_{ij}}$, for each treatment combination AB_{ij}. Plotting the survivor function over times observed in the data set provides a useful tool to understand the information provided by $S_{ij}(t)$. Figure 12.1 shows the plot.

The vertical axis gives the probability of survival up to the time given on the horizontal axis. Note that the survival functions for the two levels of B with A_1 (AB_11 and AB_12) are so close together they cannot be distinguished on Figure 12.1. Consistent with the mean survival times in the LSMEANS output shown earlier, the survival probabilities for A_1 are distinctly greater than those for A_2 whereas the level of B does not make much difference. The reference line at $S(t)=0.5$ shows the survival time for each treatment combination corresponding to a survival probability of 0.5. For example, the two survivor functions for A_1 intersect $S(t)=0.5$ between time 35 and 40; the survivor functions for A_2 appear to intersect between time 20 and 25. We could, of course, determine the exact value of t for a given $S(t)$ analytically, but this plot allows a quick visual summary of the important information.

12.3.3 Exponential Survival GLMM for Censored Data

Suppose that instead of continuing the study until the last patient succumbs, as was the case with the immediately preceding analysis that led to Figure 12.1, the study has to be concluded at a fixed time. When we plan a study, we might be given a fixed quitting time in advance and the data we have at that point is what we have to use, or we may have some flexibility but plan to conclude the study when, say, a predetermined percentage of patients have succumbed. In either case, we now have right-censored data.

We can still fit a GLMM to the censored data and obtain reasonable hazard and survivor function estimates. The following notation is used in McCullagh and Nelder (1989) and Dobson and Bennett (2008), let c be a binary variable where

$$c = \begin{cases} c = 0 \text{ if the observation is censored} \\ c = 1 \text{ if the observation is } \textit{not} \text{ censored} \end{cases} \tag{12.7}$$

If an observation is censored, it has survived at least up until the time the study stops. If an observation is uncensored, it means the patient succumbed before the study was concluded. The resulting likelihood over all patients is $[f(t)]^c + [S(t)]^{1-c}$; hence, the log-likelihood is

$$c \times \{\log[f(t)]\} + (1-c) \times \{\log[S(t)]\} = c \times \left\{\log\left[\frac{f(t)}{S(t)}\right]\right\} + \log[S(t)] \tag{12.8}$$

Using (12.6) we can reexpress (12.8) as

$$c \times \{\log[h(t)]\} - H(t) \tag{12.9}$$

For the exponential model, the log-likelihood is

$$c \times \{\log(\lambda)\} - \lambda t \tag{12.10}$$

which we note is the Poisson log-likelihood. Thus, we can regard the censoring random variable C as having a Poisson distribution with rate parameter λ. Hence, at a given time t, $E(C \mid t) = \mu_C = \lambda t$. The generalized linear model would thus use c as the response variable, the Poisson as the conditional distribution of c given the random model effects, and $\log(\mu_C) = \log(\lambda) + \log(t)$ as the link. Because (12.10) is the form of the log-likelihood the GLMM works with in the estimating equations, the model uses $\log(t)$ as an *offset*. The linear predictor for the censored model is the same as it was for the uncensored model.

To illustrate implementation of the model with censoring, suppose the study that generated Data Set 12.3 lasted 50 days only—any patient that had not succumbed as of the day denoted $Y = 51$ was classified as censored. Thus, there are two response variables for every observation: C, the censoring variable, and T, the survival time. If the patient succumbs on or before day 50, the censoring variable is $C = 1$ and T is whatever value $Y \leq 50$ was in the original data. If the patient has not succumbed, then $C = 0$ and $T = 50$—that is, the observation *is* censored at T, which is set to 50, the last day of the trial.

The GLIMMIX statements for the censored data are

```
proc glimmix method=laplace data=censoring;
  class clinic a b;
  logt=log(t);
  model c=a*b / noint d=poisson offset=logt solution;
  random intercept b / subject=clinic*a;
  lsmeans a*b / ilink;
  lsmestimate a*b 'mean surv_time for AB_11' -1 0 0 0,
        'mean surv_time for AB_12' 0 -1 0 0,
        'mean surv_time for AB_21' 0 0 -1 0,
        'mean surv_time for AB_22' 0 0 0 -1/exp;
```

The statements here use the "cell means" formulation of the linear predictor. We could also use the effects parameterization, as shown with the uncensored data. In either case, the main information comes from the LSMEANS and LSMESTIMATE statements, which are the same regardless of whether the MODEL statement uses the effects or cell means parameterization. Notice the addition of the OFFSET variable. The program statement immediately before MODEL defines the offset, and then the OFFSET option in the MODEL statement adds the offset term to the linear predictor. The RANDOM statement does not change from what we used with the uncensored data. The LSMEAN statement gives the estimates of $\log(\lambda)$ and the ILINK gives the actual values of the estimated hazard function for each treatment without the additional ODS or DATA step we had to use with the uncensored data. The LSMESTIMATE statements allow us to obtain estimates of the mean survival time—estimates that resulted from LSMEANS ILINK with the uncensored data.

The output from the LSMEANS statement appears as

a*b Least Squares Means

a	b	Estimate	Standard Error	DF	*t* Value	Pr > |*t*|	Mean	Standard Error Mean
1	1	−4.1716	0.2403	10	−17.36	<0.0001	0.01543	0.003707
1	2	−3.9693	0.2124	10	−18.69	<0.0001	0.01889	0.004011
2	1	−3.4139	0.2156	10	−15.84	<0.0001	0.03291	0.007095
2	2	−3.3194	0.2254	10	−14.73	<0.0001	0.03617	0.008152

The MEAN column shows the estimates of the hazard functions $\hat{\lambda}_{ij}$ for each AB_{ij} treatment combination. The ESTIMATE column shows $\hat{\eta}_{ij} = \log(\hat{\lambda}_{ij})$; this value is of little intrinsic interest in this discussion. The LSMESTIMATE results are as follows:

Least Squares Means Estimates

Effect	Label	Estimate	Standard Error	DF	*t* Value	Pr > |*t*|	Exponentiated Estimate
a*b	surv_time for AB_11	4.1716	0.2403	10	17.36	<0.0001	64.8180
a*b	surv_time for AB_12	3.9693	0.2124	10	18.69	<0.0001	52.9497
a*b	surv_time for AB_21	3.4139	0.2156	10	15.84	<0.0001	30.3836
a*b	surv_time for AB_22	3.3194	0.2254	10	14.73	<0.0001	27.6444

The Exponentiated Estimate column shows the estimated mean survival times, $\hat{\mu}_{ij}$, as defined previously for the analysis of the uncensored data. If desired, we could add comparisons to the LSMESTIMATE statement to obtain hazard ratios.

As with the analysis of the uncensored data, we can use these estimates to plot the survivor functions for each AB_{ij} treatment combination. The plot appears in Figure 12.2.

The parameter estimates for the censored and uncensored models are not identical, but they are sufficiently consistent that the plots in Figures 12.1 and 12.2 appear to be almost identical. We interpret them exactly the same way.

The main difference between the model and analysis in this section and the examples in generalized linear model texts or in introductory survival analysis textbooks is the

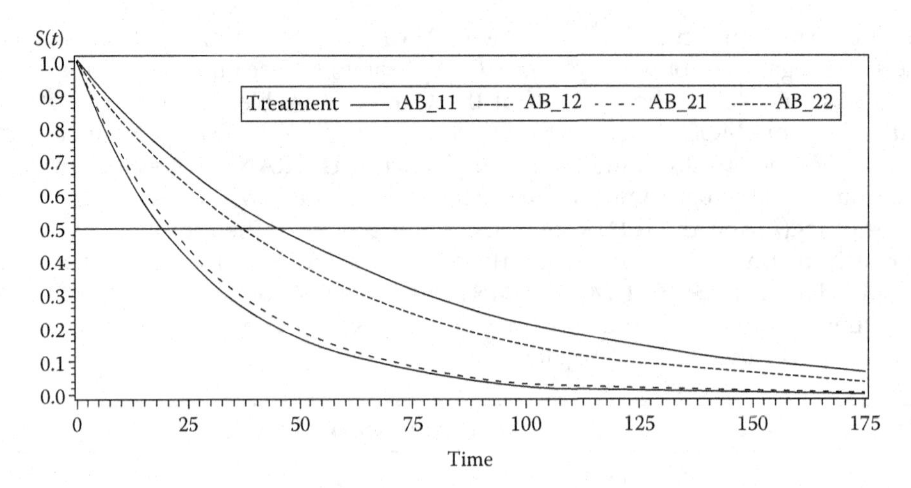

FIGURE 12.2
Plot of survivor functions for each treatment combination in Data Set 12.3 based on Poisson with offset GLMM for censored data.

inclusion of random model effects. Clearly, in these examples, they exist as a consequence of the study design. We must account for them, or we raise the same marginal vs. conditional model and correct standard error and test statistic issues we have seen for other mixed models with non-normal data.

12.4 Summary

The purpose of this section was to give reader an entry point to see the connection between GLMMs and survival analysis. The main ideas are as follows:

1. Time-to-event response variables generally must be continuous and nonnegative and are often skewed.
2. The Poisson process provides a theoretical starting point for developing time-to-event distributions.
   a. Leads directly to the exponential distribution
   b. Exponential easily generalizes to the gamma distribution (a standard exercise in probability and mathematical statistics courses)
   c. Other distributions used in survival analysis are further generalizations
3. Section 12.3 covers examples of simple time-to-event GLMMs using the exponential and gamma distributions.
   a. Use covariance inference tools to distinguish between exponential and gamma.
   b. Risk of using exponential when misspecified, similar to risk of using Poisson for overdispersed count data.
   c. Once distribution issue is settled, inference on estimable functions similar to any other GLMM.

4. Section 12.3 also introduces survival analysis–GLMM connection.
   a. Basic analysis is GLMM with exponential or gamma data, but couched in survival analysis framework
   b. Estimates expressed in terms of survival analysis terminology
   c. Censoring creates additional complication
   d. Simple censoring addressed using OFFSET; more complex censoring beyond the scope of this text
   e. Summary feature of survival analysis is the survivor function and its plot (e.g., Figures 12.1 and 12.2)

13

Multinomial Data

13.1 Overview

In Chapter 10, we considered rates and proportions. Discrete proportions—binary and binomial—can be viewed as categorical data with two response categories. In this chapter, we consider models for categorical data with more than two categories, data assumed to have a multinomial distribution.

For the multinomial distribution, we take N observations. Each observation belongs to exactly one of C mutually exclusive categories. Let π_c $(c=1,2,\ldots,C)$ denote the probability that an observation sampled at random from the population belongs to category c. Note that $\sum_{c=1}^{C} \pi_c = 1$. The multinomial distribution refers to the probability that exactly y_1 observations from our sample belong to category 1, y_2 observations belong to category 2, and so forth through y_C observations in category C, where $\sum_{c=1}^{C} y_c = N$. The p.d.f. is

$$f(y_1, y_2, \ldots, y_C) = \frac{N!}{y_1! y_2! \ldots y_C!} \pi_1^{y_1} \pi_2^{y_2} \ldots \pi_C^{y_C} \tag{13.1}$$

We can divide multinomial data into two types, *ordinal* and *nominal*. Ordinal refers to categories that have an obvious order, such as disease rating (asymptomatic, mild symptoms, moderate symptoms, severe symptoms) or approval rating (strongly approve, approve, neither approve nor disapprove, disapprove, strongly disapprove). Nominal refers to categories that have no obvious order, such as eye color, plant species, or college major.

Models for multinomial data build on the same strategy as binomial data. We motivate the link functions and the expression of an underlying process characterized by a function such as the logit or probit. *Cumulative logit* and *cumulative probit* models define link functions that, when they fit the data adequately, allow parsimonious modeling of ordinal multinomial data. *Generalized logit* and *generalized probit* models do not require ordered categories and are, therefore, suitable for nominal multinomial data.

In pre-GLMM days, ordinal categorical data were often given a numeric code—for example, asymptomatic = 0, mild = 1, moderate = 2, severe = 3—and analyzed as if the data had an approximate normal distribution. This raises two problems. First, how do we know what numeric code is "right"? Why equal spacing? Why not asymptomatic = 0, mild = 1, moderate = 2, severe = 4 or asymptomatic = 0, mild = 2, moderate = 5, severe = 10? Second, even if we do agree on a numeric code, how do we interpret the average rating? Assuming the 0-1-2-3 equally spaced code, what does an average rating of 1.5 mean? It could mean half of the responses were 1 and the other half were 2; or half 0 and half 3; or one-sixth 0, half 1, and one-third 3. There are three different ways to get 1.5, each with distinctly different meaning and consequences. If we have two treatments, what does it mean if one treatment has a mean rating of 1.5 and the other has a mean rating of 1.75? Again, we can think of many ways to get a 0.25 difference between mean ratings—ways that could tell us very different things about the treatments. The models we consider in this chapter focus attention on the multinomial probabilities, π_c, directly rather than a possibly meaningless, and certainly ambiguous, numeric code.

Section 13.2 focuses on ordinal data. Models using cumulative logits are called *proportional odds* models in categorical data analysis. Models using cumulative probits are also known in quantitative genetics as *threshold models*. Section 13.3 considers generalized logit models.

13.2 Multinomial Data with Ordered Categories

Recall that for GLMs and GLMMs with binomial data, we can visualize both the logit and probit link functions following the relationships summarized by Figure 13.1.

The right-hand vertical axis denotes the probability, π, and is the scale of reference for the cumulative plot, what we think of as the inverse link function in a GLM or GLMM: $\pi = h(\eta) = 1/(1+e^{-\eta})$ for the logit, $\Phi(\eta)$ for the probit.

We can also think of the inverse link in terms of the mass function, $\partial h(\eta)/\partial\eta$. For the probit link, this is the standard Gaussian p.d.f.; for the logit link it is not a true p.d.f., but we can visualize it the same way. This is the bell-shaped curve in the figure. The area under the curve corresponds to the probability for values of $\mathbf{X\beta}+\mathbf{Zb}=\boldsymbol{\eta}$ on the boundary point shown in the figure.

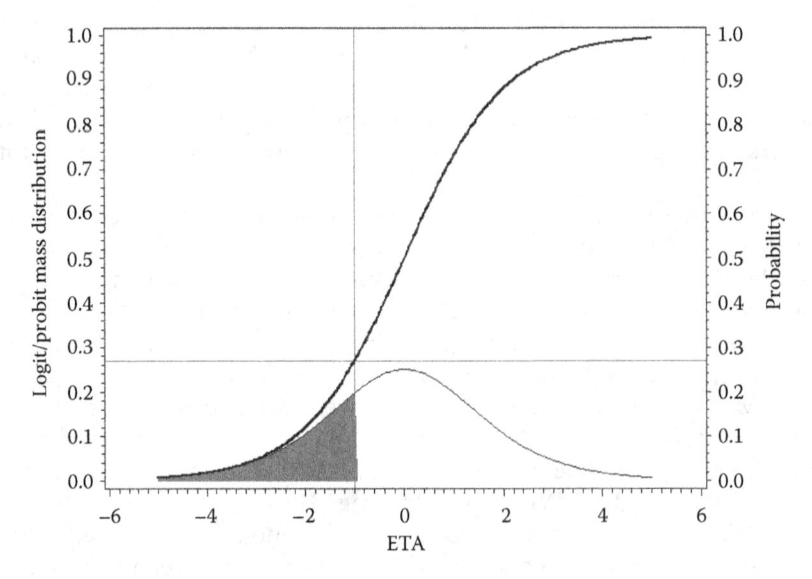

FIGURE 13.1
Relationship between $\boldsymbol{\eta}=\mathbf{X\beta}+\mathbf{Zb}$ (ETA) and probability for binomial link functions.

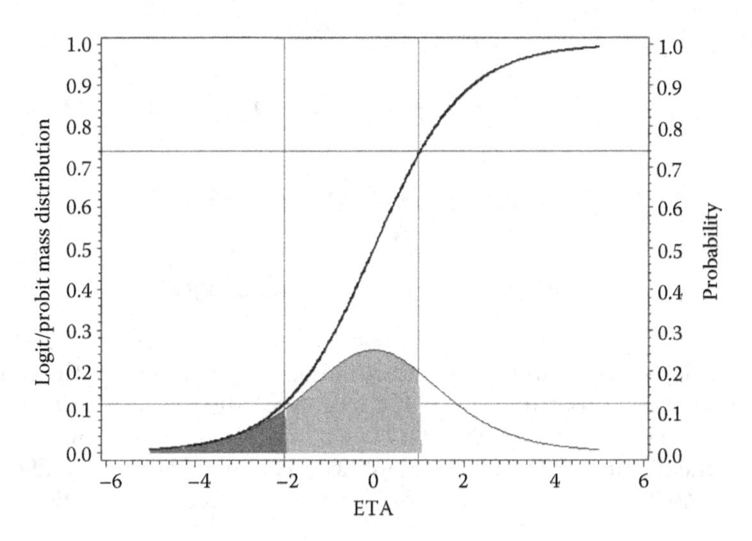

FIGURE 13.2
Plot of link vs. probability for three-category ordered multinomial.

Ordinal multinomial models extend this idea to three or more categories. Figure 13.2 shows a plot of the setup for a three-category multinomial GLM or GLMM.

Now, there are two boundaries subdividing the plot into three regions, one for each category. The lower ETA, here $\eta = -2$, shows the boundary between response category 1 and response category 2. The upper ETA, here $\eta = 1$, defines the boundary between categories 2 and 3. For the mass distribution (the bell-shaped curve), the area under the curve to the left of the lower boundary—or, alternatively, the cumulative probability referenced by the right-hand axis—corresponds to the response category 1 probability for that $\boldsymbol{\eta} = \mathbf{X}\boldsymbol{\beta} + \mathbf{Zb}$. The cumulative probability for the upper boundary—or, alternatively, the combined area of the two shaded regions under the bell-shaped curve—depicts the combined probability of categories 1 and 2.

In GLM/GLMM terms, each boundary on Figure 13.2 corresponds to a link function. Multinomial distributions with C categories require $C-1$ link functions to fully specify a model relating the response probabilities, $\{\pi_1, \pi_2, \ldots, \pi_C\}$ to the linear predictor. Two commonly used models are the cumulative logit model, also known as the proportional odds model, and the cumulative probit model, also known as the threshold model. Their link functions are as follows:

Cumulative logit

$$\eta_1 = \log\left(\frac{\pi_1}{1-\pi_1}\right) = \eta_1 + \mathbf{X}\boldsymbol{\beta} + \mathbf{Zb}$$

$$\eta_2 = \log\left(\frac{\pi_1 + \pi_2}{1-(\pi_1 + \pi_2)}\right) = \eta_2 + \mathbf{X}\boldsymbol{\beta} + \mathbf{Zb} \tag{13.2}$$

$$\eta_{C-1} = \log\left(\frac{\pi_1 + \pi_2 + \cdots + \pi_{C-1}}{1-(\pi_1 + \pi_2 + \cdots + \pi_{C-1})}\right) = \eta_{C-1} + \mathbf{X}\boldsymbol{\beta} + \mathbf{Zb}$$

Cumulative probit

$$\eta_1 = \Phi^{-1}(\pi_1) = \eta_1 + \mathbf{X\beta} + \mathbf{Zb}$$

$$\eta_2 = \Phi^{-1}(\pi_1 + \pi_2) = \eta_2 + \mathbf{X\beta} + \mathbf{Zb} \tag{13.3}$$

$$\eta_{C-1} = \Phi^{-1}(\pi_1 + \pi_2 + \cdots + \pi_{C-1}) = \eta_{C-1} + \mathbf{X\beta} + \mathbf{Zb}$$

Notice that for both the cumulative logit and cumulative probit models, the *only* parameter that changes among the link functions is the intercept. The intercepts define the boundaries between the categories when $\mathbf{X\beta}+\mathbf{Zb}=0$. Changes in $\mathbf{X\beta}+\mathbf{Zb}$ move the boundaries together so that the distance between them on the η-axis remains constant. For example, if we increase $\mathbf{X\beta}+\mathbf{Zb}$ by 0.5 units, we now have a link-to-probability relationship depicted by Figure 13.3.

Notice that the relative distance between the boundaries, that is, between link $\boldsymbol{\eta}_1$ and $\boldsymbol{\eta}_2$, remains unchanged, but because the entire set of boundaries shift to the right, the probabilities associated with each category change.

The inverse links for each model are given as follows:

Cumulative logit

$$h(\eta_1) = \frac{1}{\left(1+e^{-\eta_1}\right)} = \pi_1 \tag{13.4}$$

$$h(\eta_2) = \frac{1}{\left(1+e^{-\eta_2}\right)} = \pi_1 + \pi_2$$

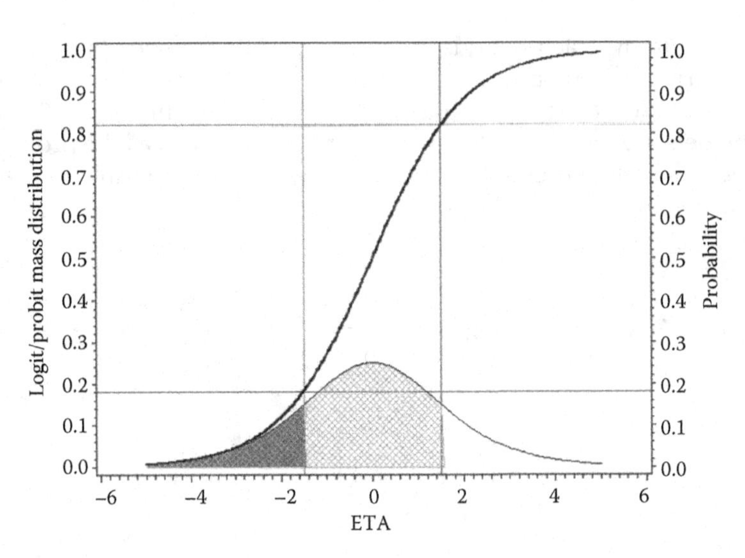

FIGURE 13.3
Link–probability relationship, linear predictor increased by 0.5.

From these, we obtain estimates of π_2 and π_3:

$$\pi_2 = h(\eta_2) - h(\eta_1)$$

$$\pi_3 = 1 - h(\eta_2)$$

Cumulative probit

$$h(\eta_1) = \Phi(\eta_1) = \pi_1 \tag{13.5}$$

$$h(\eta_2) = \Phi(\eta_2) = \pi_1 + \pi_2$$

Once we have estimates of $h(\boldsymbol{\eta}_1)$ and $h(\boldsymbol{\eta}_2)$, we can obtain $\hat{\pi}_2$ and $\hat{\pi}_3$ the same way we did for the cumulative logit model.

Example 13.1 Cumulative logit/proportional odds model

The data for this example appears in the SAS Data and Program Library as Data Set 13.1. The data are from a nested factorial design using the setup introduced in Figure 2.5—10 blocks each of size 3, 6 treatments divided into sets {0,1,2} and {3,4,5}. Each set was randomly assigned to five blocks. Each treatment was randomly assigned to a unit within the block. On each unit, N_{ij} observations were taken. The response variable is ordinal with three categories: slight, moderate, and severe, coded in this example as $C=0$, 1, and 2, respectively. The data to be analyzed are thus $\{y_{0ij}, y_{1ij}, y_{2ij}\}$, where y_{cij} denotes the number of observations in category c for treatment i in block j. A cumulative logit—a.k.a. proportional odds—GLMM for these data:

1. Linear predictor: $\eta_{cij} = \eta_c + \tau_i + b_j$, where η_{cij} denotes the cth link ($c=0,1$) for the ijth treatment–block combination, η_c is the intercept for the cth link, and τ_i and b_j denote treatment and block effects, respectively. Note that because there are three response categories, we have two link functions, one for each boundary as shown in Figure 13.2.
2. Distributions:
   a. $y_{0ij}, y_{1ij}, y_{2ij} | b_j \sim \text{Multinomial}(N_{ij}, \pi_{0ij}, \pi_{1ij}, \pi_{2ij})$
   b. b_j i.i.d. $N(0, \sigma_B^2)$
3. Links:
   a. $\eta_{0ij} = \log\left(\dfrac{\pi_{0ij}}{1-\pi_{0ij}}\right)$
   b. $\eta_{1ij} = \log\left(\dfrac{\pi_{0ij}+\pi_{1ij}}{1-(\pi_{0ij}+\pi_{1ij})}\right)$

Notice that we can easily reframe this GLMM as a cumulative probit—a.k.a. a threshold—model by replacing the links shown previously by the inverse Gaussian functions shown in (13.3).

For the cumulative logit model shown, we can use the following PROC GLIMMIX statements to implement the analysis.

```
proc glimmix data=univar_multinom;
class blk trt;
model rating(order=data)=trt/dist=multinomial solution oddsratio;
random intercept/subject=blk solution;
estimate 'c=0, t=0' intercept 10 trt 10,
  'c=1, t=0' intercept 0 1 trt 1 0,
  'c=0, t=1' intercept 1 0 trt 0 1 0,
  'c=1, t=1' intercept 0 1 trt 0 1 0,
  'c=0, t=2' intercept 1 0 trt 0 0 1 0,
  'c=1, t=2' intercept 0 1 trt 0 0 1 0,
  'c=0, t=3' intercept 1 0 trt 0 0 0 1 0,
  'c=1, t=3' intercept 0 1 trt 0 0 0 1 0,
  'c=0, t=4' intercept 1 0 trt 0 0 0 0 1 0,
  'c=1, t=4' intercept 0 1 trt 0 0 0 0 1 0,
  'c=0, t=5' intercept 1 0 trt 0 0 0 0 0 1,
  'c=1, t=5' intercept 0 1 trt 0 0 0 0 0 1/ilink;
freq y;
```

The CLASS, MODEL, and RANDOM statements appear as any other GLMM we have seen thus far. Implicit in these statements, but not obvious to first-time users, the data must have one line per block–treatment category, in this case referenced by the variables BLK, TRT, and RATING, respectively. In addition to these three variables, each data line must also contain the number of observations for each category—here referenced by the variable Y. Here is a sample of the first several lines of the data set in "GLIMMIX-ready" form:

Obs	blk	trt	rating	*y*
1	1	0	slight	1
2	1	0	modrat	4
3	1	0	severe	23
4	1	1	slight	2
5	1	1	modrat	7
6	1	1	severe	23
7	1	2	slight	4
8	1	2	modrat	7
9	1	2	severe	18

Returning to the GLIMMIX statements, (ORDER=DATA) specifies that the order in which the categories appear in the data set be treated as the ordinal categories from lowest to highest for the analysis. Notice that observations on each block–treatment combination always appear in threes and always in the order SLIGHT (for "slight"), MODRAT (for "moderate"), and SEVERE. If you do not use an ORDER option with the response variable in the MODEL statement, GLIMMIX will rearrange your categories in numeric or alphabetical order, depending on whether you enter RATING as a number or a name. The last statement in the GLMMIX program FREQ Y causes GLIMMIX to use Y as the number of observations on the corresponding RATING. If no FREQ statement appears, GLIMMIX simply counts the number of data lines per RATING per BLK per TRT. Alternatively, you could delete the Y variable but have one data line per observation. For example, you could have 1 data line for BLK 1, TRT 0, RATING SLIGHT; 4 data lines for BLK 1, TRT 0, RATING MODRAT; and 23 data lines for BLK 1, TRT 0, RATING SEVERE.

The ESTIMATE statements specify estimable functions that form the boundaries between categories for each of the six treatments. For example, the first ESTIMATE "C=0, T=0" defines $\eta_0 + \tau_0$, the boundary between the SLIGHT and MODERATE categories for treatment 0. Notice that this estimates $\log[\pi_{00}/(1-\pi_{00})]$, the logit for the probability that a member of the population receiving treatment 0 responds in category 0—that is, SLIGHT. We can take the inverse link of this value to obtain the estimated probability π_{00}. The second ESTIMATE defines $\eta_1 + \tau_0$, the linear predictor of the boundary between categories 1 and 2 for treatment 0. It estimates the logit of the cumulative probability, $\pi_{00} + \pi_{10}$, that the rating is no worse than MODERATE under treatment 0.

Selected output:

Solutions for Fixed Effects

Effect	rating	trt	Estimate	Standard Error	DF	*t* Value	Pr > |*t*|
Intercept	slight		0.3492	0.2525	8	1.38	0.2041
Intercept	modrat		1.9956	0.2641	8	7.56	<0.0001
trt		0	−2.8326	0.3725	768	−7.60	<0.0001
trt		1	−2.6716	0.3617	768	−7.39	<0.0001
trt		2	−1.4795	0.3541	768	−4.18	<0.0001
trt		3	−1.2353	0.2469	768	−5.00	<0.0001
trt		4	−0.5820	0.2360	768	−2.47	0.0139
trt		5	0				

Notice the two intercept estimates. The first is $\hat{\eta}_0 = 0.3492$; it defines the boundary between categories 1 and 2—SLIGHT and MODERATE. The second, $\hat{\eta}_1 = 1.9956$, defines the boundary between categories 2 and 3—MODERATE and SEVERE. The TRT effect estimates, or $\hat{\tau}_i$, show how far up or down the boundaries move under different treatment applications. In this case, all of the treatment effects are negative relative to treatment 5. This means treatments 0 through 4 all have a lower probability of SLIGHT and a greater probability of SEVERE than treatment 5. Notice that for proportional odds (and threshold) models, the τ_i are *not* category specific; treatment effects move the boundaries as a group.

Odds-Ratio Estimates

Trt	_trt	Estimate	DF	95% Confidence Limits	
0	5	0.059	768	0.028	0.122
1	5	0.069	768	0.034	0.141
2	5	0.228	768	0.114	0.456
3	5	0.291	768	0.179	0.472
4	5	0.559	768	0.352	0.888

The odds ratios result from taking $e^{\hat{\tau}_i}$ for treatments 0 through 4. These are the estimated odds ratios of adjacent categories for treatment *i* relative to treatment 5. Because the τ_i are not category specific, the odds ratio for SLIGHT vs. MODERATE and MODERATE vs. SEVERE are the same (hence, the name "proportional odds" model).

Type III Tests of Fixed Effects

Effect	Num DF	Den DF	*F* Value	Pr > *F*
trt	5	768	17.67	<0.0001

The F-value refers to the overall test H_0: all $\tau_i = 0$. Were H_0 true, we would conclude that all odds ratios = 1. From the aforementioned odds-ratio results, it should be obvious why F and the corresponding p-value here are what they are.

Estimates

Label	Estimate	Standard Error	DF	t Value	Pr > |t|	Mean	Standard Error Mean
c=0, t=0	−2.4834	0.2791	768	−8.90	<0.0001	0.07703	0.01984
c=1, t=0	−0.8370	0.2658	768	−3.15	0.0017	0.3022	0.05605
c=0, t=1	−2.3224	0.2643	768	−8.79	<0.0001	0.08929	0.02149
c=1, t=1	−0.6760	0.2510	768	−2.69	0.0072	0.3372	0.05609
c=0, t=2	−1.1303	0.2521	768	−4.48	<0.0001	0.2441	0.04651
c=1, t=2	0.5162	0.2485	768	2.08	0.0381	0.6262	0.05817
c=0, t=3	−0.8861	0.2510	768	−3.53	0.0004	0.2919	0.05188
c=1, t=3	0.7604	0.2503	768	3.04	0.0025	0.6814	0.05434
c=0, t=4	−0.2328	0.2391	768	−0.97	0.3305	0.4421	0.05897
c=1, t=4	1.4136	0.2460	768	5.75	<0.0001	0.8043	0.03872
c=0, t=5	0.3492	0.2525	768	1.38	0.1671	0.5864	0.06125
c=1, t=5	1.9956	0.2641	768	7.56	<0.0001	0.8803	0.02782

There are the estimates of the various $\eta_c + \tau_i$. For example, the estimate for "C=0, T=0"—SLIGHT, treatment 0, is $\hat{\eta}_0 + \hat{\tau}_0 = -2.48$. We can check the table of SOLUTIONS shown previously to verify this estimate. Taking the inverse link yields $\hat{\pi}_{00} = 1/\left(1 + e^{2.48}\right) = 0.077$. This is the estimated likelihood of an individual receiving treatment 0 having a response rating of SLIGHT. The inverse link appears in the second column from the right labeled MEAN. For "C=1, T=0," the inverse link is 0.3022. This is the estimate $\hat{\pi}_{00} + \hat{\pi}_{10}$. From this, we can isolate the probability of a MODERATE response and a SEVERE response under treatment 0. For MODERATE, $\hat{\pi}_{10} = 0.302 - 0.077 = 0.225$; for SEVERE, $\hat{\pi}_{20} = 1 - 0.302 = 0.698$.

We can continue in this fashion for the other five treatments.

For the cumulative probit, or threshold model, we simply add the option LINK=CPROBIT to the MODEL statement. For DIST=MULTINOMIAL, the cumulative logit is the default link. If we change to the cumulative probit, the output will contain all of the same items, except for the odds ratios, which are only computed under logit GLMMs. Aside from the odds ratios, we work through the analysis exactly as we do for the cumulative logit analysis.

13.3 Nominal Categories: Generalized Logit Models

Not all multinomial data come from ordered categories. Moreover, the assumption that $\mathbf{X\beta} + \mathbf{Zb}$ does not depend on category, shared by both the proportional odds and threshold models, is often too strong even when we do have ordered categories. The generalized logit model allows us to fit less-restrictive GLMs and GLMMs for these situations.

Like the cumulative logit and probit models, the generalized logit model has $C - 1$ link functions, where C denotes the number of response categories. We define one category as the reference category. This can be arbitrary, or there may be compelling logic in the context of a given study to designate a particular response category as the reference. In practice, it really does not matter which category we use as the reference as long as we follow through consistently during the analysis. Therefore, without loss in generality, suppose we use category C as the reference. Then we define the generalized logits as

$$\eta_1 = \log\left(\frac{\pi_1}{\pi_C}\right) = \mathbf{X}\boldsymbol{\beta}_1 + \mathbf{Z}\mathbf{b}_1$$

$$\eta_2 = \log\left(\frac{\pi_2}{\pi_C}\right) = \mathbf{X}\boldsymbol{\beta}_2 + \mathbf{Z}\mathbf{b}_2$$

and so forth, through

$$\eta_{C-1} = \log\left(\frac{\pi_{C-1}}{\pi_C}\right) = \mathbf{X}\boldsymbol{\beta}_{C-1} + \mathbf{Z}\mathbf{b}_{C-1}$$

Notice that we now have distinct model effects, $\boldsymbol{\beta}$ and $\mathbf{b}$, for each link function. With a little algebra, we can express the inverse link using the following system of equations:

$$\begin{bmatrix} e^{\eta_1} \\ e^{\eta_2} \\ \cdots \\ e^{\eta_{C-1}} \end{bmatrix} = \begin{bmatrix} 1+e^{\eta_1} & e^{\eta_1} & \cdots & e^{\eta_1} \\ e^{\eta_2} & 1+e^{\eta_2} & \cdots & e^{\eta_2} \\ \cdots & \cdots & \cdots & \\ e^{\eta_{C-1}} & e^{\eta_{C-1}} & & \ldots 1+e^{\eta_{C-1}} \end{bmatrix} \begin{bmatrix} \pi_1 \\ \pi_2 \\ \cdots \\ \pi_{C-1} \end{bmatrix}$$

With a little additional algebra, we can show that the general form of the inverse link is

$$\pi_c = \frac{e^{\eta_c}}{1+\sum_{c=1}^{C-1} e^{\eta_c}} \quad \text{for} \quad c = 1, 2, \ldots, C-1$$

For the reference category, once we determine $\pi_1, \pi_2, \ldots, \pi_{C-1}$, then $\pi_C = 1 - \sum_{c=1}^{C-1} \pi_c$.

Example 13.2 Generalized logit model

Let us implement the generalized logit model for Data Set 13.1. We do this mainly to show how to implement the generalized logit model with GLIMMIX and to illustrate important differences between working with the cumulative logit and generalized logit models. While there is no compelling reason *not* to use generalized logit models for ordinal data, there is also no compelling reason to use them in favor of cumulative logit models either. In practice, we use cumulative models for ordinal data and generalized logit models for nominal data.

One question that is of interest would be to see if the estimated probabilities change and how our conclusions regarding treatment effects are affected by the extra parameterization of the generalized logit. We explore this question in Section 13.4.

The following GLIMMIX statements implement the analysis:

```
proc glimmix data=univar_multinom;
class rating blk trt;
model rating(reference='severe')=trt/dist=multinomial link=glogit
solution oddsratio;
random intercept/subject=blk solution group=rating;
estimate 't=0' intercept 1 trt 1 0,
  't=1' intercept 1 trt 0 1 0,
  't=2' intercept 1 trt 0 0 1 0,
  't=3' intercept 1 trt 0 0 0 1 0,
  't=4' intercept 1 trt 0 0 0 0 1 0,
  't=5' intercept 1 trt 0 0 0 0 0 1/ilink bycat;
freq y;
```

Note several changes in the program. First, we use LINK=GLOGIT to override the default cumulative logit link and use the generalized logit instead. Also in the MODEL statement we use a REFERENCE= option to designate the reference category. If we omit the REFERENCE= option, on default GLIMMIX uses the last category in the data set. If we used the ORDER statement as we did with the cumulative logit model, GLIMMIX would still use SEVERE as the reference category because we entered it as the last category. If we omitted both REFERENCE and ORDER options, GLIMMIX would alphabetize the categories and used the last one as the reference. For these data, the default reference category would then be SLIGHT.

The BYCAT option in the ESTIMATE statement is unique to the generalized logit among the multinomial models. In conjunction with ILINK, these commands direct GLIMMIX to compute estimated probabilities of all *non-reference* categories for each treatment.

Selected results:

Solutions for Fixed Effects

Effect	Rating	trt	Estimate	Standard Error	DF	t Value	Pr > $\|t\|$
Intercept	slight		1.6495	0.3770	16	4.37	0.0005
Intercept	modrat		0.9294	0.3784	16	2.46	0.0259
Trt	slight	0	−3.9101	0.5714	755	−6.84	<0.0001
Trt	modrat	0	−2.0519	0.4873	755	−4.21	<0.0001
Trt	Slight	1	−3.6198	0.5356	755	−6.76	<0.0001
trt	modrat	1	−1.9528	0.4795	755	−4.07	<0.0001
trt	slight	2	−2.1166	0.5099	755	−4.15	<0.0001
trt	modrat	2	−0.8848	0.4830	755	−1.83	0.0674
trt	slight	3	−1.7101	0.3706	755	−4.61	<0.0001
trt	modrat	3	−0.8260	0.3813	755	−2.17	0.0306
trt	slight	4	−0.8078	0.3627	755	−2.23	0.0262
trt	modrat	4	−0.3578	0.3874	755	−0.92	0.3559
trt	slight	5	0				
trt	modrat	5	0				

Notice that for the generalized logit model, we have separate estimates of all treatment effects for each link function, as well as separate intercepts. SLIGHT refers the parameters of the first link function, η_1, and MODRAT refers to the second link function, η_2.

Estimates

Label	Rating	Estimate	Standard Error	DF	*t* Value	Pr > |*t*|	Mean	Standard Error Mean
t=0	slight	−2.2606	0.4293	755	−5.27	<0.0001	0.07294	0.02901
t=0	modrat	−1.1225	0.3070	755	−3.66	0.0003	0.2276	0.05380
t=1	slight	−1.9703	0.3804	755	−5.18	<0.0001	0.09302	0.03212
t=1	modrat	−1.0234	0.2944	755	−3.48	0.0005	0.2398	0.05350
t=2	slight	−0.4671	0.3433	755	−1.36	0.1741	0.2346	0.06177
t=2	modrat	0.04460	0.3001	755	0.15	0.8819	0.3913	0.07154
t=3	slight	−0.06061	0.3381	755	−0.18	0.8578	0.3086	0.07156
t=3	modrat	0.1033	0.3105	755	0.33	0.7393	0.3636	0.07207
t=4	slight	0.8418	0.3320	755	2.54	0.0114	0.4558	0.08105
t=4	modrat	0.5715	0.3193	755	1.79	0.0738	0.3478	0.07402
t=5	slight	1.6495	0.3770	755	4.37	<0.0001	0.5957	0.08237
t=5	modrat	0.9294	0.3784	755	2.46	0.0143	0.2899	0.07354

The MEAN column shows the estimated probabilities, that is, $\hat{\pi}_{ci}$, for each non-reference category by treatment. For example, for treatment 0, the estimated probabilities for SLIGHT and MODRAT (moderate) are $\hat{\pi}_{00} = 0.0729$ and $\hat{\pi}_{10} = 0.228$, respectively. From these, we can determine the estimated probability for SEVERE, $\hat{\pi}_{20} = 1-(0.0729+0.228) = 0.699$. We use these estimates to compute odds ratios.

Odds-Ratio Estimates

Rating	trt	_trt	Estimate	DF	95% Confidence Limits	
slight	0	5	0.020	755	0.007	0.062
modrat	0	5	0.128	755	0.049	0.334
Slight	1	5	0.027	755	0.009	0.077
modrat	1	5	0.142	755	0.055	0.364
slight	2	5	0.120	755	0.044	0.328
modrat	2	5	0.413	755	0.160	1.065
slight	3	5	0.181	755	0.087	0.374
modrat	3	5	0.438	755	0.207	0.925
slight	4	5	0.446	755	0.219	0.909
modrat	4	5	0.699	755	0.327	1.496

These odds ratios compare the odds for the labeled category vs. the reference category for treatments 0 through 4 relative to treatment 5. For example, as we will see in the following output, for treatment 0, the estimated probabilities of SLIGHT and SEVERE are $\hat{\pi}_{00} = 0.0729$ and $\hat{\pi}_{20} = 0.699$, respectively. For treatment 5, the estimated probabilities are $\hat{\pi}_{05} = 0.596$ and $\hat{\pi}_{25} = 0.114$, respectively. The estimated odds ratio is thus

$$\frac{0.0729}{0.699} \bigg/ \frac{0.596}{0.114} = 0.0199 \cong 0.02$$

what we see above under ESTIMATE for SLIGHT TRT 0 vs. TRT 5.

Type III Tests of Fixed Effects				
Effect	**Num DF**	**Den DF**	***F* Value**	**Pr>*F***
trt	10	755	9.10	<0.0001

Notice that we have 10 numerator degrees of freedom, not five as we did for the proportional odds model. This is because we now have a distinct estimate of treatment effect for each link function; we are now testing H_0: all $\tau_{1i}=0$ and all $\tau_{2i}=0$. We could, in principle, partition this *F*-statistic into components that test the treatment effect separately for each link function. The GLIMMIX CONTRAST statement has a BYCAT option that allows one to do this easily.

13.4 Model Comparison

Comparing the analysis using the cumulative logit link and the generalized logit link, we see at most negligible changes in estimated category probabilities by treatment and the significance level for the test of treatment effects. It does not appear that we gained any additional information by using the less parsimonious generalized logit model. Informally, we could use this as a basis to suggest that the proportional odds model provides an adequate fit for this particular data set. How could we evaluate this more formally?

While both of these analyses have been shown using the default PL estimation algorithm, we could have used integral approximation instead. If we rerun the two models using METHOD=QUAD, for example, we get the following results:

Fit Statistic	Cumulative Logit/Prop-Odds	Generalized Logit
−2 log Likelihood	1494.01	1496.49
AICC	1510.20	1525.03

These results involve changes in both link and the effects, fixed and random, of each model. Although, taking them in context, the fit statistics appear to support our impression that the simpler cumulative logit model provides an adequate fit for these data and the generalized logit is an unneeded complication, this is not an appropriate use of fit statistics. We need a better defined formal test.

In fit statistics stated previously, the following two questions are confounded:

1. Do we need to separate treatment and block effects for each link, that is, are the link functions $\eta_{cij}=\eta_c+\tau_i+b_j$; $c=0,1$ sufficient, or do the link functions $\eta_{cij}=\eta_c+\tau_{ci}+b_{cj}$; $c=0,1$ provide a better fit?
2. How does the choice between cumulative and generalized link functions affect inference?

We need to untangle the two and focus on the first question. To do this, we need to fit a cumulative logit model that does not assume proportional odds, that is,

1. Linear predictor: $\eta_{cij}=\eta_c+\tau_{ci}+b_{cj}$, where η_{cij} denotes the cth link ($c=0,1$) for the ijth treatment–block combination, η_c, τ_{ci}, and b_{cj} denote the intercept, treatment, and block effects, respectively, for the cth link.
2. Distributions:
   a. $y_{0ij}, y_{1ij}, y_{2ij} \mid b_j \sim \text{Multinomial}(N_{ij}, \pi_{0ij}, \pi_{1ij}, \pi_{2ij})$
   b. b_{0j} i.i.d. $N(0,\sigma^2_{0B})$; b_{1j} i.i.d. $N(0,\sigma^2_{1B})$; b_{0B} and b_{1B} mutually independent
3. Links:
   a. $\eta_{0ij}=\log\left(\dfrac{\pi_{0ij}}{1-\pi_{0ij}}\right)$
   b. $\eta_{1ij}=\log\left(\dfrac{\pi_{0ij}+\pi_{1ij}}{1-(\pi_{0ij}+\pi_{1ij})}\right)$

Once we fit the model, we can test the proportional odds assumption formally by testing H_0: $\tau_{0i}=\tau_{1i}$ for all $i=0,1,\ldots,5$ and H_0: $b_{0i}=b_{1i}$ for all $i=0,1,\ldots,5$. Because the block effects are random, we do this in two stages. First, test the equality of treatment effects, H_0: $\tau_{0i}=\tau_{1i}$. If we reject H_0, we are done. If we conclude that assuming equal treatment effects for both links is tenable, we can compare to fit of the model with separate variance components σ^2_{cB}; $c=0,1$ to the model with a single variance component σ^2_B.

We cannot fit the nonproportional odds cumulative logit model with PROC GLIMMIX, but we can with PROC NLMIXED. The program statements appear in the following and can be accessed in the SAS Data and Program Library in file Ch_13_CumLogit_NLMIXED.

The NLMIXED statements are

```
proc nlmixed data=indiv_data_multinom;
parms int_0=-1 int_1=0.5 tau_00=0 tau_10=0 tau_01=0 tau_11=0 tau_02=0
tau_12=0
  tau_03=0 tau_13=0 tau_04=0 tau_14=0 sdb0=0.1 sdb1=0.1;
eta_0=int_0+(trt=0)*tau_00+(trt=1)*tau_01+(trt=2)*tau_02+(trt=3)*tau_03
+(trt=4)*tau_04+blk_0j;
eta_1=int_1+(trt=0)*tau_10+(trt=1)*tau_11+(trt=2)*tau_12+(trt=3)*tau_13
+(trt=4)*tau_14+blk_1j;
pi_0=1/(1+exp(-eta_0));
pi_1=1/(1+exp(-eta_1))-pi_0;
pi_2=1-pi_0-pi_1;
if rating=0 then pi=pi_0;
else if rating=1 then pi=pi_1;
else pi=pi_2;
  p=(pi>0 and pi<=1)*pi+(pi<=0)*1e-8+(pi>1);
  LL=log(p);
model rating ~ general(LL);
random blk_0j blk_1j ~ normal([0,0],[sdb0*sdb0,0,sdb1*sdb1]) subject=blk;
estimate 'blk variance for link 0' sdb0*sdb0;
estimate 'blk variance for link 1' sdb1*sdb1;
contrast 'separate trt effect by link?' tau_00-Tau_10,tau_01-
  Tau_11,tau_02-Tau_12,tau_03-Tau_13,tau_04-Tau_14;
```

The CONTRAST statement "`separate trt effect by link?`" generates the test statistic for H_0: $\tau_{0i}=\tau_{1i}$. For these data, $F=0.08$ and $p>0.99$; there is no evidence to suggest unequal treatment effects by link function. Next, we modify the linear predictor from $\eta_{cij} = \eta_c + \tau_{ci} + b_{cj}$; $c = 0,1$ to $\eta_{cij} = \eta_c + \tau_i + b_j$; $c = 0,1$ by replacing the ETA_0 and ETA_1 statements with

```
eta_0=int_0+(trt=0)*tau_0+(trt=1)*tau_1+(trt=2)*tau_2+(trt=3)*tau_3+(trt=4)
*tau_4+blk_0j;
eta_1=int_1+(trt=0)*tau_0+(trt=1)*tau_1+(trt=2)*tau_2+(trt=3)*tau_3+(trt=4)
*tau_4+blk_1j;
```

and making corresponding changes in the PARMS statement. This model has equal treatment effects for the two links but separate random block effects. The AICC for this model is 1511.8. With equal random effects for the two links, the AICC is 1510.2. Hence, no evidence to suggest unequal block effects by link function.

This establishes the adequacy of the proportional odds cumulative logit model shown in Example 13.2. Readers interested in more detail and a much expanded array of tools for assessing model adequacy for multinomial data are referred to Hedeker and Gibbons (1994), Jansen (1990) and texts such as Agresti (2002) and Stokes et al. (2000).

13.5 Summary

- Analysis of multinomial data should focus on treatment effect on response categories not mean ratings.
- Mean rating only possible for ordinal data, and too much is lost aggregating the data this way.
- Conceptual basis for binomial logit and probit models extends naturally to multinomial data.
- Cumulative logit and probit models can be used for ordinal data only.
- Generalized logit and probit models may be used for any multinomial data.
- Generalized logit or probit may be unnecessary elaboration, not parsimonious, for ordinal data—this can be assessed using fit statistics and by comparing what is learned from each model.

Exercises

13.1 This problem uses file `Ch _ 13 _ Problem1.sas` found in SAS Data and Program Library.

The data are from a four treatment multisite study. There were 16 sites. At 8 sites, chosen at random, treatments 1 and 2 were observed; at 8 other sites, treatments 3 and 4 were observed. At each site, 120 people participated—60 were assigned to one of the treatments at the site while the other 60 were assigned to the other treatment. Participants were asked to rate their assigned treatment as "bad," "fair," or "good."

a. Fit the cumulative logit proportional odds model to these data. Do a complete and appropriate analysis of the data, focusing on
   i. An assessment of treatment effects
   ii. Interpretation of odds ratios
   iii. The expected probability by category for each treatment
b. Test the adequacy of the proportional odds assumption. Cite relevant evidence to support your conclusion regarding the assumption's adequacy.
c. If as a result of (b) an alternative cumulative logit model is deemed better, revise your analysis in (a) accordingly.
   If the model in (a) is adequate, (c) can be used as an exercise in using NLMIXED output instead of GLIMMIX output as a basis for a report.

13.2 This problem uses file `Ch_13_Problem2.sas` found in SAS Data and Program Library.

The data are from a blocked design with seven treatments. At each block, only three of the treatments can be observed, resulting in an incomplete block structure. Assume block effects are random. For each treatment at each block, a number of observations are taken (the numbers vary). Each observation is classified according to categories denoted 0,1, and 2. Assume these categories are nominal.

a. Fit a generalized logit model to these data.
b. Fit a reduced generalized logit model with equal model effects by link to these data (you will need to use NLMIXED or equivalent to write a program along the lines of the cumulative logit program shown in Section 13.4). Assess model fit to decide if separate treatment and/or block effects are needed.
c. Do a complete and appropriate analysis of the data focusing on treatment effect and its impact of probability by category using the model you deemed most appropriate in (b).

14

Correlated Errors, Part I: Repeated Measures

14.1 Overview

Repeated measurements are used to study changes over time or space and the effect of treatments on these changes. Repeated measures in space may occur in one or multiple dimensions. One dimension implies taking observations along a line. Two dimensions mean we take observations on a place and locate them, for example, by latitude and longitude. Adding altitude or ocean depth, we have spatial data in three dimensions. Spatial–temporal repeated measures involve both location and time of measurement.

Data collected over time are also called *longitudinal* data. From a modeling perspective, the terms *repeated measures* and *longitudinal* have interchangeable meanings, reflecting differences in linguistic history and jargon among disciplines, not any difference in definition. Hereafter, we use *repeated measures*—readers should understand that the methods and concerns we discuss apply equally to longitudinal data.

We define a *repeated measures study* as one in which units are observed at two or more planned times or places over the course of the study. The key feature common to all repeated measures is the possibility of correlation between observations on the same subject. Moreover, the magnitude of this correlation depends on the distance between pairs of observations.

In this chapter, we consider one-dimensional repeated measures only—observations in time or along a straight line in space. In Chapter 15, we consider multidimensional repeated measures. There are entire courses on time series and on spatial statistics. Chapters 14 and 15 will necessarily be surveys introducing readers to modeling with repeated measures. Interested readers wanting more depth are, of course, encouraged to read further and consider taking more advanced courses in these areas.

14.1.1 What Are Repeated Measures/Longitudinal Data

We start this section by first clarifying the distinction between *multiple measurements*, also known as *subsampling*, and true *repeated measures data*. Some examples of multiple measurements: weighing a mouse three times during the same measurement occasion, taking blood serum measurements on three samples from the same blood draw, determining the proportion of leaf area infected on three leaves taken from the same plant at the same time, and having three judges rate the taste of a food item. None of these qualify as repeated measures. We reserve the term repeated measures for observations that are taken at distinct times or places *with the express purpose of modeling change over time or space* and, usually for these kinds of studies, treatment effects on these changes. Therefore, models for repeated measures *must* have a time or space source of variation in $\mathbf{X}\boldsymbol{\beta}$, the fixed component of the linear predictor. If treatment effects are also part of the model, corresponding treatment-by-time effects must also be included—with very rare exceptions (so rare they are not worth considering here).

Examples of true repeated measures experiments include sampling a pharmaceutical product at planned storage times (e.g., after 3, 6, 9, and 12 months) to monitor changes in a stability-limiting characteristic, math achievement tests at the end of the school year from fifth through eighth grade to monitor students' progress, taking insect counts at key times during the growing season to assess their impact on a crop, measuring nutrient uptake of plant roots at various soil depths, and annual data on the number of repairs required by a component as it ages.

Repeated measures studies can be conducted in conjunction with any design—completely randomized, blocked, row–column, multilevel, etc. Recalling our discussion in Chapters 7 and 8, we can think of repeated measures as a form of treatment design that also has implications for the covariance. In addition, the "topological aspect" (using Fisher's term) of the design must account for measurements on the subjects over time.

The repeated measures and design aspects affect the way we write the linear predictor and the distributions differently depending on whether we have Gaussian or non-Gaussian data. For Gaussian data, repeated measures primarily determine how we write $\mathbf{X\beta}$ and $\mathbf{R} = Var(\mathbf{y}|\mathbf{b})$ (see Stiratelli, et al., 1984). The other aspects to the design mainly affect $\mathbf{Zb}$. For non-Gaussian repeated measures, we have two major *types* of models: the GEE-type models repeated measures through $\mathbf{X\beta}$ and working covariance embedded in $\mathbf{R}$ (see Zeger and Liang, 1986); generalized linear mixed models or GLMMs model $\mathbf{X\beta}$ much as any other repeated measures model, but account for correlation through additional terms in $\mathbf{Zb}$ with nontrivial covariance structure, an approach also used with non-linear repeated measures (see, for example, Davidian and Giltinan, 2003). We will work through examples illustrating these alternative approaches in Section 14.4.

14.1.2 Pre-GLMM Methods

In the days preceding widespread access to linear mixed model (LMM) and GLMM software, there were two predominant modeling approaches for repeated measures. One was essentially a multilevel analysis of variance model, identical to the nested factorial and split-plot examples we have seen in earlier chapters. In fact, in agricultural statistics, the approach is called "the split-plot-in-time." At the other extreme was multivariate analysis of variance (MANOVA). Rather than focusing on a single response variable, MANOVA defines a single observation as a vector with one element for each time (or place) of observation.

To make this more tangible, let us apply each of these models to Data Set 6.2, the repeated measures data set we used as part of our introduction covariance component testing in Chapter 6. The data set contains two treatments each assigned to six subjects using a completely randomized design. Observations are taken on each subject at six equally spaced times. Let y_{ijk} denote the observation on the kth subject assigned to the ith treatment at the jth time.

We saw the "split-plot-in-time" model in Chapter 6, also called the independent errors model or the compound symmetry (CS) model. In GLMM terms, the model is

1. Linear predictor:
   a. Independent errors version: $\eta_{ijk} = \alpha_{ij} + s_{ik}$.
   b. CS version: $\eta_{ijk} = \alpha_{ij}$.

c. α_{ij} denotes the treatment × time mean model. We saw in Chapter 6 that we can partition this into treatment and time components, that is, $\alpha_{ij} = \alpha + \tau_i + \gamma_j + (\tau\gamma)_{ij}$, where τ and γ denote time and treatment effects.

d. s_{ik} denote subject effects.

2. Distributions:

a. Independent error version: $y_{ijk}|s_{ik} \sim NI(\mu_{ij}, \sigma^2)$ and s_{ik} i.i.d. $N(0, \sigma_S^2)$

b. CS version: $\mathbf{y}_{ik} \mid s_{ik} \sim N\left(\mathbf{0}, \sigma_C^2 \begin{bmatrix} 1 & \rho & \rho & \rho & \rho & \rho \\ & 1 & \rho & \rho & \rho & \rho \\ & & 1 & \rho & \rho & \rho \\ & & & 1 & \rho & \rho \\ & & & & 1 & \rho \\ & & & & & 1 \end{bmatrix}\right)$, where

$$\mathbf{y}'_{ik} = \begin{bmatrix} y_{i1k} & y_{i2k} & y_{i3k} & y_{i4k} & y_{i5k} & y_{i6k} \end{bmatrix}$$

3. Link: identity

We know that for Gaussian data, as long as $\sigma_S^2 \geq 0$, the independent error and CS models are equivalent.

The MANOVA model treats the vector $\mathbf{y}'_{ik} = \begin{bmatrix} y_{i1k} & y_{i2k} & y_{i3k} & y_{i4k} & y_{i5k} & y_{i6k} \end{bmatrix}$ as a single observation. The model is

4. Linear predictor: $\boldsymbol{\eta}_{ik} = \boldsymbol{\alpha}_i$, where $\boldsymbol{\alpha}'_i = \begin{bmatrix} \alpha_{i1} & \alpha_{i2} & \alpha_{i3} & \alpha_{i4} & \alpha_{i5} & \alpha_{i6} \end{bmatrix}$

5. Distribution: $\mathbf{y}_{ik} \sim N(\boldsymbol{\mu}_i, \boldsymbol{\Sigma}_{ik})$, where $\boldsymbol{\mu}'_i = \begin{bmatrix} \mu_{i1} & \mu_{i2} & \mu_{i3} & \mu_{i4} & \mu_{i5} & \mu_{i6} \end{bmatrix}$ and

$$\boldsymbol{\Sigma}_{ik} = \begin{bmatrix} \sigma_1^2 & \sigma_{12} & \sigma_{13} & \sigma_{14} & \sigma_{15} & \sigma_{16} \\ & \sigma_2^2 & \sigma_{23} & \sigma_{24} & \sigma_{25} & \sigma_{26} \\ & & \sigma_3^2 & \sigma_{34} & \sigma_{35} & \sigma_{36} \\ & & & \sigma_4^2 & \sigma_{45} & \sigma_{46} \\ & & & & \sigma_5^2 & \sigma_{56} \\ & & & & & \sigma_6^2 \end{bmatrix}$$

6. Link: identity $\boldsymbol{\eta}_i = \boldsymbol{\mu}_i$

"Linear predictor" and "link" would not be concepts that would occur to anyone doing multivariate analysis of variance, but we present the model this way here to make the connection with contemporary linear model theory. Notice that the MANOVA model shares the same covariance with the *unstructured* (*UN*) LMM we saw in Chapter 6. In fact, for balanced completely randomized designs, they produce identical results.

With pre-GLMM SAS®, we could implement the "split-plot-in-time" and MANOVA analyses using the following PROC GLM statements:

```
proc glm data=multivar;
  class trt;
  model t1-t6=trt;
  repeated time 6/printe;
```

Notice that data must be in a different format than we use with PROC GLIMMIX (or, for that matter, any *non*-repeated measures linear model for a single response variable implemented with PROC GLM): the observations at each time must appear with a different variable name—in this case T1, T2,..., T6—on the same line for each subject. In a repeated measures experiment with two or more treatments, the analysis should begin by testing the treatment × time interaction: Does treatment affect the pattern of change over time? The output for the two analyses appears as follows:

MANOVA:

MANOVA Test Criteria and Exact F Statistics for the Hypothesis of no time*trt Effect
H = Type III SSCP Matrix for time*trt
E = Error SSCP Matrix
S = 1 M = 1.5 N = 2

Statistic	Value	F Value	Num DF	Den DF	Pr > F
Wilks' Lambda	0.33118535	2.42	5	6	0.1558
Pillai's Trace	0.66881465	2.42	5	6	0.1558
Hotelling–Lawley Trace	2.01945720	2.42	5	6	0.1558
Roy's Greatest Root	2.01945720	2.42	5	6	0.1558

Split-plot-in-time:

The GLM Procedure
Repeated Measures Analysis of Variance
Univariate Tests of Hypotheses for Within-Subject Effects

						Adj Pr > F	
Source	DF	Type III SS	Mean Square	F Value	Pr > F	G – G	H-F-L
time	5	145.7552376	29.1510475	77.73	<.0001	<.0001	<.0001
time*trt	5	6.5830678	1.3166136	3.51	0.0085	0.0261	0.0109
Error(time)	50	18.7508581	0.3750172				

The "split-plot-in-time" analysis is equivalent to the "type III tests of fixed effects" for the TIME and TIME*TRT effects. "Pr>F" is the usual p-value. The "Adj Pr>F" values are corrections due to Greenhouse and Geisser (1959) and Huynh and Feldt (1970) who were early investigators concerned with the unrealism of the independence assumption of the split-plot analysis shown earlier. Notice that the most conservative p-value in the univariate "split-plot-in-time" is 0.0261, and the p-values for the MANOVA are 0.1558. Clearly, these results are in conflict.

As it turns out, neither of these analyses are satisfactory. Historically, this is one of the issues that played a prominent role in raising linear mixed model methodology from an under-appreciated novelty in the 1970s to the mainstream linear model for Gaussian data by the mid-1990s. Here are the problems.

MANOVA is far too conservative. Worse, what we see in this example is a best case scenario. MANOVA requires the data vector to be complete—even if only one observation at any time is missing, we must throw out the entire vector. This is an indefensibly wasteful approach to data analysis. Even when there are no missing data, the MANOVA procedure over-models within-subject correlation, resulting in loss of efficiency. MANOVA's lack of power makes it an excellent way *not* to find a treatment effect even if it actually exists.

More parsimonious linear model procedures can account for correlation, control type I error, use all of the data and provide greater power.

The split-plot-in-time, a.k.a. the independent errors/CS model, has the opposite problem. When nontrivial within-subject correlation exists, the independent errors/CS model underestimates within-subject variance. This in turn yields upwardly biased test statistics for time and time × treatment effects, loss of control over type I error rates and poor confidence interval coverage.

At the beginning of Section 14.3, we look at one example to illustrate how we assess MANOVA's lack of power and CS's lack of type I error control. First, we need to review the mixed model alternatives between these two extremes. As we do this, keep in mind our goal: we need to adequately account for within-subject correlation (if we do not, we fail to control type I error), but we need to do so parsimoniously, not wastefully (if we do not, we compromise power). Compromising power often leads researchers to try to compensate by increasing replication. This means that the hidden cost of MANOVA (and, as we will see in the next section, the UN covariance model) is that it drives researchers to spend other people's money foolishly.

14.2 Gaussian Data: Correlation and Covariance Models for LMMs

In Chapter 6 we introduced a few of prominent covariance models for repeated measures LMMs. Here is a brief review. As we review, we will say "...referred to hereafter as..." for each model and then give an acronym. The acronym is the PROC GLIMMIX name, so we can minimize confusion as we work through the examples later in this chapter.

The **independent errors** or **CS** model. As we saw in Section 14.1, this is the "split-plot-in-time" model, one of the two repeated measures models widely used in the premixed-model era. Because our focus in this chapter is on within-subject correlation among repeated measurements, we will hereafter refer to this as the CS model.

With regard to correlation, CS implies that repeated measurements are equally correlated regardless of how widely separated they are in time. This is a very strong and often unrealistic assumption. If not satisfied, CS underestimates within-subject variance, and we lose control of type I error for tests involving time. The CS model *does* make sense when the times of observation are widely separated and the data are noisy relative to the strength of within-subject correlation. For example, some agricultural trials take monthly measurements over a growing season. Between measurements, weather events alone can overwhelm month-to-month correlation, making it virtually undetectable.

The unstructured covariance model, hereafter referred to as UN. The within-subject covariance matrix is identical to $\boldsymbol{\Sigma}_{ik}$ for MANOVA. As mentioned in Section 14.1, for a balanced, completely randomized between-subject design, for example, Data Set 6.2, UN LMM and MANOVA are just different ways of writing the same model. For more complicated designs, for example, when the between-subject design uses blocking, the models are similar in spirit but not quite identical.

If we have t times, the UN model requires $t+t(t-1)/2$ covariance parameters. Usually, this is overkill in terms of what we actually need to account for within-subject covariance. Nonetheless, the UN model can be a good place to start if we lack any intuition about the extent to which covariance decays with increasing distance (but we should be sure we truly lack intuition: UN should never be a substitute for thinking carefully about the problem at hand). We can also use the UN parameter estimates to create visual aids to help identify a more parsimonious model. More about this in Section 14.3.2.

The **first-order ante-dependence** model, hereafter referred to as ANTE(1). The form of the covariance matrix was given as equation (6.4) in Chapter 6. Following our notation convention—observation y_{ijk} and subject effect s_{ik}—begun in Section 14.1, the ANTE(1) process assumes that the within-subject "error" or residual term, denoted $w_{ijk} = \rho_j w_{i,j-1,k} + z_{ijk}$, where ρ_j denotes the auto-regression parameter at time j and z_{ijk} is a Gaussian random deviate such that $z_{ijk} \sim N(0, \sigma_j^2)$. Think of w_{ijk} as the discrepancy between y_{ijk} and $\hat{\mu}_{ijk}$ for the $i$kth subject at time j. At $j=1$, $w_{i,j-1,k}=0$.

The ANTE(1) model assumes that the correlation between observations at times $j-1$ and j may change with j and that the variance also changes over time.

The **first-order auto-regression** model, hereafter referred to as AR(1). The AR(1) model can be viewed as a special case of the ANTE(1) model where all $\rho_j = \rho$ and all $\sigma_j^2 = \sigma^2$. The ARH(1) model assumes $\rho_j = \rho$ but not $\sigma_j^2 = \sigma^2$.

The **Toeplitz** model, hereafter referred to as TOEP. The TOEP model is similar to the AR(1) model except that it relaxes the strict power relationship between distance and correlation. That is, the TOEP correlation between observations on the same subject two units apart in time is ρ_2 instead of ρ^2; between observations three units apart, the TOEP correlation is ρ_3 instead of ρ^3; and so forth.

Notice that the ANTE(1), AR(1), and TOEP models all assume equal spacing between times. Designs often use geometric rather that equal spacing, for example, time 1, 2, 4, and 8. We can make these observations appear to be equally spaced using the log—for example, $\log_2(1)=0$, $\log_2(2)=1$. Whether this allows us to use ANTE(1), AR(1), and TOEP or merely distorts the correlation–distance relationship causing these three models to yield poor fits is something we have to decide on a case by case basis.

One alternative, at least for AR(1), is the **spatial power** model, hereafter referred to as SP(POW). This model is similar to the AR(1), but the correlation between adjacent pairs of observations on the same subject is $\rho^{|t_j - t_{j-1}|}$, where t_j denotes that actual time the jth repeated measurement was taken. This allows observations to be unequally spaced. It also means that different subjects can have repeated measurements taken at different times. This can be extremely useful for observational or multisite studies, where the exact timing of repeated measurements is hard to control, as well as studies with relatively high rates of missing data. The SP(POW) model was designed for two-dimensional spatial modeling (see Chapter 15) but can be used for time—or space in one dimension—by giving only one coordinate. We will work through an example in Section 14.3.

14.3 Covariance Model Selection

Our task is to identify a reasonable compromise between over- and under-modeling within-subject correlations. To do this, we use many of the tools for inference on covariance components introduced in Chapter 6 and add graphical methods that can often, but not always, add insight to covariance model selection.

In Chapter 6, we went through the basics for Data Set 6.2, to which we applied the split-plot-in-time and MANOVA in Section 14.1. In that exercise, we selected the AR(1) model. In Section 14.3.2, we will consider two more advanced examples. First, however, let us take a closer look at *why* we regard covariance model selection to be an essential prerequisite before we begin inference on time and treatment effects.

14.3.1 Why Does It Matter?

At the end of Section 14.1, we said that over-modeling correlation compromises power and under-modeling compromises type I error control. To get a sense of the magnitude of the compromises we are talking about and what aspects of inference are most vulnerable, simulation is an invaluable tool.

Modern computers make it easy to implement simulations that, while not of publishable quality, are adequate to provide a good idea of what a procedure will do under the assumptions of a given modeling exercise. Writing the simulation program is also an excellent way to deepen your understanding of what you are modeling and the assumptions you are making. Once you get the hang of it, you should be able to write and run the kind of simulations we describe here in an hour or less.

Let us consider a study similar to Data Set 6.2. Instead of two treatments, we have 3. Instead of six repeated measurements (or times), we have 8. Suppose treatment 0 is a "control" and treatments 1 and 2 are experimental. The experimental treatments are supposed to provide temporary relief, so they should produce a higher "comfort index" immediately after being applied but then wear off over time. We want to know which treatments provide initial relief, how quickly they wear off, and if either of the experimental treatments is better at providing lasting relief. Suppose the actual "comfort index" means of the three treatments are

	Time							
Treatment	**1**	**2**	**3**	**4**	**5**	**6**	**7**	**8**
0	3.0	3.0	3.0	3.0	3.0	3.0	3.0	3.0
1	6.5	6.0	5.5	5.0	4.5	4.0	3.5	3.0
2	5.8	5.8	5.8	5.75	5.65	5.4	4.8	3.25

Also, suppose that the variance among subjects is $\sigma_S^2 = 0.5$ and variability within subjects follows an AR(1) auto-regression process with $\rho = 0.7$ and $\sigma^2 = 1$. These numbers are not out of line for certain clinical trials—see, for example, Section 5.2 of Littell et al. (2006). Here, CS would clearly under-model variation, and UN would over-model it. Some things we might want to test are as follows:

- The treatment × hour effect: This is a global test of treatment effect on changes over time.
- The simple effects of treatment | hour: Is there a treatment effect at hour 1, hour 2, etc? We would expect to see a big treatment effect at hour 1 and little treatment effect at hour 8 (when the treatments have worn off), and we would want to know how long the treatment effect lingers.
- The simple effect of hour | treatment: What do the treatment-specific patterns of change over time look like?

Usually, we also have specific contrasts and individual pair-wise tests we could identify, but the tests listed previously are sufficient to give us a sense of the damage over- and under-modeling might do to type I and II error control. First, let us consider the impact of over-modeling. The following table shows the rejections per 1000 simulated experiments for the correct model—AR(1)—and UN using $\alpha = 0.05$ as our criterion for rejection:

		Simple Effect of Treatment Given Time						Simple Effect of Time Given Treatment		
Covariance Model	**Trt × Time**	**2**	**3**	**4**	**5**	**6**	**7**	**0**	**1**	**2**
AR(1)	0.833	0.956	0.917	0.876	0.846	0.758	0.550	0.056	0.987	0.961
UN	0.561	0.930	0.869	0.787	0.749	0.657	0.431	0.042	0.858	0.720

The rejection rates for the correct model—the AR(1)—accurately reflect the power of each test. We see that in every case, UN falls short—often well short—of the actual power of the test if one uses a sensible method of analysis.

Author's comment I: Think of this example as role modeling. Readers are encouraged to explore different means and covariance parameters. The essential points here are (1) this is how you determine the price to be paid for over-modeling and (2) the price can be substantial.

Author's comment II: In some quarters, UN has been enshrined as standard operating procedure. I cannot resist pointing out that this should stand as a self-explanatory indictment of the wisdom of their approach.

What about under-modeling? Here, the issue is type I error control. If we set all of the means equal (without loss of generality, it does not matter what we set them to as long as they are all equal), we should see an observed rejection rate of roughly 5% for $\alpha = 0.05$. To be precise, for 1000 simulated experiments, the rejection rate should be between 3% and 7% to reflect the margin of error for a binomial with $N = 1000$ and probability $= 0.05$. The AR(1) and UN procedures are consistently within this range. CS varies between 6% and 9%. For certain contrasts that may possibly be of interest, for example, the difference in linear change over time for the experimental treatments (treatments 1 and 2) vs. the control (treatment 0), the observed rejection rate under the null hypothesis was 19% for treatment 0 vs. treatment 1 and 18.2% for treatment 0 vs. treatment 2! The loss of type I error control for under-modeling can be subtle, but for some key hypotheses it can also be drastic. *Caveat emptor.*

This should clearly establish why identifying an adequate but parsimonious covariance model is an essential first step in repeated measures analysis.

14.3.2 Covariance Model Selection Methods

Before we talk about statistical tools *per se*, we should mention that when a statistical scientist participates in a project that entails analysis of repeated measures data, an essential first step in covariance model selection should be to gain sufficient understanding of the subject matter of the data to make a first cut of potential models. Some covariance models may be implausible given the nature of the data being collected and the study design being used, whereas other covariance models may seem to be natural candidates. The statistical tools we discuss in this section are meant for use on the remaining candidate models after eliminating obviously implausible models from consideration.

We have three primary tools for selecting among candidate covariance models:

1. Plots of the covariance as a function of distance between pairs of repeated measures. Inspection of these plots is an inexact science, but certain covariance models have visual signatures. Cues in plots may suggest these signatures. If so, they can help point us in the right direction.
2. Formal likelihood ratio tests, assuming we have a nested hierarchy of candidate models. A frequently used hierarchy in repeated measures modeling, from most complex to least complex: ANTE(1) simplifies to ARH(1) which simplifies to AR(1).
3. Fit criteria such as the AIC, AICC, and BIC.

Likelihood ratio testing and fit statistics were introduced in Chapter 6. In that chapter, we applied both of these tools to Data Set 6.2. Because the design in that example is very similar to repeated measures format used in a wide variety of disciplines throughout the biological and social sciences, the model selection process demonstrated in Chapter 6 can be used as a template for most repeated measures analyses. What we add in this chapter are plots and applications of likelihood ratio and fit statistics to more complex, messier data.

Let us start with plots. A covariance plot involves obtaining an estimate of every parameter in the UN covariance matrix and plotting all of them by distance between pairs of observations. Instead of using REML to obtain these estimates—we could create a chicken-and-egg situation if we cannot get REML to converge to a solution—we compute the sums of squares and cross-products. Recall that this is the empirical covariance matrix used in the sandwich estimator. It is a cruder estimate than REML, but we can always get it.

The most straightforward way to obtain the sum of squares and cross-products estimate uses PROC MIXED. For Data Set 6.2, the statements are

```
proc mixed data=univar;
  class treatment time subj_id;
  model y=treatment|time;
  repeated/type=un sscp subject=subj_id(treatment);
  ods output covparms=cov;
```

In MIXED, the `repeated` statement replaces the `random _ residual _` statement used in GLIMMIX. The SSCP option overrides the default REML estimation algorithm. Note that the SSCP option is available in MIXED but not in GLIMMIX.

The ODS statement creates a new data set containing the UN covariance estimates, the σ_j^2, and $\sigma_{jj'}$. The following program statements compute the distance information needed to create the covariance plot:

```
data times;
  do time1=1 to 6;
    do time2=1 to time1;
      dist=time1-time2;
      output;
    end;
  end;
```

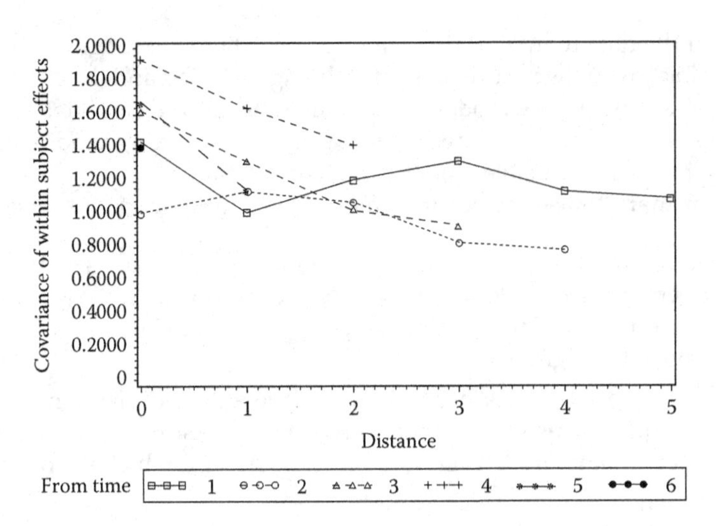

FIGURE 14.1
Within-subject covariance as a function of time between observations.

```
data covplot;
  merge times cov;
proc gplot data=covplot;
  plot estimate*dist=time2;
```

Figure 14.1 shows the resulting covariance plot.

Notice that there are several lines connecting covariance values. These lines are referred to as "From Time..." on the legend at the bottom of the plot. These connect the set $\left\{\sigma_j^2 \quad \sigma_{j,j+1} \quad \sigma_{j,j+2} \quad \ldots \quad \sigma_{j,6}\right\}$, that is, the sequence of variance and covariance terms starting at time j. The variance terms, σ_j^2, all appear on the intercept axis above distance = 0. The covariance terms $\sigma_{j,j+1}$ all appear above distance = 1. In general, covariance terms $\sigma_{j,j+d}$ appear above distance = d. We are looking for three things. First, how different are the σ_j^2? If they appear to be similar, a covariance model assuming all $\sigma_j^2 = \sigma^2$ may provide an adequate fit. Second, do the covariance terms decrease over time? If they do not, a CS model may be adequate. If they do, we need a model that accounts for decay in correlation with increasing distance. Finally, if there does appear to be decreasing covariance with distance, do the lines starting with each time appear to be parallel? Strong departure from parallel indicates unequal ρ_j, suggesting the ANTE(1)model. Weak or no departure suggests that the $\rho_j = \rho$, that is, an AR(1) model would provide adequate fit.

When we went through the example with this data set in Chapter 6, the AR(1) model, plus a random SUBJ_ID(TREATMENT), was the best fitting model using both the likelihood ratio testing and fit statistic comparison. At this point, it is informative to compare the covariance plot of the actual data in Figure 14.1 to an idealized plot of what an AR(1) covariance pattern should look like, with and without the random SUBJ_ID(TREATMENT) effect in the model. Figure 14.2 shows the idealized plot.

The two dashed lines show the signature AR(1) pattern. The solid line shows what the plot describing covariance for the SUBJ_ID(TREATMENT) random effect only, that is, $\sigma_S^2 \cong 1$ and AR(1) correlation, $\rho = 0$. We know that independent errors model is the same as a CS model—ergo, the covariance is equal for all distances. On the other hand, if the subject variance $\sigma_S^2 = 0$ and the autocorrelation $\rho > 0$, then we get the lower dashed line with

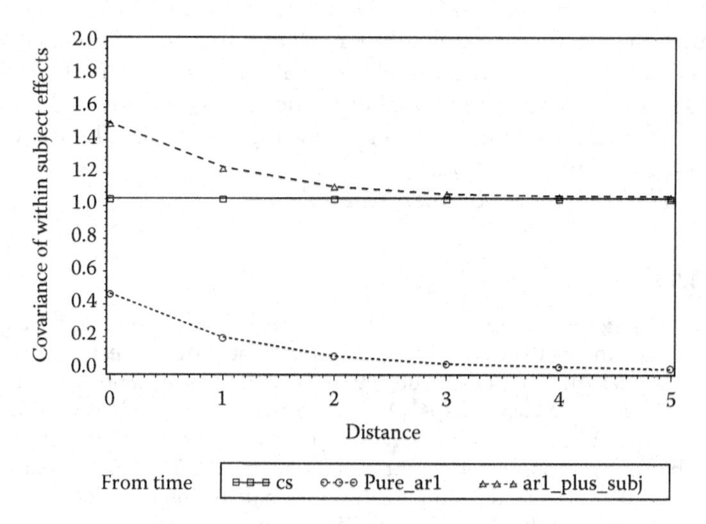

FIGURE 14.2
Idealized covariance plots by distance for CS and AR(1) models.

asymptote at 0 (because $\rho < 1$, as d increases $\rho^d \to 0$). The upper dashed line shows the covariance plot when the AR(1) correlation and the subject variance are both >0. Covariance decreases, but the asymptote is σ_S^2, not 0. Now, if we look back at Figure 14.1, we see that aside from the noise that occurs in "real" data, the AR(1) + subject random effect model is exactly what we see.

Before we move on to more complex examples, the AR(1) + subject model raises the issue of **identifiability**. We can include the AR(1) random residual structure and the subject(treatment) random effect in the same model, because each make a clear contribution to the overall covariance of the observations. This is *not* true of other models. For example, we know it is nonsense to include CS random residual and the subject(treatment) random effect in the same model: each is a different way of parameterizing the same model so including both is redundant. So CS with a random subject effect is called a **non-identifiable** covariance structures.

Less obviously, neither UN nor TOEP are identifiable from the random subject effect. To see this, consider TOEP. If we include both of the following statements

```
  random intercept/subject=subj_id(treatment);
random _residual_/type=toep subject=subj_id(treatment);
```

the resulting variance is

$$J\sigma_S^2 + \sigma_T^2 \begin{bmatrix} 1 & \rho_1 & \rho_2 & \cdots & \rho_5 \\ & 1 & \rho_1 & \cdots & \rho_4 \\ & & 1 & \cdots & \cdots \\ & & & \cdots & \rho_1 \\ & & & & 1 \end{bmatrix}. \text{ Reexpressing yields } \sigma_C^2 \begin{bmatrix} 1 & \rho_1^* & \rho_2^* & \cdots & \rho_5^* \\ & 1 & \rho_1^* & \cdots & \rho_4^* \\ & & 1 & \cdots & \cdots \\ & & & \cdots & \rho_1^* \\ & & & & 1 \end{bmatrix}$$

where the combined variance, $\sigma_C^2 = \sigma_S^2 + \sigma_T^2$, is the sum of the subject and Toeplitz variances, and ρ_d^* are the ρ_d rescaled to σ_C^2. We can write a similar reexpression for UN.

ANTE(1) occupies a gray area: technically, it is identifiable, but in practice, including random subject and ANTE(1) in the same model creates ill-conditioned matrices that play havoc with attempts to solve the estimating equations. The ANTE(1) alone can accommodate the contribution of σ_S^2 accurately enough without including the SUBJECT(TREATMENT) effect in the model, so it is best to leave it out.

Now we move on to two more complex examples.

Example 14.3.1

The data for this example appear in the SAS Data & Program Library, Data Set 14.1. There are two treatments to be compared, called treatment 0 and treatment 1. The study uses a crossover design. That is, all subjects receive both treatments during the study, but they are divided into two groups. One group of subjects receives treatment 0 first, then treatment 1; the other group receives treatment 1 first, then treatment 0. In this study, 17 subjects were observed on the former sequence, 24 subjects were observed on the latter sequence. Period 1 denotes the first half of the sequence (that is, when treatment 0 is active for the first sequence and treatment 1 is active for the second); period 2 denotes the second half.

During each treatment's application, subjects are observed at times 0 through 5, where 0 is immediately after the treatment is applied and 1, 2,…,5 are the number of time units after treating the subjects. The times are equally spaced. Also, for each subject, a baseline measurement was taken before applying the treatment.

Presume that the response variable is Gaussian for the purposes of this example. Also, assume it is *known* that changes over time are linear. A model for these data is shown as follows:

1. Linear predictor: $\eta_{ijkl}=\alpha_{ijk}+\rho_l+\beta_b X_b$, where ρ_k denotes the period effect, X_b denotes the baseline measurement, β_b denotes the regression coefficient for the baseline, and α_{ijk} denotes the treatment–time–subject model. We partition α_{ijk} according to a random coefficient model, that is, $\alpha_{ijk}=\beta_{0i}+b_{0ik}+(\beta_{1i}+b_{1ik})T_j$, where β_{0i} and β_{1i} are the intercept and slope for the ith treatment, b_{0ik} and b_{1ik} are the random intercept and slope coefficients for the kth subject on the ith treatment, and T_j denotes the time of the jth repeated measurement. Note that the random coefficients account for variation among subjects.
2. Distributions:
   a. $\begin{bmatrix} b_{0ik} \\ b_{1ik} \end{bmatrix} \sim NI\left(\begin{bmatrix} 0 \\ 0 \end{bmatrix}, \begin{bmatrix} \sigma_0^2 & 0 \\ & \sigma_1^2 \end{bmatrix}\right)$
   b. $\mathbf{y}_{ikl} \sim NI(\boldsymbol{\mu}_{ikl}, \boldsymbol{\Sigma}_{ikl})$, where $\mathbf{y}'_{ikl} = \begin{bmatrix} y_{i0kl} & y_{i1kl} & \cdots & y_{i5kl} \end{bmatrix}$ and $\boldsymbol{\mu}_{ikl}=E(\mathbf{y}_{ikl})$
3. Link: identity.

Our task is to identify the covariance structure for $\boldsymbol{\Sigma}_{ikl}$ that accounts for within-subject correlation parsimoniously.

As a first step, we can construct a covariance plot. Use the following SAS statements:

```
proc mixed data=x_over_ante1;
  class period trt t id;
    model y=period trt|t;
    repeated/type=un sscp subject=id(trt*period);
    ods output covparms=cov;
data times;
```

```
   do time1=1 to 6;
   do time2=1 to time1;
     dist=time1-time2;
   output;
   end;
  end;
data covplot;
  merge times cov;
proc gplot data=covplot;
  plot estimate*dist=time2run;
```

The SSCP option in PROC MIXED only works when all effects on the right-hand side of the MODEL statement are CLASS variables. We use a version of the model treating repeated measurements (T) as a class variable, not a linear regression. The resulting covariance plot appears in Figure 14.3.

We see some patterns we might be able to use. There appears to be a great deal of spread among the variances, that is, the points at distance = 0. The covariance plots from the first two times, the ones for which there is information at the longer distances, appear to decrease. The profiles for the shorter lines actually appear to increase—this can be an artifact of unequal variances. From the plot (you have to acquire some experience to see this) we can anticipate that ANTE(1) or ARH(1) may provide the best fit.

Now we go to the GLIMMIX statements to actually fit the model. At this point, the most practical thing to do is use fit statistics as our model-comparison criterion. The GLIMMIX statements that implement the model literally as we have written it earlier are

```
proc glimmix;
  class period trt id;
  model y=period trt t(trt) baseline/s;
  random intercept t /subject=id(trt*period);
  random _residual_/type=___ subject=id(trt*period);
```

TYPE=_____ indicates that we try candidate covariance models to find the best AICC. At this point, we need to say more about identifiability. The statement RANDOM INTERCEPT/

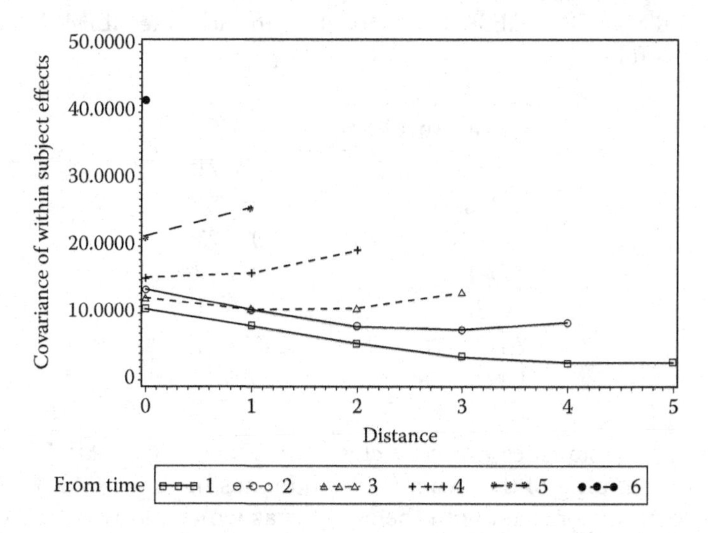

FIGURE 14.3
Covariance plot for crossover data.

SUBJECT=ID(TRT*PERIOD) defines the same random subject effect we saw in the more conventional repeated measures model earlier. It is therefore equivalent to CS. It is harder to see here, but nonetheless. To clarify this point, we run the following programs:

```
proc glimmix;
  class period trt id;
    model y=period trt t(trt) baseline/s;
    random intercept t /subject=id(trt*period);
proc glimmix;
  class period trt id;
    model y=period trt t(trt) baseline/s;
    random t /subject=id(trt*period);
    random _residual_/type=cs subject=id(trt*period);
```

The covariance parameter output is as follows:

Cov Parm	Subject	Estimate
t	id(period*trt)	1.3992
CS	id(period*trt)	8.4243
Residual		3.5245

Cov Parm	Subject	Estimate
Intercept	id(period*trt)	8.4243
t	id(period*trt)	1.3992
Residual		3.5245

We see that the CS parameter in the top output and the intercept parameter in the bottom output are identical—exactly what we have seen in other demonstrations of CS/random-subject-effect equivalence.

Therefore, with the exception of the AR(1) model, we do not include RANDOM INTERCEPT and RANDOM _RESIDUAL_ together in the same GLIMMIX program. The AICC results are as follows:

Covariance Model	AICC
CS/indep errors	3130.21
AR(1)	3062.88
ARH(1)	3035.25
TOEP	3068.67
TOEPH	3043.01
ANTE(1)	3036.71
UN	3053.80

We see that the AICC statistics are very close for ANTE(1) and ARH(1). However, the ARH(1) is both the simplest of the two models and it has the lowest AICC, so we use this model. At this point, we proceed with the analysis as we would with any other factorial treatment design with TREATMENT and T as the two factors. Our interest would logically center on comparing the intercept and slope parameters for the two treatments, that is, H_0: $\beta_{00} = \beta_{01}$ and H_0: $\beta_{10} = \beta_{11}$. We add the SAS statements

```
contrast 'equal intercepts' trt 1 -1;
contrast 'equal slopes' t(trt) 1 -1;
```

to the ARH(1) model. The result is as follows:

Contrasts

Label	Num DF	Den DF	F Value	Pr>F
Equal intercepts	1	526	1.16	0.2818
Equal slopes	1	104	4.18	0.0434

We could continue by obtaining the regression coefficient estimates, etc.

Example 14.3.2

Data for this example appear in the SAS Data & Program Library as Data Set 14.2. We have two treatments. There are 17 subjects observed on treatment 1, 24 on treatment 2. Repeated measurements are taken at times 0, 1, 2, 4, 8, 16, 32, 64, and 128. However, every subject in the data set has some times at which observations are missing. In fact, of the $(17+24)\times 9 = 369$ possible observations that could have been taken, only 101 observations actually were taken.

Assume, as in the previous example, that we can restrict our attention to linear changes over time. A model for these data would be

1. Linear predictor: $\eta_{ijk} = \beta_{0i} + b_{0ik} + (\beta_{1i} + b_{0ik})X_j$, where X_j denotes the jth measurement time for the study, that is, $X_1 = 1, X_2 = 2, X_3 = 4, \ldots, X_9 = 128$, and β and b denote fixed by-treatment and random by-subject intercept and slope coefficients
2. Distributions:
   a. $\mathbf{y}_{ik} | b_{0ik}, b_{1ik} \sim NI(\boldsymbol{\mu}_{ik}, \boldsymbol{\Sigma}_{ik})$, where $\boldsymbol{\mu}_{ik}$ and $\boldsymbol{\Sigma}_{ik}$ are as defined in Example 14.3.1
   b. $\begin{bmatrix} b_{0ik} \\ b_{1ik} \end{bmatrix} \sim NI\left(\begin{bmatrix} 0 \\ 0 \end{bmatrix}, \begin{bmatrix} \sigma_0^2 & 0 \\ & \sigma_1^2 \end{bmatrix}\right)$
3. Identity

We obtain the covariance plots much as we have in the previous examples. We leave this as an exercise. The GLIMMIX statements to implement the aforementioned model are

```
proc glimmix data=all;
class trt subject;
model y=trt t(trt)/noint s;
random intercept t/subject=subject(trt);
random _residual_/type=_____ subject=subject(trt);
```

We insert candidate covariance models in the TYPE= option of the RANDOM_ RESIDUAL_ statement. As we saw in Example 14.3.1, we must delete RANDOM INTERCEPT, but retain RANDOM T for certain structures because of identifiability issues.

We limit our attention to three candidate models. One is CS, in case we have only correlation resulting from the random subject effects. One is AR(1). However, the times of observation are not equally spaced. To account for this, our third candidate covariance model is SP(POW). The underlying premise of the SP(POW) model is identical to AR(1), but it takes into account unequal spacing in time. The GLIMMIX statements for SP(POW) are

```
proc glimmix data=all;
  class trt subject;
  model y=trt t(trt)/noint s;
  random intercept t/subject=subject(trt);
  random _residual_/type=sp(pow)(t) subject=subject(trt);
```

Notice that there is an extra term (T) after TYPE=SP(POW). This specifies the time coordinate so that the program can determine the distance between observations.

Here, we must use the AICC criterion (or one of the other fit criteria) because these models are not nested. Here are the results:

Covariance Model	AICC
CS	583.67
AR(1)	585.77
SP(POW)	586.54

From these results, it appears that evidence for nontrivial autocorrelation is lacking.

In many studies, the covariance structure may differ for two treatments—or two locations in a multisite study, or over other factors. Suppose we have reason to suspect that this is the case here. How would we know? This is one example of why it is important for the statistical scientist to have some appreciation of the subject matter attending the data. The only statistical cue might be that it occurs to you to check. However, a working knowledge of the subject matter (and good communication with the research team, knowing the right questions to ask, etc.) makes it more likely that you would have a basis for suspecting unequal variance.

The GLIMMIX statements to estimate separate covariance models for each treatment are

```
proc glimmix data=all;
  class trt subject;
  model y=trt t(trt)/noint s;
  random intercept t/subject=subject(trt);
  random _residual_/type=sp(pow)(t) subject=subject(trt) group=trt;
  covtest 'equal covariance structure?' homogeneity;
```

The statements for the AR(1) and CS models are similar except for the TYPE=option and identifiability concerns if any. The AICC comparisons are now

Covariance Model	AICC
CS	579.07
AR(1)	577.75
SP(POW)	575.96

The likelihood ratio test for homogeneity for the SP(POW) model is 8.65 with a p-value of 0.0132, suggesting that there is sufficient evidence to conclude that the treatments do indeed have different covariance parameters.

Analysis from this point proceeds like any other quantitative–qualitative factorial treatment design. The important thing is to use the heterogeneous SP(POW)

covariance model so that you can base inference on appropriate standard errors and test statistics.

14.4 Non-Gaussian Case

In Chapter 6, we used a binomial repeated measures example, Data Set 6.3, to select the appropriate GLMM covariance structure. The design was the same as Data Set 6.2; the only difference was the response variable. Recall that there were two approaches: the GEE-type model, which we also call the R-side model, and the G-side or "true GLMM." We noted that we can only do true covariance model selection with the G-side model. Nevertheless, GEE-type models are widely used for non-Gaussian repeated measures data. In this section, we simply present the GLIMMIX statement and selected elements of the analysis so we can compare and contrast. In Section 14.5, we will discuss the major issues raised by these two analyses.

14.4.1 GEE-Type Models

In Chapter 6, we saw that because GEE models are quasi-likelihood models, we must use some form of linearization method—PL in PROC GLIMMIX. We, therefore, have no basis for likelihood ratio testing or comparing fit statistics. This makes working covariance model selection problematic. As an example implementation of the GEE-type model, we use Data Set 6.3 with the AR(1) working covariance (because it was selected for the G-side GLMM—shown later). The GLIMMIX statements are

```
proc glimmix DATA=RptM_Binomial;
  class trt time id;
  model Fav/N=trt|time/ddfm=kr;
  random intercept/subject=id(trt);
  random _residual_/type=ar(1) subject=id(trt);
  lsmeans trt*time/slicediff=time oddsratio cl;
```

The example program here shows the random subject effect also included. The statements as written here yield model-based standard errors and test statistics. The PL version of the Kenward–Roger adjustment also appears here. If we want to use the sandwich estimators instead, we remove the DDFM = KR option in the MODEL statement and add EMPIRICAL = MBN to the MODEL statement.

How do the *ad hoc* KR, MBN, and uncorrected sandwich estimates compare? Here are the "type III tests of fixed effects" results:

Ad hoc KR

Type III Tests of Fixed Effects

Effect	Num DF	Den DF	F Value	Pr>F
trt	1	9.86	3.79	0.0806
Time	5	37.36	1.95	0.1084
trt*time	5	37.36	1.87	0.1234

Empirical = MBN

Type III Tests of Fixed Effects

Effect	Num DF	Den DF	F Value	Pr > F
trt	1	10	2.16	0.1722
Time	5	50	1.33	0.2667
trt*time	5	50	2.32	0.0573

Uncorrected Sandwich

Type III Tests of Fixed Effects

Effect	Num DF	Den DF	F Value	Pr > F
trt	1	10	4.05	0.0717
Time	5	50	2.97	0.0202
trt*time	5	50	22.34	<.0001

We see here the uncorrected sandwich estimator's upward bias in test statistics. As we have noted elsewhere, we want to avoid using them. On the other hand, the differences between the *ad hoc* KR and MBN corrected sandwich estimates are more nuanced. The MBN procedure provides marginally strong evidence of a treatment–time interaction, whereas KR show less compelling evidence of an interaction but more compelling evidence of a treatment main effect.

We could also look at the interval estimates. Assuming that there is an interaction, we would want to look at simple effects, for example, the time-specific treatment odds ratios.

Obs	SimpleEffect Level	trt	_trt	L_KR	KR_OR	U_KR	L_MBN	MBN_OR	U_MBN
1	Time 1	1	2	0.30518	1.09767	3.94811	0.27969	1.09767	4.30790
2	Time 2	1	2	0.23027	0.84922	3.13193	0.16650	0.84922	4.33149
3	Time 3	1	2	0.15249	0.54655	1.95891	0.09861	0.54655	3.02915
4	Time 4	1	2	0.10032	0.37267	1.38432	0.06082	0.37267	2.28343
5	Time 5	1	2	0.11745	0.44709	1.70193	0.07390	0.44709	2.70508
6	Time 6	1	2	0.02006	0.08784	0.38452	0.01904	0.08784	0.40517

The shaded columns show the lower 95% confidence limit, the odds-ratio estimate, and the upper 95% confidence limit for the *ad hoc* KR procedure. The three columns to the right show the corresponding MBN derived values. Notice that the MBN interval estimates are considerably more conservative. This is an area where more research is needed before we can make definitive statements about which method to recommend.

Finally, we could take a quick look at robustness. At this point, GEE practitioners would probably lean toward the MBN-corrected sandwich estimators, so let us use those to compare type III tests of fixed effects and time-specific odds ratios for the CS, AR(1), and UN working covariance models.

CS

Type III Tests of Fixed Effects				
Effect	**Num DF**	**Den DF**	**F Value**	**Pr>F**
trt	1	10	2.16	0.1722
Time	5	50	1.33	0.2667
trt*time	5	50	2.32	0.0573

AR(1)—shown earlier

UN

Type III Tests of Fixed Effects				
Effect	**Num DF**	**Den DF**	**F Value**	**Pr>F**
trt	1	10	2.07	0.1804
Time	5	10	1.55	0.2606
trt*time	5	10	12.05	0.0006

Time-specific odd ratios

Obs	**Simple Effect Level**	**trt**	**_trt**	**L_AR1**	**AR1_ OR**	**U_AR1**	**L_CS**	**CS_OR**	**U_CS**	**L_UN**	**UN_ OR**	**U_UN**
1	Time 1	1	2	0.27969	1.09767	4.30790	0.28564	1.10294	4.25884	0.27979	1.10294	4.34781
2	Time 2	1	2	0.16650	0.84922	4.33149	0.17375	0.85682	4.22519	0.14706	0.85682	4.99225
3	Time 3	1	2	0.09861	0.54655	3.02915	0.10498	0.56250	3.01389	0.08255	0.56250	3.83282
4	Time 4	1	2	0.06082	0.37267	2.28343	0.06490	0.38521	2.28628	0.04941	0.38521	3.00324
5	Time 5	1	2	0.07390	0.44709	2.70508	0.07710	0.45074	2.63496	0.05897	0.45074	3.44517
6	Time 6	1	2	0.01904	0.08784	0.40517	0.02074	0.09474	0.43272	0.02091	0.09474	0.42929

Here we see close agreement between the CS and AR(1) results, but several important departures for UN. The treatment–time interaction test using UN appears to be inflated. On the other hand, the odds-ratio interval estimates obtained with UN appear to be noticeably conservative. While ostensibly robust, this illustrates that GEE-type models are hardly invulnerable to working covariance misspecification.

14.4.2 GLMMs

We saw in Chapter 6 that, for Data Set 6.3, we were able to use likelihood ratio testing and fit statistic comparison to establish that the AR(1) + random subject effect model fit best. How do these results compare with the GEE? We use statistics obtain using EMPRICAL=MBN, since there is no KR option with LAPLACE and LAPLACE is the method of choice with repeated measures GLMMs. First, the GLIMMIX statements are as follows:

```
proc glimmix DATA=RptM_Binomial method=laplace empirical=mbn;
  class trt time id;
  model Fav/N=trt|time;
  random intercept/subject=id(trt);
```

```
random time/type=ar(1) subject=id(trt);
lsmeans trt*time/slicediff=time oddsratio cl;
```

The type III tests:

Type III Tests of Fixed Effects

Effect	Num DF	Den DF	F Value	Pr>F
trt	1	10	4.10	0.0703
Time	5	50	2.45	0.0464
trt*time	5	50	2.26	0.0623

The p-value for the treatment–time interaction effect was 0.0573 with the GEE. The main effects tests for the GLMM have much lower p-values. This is consistent with our discussion in Chapter 3 when we introduced the marginal and conditional distributions. Because of the skewness of the marginal distribution, the GEE tends to understate factor level difference and therefore tend, in general, toward less powerful tests.

How do the inverse-linked LSMeans compare? Recall that for the GEE, these are estimates of the marginal mean. For the GLMM, they are estimates of the binomial probability. The two sets of estimates are *not* different ways of estimating the same parameter.

Obs	Time	trt	Marg_Mean_GEE	P_Hat_GLMM
1	1	1	0.42547	0.41451
2	1	2	0.40287	0.38497
3	2	1	0.33950	0.31159
4	2	2	0.37704	0.35992
5	3	1	0.42441	0.40267
6	3	2	0.57431	0.60163
7	4	1	0.47403	0.46852
8	4	2	0.70746	0.74594
9	5	1	0.56347	0.58063
10	5	2	0.74274	0.77524
11	6	1	0.31190	0.27403
12	6	2	0.83767	0.85691

We see the signature discrepancy between the GEE's $\hat{\mu}_P$ and the GLMM's $\hat{\pi}$. Above 0.5, $\hat{\mu}_P < \hat{\pi}$; below 0.5, $\hat{\pi} < \hat{\mu}_P$. This affects the odd-ratio estimates as follows:

Obs	Simple Effec Level	trt	_trt	L_GEE	GEE_ OR	U_GEE	L_GLMM	GLMM_ OR	U_GLMM
1	Time 1	1	2	0.27969	1.09767	4.30790	0.22073	1.13106	5.79583
2	Time 2	1	2	0.16650	0.84922	4.33149	0.11404	0.80494	5.68160
3	Time 3	1	2	0.09861	0.54655	3.02915	0.05143	0.44637	3.87396
4	Time 4	1	2	0.06082	0.37267	2.28343	0.03313	0.30025	2.72091
5	Time 5	1	2	0.07390	0.44709	2.70508	0.03762	0.40140	4.28305
6	Time 6	1	2	0.01904	0.08784	0.40517	0.01065	0.06303	0.37315

The odds-ratio estimates show discrepancies induced by the marginal-conditional difference noted earlier. In addition, the 95% confidence intervals show striking dissimilarities. Here is a different version of this table, with the GLMM odds-ratio interval estimates determined from model-based statistics:

Obs	Simple Effect Level	trt	_trt	L_GEE	GEE_OR	U_GEE	L_GLMM	GLMM_ OR	U_GLMM
1	Time 1	1	2	0.27969	1.09767	4.30790	0.27857	1.13106	4.59237
2	Time 2	1	2	0.16650	0.84922	4.33149	0.19429	0.80494	3.33485
3	Time 3	1	2	0.09861	0.54655	3.02915	0.10784	0.44637	1.84758
4	Time 4	1	2	0.06082	0.37267	2.28343	0.07094	0.30025	1.27086
5	Time 5	1	2	0.07390	0.44709	2.70508	0.09370	0.40140	1.71957
6	Time 6	1	2	0.01904	0.08784	0.40517	0.01419	0.06303	0.27995

Recall that the MBN correction for the sandwich estimate was developed explicitly for the GEE. The behavior of the MBN correction has been more thoroughly researched for GEEs. This suggests that there is work to be done on the GLMM side regarding the impact on interval estimate coverage of the model-based vs. sandwich estimator approaches.

Finally, what about robustness with the GLMM? Here are the type III test results for CS and UN compared to AR(1):

CS

Type III Tests of Fixed Effects

Effect	Num DF	Den DF	F Value	Pr>F
trt	1	60	3.71	0.0589
Time	5	60	3.83	0.0044
trt*time	5	60	2.88	0.0215

AR(1)—shown earlier

UN

Type III Tests of Fixed Effects

Effect	Num DF	Den DF	F Value	Pr>F
trt	1	60	1.12	0.2950
time	5	60	0.85	0.5182
trt*time	5	60	1.77	0.1325

The CS results are model based; the UN results use EMPIRICAL=MBN. This follows from Gaussian model reasoning. In Gaussian models, we use the Kenward–Roger correction to control standard error bias. With CS Gaussian models, Kenward–Roger has no impact; it only has impact for nontrivial within-subject correlation, and its impact increases as the covariance model's complexity increases. With Laplace and non-Gaussian data, EMPIRICAL=MBN is the closest equivalent to KR. Following the Gaussian model analogy, we would expect MBN to be overkill for CS but essential for UN. Here we see familiar impact on the results: loss of type I error control for under-modeled covariance (CS) and loss of power for over-modeled covariance (UN).

While not shown here, we would see similar impact on the odds-ratio interval estimates. The difference is that with the GLMM, we have a solid basis for using the AR(1) as the parsimonious model that accounts for autocorrelation. With the GEE we have no such assurance, so robustness concerns figure more prominently.

14.5 Issues for Non-Gaussian Repeated Measures

In Section 14.4, we reviewed the major features of non-Gaussian repeated measures analysis using an example with binomial data. In this section, we review the thought process leading to the point of departure between the GEE, R-side approach and the GLMM, G-side approach. We briefly review the overarching issues of model selection and standard error and test statistic construction.

14.5.1 How Do Correlated Errors Arise? Deciding What We Are Modeling

At several points in this textbook, we have emphasized the close relationship between statistical modeling and design concepts. One of the recurring tools has been the skeleton ANOVA WWFD process. Just as we used it in Chapter 11 to illuminate the over-dispersion issue, we can use it here to shed light on how we think about within-subject correlation.

Data Sets 6.2 and 6.3 share a basic template common to the vast majority of repeated measures studies. We have seen that we can summarize this template in skeleton ANOVA terms as follows:

Topographical		Treatment		Combined	
Source	**d.f.**	**Source**	**d.f.**	**Source**	**d.f.**
		Treatment	a-1	Treatment	a-1
Subject	s-1			Subject(treatment)	s-a
		Time	t-1	Time	t-1
		Time × treatment	(a-1)(t-1)	Time × treatment	(a-1)(b-1)
Measurement (subject)	s(m-1)	"Parallels"	sm-at	Time × subject (treatment) a.k.a. residual	s(m-1)-a(b-1)
Total	sm-1	Total	sm-1	Total	sm-1

At this point in the model-building process, we focus our attention on the unit of observation, the intersection of time, treatment, and subject. If the data are Gaussian, we have a clear way to proceed. The observations are multivariate normal, and we define their correlation structure on the vector of observations at the t times on the ikth subject within treatment.

For non-Gaussian data, our modeling paths diverge at this point. We can either follow the Gaussian analogy or continue to follow the WWFD process. Following the

Gaussian analogy, we mimic the Gaussian R-side covariance matrix as closely as we can. This gives us a working covariance matrix, a quasi-likelihood, and a GEE-type model. It also anchors us to the marginal, or population-averaged, inference space. One the other hand, if we continue to follow WWFD and if the observation has a distribution belonging to the one-parameter exponential family, then we say that within-subject correlation occurs on the time × subject(treatment) random model effects. If we denote these effects as $ts(a)_{ijk}$ for the *ijk*th treatment–time–subject, then the vector $\mathbf{s}(\mathbf{a})_{ik} = \begin{bmatrix} ts(a)_{i1k} & ts(a)_{i2k} & \cdots & ts(a)_{iTk} \end{bmatrix} \sim NI(\mathbf{0}, \boldsymbol{\Sigma}_{ik})$. The G-side correlation structure $\boldsymbol{\Sigma}_{ik}$ corresponds to familiar covariance models for Gaussian distributions.

The primary ambiguity in repeated measures model building for non-Gaussian data occurs when we have a member of the two-parameter exponential family. How do we model repeated measures for beta or negative binomial or other distributions in this family? For GEE models, it is not clear how the working correlation parameters co-exist with the scale parameters intrinsic to the distribution of $\mathbf{y}|\mathbf{b}$. For G-side models, it is not clear how the random $ts(a)_{ijk}$ effect coexists with $f(\mathbf{y}|\mathbf{b})$'s scale parameter.

For Gaussian data and one-parameter exponential family members, LMM and GLMM repeated measures methodology is reasonably well understood. The area in need of further development is clearly the two-parameter exponential family.

14.5.2 Covariance Model Selection and Non-Gaussian Repeated Measures

This will be a short section, but its main points need to be made. We summarize what we have seen in the examples from Chapter 6 and in the previous section. Because GEE models are exclusively defined on quasi-likelihood, the usual tools for covariance model selection are not defined. For GEE models, we need to depend on the inherent consistency of GEE estimates and on our knowledge of the subject matter context of the observations. Fortunately, GEE working covariance models tend to give similar results if we have a reasonably plausible "in-the-ballpark" approximation.

For GLMMs defined on one-parameter exponential family members (mainly binomial proportions and Poisson counts), as long as we use integral approximation we have a well-developed set of tools (likelihood ratio tests and fit statistic comparison) at our disposal.

For two-parameter exponential family members, the situation is less clear, mainly because we have not yet clearly established how best to model repeated measures with these data.

14.5.3 Inference Space, Standard Errors, and Test Statistics

The GEE-type repeated measures model embeds within-subject correlation in a working covariance matrix. This means that we have a quasi-likelihood—strictly speaking, there does not exist a probability mechanism that could give rise to the data according to the process described by the GEE. Raudenbush (2009) notes that the GEE starts with the marginal association between response and the explanatory variables, treatment and time. Assuming this is the intended target of inference and our GEE model provides a reasonable description of the marginal association, GEE-type models are useful.

Estimates from GEE-type models are less sensitive to misspecification of the working correlation matrix than GLMMs are to misspecification of the corresponding G-side covariance model. GEE literature refers to this as "robustness." However, as Raudenbush (2009) cautions, it really just means that GEE-type models with different working correlation matrices produce similar estimates. If they all target the wrong inference space, they are not robust, just alike. This is an important point, because we know the GEE targets the marginal means, and we have seen several examples throughout this text where our inference target is a binomial probability or a count distribution rate parameter, not the mean of the marginal distribution.

GEE-type models have working correlation analogs for each of the structures we have seen for Gaussian covariance models. The one exception is the independent errors model. In the Gaussian case, CS and the random subject effect/$\mathbf{R} = \mathbf{I}\sigma^2$ models are equivalent. For GEE-type models, the latter would simply be a GLMM with no working correlation structure, so only the CS working covariance is a GEE-type model. We have seen that the results are not equivalent to the random subject effect GLMM.

While GEE estimates of $\boldsymbol{\beta}$ are consistent and robust to working covariance misspecification, standard errors of estimable functions $\mathbf{K}'\hat{\boldsymbol{\beta}}$ are not. This makes model-based standard errors and test statistics more vulnerable to model misspecification than sandwich estimators. For this reason, sandwich estimators are frequently used in conjunction with GEEs. For all but very large sample sizes, bias-corrected sandwich estimators are preferable. If model-based estimators are used, experience to date suggests that the *ad hoc* Kenward–Roger adjustment should also be used.

Repeated measures GLMMs model correlation among repeated measurements as part of the linear predictor. We motivate this the same way we motivate modeling overdispersion on the G-side—through the skeleton ANOVA-based design-to-model process. G-side covariance modeling essentially conceptualizes autocorrelation as occurring among the components of $\mathbf{Zb}$ randomly perturbing the conditional expectation on the model (link) scale.

The GLMM approach has several advantages, as we have seen throughout this text. First, we explicitly target the expectation of a member of the exponential family, for example, a binomial probability or a counting distribution rate parameter. Second, we can work with the GLMM using integral approximation. This gives us more accurate estimates of the covariance components that we could obtain using linearization methods such as PL. Perhaps more importantly, it also allows us to perform legitimate likelihood ratio tests on nested covariance models and to compute legitimate fit statistics, giving us access to all of the covariance model-selection tools we have with Gaussian data.

GLMMs may be more susceptible to covariance model misspecification, but GLMMs also allow for much more accurate selection of a reasonable-fitting covariance model. On the other hand, we have few covariance model–selection tools at our disposal with GEE-type models. As we saw in our example using Data Set 6.3, GEEs are not so robust as to preclude big discrepancies in results obtained with competing working covariance models.

As with any repeated measures modeling, GLMMs are susceptible to standard error and test statistic bias that results from using estimated covariance components. Because we obtain more accurate covariance estimates using integral approximation and covariance models are generally too complex for quadrature to handle, this leaves Laplace as the method of choice for non-Gaussian G-side repeated measures GLMMs. With Laplace, the *ad hoc* Kenward–Roger correction is undefined. This leaves bias-corrected sandwich estimators as the apparent choice. We say this with the caveat that this is an under-researched area—we need to know more about small sample behavior of inferential statistics for repeated measures GLMMs.

14.6 Summary

- Repeated measures in time occur when the same unit is observed at multiple times. "Multiple times" refers to distinct times of measurement, not multiple subsamples at the same time.
- Interest usually focuses on change in response over time and possible treatment effects on those changes. Changes over time may be characterized by regression, but regression is not essential to repeated measures analysis.
- Repeated measures design structure superficially resembles a split-plot, with between-subject effects analogous to whole-plot effects and within-subject effects analogous to the split-plot. The difference is that split-plot levels can be randomized and split-plot effects are assumed to be independent. Time cannot be randomized and repeated measurements on the same subject are correlated.
- Independence among split-plot experimental units is equivalent to compound symmetry. Repeated measures differ in that correlation among measurements on different times on the same subject typically decreases in strength with increasing distance in time between pairs of observations.
- "Covariance structure" refers to models of the relationship between covariance (or correlation) and distance between pairs of observations on the same unit. Covariance models range in complexity from CS to UN.
- In the typical repeated measures scenario, in which covariance decreases as a function of distance, CS under-models covariance and UN over-models covariance. Both are forms of covariance misspecification. Both are undesirable in their own way. Under-modeling inflates type I error rate. Over-modeling reduces power, reducing the chance of identifying beneficial treatment effects or making studies needlessly expensive.
- The first step in a repeated measures analysis is to identify the covariance structure that strikes the best balance between over- and under-modeling.
- For Gaussian data, covariance structure determines the **R** matrix of the conditional distribution $\mathbf{y}|\mathbf{b} \sim N(\boldsymbol{\mu}|\mathbf{b},\mathbf{R})$. The random model effects **b** specify the design structure for the subjects.
- For non-Gaussian data, both "R-side" and "G-side" covariance models exists.
- "R-side" models mimic the Gaussian model construction process. For a given design, the resulting linear predictor is identical. The covariance structure, however, is not. Instead, we have a working covariance structure $\phi\mathbf{V}_{\mu}^{1/2}\mathbf{P}\mathbf{V}_{\mu}^{1/2}$, where $\mathbf{V}_{\mu}$ is the diagonal matrix of variance functions, ϕ is the scale parameter, and **P** is a working correlation matrix.
- All "R-side" models are by definition quasi-likelihood models.
- "G-side" models embed the repeated measures process in within-subject unit-level effects. Within each subject, they have a multivariate normal distribution with covariance structure similar to the **R** matrix in a Gaussian LMM.
- "G-side" covariance structures are well-defined for one-parameter non-Gaussian distributions—for example, binomial and Poisson.

- Gaussian LMMs and "G-side" GLMMs have well-defined likelihoods. Therefore, likelihood-based model-selection tools, notably likelihood ratio tests and fit statistics, can be used for covariance model selection.
- Likelihood ratio testing and fit statistics are undefined for "R-side" GLMMs and should not be used.
- "G-side" GLMMs are considered more vulnerable to covariance model misspecification; "R-side" GLMMs are considered more robust. On the other hand, "G-side" GLMMs are more accurate if the covariance model is not misspecified. This characterization gives us the following metaphor. One can think of the "G-side" GLMM as a high-quality telephoto lens (very sharp when in focus but with a shallow depth of field) and the "R-side" GLMM as a lesser-quality wide-angle lens (never as sharp as a premium lens but its greater depth of field makes it capable of a useable image even when somewhat out of focus). However, see next point.
- For small samples, there is as much "data-free opinion" on this matter as there is hard evidence. Documenting actual behavior of the two in small sample settings is, at present, work in progress.
- Both "G-side" and "R-side" covariance structures are problematic for two-parameter non-Gaussian distributions, for example, negative binomial, beta and gamma. This is an active area of GLMM research.

Exercises

14.1 An experiment was conducted to compare three allergy treatments: trt_0 (a placebo) and trt_1 and trt_2 (two experimental treatments being investigated). The treatments are applied to subjects, and then measurements of "breathing comfort" are taken every 2 hours, starting at hour=0 (immediately after the treatment is applied) until hour=8 (when the treatments are assumed to have worn off).

File `ch_14_problem1.sas` contains the data for this exercise. The file also contains a GLIMMIX program that generates an interaction plot for these data (although it does not implement an adequate model).

a. Using the interaction plot, state in words what appears to be going on. In other words, what would you expect to conclude about the three treatments and how long the allergy relief they provide lasts?

b. Write a statistical model (linear predictor, distributions, link, etc.) consistent with the description of the study given previously.

c. Assuming the allowable covariance structures are limited to those listed in the following, which would be most appropriate to use to completing the formal analysis of this experiment? Cite relevant evidence to support your conclusion.

_____ CS

_____ AR(1)

_____ ANTE(1)

_____ UN

Answer the remaining questions using the analysis from the covariance model selected in (c).

d. The following sets of equations were estimated from these data. At most one of them is appropriate to report. Which, if any, is the appropriate set? Why? (Cite relevant evidence to support your answer.) Note: $\hat{\mu}_{ij}$ denotes the estimated mean at the ith treatment and the jth hour, where H denotes the hour.

   i. $\hat{\mu}_{0j} = 3.52$ $\hat{\mu}_{1j} = 5.79$ $\hat{\mu}_{2j} = 5.18$

   ii. $\hat{\mu}_{0j} = 5.21 - 0.42H$ $\hat{\mu}_{1j} = 7.48 - 0.42H$ $\hat{\mu}_{2j} = 6.88 - 0.42H$

   iii. $\hat{\mu}_{0j} = 3.47 + 0.003H$ $\hat{\mu}_{1j} = 8.36 - 0.65H$ $\hat{\mu}_{2j} = 7.69 - 0.62H$

   iv. None of the above—there is evidence of lack of fit from linear regression over hour

e. The researcher wants to compare time effect for treatment 1 vs. treatment 2 and for the average of treatment 1 and 2 vs. treatment 0. Define the required contrasts, state what H_0 is tested by each, implement them using GLIMMIX, and state the conclusions you reach.

14.2 The following defines a first-order autoregressive [AR(1)] process. Assume there are observations on the same subject at times $j = 1,2,\ldots,t$. Let w_{ij} denote the random component at time j on the ith subject. In the AR(1) process, $w_{ij} = \rho w_{i,j-1} + e_{ij}$, where e_{ij} is referred to as "white noise," we assume e_{ij} i.i.d. $N\left(0, \sigma_N^2\right)$, σ_N^2 denotes the white noise variance and $w_{i,0} = 0$.

The data step of SAS file `ch _ 14 _ problem2.sas` shows how this AR(1) process occurs in the context of a two-treatment repeated measures experiment with blocking. The new term in the program that you may not have seen is RANNOR(seed). This generates a N(0,1) deviate. "Seed" is a number you pick at random to initialize the random number generator. You set this value just once, at the beginning of a program.

a. Write the statistical model in "GLMM-appropriate form" that describes the process being simulated in this program.

b. Implement the analysis of these data consistent with your model in part (a). Using your analysis

   i. Give the estimates of all variance, covariance, or correlation parameters defined in (a).

   ii. Is there evidence that a linear regression over time would be adequate to explain any time effects? Cite relevant supporting evidence.

   iii. Is there evidence that the time effects differ for each treatment? Cite relevant supporting evidence.

   iv. Is there evidence of a treatment effect? If so, characterize it. In either case, what evidence is relevant to answering this question?

c. Do the variance/covariance/correlation estimates you obtained in b-i seem consistent with what was defined in the data step of the program? Briefly explain.

d. Suppose you did not know that these data were generated by an AR(1) process. Compute the appropriate fit statistics if you fit these data with

   - CS
   - ANTE(1)
   - UN

covariance structures. Did your analyses correctly identify your model from (a) as the "correct" model?

14.3 Modify the data step in the program from Problem 14.2 so that it generates Poisson repeated measures data.

Hint: the GLMM for Poisson repeated measures data most closely analogous to the model for Gaussian data in Problem 14.2 should suggest what to do. Remember the linear predictor works on the link-function scale. The SAS function that generates Poisson random deviates is RANPOI(seed, λ), where "seed" works as in Problem 14.2 and λ is the rate parameter on the data scale. If you have trouble, look at the SAS program for Chapter 11, problem 3, which generates Poisson data for a GLMM in a slightly different context.

a. Write the statistical model in "GLMM-appropriate form" that describes the process being simulated in this program.

b. Implement the analysis of these data consistent with your model in part (a). [this means you will need to successfully compute the simulated data]. Using your analysis

   i. Give the estimates of all variance, covariance, or correlation parameters defined in (a).

   ii. Is there evidence that a linear regression over time would be adequate to explain any time effects? Cite relevant supporting evidence.

   iii. Is there evidence that the time effects differ for each treatment? Cite relevant supporting evidence.

   iv. Is there evidence of a treatment effect? If so, characterize it. In either case, what evidence is relevant to answering this question?

c. Write the GEE (R-side) model that comes as close as possible to your model in (a).

d. How do you know the correct answer for (a) cannot be a GEE model?

e. Suppose you did not know that these data were generated by an AR(1) process. Compute the appropriate fit statistics if you fit these data with

- CS
- ANTE(1)
- UN

covariance structures. Did your analyses correctly identify your model from (a) as the "correct" model?

14.4 Modify the data step in the program from Problem 14.3 so that it generates binomial repeated measures data.

Hint: Remember the linear predictor works on the link-function scale. The SAS function that generates binomial random deviates is RANBIN(seed,N,π).

a. Write the statistical model in "GLMM-appropriate form" that describes the process this program simulates.

b. Implement the analysis of these data consistent with your model in part (a). (This means you will need to successfully compute the simulated data.) Using your analysis

i. Give the estimates of all variance, covariance, or correlation parameters defined in (a).
ii. Is there evidence that a linear regression over time is *adequate* to explain any time effects? Cite relevant supporting evidence.
iii. Is there evidence that the time effects differ for each treatment? Cite relevant supporting evidence.
iv. Is there evidence of a treatment effect? If so, characterize it. In either case, what evidence is relevant to answering this question?

c. Write the GEE (R-side) model that comes as close as possible to your model in (a).
d. How do you know the correct answer for (a) cannot be a GEE model?
e. Suppose you did not know that these data were generated by an AR(1) process. Compute the appropriate fit statistics if you fit these data with

- CS
- ANTE(1)
- UN

covariance structures. Did your analyses correctly identify your model from (a) as the "correct" model?

15

Correlated Errors, Part II: Spatial Variability

15.1 Overview

In Chapter 14, we considered repeated measurements in one dimension—measurements over time or in a single spatial direction, such as depth. In this chapter, we consider models for observations taken in two dimensions, points on a grid or map referenced by row and column, latitude and longitude, or, to use standard spatial terminology, "easting" and "northing."

As mentioned at the beginning of Chapter 14, correlated error methods extend into three and higher dimensions, such as spatial–temporal models that conceivably could involve easting, northing, altitude, ocean depth, and time. There are entire courses (indeed course *sequences*) devoted to spatial statistics. Models with more than two dimensions are beyond the scope and purpose of this book. Interested readers are referred to spatial statistics textbooks such as Cressie (1993), Isaaks and Srivastava (1989), Journel and Huijbregts (1978), Schabenberger and Gotway (2004), and Cressie and Wikle (2011). Our goal here is simply to provide the students and readers new to GLMMs with a basic introduction to linear models with spatially correlated observations.

15.1.1 Types of Spatial Variability

We can divide spatial statistics into two broad classes: description and adjustment. *Description* refers to applications whose goal is the construction of pictures or maps. Examples include GIS (geological information systems) and satellite data. Spatial statistics are used to predict response at unobserved locations in the proximity of data points; these predictions help refine or improve the resolution of the maps or pictures. *Adjustment* refers to applications similar to those introduced in Chapter 14. With adjustment, the main goal is to improve the accuracy of inferential statistics in comparative studies. Often, estimates of treatment means and differences can be distorted by local variation. Employing multidimensional covariance structures, much like we did for repeated measures in Chapter 14, provides an effective and efficient way to account and correct for localized spatial effects.

In this chapter, we limit our focus to models for *adjustment*. Most of the linear models we have considered throughout this text have dealt with estimation and inference about estimable functions. Models for adjustment continue this theme. Models for description take us into territory best reserved for more specialized books and courses on spatial statistics *per se*.

Like repeated measures, spatial models' main premise is that proximate observations are correlated and that the magnitude of this correlation decays with increasing distance.

Much of the early work in spatial statistics occurred in geology, particularly in mining applications, and as a result, the terminology and many of the covariance models draw on roots in geostatistics. Here, we give a very quick tour of the main ideas, then translate them into forms amenable to GLMM application.

A fundamental geostatistical tool is the semivariogram, defined as

$$\gamma(d)=\left(\frac{1}{2}\right)E\left[y\left(loc_i\right)-y\left(loc_j\right)\right]^2 \tag{15.1}$$

where

$\gamma(d)$ denotes the semivariogram
d denotes distance
loc_i and loc_j denote any two positions distance d apart
$y(loc_i)$ and $y(loc_j)$ denote observations at loc_i and loc_j, respectively

In the two-dimensional spatial models we consider in this chapter, loc_i and loc_j must each be referenced by two variables that are, in essence, Cartesian coordinates, for example, row and column, latitude and longitude, and easting and northing. If we denote $loc_i=(r_i,c_i)$ and $loc_j=(r_j,c_j)$, where (r,c) gives the row–column coordinates of loc, then distance $d=\sqrt{\left(r_i-r_j\right)^2+\left(d_i-d_j\right)^2}$. Figure 15.1 shows the general form of the semivariogram.

Notice that the semivariogram increases linearly at smaller distances then levels off at greater distances. Smaller semivariogram values imply greater spatial dependence. Where the semivariogram is increasing, observations are correlated. The *range* is the distance at which the semivariogram plateaus. For distances exceeding the range, observations are effectively uncorrelated. The *sill* corresponds to the variance among independent observations. The *nugget* measures highly localized variance. In mining exploration, for example, a probe might not detect any ore, whereas a probe almost immediately adjacent might hit a rich vein of ore, hence the name "nugget effect." Covariance models capture this by adding an extra dispersion parameter at distance zero.

The semivariogram depicted in Figure 15.1 is called *isotropic* because it does not distinguish between distances in different directions. *Anisotropic* spatial models allow correlation to be stronger in one direction than another. For example, if spatial correlation results from prevailing winds that are predominantly west to east, then we might expect stronger

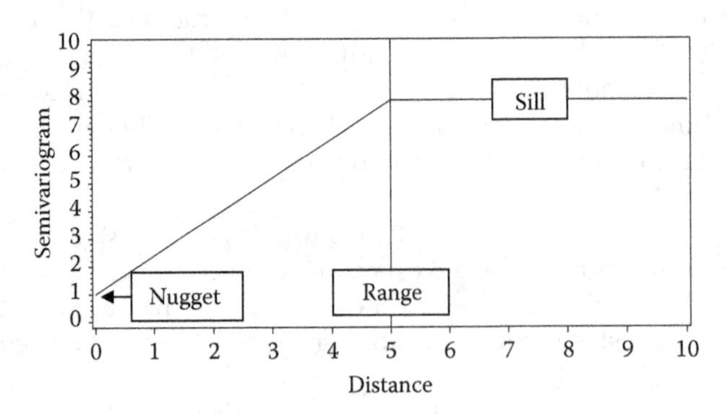

FIGURE 15.1
Generic form of semivariogram.

correlations in this direction that we would in a north–south direction. In this chapter, we focus exclusively on isotropic spatial correlation.

As we will see in Section 15.2, the semivariogram can provide a useful diagnostic to help choose an appropriate covariance model, but it does not fit the linear model framework. We need to express covariance in terms that we can capture in the **R** matrix for Gaussian linear models and either the working correlation matrix for GEE-type models or G-side covariance for true GLMMs with non-Gaussian data. If we denote by $\rho(d)$ the correlation between pairs of observations a distance d apart, then we can use the same approach that we did in Chapter 14. That is, for Gaussian models, the ijth element of the **R** matrix has the general form $r_{ij}=\sigma^2\,\rho(d_{ij})$, where d_{ij} is the distance between the pair of observations corresponding to the ijth element of the **R** matrix. For GEE-type models, we substitute $\rho_{W,ij}=\phi\rho(d_{ij})$, where $\rho_{W,ij}$ is the ijth element of the working covariance matrix and ϕ is the working covariance scale parameter.

Common spatial covariance models are

- *Exponential*

$$\rho(d)=e^{-(d/\alpha)} \quad \alpha>0 \tag{15.2}$$

- *Gaussian*

$$\rho(d)=e^{-(d/\alpha)^2} \quad \alpha>0 \tag{15.3}$$

- *Spherical*

$$\rho(d)=\begin{cases}1-1.5\left(\dfrac{d}{\alpha}\right)+0.5\left(\dfrac{d}{\alpha}\right)^3 & \text{where } 0\le d<\alpha\\[2ex] 0 & \text{where } d\ge\alpha\end{cases} \tag{15.4}$$

Each model depends on a spatial parameter α, but α acts on each model somewhat differently. Figure 15.2 shows the relationship between correlation as a function of distance, d, for various α for each model.

Notice that for the spherical model, the spatial parameter α precisely defines the range. For the exponential and Gaussian, correlation approaches zero asymptotically. Each plot has a horizontal bar at $\rho(d)=0.05$. Below this value, we regard correlation as negligible to the point that we consider the observations to be essentially independent. We call the distance d at which $\rho(d)=0.05$ the *effective range*.

> *Disclaimer*: Our use of $\rho(d)=0.05$ as the effective range is for the purposes of this discussion. In practice, effective range is a matter of judgment and, even then, spatial experts may differ.

For the exponential model, for example, $\alpha=0.5$, the effective range is just greater than $d=1$, meaning that for the exponential model with $\alpha=0.5$ there is no consequential spatial

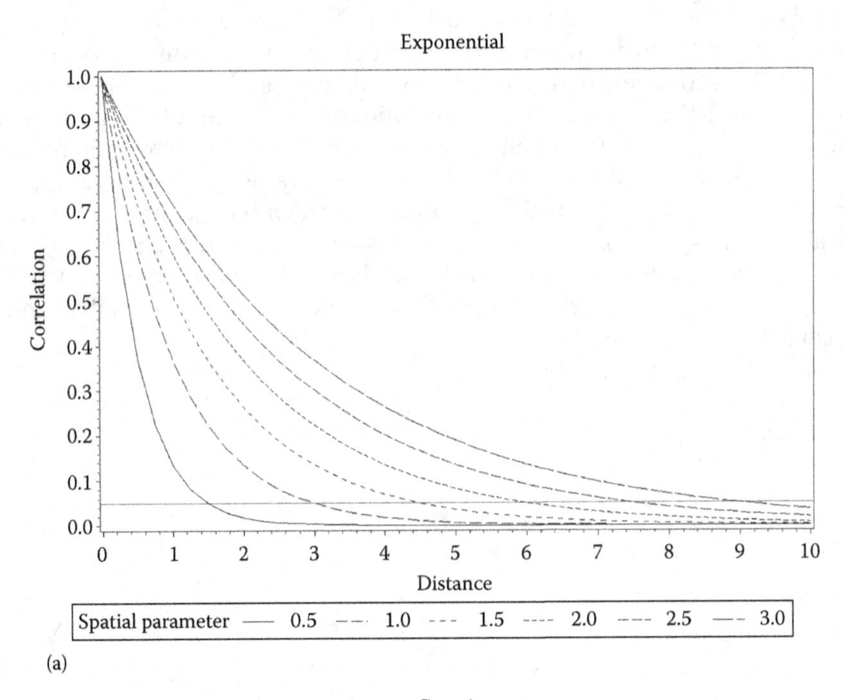

(a)

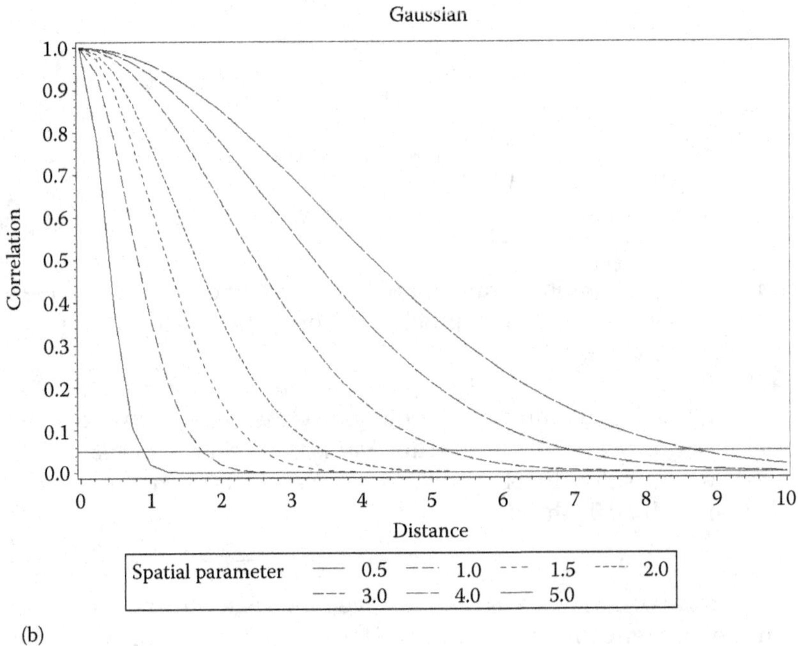

(b)

FIGURE 15.2
Correlation as a function of distance: (a) exponential spatial function, (b) Gaussian spatial function.

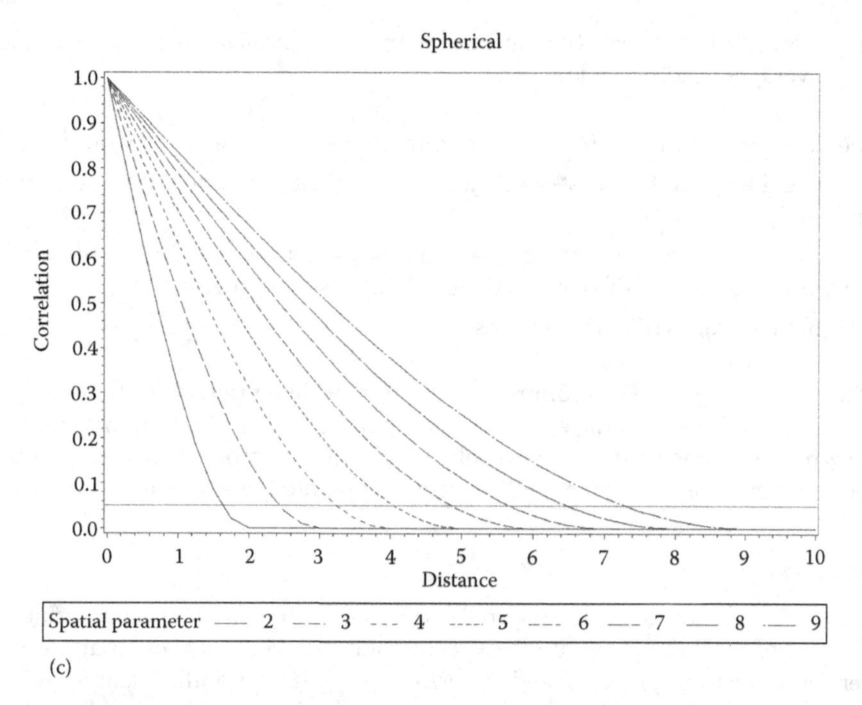

FIGURE 15.2 (continued)
Correlation as a function of distance: (c) spherical spatial function.

correlation. On the other hand, for $\alpha = 2$, the effective range is just above $d = 4$—observations 4 units apart or less are correlated, and the analysis must account for this correlation in a meaningful way.

What are the consequences of failing to account for spatial correlation? As we saw for repeated measures in Chapter 14, failure to account for spatial variability distorts standard errors and plays havoc with type I error control and confidence interval coverage. Unlike repeated measures, where the estimated means themselves were relatively unaffected, spatial variability can substantially distort estimated means as well. We will see vivid demonstrations of such distortion in the examples that appear in Sections 15.2 through 15.4.

15.1.2 Pre-GLMM Methods

Before proceeding to the examples in the next sections, we might ask how spatial variation was handled before geostatistics developed semivariogram models and before linear model methodology advanced to the point that it could incorporate spatial models into the correlated error toolbox. Essentially there were two approaches—*nearest-neighbor adjustment* and blocking. While spatial covariance modeling has largely supplanted nearest-neighbor adjustment, blocking, properly done, can still be an effective tool. We should also point out that blocking, *im*properly done, can exacerbate problems associated with spatial variability.

15.1.2.1 Nearest-Neighbor Adjustment

Nearest-neighbor adjustment (NNA) was introduced by Papadakis (1937). Readers are also referred to articles by Atkinson (1969) and Bartlett (1978). NNA can be understood as a form of analysis of covariance. The basic idea is to compute averages of observations adjacent to each unit of observation and use them as covariates. There are several ways to take these averages. These include

- The average of the four observations immediately east, west, north, and south
- The average of the two east–west neighbors and the two north–south neighbors separately
- The average of the surrounding observations, the four east–west–north–south and the four diagonal, northwest–northeast–southeast–southwest
- Weighted averages of various types

All of these methods share the common problem that data are used both as observations and as covariates. This introduces problems of bias, accurate determination of standard errors, degrees of freedom, etc. Once correlated error mixed model methodology advanced to the point where it became practical, it largely supplanted nearest-neighbor adjustments.

15.1.2.2 Blocking

Blocking can be regarded as the original tactic for dealing with spatial variability. The essential idea of blocking is to divide experimental units into subsets that are internally homogeneous. For example, in a field experiment, spatial variability nearly always exists to some extent in the form of local variation in parts of the field. How localized this variation is depends on the geology of the area. Blocking works when we subdivide the field so that local variability is essentially nonexistent *within* blocks—all or most of it should occur *between* blocks. Blocks *should* be different—units within blocks should *not*.

Blocking fails when researchers attempt to put too many treatments in the same block. In design, this is sometimes called "convenience blocking." Mead (1988) wryly—with tongue firmly in cheek—observed that it must be a beneficent deity who accommodatingly rearranges the physical world so that natural block size always corresponds to the number of treatments in one's experiment. Of course we know it does not—experiments with a large number of treatments cannot possibly be run in plausibly homogeneous complete blocks (rules of thumb are difficult, but certainly more than eight treatments begin to get into sketchy territory, and more than 16 are certainly too many, *no exceptions*).

Incomplete block designs were developed explicitly to allow experiments with large numbers of treatments to use blocking effectively. "Natural block size" is essentially a spatial variability concept. The first line of defense for spatial variability should *always* be to divide the area where the experiment is to be conducted into blocks sufficiently small to be internally homogeneous—and use incomplete block designs when the number of treatments exceeds the natural block size.

15.2 Gaussian Case with Covariance Model

In this section and in Section 15.3, we consider comparative studies with Gaussian data and spatial variability. Within the mixed model framework, we have two viable ways to account for spatial effects. The first, considered in this section, uses geostatistical covariance models for Var($\mathbf{y} \mid \mathbf{b}$) = $\mathbf{R}$, much as we used serial covariance $\mathbf{R}$ matrices for repeated measures in Chapter 14. The second, presented in Section 15.3, uses smoothing splines, first discussed in Chapter 7, to model an underlying response surface. The spline approach follows from the idea that local variability manifests itself as pattern in the residuals; if we can capture that pattern, then we can attribute individual deviation from that pattern to the treatment or explanatory factor level received by that observation, thereby giving us an accurate picture of the treatment effects or explanatory variable impact.

To introduce the covariance model approach, consider the following example. The data set in this example appears as Data Set 15.1 in the SAS Data and Program Library. The data are from a study to compare the response of 48 treatments. Such studies are common in plant breeding, for example, to screen large numbers of test varieties. One example of such a trial from Stroup et al. (1994) is discussed in Section 15.3.

The study was conducted on a 12×12 grid of plots. The grid was divided into three 12×4 complete blocks. The randomized complete block (RCB) study design should immediately raise red-flag, "suspect trouble" alarms: 48 treatments are far too many to allow much hope of within-block homogeneity. It is very unlikely that blocking accomplished its intended purpose. Confirmation comes when we implement the standard analysis for the RCB using the linear predictor, $\eta_{ij} = \eta + \tau_i + b_j$, with the random-block effect b_j assumed i.i.d. $N(0, \sigma_B^2)$. Not surprisingly, the solution for $\hat{\sigma}_B^2$ is negative and set to zero. With a large number of treatments, this is the classic symptom that blocking has failed and, in this context, a strong indication that we have unaccounted for spatial variability.

Our task at this point is to (1) determine a covariance structure that adequately accounts for the spatial variability in our data and (2) reestimate treatment means and associated inferential statistics using the spatial covariance model.

15.2.1 Covariance Model Selection

We can use semivariograms to help us identify a suitable spatial covariance model. Regardless of which covariance model we use, the $\mathbf{X}\boldsymbol{\beta}$ part of our linear predictor, in this case the treatment component $\eta + \tau_i$, will not change. We compute residuals from the treatment model and use them to construct empirical semivariograms. We can do this for the exponential, Gaussian, and spherical structures given in (15.2) through (15.4)—and any other spatial covariance structure we deem plausible. While not precisely equivalent to fitting these structures in the mixed model itself, they do give us a direction in which to proceed.

We can use the following SAS statements:

```
proc glimmix data=ch15_ex1;
  class trt;
  model Y= trt;
  output out=leftovers residual=r;
ods graphics on;
proc variogram data=leftovers plots=(semivar(all unpack));
  var r;
  model form=sph;
  coordinates xc=lat yc=lng;
  compute lagd=0.5 maxlag=16;
```

The GLIMMIX procedure fits the treatment model and uses the OUPUT statement to create a new data set with the residuals referenced as variable *R*. The VARIOGRAM procedure creates semivariogram plots using the PLOTS option in the PROC statement. The MODEL statement specifies that we want to evaluate the spherical semivariogram's fit. We change the option in the MODEL statement to FORM=EXP for the exponential and FORM=GAU for the Gaussian semivariogram. The COORDINATES statement identifies the variables we use in the data set to locate each observation in space. The COMPUTE statement specifies the distances we want to use in the semivariogram. There is a bit of trial and error associated with the COMPUTE statement; LAGD=0.5 and MAXLAG=16 causes the procedure to use all possible distances between pairs of observation up to about halfway across the diagonal length of the 12×12 grid on which the design was placed. Beyond this distance, there are too few pairs of observations to get a meaningful estimated semivariogram value.

Figure 15.3 shows the observed semivariogram with the estimated spherical semivariogram superimposed.

The exponential and Gaussian semivariograms are not shown because, as we shall see, the spherical structure fit the best. Notice that the spherical semivariogram has an intercept at approximately 2.5. This indicates a possible nugget effect. As we mentioned earlier, the semivariograms obtained with residuals from fitting $\mathbf{X}\boldsymbol{\beta}$ point the way, but do not necessarily give the same estimates we obtain from fitting the full model with PROC

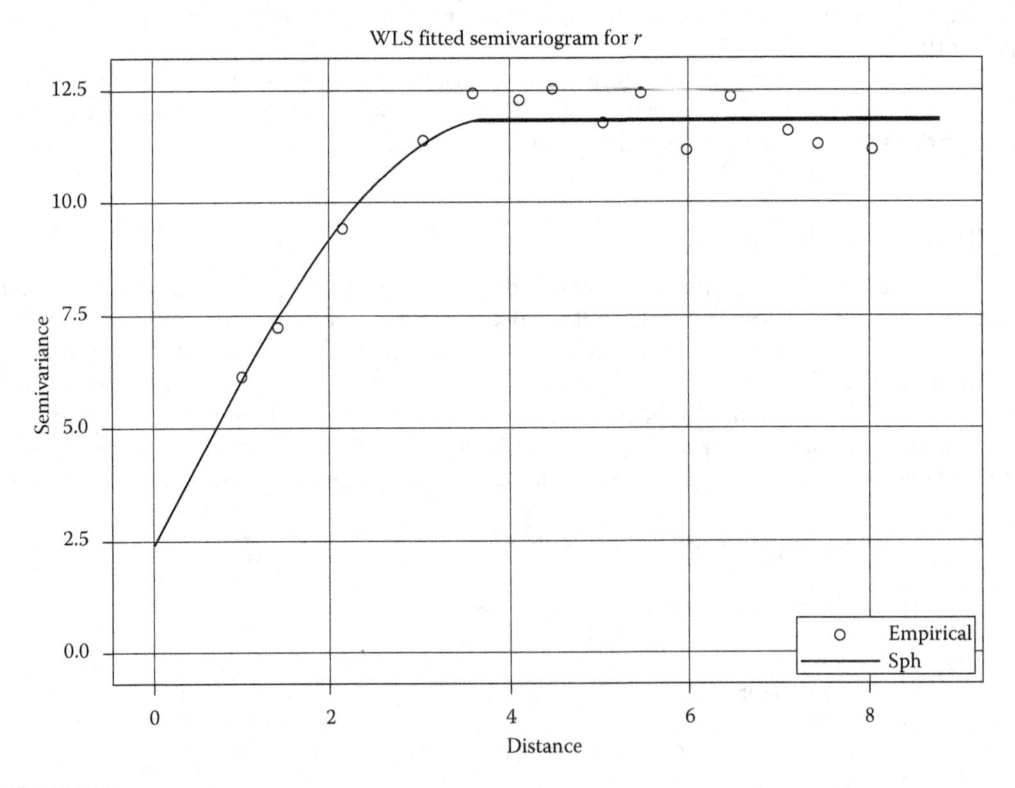

FIGURE 15.3
Observed (empirical) semivariogram with fitted spherical model superimposed.

MIXED or PROC GLIMMIX. MIXED allows us to include a nugget effect in the covariance structure. The model specification is thus

- Linear predictor: $\eta_i = \eta + \tau_i$
- Distribution: $\mathbf{y} \sim N(\boldsymbol{\mu}, \mathbf{R})$; $\left[r_{ij}\right] = \sigma_N^2 + \sigma^2 f\left(d_{ij}, \alpha\right)$, where σ_N^2 is the nugget effect variance, d_{ij} is the distance between observations i and j, and α is the spatial parameter
- Link: identity

The PROC MIXED statements are as follows:

```
proc mixed data=asademo maxiter=150 scoring=150;
  class trt;
  MODEL Y= trt;
  repeated/subject=intercept type=sp(sph)(lat lng) local;
  parms (3.8)(9.4)(2.5);
run;
```

Notice that in MIXED, REPEATED replaces the RANDOM _RESIDUAL_ statement to specify the structure of the **R** matrix. TYPE=SP specifies the various spatial models; (LAT LNG) locates the observations so distance between pairs of observations can be determined. The LOCAL option specifies the nugget effect. The MAXITER option in the PROC statement causes the procedure to run longer than the default cutoff of 50 iterations. REML estimation of spatial covariance can be sensitive to starting values. With default starting values we often see convergence, but convergence to nonsense estimates. Using starting values based on a preliminary assessment of the semivariogram is strongly recommended. Hence, the PARMS statement, which specifies starting values for the estimated range, sill and nugget (corresponding to α, σ^2, and σ_N^2, respectively) from the VARIOGRAM procedure. If convergence issues occur with Newton–Raphson, try overriding it with the SCORING option, as shown here in the PROC statement.

In this case, we obtain convergence with Fisher scoring after 122 iterations. When we do, the nugget effect variance estimate is $\hat{\sigma}_N^2 = 0.000259$. The AICC fit criterion with the nugget effect is 515; without the nugget effect, AICC = 512.9.

We also fit the exponential and Gaussian semivariogram models and the RCB model. Also, it turns out that there are in fact three 4×4 incomplete blocks embedded within each 4×12 complete block grid. The fit criteria for these various approaches are

Random Model/Covariance Component	**AICC**
Spherical, no nugget	512.9
Exponential	521.9
Gaussian	530.9
Incomplete block	588.8
Complete block	597.9

These results underline the ineffectiveness of blocking with large block sizes and confirm the spherical no-nugget covariance structure as the model of choice. We now proceed to inference on the treatment effects.

15.2.2 Impact of Spatial Variability on Inference

Consistent with our covariance model selection in the previous section, we use the model as described earlier except we drop the nugget effect. Thus, the ijth element of the **R** matrix is $[r_{ij}] = \sigma^2[1 - 1.5(d_{ij}/\alpha) + 0.5(d_{ij}/\alpha)^3]$ when $d_{ij} < \alpha$ and zero otherwise.

We can implement the analysis either using the MIXED statements shown previously, deleting LOCAL, or using the following PROC GLIMMIX statements:

```
proc glimmix data=ch15_example1;
  class trt;
  model y= trt/ddfm=kr;
  random _residual_/subject=intercept type=sp(sph)(lat lng);
  parms (4)(12);
  lsmeans trt;
```

The covariance parameter estimates appear as

Covariance Parameter Estimates			
***Cov* Parm**	**Subject**	**Estimate**	**Standard Error**
SP(SPH)	Intercept	4.1214	0.2483
Residual		14.0107	2.1665

SP(SPH) corresponds to the spatial parameter, $\hat{\alpha} = 4.12$, that is, the range is just over 4 units distance. We can refer to Figure 15.2c to visualize the strength of correlation for pairs of observations 4 units or less apart. RESIDUAL tells us $\hat{\sigma}^2 = 14.01$.

Table 15.1 shows the estimated treatment means, that is, $\hat{\mu}_i = \hat{\eta} + \hat{\tau}_i$, which we obtain from the spatial model and from the original RCB analysis. In the interest of space, only the 10 greatest means using the spatial model are shown. In addition, the rankings of these means under the RCB analysis are also shown.

Notice that several of the top 10 means under that spatial model did not even appear in the top 10 using the RCB model. For example, the highest ranked mean, from treatment 45, was ranked 10th by the randomized block. Without spatial adjustment, its estimated mean was 19.79, with spatial adjustment, $\hat{\mu}_{45} = 23.17$. The highest ranking mean under the

TABLE 15.1

Estimated Top 10 Treatment Means under Spatial Model Compared with RCB Means

Obs	rcb_rank	spatial_rank	TRT	mean_rcb	mean_spatial
1	10	1	45	19.7938	23.1658
2	4	2	47	21.0693	23.0594
3	20	3	43	18.4296	20.8899
4	7	4	34	20.2231	19.9564
5	2	5	46	22.9891	19.9364
6	13	6	48	19.3958	19.6969
7	6	7	42	20.5483	19.5630
8	15	8	35	19.2219	19.3637
9	25	9	41	17.3714	19.1306
10	5	10	29	20.7532	18.9632

RCB analysis was treatment 30: before adjustment, its estimated mean was 23.14 and after adjustment, $\hat{\mu}_{30} = 18.73$.

Why the large discrepancies? Inevitably, with large, internally heterogeneous blocks, some treatments will be randomly assigned consistently to the "worst" parts of the blocks—that is, the part of the block that tends to produce the lowest y_{ij}. By the luck of the draw, the opposite will happen to other treatments. When the blocks are too big, randomization cannot overcome this (and, by the way, nonrandom "expert judgment" assignment *will* make things worse, not better). The only ways around this are (1) design more intelligently, meaning *plan to use incomplete block designs* (the preferred approach) or (2) if it is too late—meaning the data were collected before competent design planning was done—use spatial adjustment.

What about the impact on standard errors? For the randomized block analysis for these data, the standard error of a treatment difference was 2.36. For the spatial analysis, the standard error depends on the average distance between the units where the treatments being compared were placed: For these data, the standard error of a difference ranged between 1.48 and 1.63. We can show by simulation that the standard errors from the spatial analysis closely reflect the observed sampling distribution of treatment differences, whereas the standard errors from the RCB analysis are inflated. This translates to lost power and precision.

15.3 Spatial Covariance Modeling by Smoothing Spline

The exponential, Gaussian, and spherical correlation functions are all examples of *parametric* spatial models because they describe the correlation–distance relationship in terms of a parameter. Often, spatial variability arises from natural causes not easily captured by the parametric functions, $\rho(d)$, discussed in the previous section. In these cases, smoothing splines provide an alternative that adapts to observed local variation rather than imposing a formal structure.

As an example, we consider a plant breeding trial from Alliance, NE, that first appeared in Stroup et al. (1994). The trial had 56 wheat varieties in a RCB design with four blocks. For the same reasons discussed in the previous section, the standard RCB analysis was compromised by within-block heterogeneity. Stroup et al. used nearest-neighbor analysis. Littell et al. (2006) reanalyzed the data using a spherical covariance structure with a nugget effect. The nearest-neighbor and spherical covariance analyses yielded similar results. While both methods were effective *numerically* in recovering accurate inference, neither provided a compelling scientific interpretation of the spatial effects.

In reality, the local variation in the Alliance data was caused by winter kill, a problem that occurs with winter wheat when the plants have inadequate snow cover and are exposed to sustained freezing wind chill. One part of the field at Alliance was on a rise more exposed to the wind, and another part was in a low-lying, protected area. The arrangement of the blocks did not correspond to the exposed and protected areas—the perfect recipe for pronounced spatial effects to occur.

We can use smoothing spline methodology (see Ruppert et al. [2003] for an introduction to smoothing splines) to model the underlying pattern of exposed and protected areas in the field. Splines have the advantage of being able to adapt to rough irregular contours typical of geological entities.

The following statements implement the initial analysis:

```
proc glimmix data=nin.alliance;
  class Entry;
  model Yield=Entry;
  random latitude longitude/type=rsmooth;
  lsmeans entry/diff=control('BUCKSKIN');
  output out=gmxout pred=p;
  ods output lsmeans=lsm1;
  id entry latitude longitude _zgamma_;
proc means data=gmxout; var _zgamma_;
run;
```

In the LSMEANS statement, the DIFF option defines ENTRY (wheat variety) "BUCKSKIN" as the control treatment. This follows from the Stroup et al. (1994) and Littell et al. (2006) discussions. At the time this wheat trial was conducted, BUCKSKIN was a variety known to be a high-yielding benchmark; its mediocre mean yield in the RCB analysis despite it being observed in the field outperforming all varieties in the vicinity was one symptom that the RCB analysis was giving nonsense results.

Note the form of the random statement: we use LATITUDE and LONGITUDE as random-model effects. TYPE=RSMOOTH invokes a smoothing spline procedure to create a response surface of the residuals from the treatment linear predictor $\eta + \tau_i$ defined by the MODEL statement. We call smoothing spline methods used in this context *radial smoothing*, hence the option TYPE=RSMOOTH. Figure 15.4 shows the resulting surface.

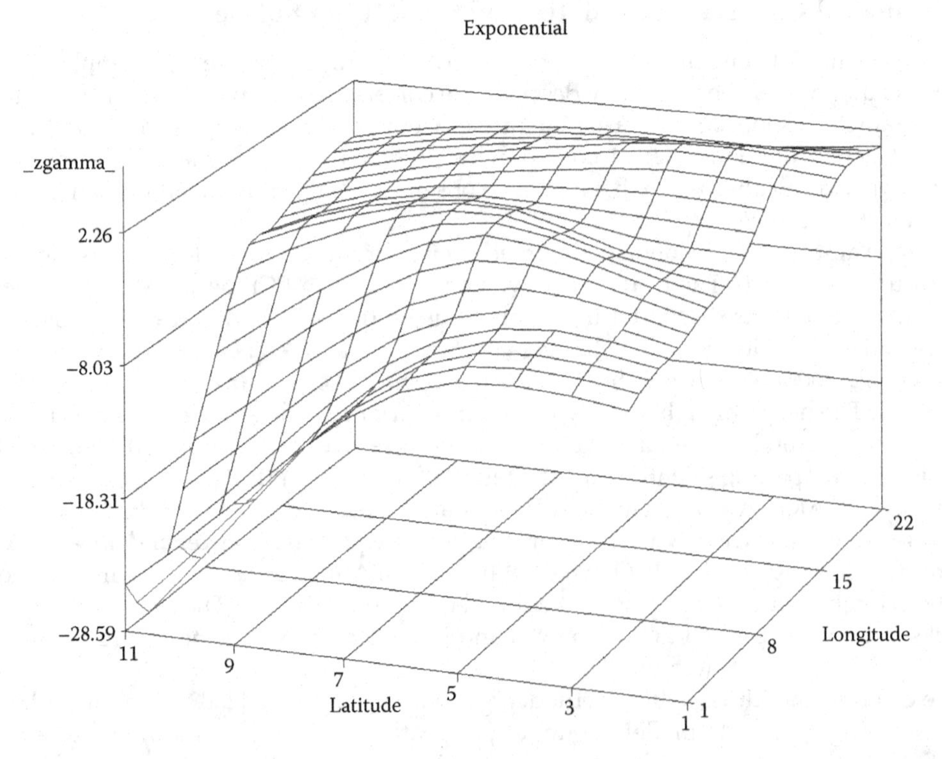

FIGURE 15.4
Plot of residual response surface, Alliance data, "naive" analysis.

The variable _ZGAMMA_ denotes the residuals forming the spatial smoothing spline response surface.

Notice that the area in the foreground-left, corresponding to higher LATITUDE and lower LONGITUDE, has noticeably lower residual values. This corresponds to the area exposed to winter kill. Also notice that the residuals are not centered about zero; this does not affect *relative differences* among the estimated treatment means adjusted for spatial effects, but all treatment means will be shifted up or down depending on whether the sum of the _ZGAMMA_ is positive or negative. The OUTPUT and PROC MEANS statements together obtain the mean—here the mean of ZGAMMA_ is −4.46.

We center the _ZGAMMA_ values by adding a response surface to the MODEL statement. This, in effect, defines a random coefficient model to account for variation over LATITUDE and LONGITUDE. We start by adding a second-order polynomial. Our new MODEL statement is

```
model Yield=Entry lat|lat|lng|lng@2;
```

This reduces the mean _ZGAMMA_ to 0.412. By trial and error, with polynomial regression we reach a point of diminishing returns by adding cubic terms, that is, with MODEL statement

```
model Yield=Entry lat|lat|lat|lng|lng|lng;
```

This yields a mean _ZGAMMA_ of 0.34. We can do a bit better with a smoothing spline in the linear predictor. The GLIMMIX statements are now

```
effect sp_rs=spline(lat lng);
model Yield=Entry sp_rs;
random latitude longitude/type=rsmooth;
```

Now we have a true smoothing spline random coefficient model. The mean _ZGAMMA_ for this model is 0.21. This is as close to centered as we can get with these data.

As with Example 15.1, we show the adjusted means and the unadjusted RCB estimated means for the spatially adjusted top 12 yielding entries.

Entry	rank_ rsmooth	rank_ sp_sph	rank_rcb	mean_ rsmooth	mean_ sp_sph	mean_rcb
BUCKSKIN	1	1	28	34.6874	32.8656	25.5625
NE85556	2	2	23	29.7090	28.8507	26.3875
NE83498	3	4	6	29.2353	28.5911	30.1250
NE87619	4	3	2	29.2301	28.7235	31.2625
NE86503	5	7	1	28.4106	26.9155	32.6500
NE86507	6	10	39	28.3175	26.4693	23.7875
REDLAND	7	5	4	28.2672	28.0658	30.5000
CENTURK 78	8	12	5	28.0781	26.3294	30.3000
NE86606	9	11	8	27.8901	26.4019	29.7625
KS831374	10	8	37	27.7967	26.7279	24.1250
NE87613	11	9	10	27.6394	26.6207	29.4000
NE86527	12	14	44	27.4973	25.8116	22.0125

The columns RSMOOTH denote the spatially adjusted means using the smoothing spline method we just described. The columns SP_SPH denote the spherical covariance plus nugget effect shown in Littell et al. (2006). The columns RCB denote the original RCB analysis. The discrepancies between the two spatial adjustment methods are comparatively minor, but the rankings from the RCB analysis show the characteristic distortion caused by spatial variability.

15.4 Non-Gaussian Case

We now consider spatial adjustment models for non-Gaussian data. As with the repeated measures for one dimension discussed in Chapter 14, there are two approaches: GEE-type, or R-side, spatial correlation models and G-side GLMM spatial models.

To illustrate, we use binomial data from a study with spatial variability first reported by Gotway and Stroup (1997). The objective of the trial was to compare wheat varieties for their resistance to damage from Hessian flies. There were 16 varieties conducted as a 4×4 lattice design. The layout of the design is shown in Figure 15.5.

Lattice designs can be regarded as a RCB design (the larger, 16-plotquarters of the field separated by the solid white boundaries) or as incomplete block designs, where the incomplete blocks are the four-plot, 2×2 grids that subdivide each complete block into quarters. The main rationale for incomplete blocks designs was the potential existence of localized variability whose magnitude might cause excessive within-block heterogeneity for complete blocks of size 16.

As mentioned earlier in this chapter, incomplete blocks and spatial correlation models can be regarded as alternative ways of accounting for highly localized spatial variability. This said, we consider four models in this section:

1. RCB
2. Incomplete block
3. GEE—"R-side" spatial working correlation
4. GLMM—"G-side" spatial correlation

The response variable for this study was plant condition: damaged or undamaged. In each plot, N_{ij} plants were sampled and y_{ij}, the number of plants showing Hessian fly damage, was

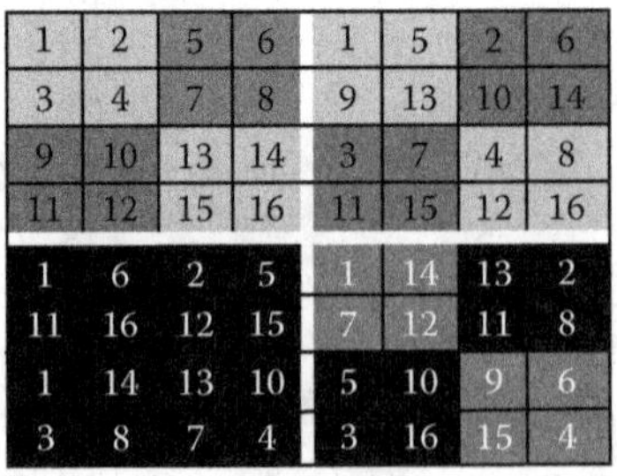

1	2	5	6	1	5	2	6
3	4	7	8	9	13	10	14
9	10	13	14	3	7	4	8
11	12	15	16	11	15	12	16
1	6	2	5	1	14	13	2
11	16	12	15	7	12	11	8
1	14	13	10	5	10	9	6
3	8	7	4	3	16	15	4

FIGURE 15.5
Plot layout of resistance to Hessian fly variety trial.

determined. Within each plot, assuming each plant represents an independent Bernoulli trial, $y_{ij} \mid plot_{ij} \sim \text{Binomial}(N_{ij}, \pi_{ij})$, where π_{ij} denotes the probability of a damaged plant in the ith variety, jth block.

We will illustrate all four models using the logit link, $\eta_{ij} = \log[\pi_{ij}/(1 - \pi_{ij})]$. Alternatively, we could use the probit model; the principles discussed here apply equally. For all four models, the fixed-effect component of the linear predictor is $\eta + \tau_i$, where η denotes the intercept and τ_i denotes the effect of the ith variety on the logit scale. The only distinction between each model is the form of the linear predictor with regard to the random effects and the assumptions associated with them. The following completes the specification of the two blocked models.

15.4.1 Randomized Complete Block Model

The linear predictor for the RCB model is

$$\eta_{ij} = \mu + \tau_i + r_j \tag{15.5}$$

where r_j is the effect of the jth complete block, $j = 1, 2, 3, 4$. The block effects are assumed i.i.d. $N\left(0, \sigma_R^2\right)$.

15.4.2 Incomplete Block Model

The linear predictor for the incomplete block model is

$$\eta_{ij} = \mu + \tau_i + b_j \tag{15.6}$$

where b_j is the effect of the jth incomplete block, $j = 1, 2, \ldots, 16$. The block effects are assumed i.i.d. $N\left(0, \sigma_B^2\right)$.

15.4.3 GLIMMIX Statements

The GLIMMIX statements for these two models are as follows.

15.4.3.1 RCB

```
proc glimmix data=HessianFly method=LaPlace;
  class entry rep;
  model Yij/Nij=entry;
  random intercept/subject=rep;
```

Note that for this data set, REP identifies the complete blocks. Also note the use of METHOD=LAPLACE. This allows us to use information criteria fit statistics, such as AICC, to compare models. Alternatively, we could also use METHOD=QUADRATURE.

15.4.3.2 Lattice Incomplete Blocks

```
proc glimmix data=HessianFly method=LaPlace;
  class entry block;
  model Yij/Nij=entry;
  random intercept/subject=block;
```

The only difference between the RCB and incomplete block statements is that BLOCK, which identifies the incomplete blocks, replaces REP in the CLASS and SUBJECT statements. Relevant output appears later, after we describe the R-side and G-side spatial models and their accompanying GLIMMIX statements.

15.4.4 GEE-Type "R-Side" Spatial Correlation Model

This is the model that appears in Gotway and Stroup (1997). The linear predictor is

$$\eta_{ij} = \mu + \tau_i$$

Note that block effects do not appear in the model. Instead, we account for variation among plots over and above treatment with a working correlation matrix. As with Gaussian data, spatial working correlation models for normal data are typically borrowed from geostatistics. Gotway and Stroup used the spherical working correlation structure. Letting $\mathbf{P}_W$ denote the working covariance matrix, we write the element corresponding to any pair of observations y_{ij} and $y_{i'j'}$ as

$$\rho_W\left(y_{ij}, y_{i'j'}\right) = \begin{cases} \phi\left[1 - 1.5\left(\dfrac{d}{\alpha}\right) + 0.5\left(\dfrac{d}{\alpha}\right)^3\right] & \text{for } d < \alpha \\ 0 & \text{for } d \geq \alpha \end{cases} \tag{15.7}$$

where
 ϕ denotes a scale parameter
 d denotes distance between the two observations
 α denotes the *range*

We can interpret the range parameter as the distance within which pairs of observations behave as if they are correlated and beyond which observations behave as if they are independent. In this sense, the range has the same interpretation in the working correlation as it does in a true correlation model for Gaussian data.

The characteristics described in Chapter 14 for the GEE-type R-side repeated measures models apply to GEE-type R-side spatial models as well. They target the mean of the marginal distribution, whereas the G-side model presented later targets the actual binomial probability parameter of the conditional distribution—a distinction we have seen in various forms throughout this textbook. The range parameter's interpretation is similar to that of the range parameter for the G-side model, but the scale parameter is not a true variance component, so we cannot interpret it as such.

The GLIMMIX statements for the GEE-type spatial model are

```
proc glimmix data=HessianFly;
  class entry;
  model Yij/Nij=entry/ddfm=kr;
  random _residual_/type=sp(sph)(lat lng) subject=intercept;
  parms (3) (1);
```

There are three primary differences between these statement and those for the two block designs. First, the RANDOM _RESIDUAL_ statement replaces the RANDOM statement.

This is similar to other GLIMMIX programs that define a working correlation structure. The syntax TYPE=SP(SPH) specifies the spherical working correlation model. The variables LAT and LNG give the location of each plot. Notice that the syntax here is identical to that used with the spatial covariance structures, shown in Section 15.2 for Gaussian data. The second difference: because we have a RANDOM _RESIDUAL_ statement, we cannot use METHOD=LAPLACE. Recall that all working covariance structures define quasi-likelihood, not a true likelihood. We must use the pseudo-likelihood computing algorithm. Finally, in this program, we use the DDFM=KR option in the MODEL statement. We cannot use the Kenward–Roger option with LAPLACE, but we can here—and it is *probably* advisable to do so. "Probably," because simulation studies to date tentatively suggest that the approximation does work effectively for spatial GEEs, but the emphasis is on "tentatively suggest"—more work is needed in this area.

Note the use of the PARMS statement to specify starting values for the range parameter (α) and the scale parameter (ϕ), respectively. As with spatial models for Gaussian data, spatial working covariance models tend to be sensitive to starting values. Default starting values can result in, to quote Oscar Kempthorne's colorful phrase, "obviously rotten" covariance parameter estimates. Our logic for the starting values is shown here: (1) if the data truly follow a binomial distribution without spatial variability, there would be no RANDOM _RESIDUAL_ statement and, hence, no scale parameter. Setting ϕ to 1 is equivalent to not having a scale parameter. The range parameter was chosen because there were normally distributed data also analyzed from this trial (e.g., yield) that produced range estimates in the vicinity of three. In the absence of such information, one could use the size of the complete blocks (side length = 3 as measured from plot-center to plot-center, diagonal = $\sqrt{18}$, for example) as a rough guideline.

15.4.5 "G-Side" Spatial Correlation Model

This is a G-side version of Gotway and Stroup model. The linear predictor is

$$\eta_{ij} = \mu + \tau_i + p_{ij}$$

where p_{ij} denotes a random effect of the ijth plot, that is, the ijth treatment × block combination as depicted in Figure 15.5. If we work though the WWFD skeleton ANOVA process, p_{ij} corresponds to the block-by-treatment effect, which we know from our discussions in Chapters 10, 11, and 14 is reserved for "residual" with Gaussian data, but can appear in the linear predictor for one-parameter members of the exponential family—for example, binomial and Poisson data. We assume $\mathbf{p}$ the vector of p_{ij} effects to be random and have a $N(\mathbf{0}, \mathbf{\Sigma}_p)$ distribution. As with the R-side model (15.7), let d equal the distance between a given pair of plots. The covariance between a given pair of p_{ij} effects depends on the distance between them. For comparability, assume a spherical covariance model:

$$\text{Cov}\left(p_{ij}, p_{i'j'}\right) = \begin{cases} \sigma^2\left[1 - 1.5\left(\dfrac{d}{\alpha}\right) + 0.5\left(\dfrac{d}{\alpha}\right)^3\right] & \text{for } d < \alpha \\ 0 & \text{for } d \geq \alpha \end{cases} \quad (15.8)$$

where $\text{Cov}\left(p_{ij}, p_{i'j'}\right)$ denotes the element of $\mathbf{\Sigma}_p$ corresponding to the ijth and $i'j'$th pair of plots. The GLIMMIX statements for the G-side model are

```
proc glimmix data=HessianFly method=LaPlace;
  class id entry;
  model Yij/Nij=entry/ddf=48;
  random id/type=sp(sph)(lat lng) subject=intercept;
  parms (1)(3);
```

The variable ID identifies each plot. The G-side model uses a RANDOM statement, not RANDOM _RESIDUAL_. Except for deleting _RESIDUAL_, the statement is otherwise identical to the statement specifying spatial covariance in the R-side model (15.7). Also, we use the same values in the PARMS statement, for much the same reasons. Note, however, that the order of the parameters is reversed when you use the G-side formulation. Finally, because this is a G-side model, we can use the LAPLACE method. This will allow us to compare fit statistics with the blocked designs. While not shown here, we could also try other spatial models to see how their fit compares with the spherical covariance structure.

15.4.5.1 G-Side Spatial Radial Smoothing Model

Instead of using a parametric covariance model, we can use the smoothing spline approach on non-Gaussian data. We introduced this approach for Gaussian data with the Alliance wheat trial example in Section 15.3. The approach for non-Gaussian data is similar. As with the Alliance data, we can start either by fitting a polynomial response surface or a smoothing spline response surface and then use radial smoothing to fit remaining spatial variability. As with Gaussian data, radial smoothing is a G-side random effect. For the Hessian fly data, the two possible approaches use the following GLIMMIX statements:
Smoothing spline response surface:

```
proc glimmix data=HessianFly method=laplace;
  class entry;
  effect sp_rs=spline(lat lng);
  model y/n=Entry sp_rs;
  random lat lng/type=rsmooth;
```

Polynomial response surface:

```
proc glimmix data=HessianFly method=laplace;
  class entry;
  model y/n=entry lat|lat|lat|lng|lng|lng@3;
  random lat lng/type=rsmooth;
```

The main difference between these programs is the use of the EFFECT definition and the created variable SP_RS in the MODEL statement for the first program and the third-order polynomial response surface defined on LAT and LNG, the spatial locator variables, in the second program. As with the Alliance data, the third-order polynomial results from trial and error whose goal is to center the residual about zero.

15.4.5.2 Relevant Output

We start by comparing the complete and incomplete blocks. Focus first on the fit statistics and the variance component estimate for block. If there is nonnegligible spatial variability,

Fit Statistics	
−2 Log likelihood	269.96
AIC (smaller is better)	303.96
AICC (smaller is better)	317.27

Covariance Parameter Estimates

Cov Parm	Subject	Estimate	Standard Error
Intercept	rep	0.000968	0.01981

FIGURE 15.6
Fit statistic and variance estimates for RCB model.

there should be evidence of within-block heterogeneity in the complete block analysis (15.5). Evidence takes two forms: (1) the information criteria for the incomplete block analysis (15.6) should be substantially better than for the complete block analysis and (2) the block variance component for the complete block model should be noticeably lower than the block variance component for the incomplete block model (because there is more variation *within* blocks than *among* blocks in the complete block model).

Figures 15.6 and 15.7 shows the relevant output for this comparison.

The AICC shows that the incomplete block model provides a better fit than the complete block models and, consistent with this result, the incomplete block variance component estimate, $\left(\hat{\sigma}_B^2 = 0.426\right)$, is noticeably greater than the complete block variance component estimate, $\left(\hat{\sigma}_R^2 = 0.000968\right)$. The latter is more of an observation—there is no need to do a formal test. It is a matter of the estimates being consistent with what one would except if spatial variation is present using the criteria described earlier.

This establishes that given a choice between the RCB and lattice incomplete block models, choosing the incomplete block analysis will give us a more accurate picture of entry effects and differences. The next question is how does the lattice model (15.6) compare to the spatial model? We start with (15.8), the G-side model, because it allows computation of the information criteria and, hence, a comparison of model fit. Figure 15.8 shows the fit statistics for (15.8).

The AIC for (15.8) is 286.7 versus 283.3 for the incomplete block model (15.6) indicating a *slightly* better fit than the spatial model. The AICC are 301.9 for the spatial model versus 296.6 for the incomplete block model, strengthening the case for using the incomplete block model.

Fit Statistics	
−2 Log likelihood	249.33
AIC (smaller is better)	283.33
AICC (smaller is better)	296.64

Covariance Parameter Estimates

Cov Parm	Subject	Estimate	Standard Error
Intercept	block	0.4262	0.2205

FIGURE 15.7
Fit statistic and variance estimates for lattice incomplete block models.

Fit Statistics	
−2 Log likelihood	250.71
AIC (smaller is better)	286.71
AICC (smaller is better)	301.91

Covariance Parameter Estimates			
Cov Parm	Subject	Estimate	Standard Error
Variance	Intercept	0.5111	0.2345
SP(SPH)	Intercept	3.2256	0.5925

FIGURE 15.8
Fit statistics and covariance parameter estimates for G-side spherical spatial model.

There is information in the spatial analysis that might be useful for its own sake. In particular, the estimated range parameter ($\hat{\alpha} = 3.23$) gives an indication of the extent of local correlation. This could be used to assess the extent of competition effects or other "nearest-neighbor" effects, or one could use it for planning future experiments (see Chapter 16 on power and planning).

The two radial smoothing models produce AICC fit statistics of 345.1 for the first version, using the smoothing spline response surface and 313.8 for the second version, using the third-order polynomial response surface. The AIC fit statistics are 298.7 and 279.6, respectively. Of the two radial smoothing models, the one based on the third-order polynomial response surface is clearly preferable. However, given the AICC statistic, there is not a compelling case for using the radial smoothing model rather than the spatial-spherical or lattice incomplete block model.

How does the G-side model (15.8) compare to the R-side, GEE-type spatial model (15.7)? Figure 15.9 shows the corresponding results for (15.7).

The range estimate $\hat{\alpha} = 3.33$ is close to the G-side estimate from (15.8). However, it is important to stress that because these two models make different assumptions about how the spatial process is conceptualized, strictly speaking these two estimates are not comparable. That said, they both address "range," so in this case it is reassuring that they are close. The "Residual" is an estimate of the scale parameter, $\hat{\phi} = 3.96$. One might be tempted to conclude that because it is $\gg 1$, this is evidence of overdispersion. However, keep in mind that the generalized χ^2/df is not a legitimate fit statistic. It is probably better to think of it as consistent with the earlier results for the incomplete block and spatial

Fit Statistics	
−2ResLogPseudo-Likelihood	164.03
Generalized Chi-Square	190.22
Gener. Chi-Square / DF	3.96

Covariance Parameter Estimates			
Cov Parm	Subject	Estimate	Standard Error
SP(SPH)	Intercept	3.3257	0.8276
Residual		3.9629	1.2027

FIGURE 15.9
Fit statistics and working covariance parameter estimates for GEE-type spherical-spatial model.

TABLE 15.2

F-Values for Test Entry Effects, Hessian Fly Data Models

Model	F_{entry}
RCB	6.81
Lattice incomplete block	6.02
Spatial—spherical G-side	4.67
Spatial—spherical R-side	4.43
Spatial—RSmooth—spline(lat lng)	7.32
Spatial—RSmooth—third-order poly	6.09

G-side models versus the complete block model: that some form of within-complete-block heterogeneity exists, and we must account for it either via spatial models or an incomplete block model.

Finally, how do the tests of variety (entry) effects and the estimated probability of a damaged plant compare for the various models? Table 15.2 shows the *F*-statistic for the overall test of entry effects, H_0: all $\tau_i = 0$.

All of the *F*-values have a *p*-value <0.0001. From the model fitting comparisons, the results of choice for research reporting purposes would be the lattice incomplete block or the spatial-spherical model, depending on whether the range associated with the spatial model was also of intrinsic interest. Here, our conclusion about entry effect is the same regardless of which model we use. Obviously, *F*-values *are* affected by model choice, so for other data sets, our decision might be more critical than it is here.

Table 15.3 shows the estimated means and their standard errors on the data scale for the lattice incomplete block model and the G-side and R-side-GEE-type spatial-spherical models.

TABLE 15.3

Data-Scale Entry Mean Estimates and Standard Errors, Lattice, and Spatial-Spherical Models

Obs	mean_lattice	mean_sph_G	mean_sph_R	se_lattice	se_sph_G	se_sph_R
1	0.87085	0.89801	0.90735	0.05199	0.05008	0.06900
2	0.83058	0.85229	0.85354	0.06070	0.06381	0.08176
3	0.77710	0.76734	0.72411	0.07205	0.08540	0.10081
4	0.61178	0.61110	0.59219	0.08624	0.10170	0.09785
5	0.72988	0.77710	0.66958	0.08996	0.09074	0.11495
6	0.65601	0.63554	0.57892	0.08964	0.10660	0.11062
7	0.71518	0.69781	0.67276	0.08356	0.10200	0.10699
8	0.50225	0.49093	0.52299	0.09639	0.11436	0.11534
9	0.53339	0.54694	0.49078	0.08775	0.10548	0.10348
10	0.61161	0.69421	0.59600	0.10444	0.10705	0.12697
11	0.46907	0.42685	0.37933	0.09185	0.11108	0.10733
12	0.60383	0.59947	0.62704	0.09153	0.10995	0.11158
13	0.07602	0.09052	0.09101	0.04102	0.05266	0.07472
14	0.31576	0.29814	0.38430	0.09177	0.10310	0.11834
15	0.35562	0.30002	0.26623	0.08442	0.09444	0.09464
16	0.13007	0.12889	0.13437	0.05637	0.06280	0.07602

The lattice and G-side spatial models provide estimates of the binomial probability, π_i, from the conditional model, whereas the R-side, GEE-type model provides estimates of the marginal mean. These are distinctions we have seen throughout this textbook. For most entries, the three models produce similar estimates and standard errors. However, for entries 4, 5, 10, 11, 14, and 15, there are discrepancies large enough to give us pause. The discrepancies result from the by-now familiar conditional versus marginal distribution issue we first encountered in Chapter 3. For the G-side and R-side spatial models, our trade-off is between the (in theory at least) more robust R-side estimates and the more accurate but more vulnerable to model misspecification G-side estimates.

Given these tradeoffs, the slightly better AICC fit statistic for the lattice incomplete block model, and the fact that nonspatial models for well-designed blocked experiments are harder to misspecify, the case for the incomplete block model appears to be even more compelling.

This anticipates a theme we will develop more forcefully in Chapter 16. The spatial modeling we have seen in this chapter has largely been "data repair"—that is, attempts to recover useful information from ill-conceived designs. Specifically, these examples have featured inappropriate use of a complete block design when an incomplete block design would have been a better choice. The Hessian fly example shows the approach one *should* take—a well-conceived incomplete block design was used, and it turned out to provide the analysis of choice.

15.5 Summary

1. The basic approach described in this chapter is a two-dimensional extension of the approach presented in Chapter 14.
2. Our primary concern in this chapter is inference on estimable functions $\mathbf{K}'\boldsymbol{\beta}$ in the presence of spatial covariance.
3. Spatial covariance generally means correlation between pairs of observations decreases with increasing distance.
4. Common functions used to relate covariance to distance originated in geostatistics.
5. The primary covariance-to-distance tool in geostatistics is the semivariogram. Geostatistical semivariograms can be expressed as mixed model covariance functions and vice versa.
6. As with repeated measures GLMMs, there is a "G-side/R-side issue."
    a. For Gaussian data, the covariance function is embedded in $\mathbf{R}$ of $\mathbf{y} \mid \mathbf{b} \sim N(\boldsymbol{\mu} \mid \mathbf{b}, \mathbf{R})$.
    b. For data with one-parameter distributions, for example, binomial and Poisson, G-side models add a unit-level effect to the linear predictor with a multivariate normal distribution defined by the covariance function; R-side models embed the covariance model in the quasi-variance as a working correlation.
    c. For non-Gaussian two-parameter distributions, modeling spatial correlation is an open area of research.

7. Radial smoothing is an alternative to parametric covariance models. Again, application is straightforward for Gaussian LMMs and G-side GLMMs with one-parameter distributions. Little work has been done with two-parameter distributions along these lines.

Exercise

This problem uses data provided in SAS file `Ch_15_Problem1_12x10_Grid.sas`.

The data are from a survey conducted on a 12×10 grid. Latitude (or north–south location) is referred to as "northing" and longitude (or east–west location) is referred to as "easting" in the data set. Each of the 120 locations has been treated with one of four experimental treatments designed to decrease the number of "undesirable pest intrusions." Note that this means that the response variable is a *count*.

The SAS file contains the research team's first attempt at analyzing their data. You can see that they have fit a one-way ANOVA model assuming that the natural log of the count has an approximate normal distribution (actually, they used log(count+0.1) so that observations with zero counts were not eliminated from the data set. Their supervisors have told them that this is unacceptable: they must fit a generalized linear model with a legitimate "count distribution" and, if warranted, account for spatial correlation in the data. They heard that you are taking generalized linear statistical modeling, so they have come to you for help. To do this, implement the following four analyses:

1. GLM assuming that the counts have a Poisson distribution
2. GLM assuming that the counts have a negative binomial distribution
3. GEE-type GLM assuming an exponential spatial working covariance model
4. GLMM assuming a exponential spatial correlation among the observations

Note that the fixed component of all models is $\eta_i = \eta + \tau_i$, where τ_i denotes the ith treatment effect. Also, assume that the choice of exponential correlation/covariance model is based on knowledge of the discipline and does not need to be second-guessed.

Compare and contrast the results of these analyses, including the research team's first attempt, focusing on the following:

- Adequacy of model fit using appropriate chi-square-based goodness of fit statistics, information criteria, and relevant information in the iteration history
- For the two spatial analyses, evidence (or lack thereof) of spatial correlation—this could tell researchers something about how "natural pest combatants" behave or spread
- Mean count estimates and standard errors *on the data scale*
- Ranking of treatment means and significant differences among means; given the criteria stated previously, which treatment(s) would be considered "best"

On which analysis would you recommend that the team base its final report? Why?

16

Power, Sample Size, and Planning

16.1 Basics of GLMM-Based Power and Precision Analysis

Throughout this textbook, we have seen the intimate connection between modeling and design principles. The study design and response variable distribution define the processes a model should describe. Many, perhaps most, "modeling problems" are, at least in part, design problems—either a flawed design or faulty application of design principles to model construction. Usually, we have thought of this as a one-direction process—how do we write a plausible model given the description of the design? In this chapter, we reverse the process. Given a set of objectives and a model that will allow us to address these objectives, how can we use generalized linear mixed model (GLMM) theory and methods to help determine an adequate study design?

At first, a power and planning chapter may seem out of place in a modeling book. Is not modeling about analysis—estimation and inference—something we do *after* data collection? Does not design belong in a *design* textbook, not a *modeling* textbook? However, if anything should be clear from the first 15 chapters, it is that design and modeling are not two separate topics but rather two sides of the same coin. With the GLMM, we have vastly more sophisticated modeling tools than we did a generation ago when ordinary least squares defined the state of the art. However, no matter how sophisticated the model, it cannot overcome inadequate data or a badly conceived study design. If we are going to realize the GLMM's potential, we need appropriately designed studies.

More importantly, conventional wisdom about design derives largely from ordinary least squares thinking. Design *principles* have not changed, but viewing them through the GLMM lens provides a much clearer look at design requirements. This raises three overriding concerns that provide the main motivation for this chapter:

1. *Power* and *sample size* are misnomers. The same sample size can yield very different power characteristics depending on *how* we deploy the sample size through the study design. GLMM-based power analysis allows us to accurately assess competing designs. With conventional power analysis, opportunities to greatly improve the accuracy or efficiency of a design as well as critical design flaws often go unnoticed.
2. Design strategies that make sense for Gaussian data are often inappropriate—even catastrophically inappropriate—for non-Gaussian data.
3. Correlated errors affect power and precision. Using GLMM-based power and precision analysis, it *is* possible to take correlated errors into account at planning time. Moreover, failure to do so leads to inaccurate conclusions regarding design requirements.

The main objectives of this chapter, then, are to present the essentials of GLMM-based power and precision analysis, to show how to employ them to assess competing designs, and to tailor designs to the requirements of the primary response variable and the anticipated covariance structure.

16.1.1 Essential GLMM Theory for Power and Precision Analysis

GLMM-based power and precision analysis rely on applying inference results for estimable functions developed in Chapter 5. In this context, *power* = *P*{reject H_0: $\mathbf{K'\beta} = \boldsymbol{\psi}_0 \mid H_0$ false}. "H_0 false" usually means "treatment effect exists" or "explanatory variable predicts response." Usually we do a study because we suspect a treatment effect does exist or that response can be predicted by the explanatory variable. Therefore, we can understand power as the likelihood of being able to declare our suspected treatment effect or predictor to be statistically significant when the effect does in fact exist.

Precision analysis refers to estimable functions with vector **k**. The general form of the confidence interval is $\mathbf{k}'\hat{\boldsymbol{\beta}} \pm (t \text{ or } Z)_{\alpha/2} \times s.e.(\mathbf{k}'\hat{\boldsymbol{\beta}})$: assessing precision means finding the expected width of the confidence interval for a proposed design.

The key results for power and precision analysis are (5.18) and (5.19). That is,

- For scalar **L**, $t = (\hat{\psi} - \psi_0)/\sqrt{\mathbf{L}'\hat{\mathbf{C}}\mathbf{L}} \sim$ approximately t_{ν_2}

- $\mathbf{L}'\begin{bmatrix}\hat{\boldsymbol{\beta}} \\ (\hat{\mathbf{b}} - \mathbf{b})\end{bmatrix}\left(\mathbf{L}'\hat{\mathbf{C}}\mathbf{L}\right)^{-1}\begin{bmatrix}\hat{\boldsymbol{\beta}}' \\ (\hat{\mathbf{b}} - \mathbf{b})'\end{bmatrix}'\mathbf{L} \Big/ \text{rank}(\mathbf{L}) \sim$ approximately $F_{\nu_1,\nu_2,\varphi}$

where $\boldsymbol{\psi} = \mathbf{K'\beta} + \mathbf{M'b}$, $\mathbf{L}' = [\mathbf{K}' \; \mathbf{M}']$, **C** is the generalized inverse of the left-hand side of the PL Estimating Equations (4.35) and φ denotes the non-centrality parameter. In power and precision analysis, we usually set $\mathbf{M} = \mathbf{0}$ and focus on $\mathbf{K'\beta}$. When we do, the non-centrality parameter reduces to

$$\varphi = (\mathbf{K'\beta} - \boldsymbol{\psi})'\left[\mathbf{K}'(\mathbf{X'WX})^{-}\mathbf{K}\right]^{-1}(\mathbf{K'\beta} - \boldsymbol{\psi}) \tag{16.1}$$

where

$$\mathbf{W} = (\mathbf{DVD})^{-1} = \left(\mathbf{D}\mathbf{V}_\mu^{1/2}\mathbf{A}\mathbf{V}_\mu^{1/2}\mathbf{D}\right)^{-1} \tag{16.2}$$

D, $\mathbf{V}_\mu^{1/2}$, and **A** were defined in Chapter 4. For GEE-type models, we use the working covariance matrix, $\mathbf{A}_W$, in place of **A**.

We use (16.1) to implement power analysis as follows. Under H_0, we know $\varphi = 0$, meaning the *F*-statistic has a central *F*-distribution. When H_0 is not true, $\varphi > 0$. By inspection, we can see several ways to affect the non-centrality parameter:

- The treatment effect, $\mathbf{K'\beta} - \boldsymbol{\psi}$
- The design (determines **X** and affects **V** through **Z**)
- The sample size (determines the number of rows of **X** and **Z**)

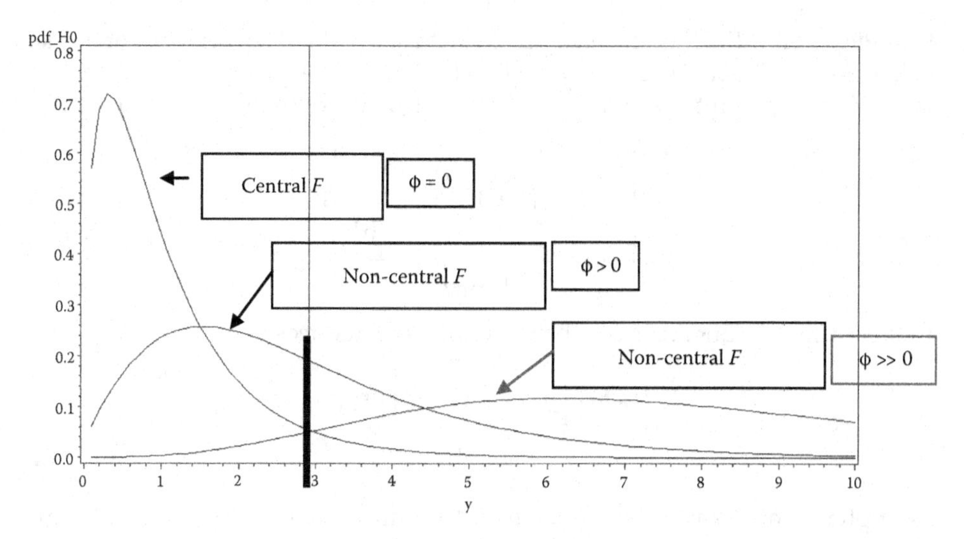

FIGURE 16.1
Central and noncentral F-distributions in an F-test.

- The variance, $\mathbf{V}$
- The distribution (determines $\mathbf{V}_{\mu}^{1/2}$ and $\mathbf{D}$)
- The covariance or working covariance structure (determines $\mathbf{A}$ or $\mathbf{A}_W$)

All power analyses require us to specify the design we wish to evaluate, treatment effect, the components of variances, and some reasonable idea of their magnitude. By "treatment effect" we usually mean the minimum treatment effect considered scientifically important—"clinically relevant" to borrow terminology from pharmaceutical statistics. Figure 16.1 illustrates how we assess power.

Notice that as φ increases, the noncentral F's probability mass shifts to the right. The solid black vertical line (in this case just below three on the horizontal axis) is the "critical value" of F, that is, the value of F above which we reject H_0 for specified α. Thus, the area to the right of the critical value under the noncentral F gives us the power of the test.

16.1.2 Using SAS PROC GLIMMIX to Implement a Power Analysis

We can implement GLMM-based power analysis in SAS using a three step process. This process first appeared applied to LMs in Littell (1980), Lohr and O'Brien (1984), and O'Brien and Lohr (1984). Stroup (1999, 2002) extended the process to linear mixed models (LMMs). Littell et al. (2006) described this process for LMMs using PROC MIXED. In this chapter, we extend these methods to the GLMM using PROC GLIMMIX. The steps are as follows:

1. Create an *exemplary* data set. That is, create a data set whose structure is identical to the data set that would be used to analyze the data if we use the design under evaluation. The only difference between the exemplary data set and the "live" data set is that we use expected values instead of actual data. Denoting the expected value, $E(y|b)$ for the ith treatment or predictor variable level by μ_i, we choose values of μ_i to reflect the minimum differences considered important (e.g., "clinically relevant") in the context of the study.

2. Run PROC GLIMMIX on the exemplary data set with all covariance components held constant. We will see how to do this in the following examples. For each estimable function of interest, GLIMMIX computes the approximate *F*-statistic

$$\frac{\mathbf{L}'\begin{bmatrix}\hat{\boldsymbol{\beta}}' \\ (\hat{\mathbf{b}}-\mathbf{b})\end{bmatrix}\left(\mathbf{L}'\hat{\mathbf{C}}\mathbf{L}\right)^{-1}\begin{bmatrix}\hat{\boldsymbol{\beta}}' \\ (\hat{\mathbf{b}}-\mathbf{b})'\end{bmatrix}\mathbf{L}}{\operatorname{rank}(\mathbf{L})}$$

In the estimable function case, the approximate *F* reduces to

$$\frac{\left\{(\mathbf{K}'\boldsymbol{\beta})'\left[\mathbf{K}'(\mathbf{X}'\mathbf{W}\mathbf{X})^{-}\mathbf{K}\right]^{-1}\mathbf{K}'\boldsymbol{\beta}\right\}}{\operatorname{rank}(\mathbf{K})}$$

Multiplying this term by rank(**K**), that is, the numerator degrees of freedom, gives us the non-centrality parameter under the proposed design, best-guess covariance components, and minimum important ("clinically relevant") treatment/predictor effect.

For precision analysis, GLIMMIX computes the standard errors for all estimable functions of interest using the ESTIMATE, LSMESTIMATE, or LSMEANS statements.

3. Use SAS probability functions to determine the critical value of *F* and the area to the right of the critical value under the noncentral *F* for each $\mathbf{K}'\boldsymbol{\beta}$ of interest.

Example 16.1

Suppose we have three treatments, a "control" and two experimental treatments. Denote them TRT 0, 1, and 2, respectively. Suppose that the control's mean is known to be $\mu_0 = 20$ and suppose that the two experimental treatments have been designed with the objective of increasing the mean response to at least 25. In other words, $\mu_i - \mu_0 = 5$; $i = 1, 2$ is the target minimum difference. Suppose that the response variable is known to be Gaussian and from past experience $\sigma^2 \cong 9$. The research team wants to know if a completely randomized design with five observations per treatment will provide adequate power. Here is how we use GLMM-based power analysis to answer this question:

Step 1. Create the exemplary data set

Use the SAS statements

```
data power_example_1;
  input trt mu;
  do obs=1 to 5;
    output;
  end;
datalines;
0 20
1 25
2 25
;
```

We use the variable MU to set the three treatment means: $\mu_0 = 20$ and $\mu_1 = \mu_2 = 25$. In many power analyses, we might set μ_2 to some value other than 25 to see how changing μ_i affects power. Here, let us use these values. The statement DO OBS=1 TO 5 in conjunction with the OUTPUT statement creates the five replicate observations. The exemplary data set is a completely randomized design unless we explicitly define a blocking structure.

Step 2. The PROC GLIMMIX step

Use the statements

```
proc glimmix data=power_example_1;
  class trt;
  model mu=trt;
  parms (9)/hold=1;
  contrast 'control vs experimental' trt 2 -1 -1;
  contrast 'control vs exp 1' trt 1 -1 0;
  contrast 'control vs exp 2' trt 1 0 -1;
  contrast 'exp 1 vs exp 2' trt 0 1 -1;
  lsmeans trt/diff cl;
  ods output tests3=F_overall contrasts=F_contrasts;
```

Note the response variable MU. The PARMS statement sets the variance. In this case, we only have one variance—the default distribution is GAUSSIAN and there is no RANDOM statement, so the sole variance component must be σ^2. The value (9) sets σ^2 and HOLD=1 overrides the default REML estimation process and fixes $\sigma^2=9$. The CONTRAST statements define comparisons *possibly* of interest. They are here for demonstration purposes; it is unlikely that you would use all of them in a real-life power assessment. The ODS OUTPUT statement creates new data sets with the *F*-values for the contrasts and the type III tests of overall TRT effect. We will use these data sets in Step 3.

The LSMEANS statement gives standard errors and confidence limits for treatment means and differences. Use these for an analysis of precision. We could add additional ESTIMATE and LSMESTIMATE statements if desired.

Step 3. Compute power

Use the statements

```
data power;
  set F_overall F_contrasts;
  nc_parm=numdf*Fvalue;
  alpha=0.05;
  F_Crit=Finv(1-alpha,numdf,dendf,0);
  Power=1-probF(F_crit,numdf,dendf,nc_parm);
proc print data=power;
  run;
```

The DATA and SET statements create a new data set combining the contrast and type III test information. NC_PARM defines the non-centrality parameter and ALPHA defines the α-level. FINV and PROBF are SAS probability functions to evaluate the *F*-distribution. Once these statements have been executed, we print the results. Here, we have

Obs	Effect	NumDF	DenDF	FValue	ProbF	Label	nc_parm	alpha	F_Crit	Power
1	trt	2	12	4.63	0.0323		9.25926	0.05	3.88529	0.66605
2		1	12	9.26	0.0102	control vs experimental	9.25926	0.05	4.74723	0.79750
3		1	12	6.94	0.0218	control vs exp 1	6.94444	0.05	4.74723	0.67750
4		1	12	6.94	0.0218	control vs exp 2	6.94444	0.05	4.74723	0.67750
5		1	12	0.00	1.0000	exp 1 vs exp 2	0.00000	0.05	4.74723	0.05000

The last column gives the POWER for each test. Often, research teams use 0.8 as the minimum power targeted for the design. By this criterion, the design is adequate (just barely) for the CONTROL VS EXPERIMENTAL contrast, but not for anything else. Notice the power for EXP 1 VS EXP 2. The means $\mu_1 = \mu_2$ in the exemplary data set, so the number we get here is the power when the means are equal, that is, when H_0 is true. By definition, this is α.

To determine how many observations per treatment are required to obtain power = 0.8, we change the DO OBS statement and rerun all three steps. For DO OBS = 1 TO 7, we get

Obs	Effect	NumDF	DenDF	FValue	ProbF	Label	nc_parm	alpha	F_Crit	Power
1	trt	2	18	6.48	0.0076		12.9630	0.05	3.55456	0.84969
2		1	18	12.96	0.0020	control vs experi-mental	12.9630	0.05	4.41387	0.92543
3		1	18	9.72	0.0059	control vs exp 1	9.7222	0.05	4.41387	0.83845
4		1	18	9.72	0.0059	control vs exp 2	9.7222	0.05	4.41387	0.83845
5		1	18	0.00	1.0000	exp 1 vs exp 2	0.0000	0.05	4.41387	0.05000

In this way, we determine that seven observations per treatment are required for this study to achieve its objective with *power* ≥ 0.8.

Example 16.2

Consider the last example but suppose the response variable is binomial, not Gaussian. We will complicate this example later in the chapter, but, for now, let us just say that the "control" treatment has a known probability of a favorable outcome of $\pi_0 = 0.15$ and we have reason to believe that the experimental treatments can improve this probability at least to $\pi_i = 0.25$ and perhaps as much as $\pi_i = 0.35$. We would like to know how many subjects per treatment we have to observe for a study to have adequate power.

We need to modify the programs as follows:

1. The exemplary data set needs to define π and N instead of μ.
2. If we have a straight binomial GLM with no random effects, we have a known scale parameter and, hence, an χ^2 rather than F-statistic for testing H_0.

The resulting SAS statements are

Step 1

```
data power_example_1;
    input trt p;
    N=250;
    mu=N*p;
  datalines;
  0 0.15
  1 0.25
  2 0.35
  ;
```

Note that N replaces OBS in determining replication for the binomial GLM. *This will change* as we consider more realistic scenarios for binomial data in Section 16.3.

Step 2

```
proc glimmix data=power_example_1;
  class trt;
  model mu/N=trt/chisq;
  contrast 'control vs experimental' trt 2 -1 -1/chisq;
  contrast 'control vs exp 1' trt 1 -1 0/chisq;
  contrast 'control vs exp 2' trt 1 0 -1/chisq;
  contrast 'exp 1 vs exp 2' trt 0 1 -1/chisq;
  lsmeans trt/diff cl;
  ods output tests3=F_overall contrasts=F_contrasts;
```

Notice that we use MU as the binomial response in defining the MODEL statement. Also, notice the use of CHISQ options in the MODEL and CONTRAST statements. The program leaves the output data set names alone; we could change the names, say, to CHISQ_OVERALL and CHISQ_CONTRASTS, but it would be for aesthetic reasons only. The names we have here are fine as long as we know what they refer to.

Step 3

```
data power;
  set F_overall F_contrasts;
  nc_parm=numdf*Fvalue;
  alpha=0.05;
  F_Crit=Cinv(1-alpha,numdf,0);
  Power=1-probchi(F_crit,numdf,nc_parm);
proc print data=power;
run;
```

Notice the use of CINV and PROBCHI here. This is because we are evaluating test statistics with χ^2, not F distributions.

The results are as follows:

Obs	Effect	NumDF	DenDF	ChiSq	FValue	Label	nc_parm	alpha	F_Crit	Power
1	trt	2	0	25.61	12.81		25.6119	0.05	5.99146	0.99706
2		1	0	18.66	18.66	control vs experimental	18.6607	0.05	3.84146	0.99086
3		1	0	7.67	7.67	control vs exp 1	7.6743	0.05	3.84146	0.79111
4		1	0	25.42	25.42	control vs exp 2	25.4209	0.05	3.84146	0.99897
5		1	0	5.91	5.91	exp 1 vs exp 2	5.9100	0.05	3.84146	0.68121

Notice the DenDF output, equal to 0. GLIMMIX will compute F-values even for GLMs with known scale parameters, but it has no denominator degrees of freedom on which to evaluate these statistics. This is why we invoke the CHISQ option. Note that this is the *only* time we use χ^2 in GLMM-based power analysis.

Question for Students: Both the CONTROL VS EXP 1 and EXP 1 VS EXP 2 contrasts have probabilities that differ by 0.10. That is, $\pi_1 - \pi_0 = 0.25 - 0.15 = \pi_2 - \pi_1 = 0.35 - 0.25 = 0.10$. Why, then, does the power differ for these two contrasts? How could you vary the design to equalize them?

16.2 Gaussian Example

In Chapter 2, Figure 2.5, we first encountered a design with 10 blocks each of size 3 and 6 treatments. We revisited this design in Chapter 8. We have referred to it variously as a disconnected design, a nested factorial design, and a split-plot.

If we take a step back, the topographical layout (10 blocks with 3 units each) and the treatment layout (6 treatments) provide a setting to demonstrate how power analysis should be used. In addition, this example illustrates what GLMM-based power analysis can provide that conventional, pre-GLMM-based power analysis cannot. Specifically, we have a total sample size of 30 experimental units available to us and we need to allocate them among six treatments. Depending on the design we use, we can drastically alter the power characteristics for various estimable functions of possible interest. There are two take-home messages you should retain from this example:

- *To statistical scientists*: The *earlier* you can be involved in *planning* a research study, the more beneficial your impact will be.
- *To consumers of statistical methods*: If you see the role of power analysis as *merely* to calculate the needed sample size *after* you have determined the design, you have *forfeited* most of the value of statistics in planning your research.

This said, let us begin the example.

There are a number of ways we could allocate 6 treatments to our 30 units of observation. Four common ways are

Block	Balanced Incomplete Block (BIB)			Control vs Experimental (CVT)			Nested Factorial (DNF)			"Rep"	Randomized Complete Block		
1	0	1	2	0	1	2	0	1	2	1	0	1	2
2	0	1	3	0	1	3	3	4	5		3	4	5
3	0	2	4	0	1	4	0	1	2	2	0	1	2
4	0	3	5	0	1	5	3	4	5		3	4	5
5	0	4	5	0	2	3	0	1	2	3	0	1	2
6	1	2	5	0	2	4	3	4	5		3	4	5
7	1	3	4	0	2	5	0	1	2	4	0	1	2
8	1	4	5	0	3	4	3	4	5		3	4	5
9	2	3	4	0	3	5	0	1	2	5	0	1	2
10	2	3	5	0	4	5	3	4	5		3	4	5

The first is a balanced incomplete block, hereafter referred to as a BIB design. Technically this is an unbalanced design in that sense that if you fit treatment first, then block, you get different sequential (type I) hypotheses of treatment effect than if you fit block then treatment. Refer to our Chapter 7 discussion of sequential hypotheses. We call the BIB "balanced" because every pair of treatments is observed together in the same block an equal number of times (in this case, 2) and the standard errors for all pair wise treatment comparisons are equal.

The second design makes sense if treatment 0 is a "control" or reference treatment. Hereafter, we refer to the design as CVT. We observe the control in every block. This allows us direct comparisons between the control and each of the other treatments in four blocks, but only pairs each of the other treatments in the same block once.

The third design is our disconnected/nested factorial (depending on how you look at it) design from Figure 2.5. Hereafter, we refer to this design as DNF. The blocks are rearranged from Figure 2.5 to emphasize its connection with the fourth and final design.

The final design is a randomized complete block, hereafter referred to as the RCB. We get this design by taking the third design, the DNF, and, in essence, disregarding the natural block size in order to get complete blocks. We do this to emphasize a point. The role of blocking in design of experiments is to create subsets of experimental units that are internally homogeneous. The idea is to minimize the observational unit variance—the variance among units within a block—in order to maximize precision. Blocking is most effective when the block size corresponds to natural variation. However, many researchers practice "convenience blocking," that is, letting the number of treatments determine the block size regardless of the actual variability. Often researchers do this because they do not know any better—design education often ends with the RCB design. Also, the way efficiency is taught in ordinary least squares driven design courses, it is easy to get the mistaken idea that complete block designs are always more efficient (they are not, as we shall see).

We can now apply power analysis to each of these four designs. For each design, we create an exemplary data set according to the design description given earlier. Suppose that we do this power analysis for the following treatment means:

Set	**Treatment: μ_{Trt}**		
1	0: 20	1: 18.75	2: 17.5
2	3: 20	4: 20	5: 20

In addition, suppose that, *assuming the natural block size is respected*, the variance components are $\sigma_B^2 = 0.5$ for block and $\sigma_U^2 = 1$ for units within blocks. These variance components apply to the BIB, CVT, and DNF design.

For the RCB, we need to adjust the variance components to take into account the fact the natural block sizes were disregarded in forming this design. To do this, denote the linear predictor for the BIB, CVT, and DNF designs by

$$\eta_{ij} = \eta + \tau_i + b_j$$

and the linear predictor for the RCB by

$$\eta_{ik} = \eta + \tau_i + r_k$$

The covariance matrix is given by $Var(\mathbf{y}) = \mathbf{V} = \mathbf{Z}_b\mathbf{Z}_b'\sigma_B^2 + \mathbf{I}\sigma_U^2$, where $\mathbf{Z}_b$ denotes the design matrix associated with the b_j. Now, for the RCB, its block variance, $Var(r_k) = \sigma_R^2$, can be determined by taking the expected value of the quadratic form that isolates σ_R^2. Here, we exploit a trick from the analysis of variance.

$$\sigma_R^2 = \frac{1}{6}\left[EMS(blk) - EMS(error)\right]$$

$$EMS(blk) = \text{trace}\left(\left\{\left[\mathbf{X}(\mathbf{X}'\mathbf{X})^{-}\mathbf{X}' - \mathbf{X}_T(\mathbf{X}_T'\mathbf{X}_T)^{-}\mathbf{X}_T'\right]/\text{rank}(\mathbf{X}_{blk})\right\}\mathbf{V}\right)$$

$$EMS(error) = \text{trace}\left(\left\{\left[\mathbf{I} - \mathbf{X}(\mathbf{X}'\mathbf{X})^{-}\mathbf{X}'\right]/\left[N - \text{rank}(X)\right]\right\}\mathbf{V}\right)$$

where

$\mathbf{X}_T$ denotes the fixed effects design matrix with columns corresponding to η and the τ_i

$\mathbf{X}_{Blk}$ denotes the design matrix for complete blocks, corresponding to the r_k and $\mathbf{X} = \begin{bmatrix}\mathbf{X}_T & \mathbf{X}_{Blk}\end{bmatrix}$

Here, when $\sigma_B^2 = 0.5$ and $\sigma_U^2 = 1$, the block and error variance for the RCB are $\sigma_R^2 = 0.2$ and $\sigma_E^2 = 1.3$, respectively.

For the BIB, the statements for the exemplary data set and the GLIMMIX step are

```
/* Step 1: create exemplary data set */
data bib;
   input block @@;
    do plot=1 to 3;
      input trt @@;
      mu=20-1.25*((trt=1)+2*(trt=2)); output;
    end;
datalines;
  1 0 1 2
  2 0 1 3
  3 0 2 4
  4 0 3 5
  5 0 4 5
  6 1 2 5
  7 1 3 4
  8 1 4 5
  9 2 3 4
 10 2 3 5
 ;

/* Step 2: GLIMMIX step */
proc glimmix data=bib;
  class block trt;
  model mu=trt;
  random intercept /subject=block;
  parms (0.5)(1)/hold=1,2;
  lsmeans trt/diff;
  contrast 'Set x 0 v 2' trt 1 0 -1 -1 0 1;
  contrast 'Trt_0 v Trt_1' trt 1 -1 0;
  contrast 'Trt_0 v Trt_2' trt 1 0 -1 0;
  contrast 'Trt_1 v Trt_4' trt 0 1 0 0 -1 0;
  contrast 'Trt_2 v Trt_5' trt 0 0 1 0 0 -1;
  ods output contrasts=c;
```

The contrasts here represent one possible set of comparisons that might be of interest. If we take SET as defined by the DNF design, we might want to compare treatments 1.25 and

2.5 units different within the same set (treatment 0 versus treatment 1 and 2, respectively) and across sets (treatment 1 versus treatment 4 and treatment 2 versus treatment 5, respectively). We might also want to see if the simple effect of the first and last treatments within each set differs, that is, is there a "SET by level 0 versus 2 within each set" interaction.

The statements for the other two incomplete block designs, CVT and DNF, will differ only by substituting the block–treatment design structure after the DATALINES statement. The RCB program statements will be different:

```
/* Step 1: create exemplary data set */
data rcb;
  do rep=1 to 5;
    do trt=0 to 5;
      mu=20-1.25*((trt=1)+2*(trt=2)); /* MU is value to response diffs to
      be assesed in power analysis */
      output;
    end;
  end;

/* Step 2: GLIMMIX step */
proc glimmix data=rcb noprofile;
  class rep trt;
  model mu=trt;
  random intercept /subject=rep;
  parms (0.20)(1.299)/hold=1,2;
  contrast 'Set x 0 v 2' trt 1 0 -1 -1 0 1;
  contrast 'Trt_0 v Trt_1' trt 1 -1 0;
  contrast 'Trt_0 v Trt_2' trt 1 0 -1 0;
  contrast 'Trt_1 v Trt_4' trt 0 1 0 0 -1 0;
  contrast 'Trt_2 v Trt_5' trt 0 0 1 0 0 -1;
ods output contrasts=c;
```

Notice that the variance component σ_E^2 has been set to 1.299 instead of 1.3. Occasionally, when you hold the variance components constant, GLIMMIX will give you an error message "`ERROR: Values given in PARMS statement are not feasible.`" The error is spurious, an artifact of GLIMMIX's internal architecture when the variance components are held constant. The "fix" is to perturb one or both of the variance components slightly. In this case, resetting σ_E^2 from 1.3 to 1.299 did the trick. This has minimal impact on power. With each new release, GLIMMIX is less susceptible to this problem, but it does occasionally happen.

Running the power program for each design, we can now assess power and compare designs. Figure 16.2 shows the results.

We see that, as expected, the CVT design does well for comparisons of treatment 0 versus others at the expense of comparison not involving treatment 0. The DNF design does well with the interaction and with comparisons within SET but it does so at the expense of simple effects across SET. The BIB, by keeping pair wise precision equal, does better across SET but at the expense of within SET and interaction comparisons. Because of within-block heterogeneity, the RCB does not compete well at all. Convenience blockers sit up and take notice!

We could examine the impact of a change in the variance among blocks. For example, suppose $\sigma_U^2 = 1$ does not change, but the variance among natural blocks of size 3 does, for example, $\sigma_B^2 = 5$. Then we can recalculate the complete block variance components: they

Obs	Label	Power_BIB	Power_CVT	Power_DNF	Power_RCB
1	Set x 0 v 2	0.68852	0.65889	0.74678	0.64563
2	Trt_0 v Trt_1	0.41127	0.46998	0.45928	0.37881
3	Trt_0 v Trt_2	0.93345	0.96396	0.95961	0.90947
4	Trt_1 v Trt_4	0.41127	0.33239	0.32927	0.37881
5	Trt_2 v Trt_5	0.93345	0.86107	0.85741	0.90947

FIGURE 16.2
Power comparison for 10×3 6 trt designs with $\sigma_B^2 = 0.5$.

Obs	Label	Power_BIB	Power_CVT	Power_DNF	Power_RCB
1	Set x 0 v 2	0.68852	0.62040	0.74678	0.26514
2	Trt_0 v Trt_1	0.41127	0.44611	0.45928	0.15605
3	Trt_0 v Trt_2	0.93345	0.95340	0.95961	0.46871
4	Trt_1 v Trt_4	0.41127	0.30481	0.11822	0.15605
5	Trt_2 v Trt_5	0.93345	0.82400	0.32927	0.46871

FIGURE 16.3
Power comparison for 10×3 6 trt designs with $\sigma_B^2 = 5$.

are now $\sigma_R^2 = 2$ and $\sigma_E^2 = 4$. Notice that as the variance among natural blocks increases, the heterogeneity of "convenience" complete blocks increases drastically. The revised power comparisons appear in Figure 16.3.

The within SET and interaction comparisons for the DNF design, the treatment 0 versus other treatment comparison for the CVT design, and all comparison for the BIB design are at most slightly affected. Increasing the block variance adversely affects all other comparisons. The impact on the RCB is particularly devastating.

One lesson from this exercise should be clear. If a statistical scientist had been brought in at the end, to run a pro forma power analysis once the research team had already decided on a RCB design without any statistical input, we would end up either with an underpowered study or a needlessly costly study if the research team compensates for the RCB's imprecision by simply increasing sample size. Of such sloppiness, intractable research budget deficits are born.

This example also makes clear the importance of the variance—or more to the point, the *variances*—in accurately assessing the relative advantages of competing designs. This raises the question: How do we obtain reasonable values of the variance or variances involved in a proposed design? The answer to this question is as varied as the disciplines that use designed studies to answer questions with data. However, broadly speaking, we can think in terms of four general approaches:

1. *Published results from similar studies*: For example, if scientific publications are doing their job, the estimated variance and covariance components should be documented and easily accessible to readers. When this is being done, those designing future studies have a sound basis for planning. To the extent that journals are not doing a good job along these lines, better documentation of variability represents an opportunity to improve.

2. *Pilot studies*: If one has the luxury of doing a pilot study, one can use the variance estimates as a basis for future planning. This approach has two drawbacks, however. First, pilot studies add time and expense to studies. Second, as we saw in Chapter 6, variance estimates from small studies tend to have relatively wide confidence intervals. Pilot studies therefore mean longer and more expensive studies planned with at best approximate information about the components of variance and covariance.
3. *Institutional memory*: Organizations that are heavily involved in research—for example, agricultural experiment stations, medical schools, and government and private sector research and development entities—conduct many similar studies over long periods of time. Information from previous studies should allow researchers to anticipate the variance of future, similar studies.
4. *Six sigma*: If all else fails, those planning the study can use the fact that with the Gaussian distribution, the probability is 0.99 that an observation randomly drawn from the population will be within ± 3 standard deviations of the mean. Determine an approximate range by asking what are the lowest and highest values of the response variable one will plausibly observe. Then use the fact that $\sigma \cong range/6$ and hence $\sigma^2 \cong (range/6)^2$.

16.3 Power for Binomial GLMMs

In Example 16.2, we introduced power analysis for a simple design with a binomial response. We now look at a more realistic example. Take the same three treatments (0, 1, and 2) and the same target probabilities ($\pi_0 = 0.15$, $\pi_1 = 0.25$, and $\pi_2 = 0.35$) as in Example 16.2, but suppose that the treatments are to be observed in five blocks. These could be different pens in a trial with animals, different clinics in a multisite clinical trial, different schools in educational research, etc. We now have to take variability among blocks into account in our power analysis. This raises two questions. First, will the 250 subjects that gave us adequate power in Example 16.2 still give us the desired power if we divide them evenly among the five blocks? Second, how do we account for variation among blocks and how will this affect power?

We answer the second question first. Recall from our discussions in Chapters 10 and 14 that for binomial data, variance among blocks means variance in the log-odds for logistic GLMMs and variance among the probits for probit GLMMs. The block variance, σ_B^2, refers to variance of the log-odds or probits among blocks averaged over all treatments. The block–treatment variance, σ_{BT}^2, refers to variance among the log-odds-ratios or their probit equivalents, depending on the link function we use.

As with the Gaussian case, the variance of random model effects substantially affects the power and precision of studies with non-Gaussian response variables. As we will see in the next section, the impact of variance components can be at least as dramatic as it is for Gaussian data. Therefore, we need ways to reasonably anticipate variance for planning in the non-Gaussian case. The key reality with non-Gaussian models is that variance is expressed on the model scale but subject matter researchers tend to think on the data scale. Recall that with Gaussian responses, the last resort for approximating variance for power and precision analysis was the "six-sigma" trick, that is, $\sigma^2 \cong (range/6)^2$. For binomial data, the range relevant to the six-sigma approximation is the difference between the maximum and minimum plausible *logit*. Thus, a conversation to elicit a variance for use in a power analysis could begin by asking the researcher what the lowest and highest plausible probability of an outcome of interest at the various locations might be, given that there *will* be variation among locations. Denoting the lowest and highest plausible probability as π_L and π_H, respectively, we then calculate the logits $l_L = \log[\pi_L/(1-\pi_L)]$ and $l_H = \log[\pi_H/(1-\pi_H)]$ and finally the range $l_H - l_L$. Now we can use the six-sigma approximation to obtain the variance on the model scale.

Example 16.3

Let us start by supposing that the log-odds vary among locations, but the variability among the odds ratios is negligible and thus $\sigma^2_{BT} \cong 0$. Suppose we have had the "six-sigma approximation" conversation with the researchers and determined that $\sigma^2_B = 0.1$. We modify the program statements we used for Example 16.1 so that the exemplary data set has five blocks with each treatment receiving 50 subjects per block and we add a RANDOM statement to the GLIMMIX step to account for σ^2_B. The statements for the exemplary data and GLIMMIX steps are

```
data power_example_16C1;
  input trt p;
  N=50;
  mu-N*p;
  do block=1 to 5;
    output;
  end;
datalines;
0 0.15
1 0.25
2 0.35
;

proc glimmix data=power_example_16C1;
  class trt block;
  model mu/N=trt;
  random intercept/subject=block;
  parms (0.1)/hold=1;
  contrast 'control vs experimental' trt 2 -1 -1;
  contrast 'control vs exp 1' trt 1 -1 0;
  contrast 'control vs exp 2' trt 1 0 -1;
  contrast 'exp 1 vs exp 2' trt 0 1 -1;
  ods output tests3=F_overall contrasts=F_contrasts;
```

Since this is now a GLMM, we use approximate *F*-statistics to assess power. The statements for the third step are now identical to those we use for power analysis with Gaussian data:

```
data power;
  set F_overall F_contrasts;
  nc_parm=numdf*Fvalue;
  alpha=0.05;
  F_Crit=Finv(1-alpha,numdf,dendf,0);
  Power_F=1-probF(F_crit,numdf,dendf,nc_parm);
```

The results are as follows:

Obs	Label	Power
1		0.96552
2	control vs experimental	0.96418
3	control vs exp 1	0.68070
4	control vs exp 2	0.99215
5	exp 1 vs exp 2	0.56971

Compare these to the results for 250 subjects per treatment without block distinctions. Recall that the contrast CONTROL VS EXP 1 is our target objective—can an experimental treatment increase the probability of a favorable outcome from 0.15 to 0.25? Without blocking, our power was 0.791. Now, the power is just a little above 0.68. This is why it is critically important to take variance among blocks into account when doing power analysis on multi-block designs.

How would we obtain adequate power with this design? We could increase the number of subjects per block. For example, if we increase N from 50 to 70, the power for the CONTROL VS EXP 1 contrast increases to 0.818. Alternatively, we could increase the number of blocks. If we increase to 6 blocks while keeping subjects per block per treatment at $N=50$, power for the CONTROL VS EXP 1 contrast using the Wald statistic is 0.78. With 7 blocks, the power is 0.852. In general, increasing the number of blocks increases power more efficiently than increasing the number of subject per block. This becomes increasingly true as the magnitude of the variance component increases. Here, $\sigma_B^2 = 0.1$, a relatively modest variance component. Also, this example naively assumes that $\sigma_{BT}^2 = 0$. What happens when it is not?

Example 16.4

What happens when the log-odds-ratio also varies among blocks? We know this happens in multisite trials: Certain treatments can be relatively more effective at certain sites for a variety of reasons. Suppose, for example, that the variance among log-odds-ratios $\sigma_{BT}^2 = 0.05$. We create the exemplary data set exactly as we did in Example 16.3. We modify the GLIMMIX step to include σ_{BT}^2 by revising the RANDOM and PARMS statements. The new statements are

```
random intercept trt/ subject=block;
parms (0.1) (0.05)/ hold=1,2;
```

The results, assuming seven blocks and $N=50$, are as follows:

Obs	Label	Power
1		0.98201
2	control vs experimental	0.97952
3	control vs exp 1	0.72666
4	control vs exp 2	0.99578
5	exp 1 vs exp 2	0.57491

The addition of σ_{BT}^2 reduces the power. Increasing the number of subjects per treatment per block to $N=65$ increases power for the CONTROL VS EXP 1 contrast to 0.801. Alternatively, increasing the number of blocks to eight increases power to 0.792. Increasing N means we need 455 total subjects per treatment; increasing the number of blocks means we need 400 subjects per treatment.

What happens if σ^2_{BT} increases, say to 0.25? Then for seven blocks and $N = 50$, we have

Obs	Label	Power
1		0.77204
2	control vs experimental	0.79705
3	control vs exp 1	0.42539
4	control vs exp 2	0.88026
5	exp 1 vs exp 2	0.28876

Under this scenario, we need 16 blocks with $N = 50$ to achieve power of 0.804. If we stay with seven blocks, even $N = 50,000$ only increases power to just short of 0.59!

Example 16.5

We now introduce a new twist. Suppose as we plan the study, the research team informs us that they want to add one more experimental treatment and, at any given block, at most 200 subjects can participate. Also, we cannot have more than two treatments per block. We encounter such restrictions if blocks correspond to sites and each site has logistical limitations. Borrowing the strategies we explored in Section 16.2, Figure 16.4 shows three design alternatives we might consider.

The first design is a BIB design with each pair of treatments observed together at one site. The second design places the control treatment at every site and allows a direct comparison with each of the experimental treatments at two sites. Call this the "control versus experimental treatment" or CVT design. The last design creates complete blocks by pairing individual sites. As we saw in Section 16.2, this will distort the variance components. Variance among blocks will be less than variance among sites, and the block-by-treatment variance will be greater than the site-by-treatment variance.

For the CVT design, if we assign 50 subjects at each site to the control treatment and 150 subjects to each experimental treatment, we will have 300 subjects total per treatment. For the BIB and RCB, assigning 100 subjects to each treatment at each site will also result in 300 subjects total per treatment.

If we start with the same variance assumptions we used in Example 16.4, we have the following variance components:

Source	BIB and CVT	RCB
Sites/blocks	$\sigma^2_S = 0.1$	$\sigma^2_B = 0.012$
Site × treatment/ block × treatment	$\sigma^2_{ST} = 0.05$	$\sigma^2_{BT} = 0.094$

Site	BIB		CVT		Complete Block		RCB	
1	0	1	0	1	1	1	0	1
2	0	2	0	1		2	2	3
3	0	3	0	2	2	3	0	1
4	1	2	0	2		4	2	3
5	1	3	0	3	3	5	0	1
6	2	3	0	3		6	2	3

FIGURE 16.4
Alternative designs for multisite study with binomial response.

Programming for the RCB only requires modification to the statements used in Example 16.4 for the number of blocks, number of treatments, number of subjects per treatment per block (N), and the variance components. For the BIB and CVT, creating the exemplary data set requires combining elements of the statements used in Section 16.2 with statements for the binomial. For example, here are the statements to create the BIB exemplary data set:

```
data bib;
input loc @@;
n_subj=100;
  do g=1 to 2;
    input trt @@;
      do r=1 to 1; /* R=# multiples of locations used to increase power */
      location=(r-1)*6+loc;
      pi=0.15*(trt=0)+0.25*(trt=1)+0.30*(trt=2)+0.35*(trt=3);
      mu=n_subj*pi;
      output;
    end;
end;
datalines;
1 0 1
2 0 2
3 0 3
4 1 2
5 1 3
6 2 3
;
```

Here, we set $\pi_2=0.30$ and $\pi_3=0.35$. This allows us to assess the target difference, $\pi_0=0.15$ versus $\pi_i=0.25$; $i=1, 2, 3$, using the contrast CONTROL VS EXP 1, but also assess power for a 0.15 unit difference (CONTROL VS EXP 2) and between probabilities of 0.25 and 0.35 (EXP 1 VS EXP 3). The DO loop DO R=1 TO R will allow us to increase the number of sites (called LOC in this program) if it turns out that 6 sites fails to provide adequate power. Note that if we do need to increase replication, we need to do it in multiples of six sites to preserve balance. The exemplary data set statement for CVT are the same except for the statements after DATALINES describing the assignment of treatments to sites, denoted LOC in these statements.

Here are the results for the three designs:

Obs	Label	Power_BIB	Power_CVT	Power_RCB
1	c vs e1	0.29992	0.31127	0.33568
2	c vs e2	0.50426	0.52659	0.57355
3	c vs e3	0.68438	0.71175	0.76916
4	e1 v e3	0.21298	0.14090	0.23040

Here, the inherent efficiency of the RCB more than offsets the increase in σ_{BT}^2 relative to σ_{ST}^2. However, for the contrast C vs E1 (control versus experimental treatment 1), the primary target for power, none of the designs provide adequate power. Increasing the RCB to 8 blocks—16 sites—yields power=0.809 for the C vs E1 contrast. If we increase to 12 blocks—24 sites—then power=0.80 for the E1 V E3 contrast.

What happens if the variance among sites, σ_S^2, increases? We know this increases the penalty for disregarding site-to-site variance when we create complete blocks, so the efficiency of the RCB design should suffer relative to the two incomplete block designs. Suppose $\sigma_S^2 = 0.25$ while σ_{ST}^2 remains 0.05. Then, for the RCB, $\sigma_B^2 = 0.083$ and $\sigma_{BT}^2 = 0.217$. Now, assuming the original six sites, the comparison among the three designs yields

Obs	Label	Power_BIB	Power_CVT	Power_RCB
1	c vs e1	0.28485	0.28824	0.23580
2	c vs e2	0.47977	0.48973	0.40817
3	c vs e3	0.65660	0.67118	0.58289
4	e1 v e3	0.20044	0.12678	0.16175

Notice that power for the RCB suffers relative to the incomplete blocks designs. Under this scenario, the BIB would be the design of choice. We would need to increase replication in multiples of 6 sites in order to keep the study design balanced. Running two sets of six sites gives us *power* = 0.705 for our target C VS E1 contrast. Three sets of six sites gives us power = 0.894 for C VS E1 and *power* = 0.721 for E1 V E3.

The following examples underline two extremely important points:

1. Questions of sample size *cannot* be addressed independently of questions regarding design. For the same sample size, power characteristics can change substantially with different designs.
2. Questions of design *cannot* be addressed without taking *all* variance components into account. A design that is efficient under one set of variance assumptions may be extremely *in*efficient if these assumptions change.

Finally, the power analyses we have worked through in Sections 16.2 and 16.3 are possible *only* with GLMM thinking and GLMM-based power analysis. We can verify using simulation that the power analysis results obtained using the three step process demonstrated here are accurate.

16.4 GLMM-Based Power Analysis for Count Data

Power analysis for other non-Gaussian distributions proceeds along much the same lines as it does for binomial data. We define the exemplary data set in terms of the expected value of the observations on the *data scale* and we specify variance components according to anticipated variance on the *model scale*.

Count data introduces a new issue that we did not have to address with binomial data: What conditional distribution of $\mathbf{y} \mid \mathbf{b}$ do we assume? We have seen that the Poisson and the negative binomial are two prominent and widely used probability models for counting processes. In standard power and sample size analyses, we need some knowledge of the minimum treatment or explanatory variable effect considered important (a.k.a. the minimum clinically relevant difference), sources of random variations, and the magnitude of their variance components. With count data, we add the conditional distribution—do we base power requirements on the Poisson or negative binomial assumption, or on some other distribution suggested by our counting process?

In this section, we compare the 4 designs with 10 blocks, 3 units per block, and 6 treatments we considered in Section 16.2, but here we assume we have count data. The four designs are the BIB, the CVT (control versus the five experimental treatments), the DNF (for disconnected/nested factorial), and the RCB (randomized complete block—formed by pairing the incomplete blocks of the DNF design to create "replications" of size six containing all treatments).

Example 16.6 Power analysis assuming no overdispersion

Let us begin by assuming we have a Poisson process with no overdispersion. The expected counts for this example are

Set	Treatment: λ_i	Treatment: λ_i	Treatment: λ_i
0	0: $\lambda_0 = 5$	1: $\lambda_1 = 10$	2: $\lambda_2 = 15$
1	3: $\lambda_3 = 5$	4: $\lambda_4 = 5$	5: $\lambda_5 = 5$

The λ_i denote the expected counts for each treatment. Implicitly, 5 and 10 count differences represent the minimum considered relevant or important.

A possible scenario might be that Set 0 represents a strain of animal or plant that is susceptible to some pest whereas Set 1 is resistant. The different treatments within each set are additional inoculants to protect the plant or animal from the pest. The resistant stain is thought to do about equally well regardless of treatment, but the susceptible strain shows higher pest counts with some treatments than with others. We might imagine that treatment 0 is the current standard treatment: the susceptible strain in conjunction with an additional treatment that does reduce the pest counts, but only at the cost of other undesirable side effects. We will assume that the same contrasts used in Section 16.2 represent a spectrum of objectives adequate to give us a sense of how the designs compare with respect to power and whether the 10 blocks provide adequate sample size.

We define the exemplary data set exactly as we did for Gaussian data. The *only* difference is the expected counts. We replace the μ_i from the Gaussian examples with the λ_i given earlier. The GLIMMIX step differs in two ways. First, we need to specify the Poisson distribution. Second, we need to specify only one variance component. For a blocked design assuming $\mathbf{y} \mid \mathbf{b} \sim$ Poisson and no overdispersion, the linear predictor is $\log(\lambda_{ij}) = \eta + \tau_i + b_j$ where the block effects are random and we assume b_j i.i.d. $N\left(0, \sigma_B^2\right)$. In this example, suppose $\sigma_B^2 = 0.20$. For the RCB, we have the same distortion of block effect induced by pairing natural blocks to form incomplete blocks. Determining σ_R^2, the replication variance we use in lieu of σ_B^2, is not quite as straightforward as it was in Section 16.2 because the within-block variance comes entirely from the rate parameter of the Poisson. One reasonable way to approximate σ_R^2 is to use simulation. If we generate 1000 simulated experiments according to the DNF design but analyze the data using the RCB model, we should get a reasonable approximation of σ_R^2. In the run we use for this example, $\sigma_R^2 \cong 0.07$.

Here is how the four designs compare:

Obs	Label	Power_BIB	Power_CVT	Power_DNF	Power_RCB
1	Set x 0 v 2	0.73457	0.71935	0.80626	0.81636
2	Trt_0 v Trt_1	0.67688	0.80313	0.75705	0.76786
3	Trt_0 v Trt_2	0.97967	0.99786	0.99372	0.99474
4	Trt_1 v Trt_4	0.68665	0.55980	0.41358	0.76786
5	Trt_2 v Trt_5	0.97967	0.93575	0.80626	0.99474

The designs perform as we would expect. The BIB design's balance across all treatment pairs yields power that depends primarily on the magnitude of the difference being compared. Unlike the Gaussian case, power is not exactly equal for differences of the same magnitude because the contribution of λ_i to within-block variance. The CVT does well for comparisons between treatment 0 and the others, but suffers for comparisons between other treatments. The DNF excels for set-by-treatment interaction and comparisons between treatments within sets, but suffers for comparisons across sets. The RCB shows the same relative strengths as the BIB, except that power is greater in every case. This results from the RCB's greater inherent efficiency relative to the BIB, *all else being equal.*

The RCB *appears* to be the best design to use here. However, appearances can be deceiving. "All else being equal" brings up a problem with the power assessment for the RCB.

In fact, all else is *not* equal with the RCB. By disregarding natural block size, we actually do have within-block heterogeneity with the RCB, but not with the other designs. This means that under the assumptions we have made with this scenario, we did not include a block-by-treatment term in the linear predictor. However, we should include a rep-by-treatment term in the linear predictor to account for within-block heterogeneity. Its linear predictor should be $\log(\lambda_{ik}) = \eta + \tau_i + r_k + (rt)_{ik}$, where $(rt)_{ik}$ i.i.d. $N(0, \sigma_{RT}^2)$. In this scenario, we use simulation as previously described to determine that $\sigma_{RT}^2 \cong 0.06$. The revised power comparison is

Obs	Label	Power_BIB	Power_CVT	Power_DNF	Power_RCB
1	Set x 0 v 2	0.73457	0.71935	0.80626	0.68957
2	Trt_0 v Trt_1	0.67688	0.80313	0.75705	0.62393
3	Trt_0 v Trt_2	0.97967	0.99786	0.99372	0.96365
4	Trt_1 v Trt_4	0.68665	0.55980	0.41358	0.62393
5	Trt_2 v Trt_5	0.97967	0.93575	0.80626	0.96365

Taking the impact of within-block heterogeneity into account, we now see that the RCB is decidedly *not* the design of choice here.

To repeat a point that cannot be made enough, when we plan studies for which the primary response variable is non-Gaussian, power, sample-size, and design assessment *must* use GLMM-based methods. Conventional power and sample size methodology simply misses nuances that have a substantial impact on our results. In an era of tightening research budgets, these considerations become increasingly important.

Example 16.7 Power analysis assuming overdispersion

We know that if we expect Var($\mathbf{y} \mid \mathbf{b}$) to be greater than $\lambda = E(\mathbf{y} \mid \mathbf{b})$, then we know that we cannot use the power analysis we did in the previous section. We have two choices. We can add a block-by-treatment term to the linear predictor, as we did for the RCB design, or we can use a different distribution.

Power analysis including a block-by-treatment term with the Poisson is similar to what we did with the RCB in the previous example and what we did with the binomial examples in Section 16.3. No need to go over old ground. In this example, we illustrate power analysis with the negative binomial.

In practice, how would we know whether to assume negative binomial or Poisson plus block-by-treatment effect? As with any power analysis, the more we know about our response variable, the better. Ideally, previous work with our primary response variable should have shed *some* light on its likely distribution. If we lack background information, one alternative might be to do the power analysis both ways and protect ourselves by going with the worst-case scenario.

For the negative binomial, in addition to the expected counts, we also need the scale parameter. We know that for the negative binomial, $\text{Var}(\mathbf{y} \mid \mathbf{b}) = \lambda(1 + \phi\lambda)$. To do a power analysis, we need to have enough of an idea of the mean–variance relationship to use a plausible value of ϕ. Suppose $\phi = 0.25$. This means that for treatments 0, 3, 4, and 5, the variance should be approximately $5(1 + 0.25 \times 5) = 11.25$; treatment 1 should have an approximate variance of $10(1 + 0.25 \times 10) = 35$; for treatment 2, variance should be roughly $15(1 + 0.25 \times 15) = 71.25$. If this accurately reflects the mean–variance relationship, we can use $\phi = 0.25$ for our power analysis. If not, try other values of ϕ until we find the most plausible.

Let us proceed, assuming $\sigma_B^2 = 0.2$, as before, and $\phi = 0.25$. We can use these values for the BIB, CVT, and DNF designs, but not the RCB. For the RCB, we have to recalibrate the block variance and the scale parameter to account for within-block heterogeneity. Here, using simulation, approximate values are $\sigma_B^2 \cong 0.08$ and $\phi \cong 0.37$. Here is how the four designs compare:

Obs	Label	power_BIB	Power_CVT	Power_DNF	Power_RCB
1	Set x 0 v 2	0.38683	0.36976	0.43204	0.35826
2	Trt_0 v Trt_1	0.33020	0.39439	0.37054	0.30438
3	Trt_0 v Trt_2	0.68808	0.78674	0.74992	0.64423
4	Trt_1 v Trt_4	0.33114	0.26677	0.26508	0.30438
5	Trt_2 v Trt_5	0.68808	0.57620	0.57019	0.64423

The relative between-design comparisons are similar to the Poisson-based power analysis. The striking result here is that power for all comparisons is sharply lower than it was assuming the Poisson with no overdispersion.

This leaves us with one take-home message for power analysis with count data. It is *unwise* to base your assessment of design requirements on the assumption that you have Poisson data with no overdispersion unless you are very, very sure that this assumption is justified. This is one situation for which it is easy to make a catastrophic misjudgment of design requirements.

16.5 Power and Planning for Repeated Measures

The methods illustrated so far in the chapter can be used for power analysis and planning repeated measures studies, at least for Gaussian data or response variables whose distribution belongs to one-parameter exponential family. This includes repeated measures over time with serial correlation as well as studies with spatial variability. Williams (1952) discussed design in the presence of correlated errors a half-century ago. Stroup (2002) discussed LMM-based power analysis for spatial data. We focus in this chapter on methods for one-dimensional repeated measures. Using the same methods we consider here, extending spatial power analysis to non-Gaussian data—at least to the binomial and Poisson—is straightforward.

The more difficult questions concern the two-parameter exponential family, for example, repeated measures data assumed to have a negative binomial or beta distribution. We begin this chapter presenting methods for Gaussian and one-parameter exponential data and end by discussing where we are—and where we have yet to go—with two-parameter exponential family distributions.

16.5.1 Straightforward Cases: Gaussian and One-Parameter Exponential Family

For Gaussian, binomial, and Poisson data, power analysis is no different, in principle, from anything we have considered so far in this chapter. We need an exemplary data set whose structure corresponds to the proposed design and whose time-by-treatment means reflect our assessment of important—a.k.a. "clinically relevant"—treatment effects on change over time. For the GLIMMIX step, we need to specify the type of covariance structure we anticipate will account for within-subject correlation among repeated measurements and we need reasonable values of what we anticipate the covariance parameters will be. With this information, power analysis proceeds like it would for any other factorial treatment design.

As an example, suppose we want to compare a control and two experimental treatments. The initial effect of the experimental treatment is to increase the response (assume greater is better). However, the response wears off over time. Our objectives might include the following:

1. Initially, for example, right after application, do the experimental treatments show a "clinically relevant" difference in response relative to the control?
2. Are the two experimental treatments different from each other initially?
3. Do the two experimental treatments' responses decay at the same rate or does one decay faster than the other?
4. How long does each experimental treatment remain effective relative to the control?

Suppose we initially decide we will do hourly measurements for 8h and the expected pattern of responses is as follows:

	Hour							
Treatment	1	2	3	4	5	6	7	8
0 = "control"	3	3	3	3	3	3	3	3
1	6.5	6	5.5	5	4.5	4	3.5	3
2	5.5	5.5	5.5	5.5	5.5	5.25	4.5	3.5

Suppose from previous experience, we know that within-subject correlation typically follows an AR(1) process with $\rho = 0.6$ and $\sigma_W^2 = 4$.

We structure the exemplary data set to define all of the treatment-by-time means and provide for the ability to vary the number of subjects per treatment so we can explore sample size requirements. Here is one example of statements defining the exemplary data set:

```
data rptM_power;
  input trt @@;
  n_subjects=6;
    do hour=1 to 8;
      input mu @@;
      do subj_id=1 to n_subjects;
        output;
    end;
  end;
```

```
datalines;
0   3   3   3   3   3    3   3   3
1 6.5   6 5.5   5 4.5    4 3.5   3
2 5.5 5.5 5.5 5.5 5.5 5.25 4.5 3.5
;
```

The GLIMMIX step uses the following statements:

```
proc glimmix data=rptM_power noprofile;
  class trt hour subj_id;
  model mu=trt|hour;
  random _residual_/type=ar(1) subject=subj_id;
  parms (4)(0.6)/hold=1,2;
  lsmeans trt*hour/slice=(trt hour);
  ods output tests3=overall_F slices=simple_effects;
run;
```

We could, of course, add CONTRAST statements for more specialized tests. We leave this as an exercise for the reader. The results of this run appear as

Obs	Effect	trt	hour	NumDF	DenDF	F_crit	power
1	trt	.	.	2	10	4.10282	0.68439
2	hour	.	.	7	35	2.28524	0.66794
3	trt*hour	.	.	14	70	1.83568	0.19655
4	trt*hour	0	_	7	70	2.14348	0.05000
5	trt*hour	1	_	7	70	2.14348	0.55563
6	trt*hour	2	_	7	70	2.14348	0.29120
7	trt*hour	_	1	2	70	3.12768	0.79183
8	trt*hour	_	2	2	70	3.12768	0.68714
9	trt*hour	_	3	2	70	3.12768	0.58576
10	trt*hour	_	4	2	70	3.12768	0.50730
11	trt*hour	_	5	2	70	3.12768	0.46512
12	trt*hour	_	6	2	70	3.12768	0.38343
13	trt*hour	_	7	2	70	3.12768	0.19507
14	trt*hour	_	8	2	70	3.12768	0.06857

We see that, in general, this study would be underpowered with only six subjects per treatment. As we have done with other examples, we could change the number of subjects in the exemplary data set and repeat the analysis until we find the number of subjects yielding acceptable power.

We can modify the exemplary data set and the GLIMMIX steps in a number of ways to explore design alternatives. We could, for example, see what would happen if our within-subject covariance model is different. We could change the number of times, or even the spacing of times. If we change the times to unequal spacing, for example, 1, 2, 4, and 8 h, we would need to make sure we also change the covariance model type to one that is appropriate for unequally spaced times, for example, SP(POW).

For the Poisson and binomial, the approach is much the same. In the exemplary data set, we replace means with expected MU/N for binomial data or expected COUNT for Poisson data, just as we have done in Sections 16.3 and 16.4. We can specify the covariance model using either a GEE working correlation structure or a G-side GLMM. For several reasons, the G-side GLMM alternative is better. First, keep in mind that the conditional distribution of $\mathbf{y}|\mathbf{b}$, not $\mathbf{y}$, is binomial or Poisson, so when you specify a binomial probability or a Poisson rate, you are probably thinking in terms of the conditional, not the marginal distribution. Also, the covariance components on the G-side reflect an actual probability distribution affecting the link function on the model scale. Who knows what the working covariance parameters mean? This does not mean you cannot do power analysis using the GEE-type model. It just means that you are much more at the mercy of trusting results others have got with similar data and less able to think through what might be reasonable "guess-timates" of covariance parameters.

16.5.2 On the Frontier: The Two-Parameter Exponential Family

Suppose we plan to do a repeated measures study with count data. Further, suppose that in our discipline, count data that follow a Poisson distribution are rare, but count data that fit the negative binomial are commonplace.

Following the script from Section 16.5.1, the GLIMMIX step for a power analysis for our count data would be as follows:

GEE-type model:

```
proc glimmix initglm;
  class trt time subj_id;
  model mu=trt|time/d=negbin;
  random _residual_/type=ar(1) subject=subj_id(trt);
  parms (1)(0.6)/hold=1,2;
```

GLMM G-side model:

```
proc glimmix initglm;
  class trt time subj_id;
  model mu=trt|time/d=negbin;
  random time/type=ar(1) subject=subj_id(trt);
  parms (0.2)(0.6)/hold=1,2;
```

Now we have a problem. With the binomial and the Poisson, if we use a GEE-type model, setting the working correlation scale parameter to 1, as we have in the PARMS statement here, corresponds to assuming that we have no overdispersion. On the other hand, the variance in the G-side model needs to reflect a reasonable idea of time-to-time variance in the response on the link, or model scale. Either way, with one-parameter exponential distributions, these parameters have an assignable and understandable meaning.

With two-parameter distributions like the negative binomial, however, the covariance model's scale or variance parameter is in direct competition with, and not clearly identifiable from, the scale parameter of the distribution, in this case the negative binomial ϕ. We saw in Chapter 11 that we do not put the last term of the skeleton ANOVA in the linear predictor *and* assume a negative binomial distribution, because the two are not identifiable. In repeated measures, the covariance model is in fact the last term in the skeleton ANOVA with autocorrelation as well as variance. This compounds identifiability problems.

To give an idea of the magnitude of this issue, consider the following power analysis with the negative binomial. Here are the exemplary data and GLIMMIX statements for both a GEE-type and G-side GLMM AR(1) model:

```
data mean;
  input drug @@;
   do hour=1 to 8;
     input mu @@;
      mu=(drug=0)*1.5*mu+(drug>0)*(6*mu-12);
      do patient=1 to 6;
        output;
     end;
   end;
datalines;
0  3    3    3    3    3    3    3    3
1  6.50 6.00 5.50 5.00 4.50 4.00 3.50 3.00
2  5.80 5.80 5.78 5.75 5.65 5.40 4.80 3.25
;

/* GEE-type model */
proc glimmix data=mean initglm;
  class drug hour patient;
  model mu=drug|hour/d=negbin ddf=10,70,70;
  random intercept/subject=patient(drug);
  random _residual_/type=ar(1) subject=patient(drug);
  parms (0.5)(1)(0.6)(0.5)/hold=1,2,3,4;;
  lsmeans drug*hour/slice=(drug hour) ilink;
  ods output tests3=f_overall slices=f_slices;

/* GLMM G-side model */
proc glimmix data=mean initglm;
  class drug hour patient;
  model mu=drug|hour/d=negbin ddf=10,70,70;
  random intercept/subject=patient(drug);
  random hour/type=ar(1) subject=patient(drug) gcorr v;
  parms (0.5)(0.15)(0.6)(0.5)/hold=1,2,3,4;;
  lsmeans drug*hour/slice=(drug hour);
  ods output tests3=f_overall slices=f_slices;
```

Notice the second parameter in the PARMS statement. This is the working correlation scale parameter in the GEE model and the within-subject AR(1) variance component for the G-side model. It makes sense to see it equal to 1 for the GEE, so that when it is multiplied by the negative binomial scale parameter—the last parameter listed in the PARMS statement—we just get the negative binomial scale parameter back. It is not entirely clear *what* we set it to for the GLMM G-side model. Here, we set the G-side variance (second parameter in the PARMS list) to 0.15—a result of trial-and-error to get the GEE- and GLMM-based power for the overall type III tests and slices to equate—which they only do approximately.

After the power analysis, a simulation was done. As we have noted many times in this text, the G-side GLMM actually does describe a probability process that can be simulated, whereas the GEE does not. With this in mind, data were simulated using the parameters given for the G-side GLMM. The simulated data were then analyzed using the two models. For the GLMM G-side model, the Laplace method was used. One run used model-based standard errors and test statistics; the other used empirical, MBN-corrected inferential statistics. The GEE model, of course, had to use PL. KR-model-based and MBN-empirical inferential statistics were computed. The following table summarizes the power for the two power analyses and the observed rejection rates for the two models using the different inferential statistics:

	Power Analysis		GLMM Run		GEE Run	
Effect	**GLMM**	**GEE**	**Model-Based**	**Empirical-MBN**	**KR-Model-Based**	**Empirical-MBN**
Trt	0.718	0.643	0.814	0.464	0.508	0.624
time	0.662	0.728	0.810	0.322	0.487	0.403
trt × time	0.309	0.353	0.507	0.190	0.281	0.204
hour | trt 0	0.050	0.050	0.094	0.052	0.111	0.070
hour | trt 1	0.701	0.671	0.826	0.577	0.692	0.410
hour | trt 2	0.448	0.644	0.555	0.334	0.440	0.208
trt | hour 1	0.750	0.803	0.792	0.415	0.515	0.516
trt | hour 2	0.714	0.769	0.766	0.407	0.484	0.495
trt | hour 3	0.674	0.730	0.729	0.336	0.428	0.456
trt | hour 4	0.632	0.689	0.708	0.324	0.405	0.438
trt | hour 5	0.584	0.641	0.660	0.292	0.371	0.399
trt | hour 6	0.521	0.575	0.596	0.235	0.296	0.339
trt | hour 7	0.403	0.449	0.447	0.178	0.232	0.271
trt | hour 8	0.095	0.101	0.140	0.025	0.036	0.073

On the whole, this is mainly a ringing endorsement for "we still have work to do." The model-based GLMM statistics clearly inflate type I error rates. On the other hand, all of the bias-corrected inferential statistics clearly overcorrect, adversely affecting power.

This suggests that at this point in the development of the GLMM, for count data, we are better off with the Poisson distribution where we are on firmer ground. We know that the covariance structures are simply correlated-error versions of including the skeleton ANOVA's last term in the linear predictor. Furthermore, we know that this is an effective way to account for overdispersion with count data.

Unfortunately, while this may provide a workable interim answer for count data, it does not address continuous rate (beta), time-to-event (exponential), and other kinds of data whose distributions belong to the exponential family and have nontrivial scale parameters. Obviously, assuming normality and returning to pre-GLMM practices is demonstrably the wrong answer. This has been a running theme throughout this textbook. So we end at the frontier. The GLMM is still a very active area of research. This is one of the frontier areas.

16.6 Summary

1. The most important contributions statistical science makes to research occur *before* data collection.

2. If anything, the greater sophistication and flexibility of GLMMs make design issues even more exacting than they are with Gaussian data and conventional LMs.
3. As we have seen in the first 15 chapters of this textbook, the serious problems that can occur with GLMMs are disproportionately likely to occur with underpowered studies.
4. Conventional wisdom about design has largely accumulated in association with requirements for Gaussian data. While the principles remain the same, the particulars can turn out very differently for non-Gaussian data.
5. Power and precision depend on the non-centrality parameter of the t, χ^2, or F distribution depending on the application. These in turn depend on
    a. The proposed design
    b. The number of units at the various levels of the design
    c. The covariance parameters
    d. The minimum relevant difference associated with treatment comparisons of interest
    e. The conditional distribution of $\mathbf{y}|\mathbf{b}$
6. The probability distribution method uses GLMM theory to assess power via the following steps:
    a. Create the exemplary data
    b. Run GLIMMIX
    c. Compute non-centrality parameters from GLIMMIX output
    d. Determine power using probability functions
7. What we know about the probability distribution method's accuracy is stated as follows:
    a. Gaussian LMMs: known to be accurate
    b. One-parameter (binomial and Poisson) distributions: appears to be accurate for G-side GLMMs, including repeated measures
    c. Two-parameter distributions:
        i. Appears to be accurate for G-side non-repeated measures GLMMs with negative Binomial, beta and gamma.
        ii. Repeated measures GLMMs for two-parameter non-Gaussian distributions are an issue.
    d. When in doubt, verify results of probability distribution method by simulation
8. Planning for R-side modeling:
    a. In principle, probability distribution method can be used.
    b. Verification an issue: since R-side models are quasi-likelihood, by definition you cannot simulate data using the R-side model you want to verify.

Exercises

16.1 A researcher wants to compare 6 treatments (a 3×2 factorial). The experiment is to be conducted on a 2×15 grid of experimental units. The gradient runs left–right—that is, experimental units in different rows in the same column tend to be homogeneous, whereas units in different columns tend to be heterogeneous. Treatment combination A_1B_1 is considered the "control" and the objective is to compare each treatment against the control. The following designs are under consideration:

a.

that is, complete blocks arranged as shown previously. If you do this, the variance among blocks is expected to be $\sigma_B^2 = 10$ and the variance within blocks is expected to be $\sigma_e^2 = 4$.

b.

that is, five complete blocks as shown previously. Within each complete block, assign a level of factor A to exactly one column per block. Within each column, assign a level of B to the upper row and the other to the lower row. If you do this, the expected variance among blocks is $\sigma_B^2 = 10$, the variance among columns within a block is expected to be $\sigma_{C(B)}^2 = 3.75$, and the variance among rows within a column is expected to be $\sigma_e^2 = 1.1$.

c.

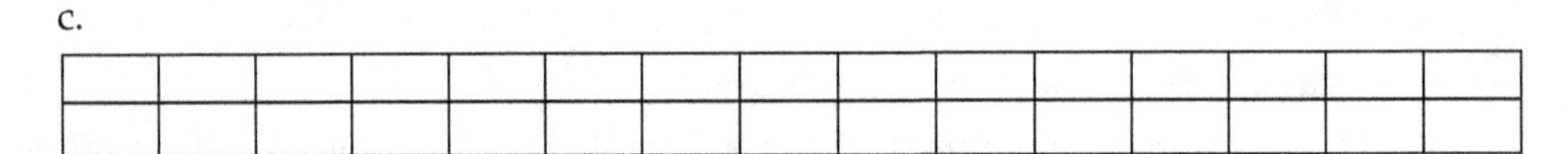

that is, 15 incomplete blocks, each with a pair of treatments. (Each block should have exactly one of the $\binom{6}{2}$ possible pairs of treatments.) If you do this, the variance among blocks is expected to be $\sigma_B^2 = 9$ and the variance within blocks is expected to be $\sigma_e^2 = 1.1$.

d.

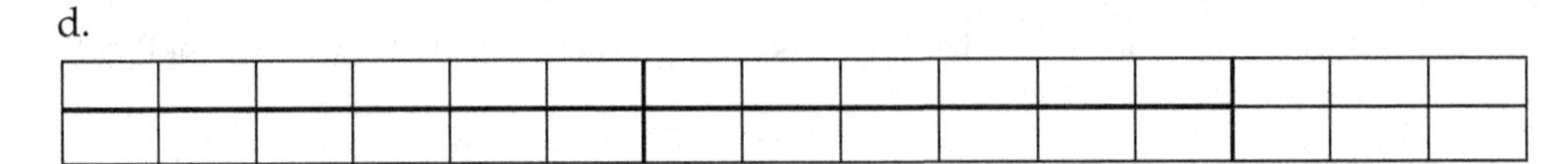

that is, five complete blocks as shown previously. If you do this, the variance among blocks is expected to be $\sigma_B^2 = 7.5$ and the variance within blocks is expected to be $\sigma_e^2 = 14$.

For each proposed design, find the power to find a three-unit difference between control and any given experimental treatment significant at the $\alpha=0.05$ level.

16.2 A researcher wants to compare the likelihood of a favorable outcome under two treatments. Treatment 2 is a "control." From previous experience, favorable outcomes occur with probability $\pi_2=0.7$. Treatment 1 is an experimental treatment—claimed to increase the probability of a favorable outcome to $\pi_1=0.8$.

a. How many observations per treatment (assuming equal N per treatment) would it take to obtain power $=0.90$ for testing H_0: $\mu_1=\mu_2$ at the $\alpha=0.05$ level?

b. The researcher's proposed design is to split the subjects per treatment in (a) equally among five locations. Assume that $\sigma_L^2=0.125$ and $\sigma_{TL}^2=0.08$. Determine the power for testing H_0: $\mu_1=\mu_2$ at the $\alpha=0.05$ level. *Hint:* you should find that it is inadequate (i.e., well below the desired power of 0.90).

c. When confronted with the results in (b), the researcher suggests increasing the number of subjects per treatment at each of the five locations by a factor of 10 (e.g., if it was 100 subjects per location, she suggests using 1000). Determine the power for H_0: $\mu_1=\mu_2$ at $\alpha=0.05$ for this design.

d. Instead of increasing the number of subjects at each location, an alternative strategy would be to increase the number of locations from 5 to 20 and divide the number of subject per treatment from (a) equally. Determine the power under these circumstances.

e. If the number of locations is held at 5, how many subjects per treatment per location are required for power $=0.9$ and $\alpha=0.05$?

f. On the other hand, if the number of subjects per treatment per location is held at 20, how many locations are required for power $=0.9$ and $\alpha=0.05$?

16.3 A researcher wants to determine the relationship between a response Y and a predictor variable X. The suspected relationship is linear, specifically $Y=10+0.25X$ for $0\leq X\leq 10$. Assume the errors have a normal distribution with iid $N(0, \sigma^2)$. Assume that you have 12 observations to "spend." Determine the power for H_0: $\beta_1=0$, where β_1 is the slope coefficient, if the true slope is $\beta_1=0.25$, when $\alpha=0.05$, for the following designs:

a. Six observations at each of two levels of X (0 and 10)

b. Four observations at each of three levels (0, 5, 10)

c. Like (b) except geometrically spaced levels (0, 1, 10)

d. Two observations at each of six equally spaced levels (0, 2, 4, 6, 8, 10)

e. Two observations at each of six geometrically spaced levels (0, 0.625, 1.25, 2.5, 5.0, 10.0)

16.4 A researcher wants to compare four treatments forming a 2×2 factorial. There are two primary response variables. One is Gaussian, the other is binomial. Factor A is a treatment and factor B is a catalyst that essentially enables the treatment to work. The two levels of factor A are "standard treatment" and "experimental, new and improved treatment." The two levels of factor B are "catalyst" or "no catalyst."

For the Gaussian variable, the catalyst is supposed to reduce the undesirable contaminant. For the standard treatment, the catalyst reduces the response 20%—from

50 to 40. It is believed that the experimental treatment used with the catalyst will reduce the response by 40%—from 50 to 30.

For the binomial variable, the catalyst is supposed to increase the likelihood of favorable response. Without the catalyst, both treatments are thought to have a 0.20 chance of favorable response; with the catalyst, the standard treatment has a 0.35 chance of favorable response, and it is believed that the new treatment with the catalyst will have a 0.50 chance of a favorable response.

The experiment is to be conducted on 6×2 grids. The grids have substantial variability lengthwise, suggesting that a natural blocking pattern for the grid would be six blocks of size 2. Three possible designs are being contemplated, as pictured in the following:

Design 1:

block 1	**block 2**	**block 3**	**block 4**	**block 5**	**block 6**
1	1	1	2	2	3
2	3	4	3	4	4

Design 2:

block 1	**block 2**	**block 3**	**block 4**	**block 5**	**block 6**
1	3	1	3	1	3
2	4	2	4	2	4

Design 3:

block 1		**block 2**		**block 3**	
1	3	1	3	1	3
2	4	2	4	2	4

The number shown in each cell corresponds to the allocation of treatments to blocks (with appropriate randomization, of course!). Here is a summary of how the treatment number corresponds to the factorial combination and other pertinent information:

		Factor B	
		Catalyst	**No Catalyst**
Factor A	Standard	Treatment: 1 Binomial π: 0.35 Gaussian: 40	Treatment: 2 Binomial π: 0.20 Gaussian: 50
	New, experimental	Treatment: 3 Binomial π: 0.50 Gaussian: 30	Treatment: 4 Binomial π: 0.20 Gaussian: 50

The needed variance information appears as follows:

		Response Variable/ Distribution	
Design	**Variance Component**	**Gaussian**	**Binomial**
Design 1	Block variance	$\sigma^2_{BLK} = 20$	$\sigma^2_{BLK} = 0.10$
	Second variance component	$\sigma^2_{Y\|BLK} = 15$	$\sigma^2_{Unit(blk)} = 0.05$
			a.k.a. $\sigma^2_{blk*trt}$
Design 2	Block variance	$\sigma^2_{BLK} = 20$	$\sigma^2_{BLK} = 0.10$
	Second variance component	$\sigma^2_{Y\|BLK} = 15$	$\sigma^2_{Unit(blk)} = 0.05$
			a.k.a. $\sigma^2_{blk*trt}$
Design 3	Block variance	** see part (b)	$\sigma^2_{BLK} = 0.06$
	Second variance component	** see part (b)	$\sigma^2_{Unit(blk)} = 0.30$
			a.k.a. $\sigma^2_{blk*trt}$

The primary contrasts of interest are

- $A \times B$ interaction $= \mu_{trt1} - \mu_{trt2} - \mu_{trt3} + \mu_{trt4}$
- Simple effect of catalyst|new treatment $= \mu_{trt3} - \mu_{trt4}$

a. For each design *x* response variable/distribution combination, determine the power for each contrast (use $\alpha = 0.05$) if only one of the 6×2 grids shown earlier is available. For the binomial response, without blocking, 420 binomial observations per treatment would be required for 80% power. Since there are three blocks per treatment, do all of your power assessment for the binomial variable with $N = 140$.

b. For the Gaussian/Complete block (design 3) combination, determine the variance components that result from disregarding the natural block size and use those variance components to determine power. Methods for determining the conversion are shown in Section 16.2.

c. You will find power to be inadequate for most, possibly all, of the scenarios in part (a). For each design *x* response variable/distribution combination, determine the number of replicate grids you would need to achieve 80% power for these objectives. To keep it simple, grids are NOT divisible. You can augment your power program by adding DO GRID=1 to *g*, where *g* is the number of replicate grids. Then make sure you substitute BLK(GRID) wherever you refer to BLK in your program for part (a).

16.5 Verify the results of your power assessments in Problem 16.4 by simulation.

Note: this could turn into a *very long* exercise. Suggested alternative: identify the most promising design from Problem 16.4 and use simulation to verify it. Do this under the Gaussian and binomial scenarios.

Include the RCB alternatives in the simulation by taking the same data but labeling blocks 1 and 2 as "rep 1," blocks 3 and 4 as "rep 2," and blocks 5 and 6 as "rep 3." This way you can determine the accuracy of your calculations in problem 4 regarding the impact of disregarding natural block size on the variance components in question as well as the impact on power.

Appendices: Essential Matrix Operations and Results

Introduction to the Appendices

Modern linear model theory depends heavily on matrix algebra and associated distribution theory for matrices. We express models in matrix form and develop estimating equations using matrix operations. We formulate estimable and predictable functions, the basis of generalized linear mixed model (GLMM) inference, as matrix operations. Subsequent development of their expectation, sampling distribution and resulting test statistics, and interval estimates all depend on matrix distribution theory. These appendices provide a summary of matrix syntax, notation, and essential operations and distribution theory. The material in Appendix A appears throughout the book, from Chapter 1 onward. The material in Appendix B is referred to mainly in Chapters 5 and 6 where the primary results for GLMM estimation and inference are developed. Readers seeking more detail are referred to texts on matrix algebra for statistics such as Harville (1997) and Schott (2005).

Appendix A: Matrix Operations

A.1 Notation

A *matrix* is an array of numbers with r rows and c columns. a_{ij} denotes the element in row i, $i = 1, 2, \ldots, r$; column j, $j = 1, 2, \ldots, c$. Matrices are denoted by boldface, uppercase characters. For example, $\mathbf{A} = \begin{bmatrix} a_{11} & a_{12} & a_{13} \\ a_{21} & a_{22} & a_{23} \end{bmatrix}$ is a 2×3 matrix, that is, a matrix with two rows and three columns.

A *vector* is a matrix with $c = 1$, that is, one column. Vectors are denoted by boldface, lowercase characters, that is, $\mathbf{a} = \begin{bmatrix} a_{11} \\ a_{21} \\ a_{31} \\ a_{41} \end{bmatrix} = \begin{bmatrix} a_1 \\ a_2 \\ a_3 \\ a_4 \end{bmatrix}$ is a 4×1 vector. Note that technically the elements of the vector are a_{i1}, but because $j = 1$ only, the second subscript is typically regarded as understood, thus the notation a_i suffices.

Note that a matrix is composed of c vectors, that is, each column is a vector. For example, we can denote the aforementioned matrix as $\mathbf{A} = \begin{bmatrix} \mathbf{a}_1 & \mathbf{a}_2 & \mathbf{a}_3 \end{bmatrix}$, where $\mathbf{a}_1 = \begin{bmatrix} a_{11} \\ a_{21} \end{bmatrix}$, $\mathbf{a}_2 = \begin{bmatrix} a_{12} \\ a_{22} \end{bmatrix}$, etc.

A *transpose* rearranges a matrix so that the rows become columns and vice versa. In this text book, the transpose is denoted "A prime," for example, if $\mathbf{A} = \begin{bmatrix} 2 & 7 & -2 \\ 9 & -2 & 12 \end{bmatrix}$, then its transpose is $\mathbf{A}' = \begin{bmatrix} 2 & 9 \\ 7 & -2 \\ -2 & 12 \end{bmatrix}$. Note that some writers denote the transpose as $\mathbf{A}^T$.

The transpose of a vector is denoted $\mathbf{a}'$. Following our convention, $\mathbf{a}'$ will always be $1 \times r$, for example, $\mathbf{a}' = \begin{bmatrix} a_1 & a_2 & a_3 & a_4 \end{bmatrix}$.

It follows that if $\mathbf{A} = \begin{bmatrix} \mathbf{a}_1 & \mathbf{a}_2 & \mathbf{a}_3 \end{bmatrix}$ then $\mathbf{A}' = \begin{bmatrix} \mathbf{a}_1' \\ \mathbf{a}_2' \\ \mathbf{a}_3' \end{bmatrix}$.

A.1.1 Functions of Vectors and Matrices

- If $\mathbf{a}$ is a $r \times 1$ vector and $f(a_i)$ is a function, then $f(\mathbf{a})$ is a $r \times 1$ vector with the function applied element-wise to each element of $\mathbf{a}$, for example, $f(\mathbf{a}) = \begin{bmatrix} f(a_1) \\ f(a_2) \\ f(a_3) \\ f(a_4) \end{bmatrix}$.
- If $\mathbf{A}$ is a $r \times c$ matrix, then $f(\mathbf{A})$ is an $r \times c$ matrix whose ijth element is $f(a_{ij})$.
- A common use of this vector and matrix functions is as a shorthand notation for the likelihood or log-likelihood of a parameter vector, the p.d.f. of a vector of observations of the link function of a mean vector. For example, if $\boldsymbol{\theta}$ is the $n \times 1$ vector of natural parameters for n observations, then $\ell(\boldsymbol{\theta})$ is the $n \times 1$ vector of log-likelihood functions, where $\ell(\theta_i)$ is the log-likelihood for the ith observation and hence the ith element of $\ell(\boldsymbol{\theta})$. Similarly, if $\boldsymbol{\mu}$ is the vector of expected values, then the vector $\boldsymbol{\eta} = g(\boldsymbol{\mu})$ denotes the link function.
- A function applied to a transpose is denoted $f(\mathbf{a}')$ for a vector function or $f(\mathbf{A}')$ for a matrix function, for example, $f(\mathbf{a}') = \begin{bmatrix} f(a_1) & f(a_2) & f(a_3) & f(a_4) \end{bmatrix}$. This is to avoid confusion with $f'(\mathbf{a})$ where the prime applied to the function operator denotes a derivative (see Appendix A.2).

A.2 Matrix Derivatives

Derivatives play an essential role in developing linear model estimating equations. The "gold standard" of linear model estimation is maximum likelihood. Hence, the primary mathematical operation to obtain estimations involves taking the derivative of the log-likelihood (or residual log-likelihood in the case of covariance component estimation), setting it to zero, and solving. As shown earlier, the joint log-likelihood of a set of data is commonly expressed as a vector function, for example, $\ell(\boldsymbol{\theta})$.

The simplest derivative involves forming a vector of the first derivatives of the elements of the vector function. For example, consider the log-likelihood vector $\ell(\boldsymbol{\theta}) = \begin{bmatrix} \ell(\theta_1) \\ \ell(\theta_2) \\ \ell(\theta_3) \\ \ell(\theta_4) \end{bmatrix}$. The first derivative vector is $\begin{bmatrix} \frac{\partial \ell(\theta_1)}{\partial \theta_1} \\ \frac{\partial \ell(\theta_2)}{\partial \theta_2} \\ \frac{\partial \ell(\theta_3)}{\partial \theta_3} \\ \frac{\partial \ell(\theta_4)}{\partial \theta_4} \end{bmatrix}$. The matrix derivative notation is $\partial \ell(\boldsymbol{\theta})/\partial \boldsymbol{\theta}$.

- The *transpose* of the vector derivative is denoted by assigning a transpose operator to the denominator vector. For example, $\frac{\partial \ell(\boldsymbol{\theta})}{\partial \boldsymbol{\theta}'} = \begin{bmatrix} \frac{\partial \ell(\theta_1)}{\partial \theta_1} & \frac{\partial \ell(\theta_2)}{\partial \theta_2} & \frac{\partial \ell(\theta_3)}{\partial \theta_3} & \frac{\partial \ell(\theta_4)}{\partial \theta_4} \end{bmatrix}$

The following results are derived from this general result:

- $\frac{\partial (\mathbf{a}'\mathbf{x})}{\partial \mathbf{x}} = \mathbf{a}$
- $\frac{\partial (\mathbf{a}'\mathbf{x})}{\partial \mathbf{x}'} = \mathbf{a}'$

Using these results, we can also show

- $\frac{\partial (\mathbf{x}'\mathbf{A}\mathbf{x})}{\partial \mathbf{x}} = (\mathbf{A} + \mathbf{A}')\mathbf{x}$
- $\frac{\partial^2 (\mathbf{x}'\mathbf{A}\mathbf{x})}{\partial \mathbf{x} \partial \mathbf{x}'} = \mathbf{A} + \mathbf{A}'$

Moreover, if **A** is symmetric, then

- $\frac{\partial (\mathbf{x}'\mathbf{A}\mathbf{x})}{\partial \mathbf{x}} = 2\mathbf{A}\mathbf{x}$
- $\frac{\partial^2 (\mathbf{x}'\mathbf{A}\mathbf{x})}{\partial \mathbf{x} \partial \mathbf{x}'} = 2\mathbf{A}$

$$\mathbf{B} = \left[b_{ij} \right] \quad n \times p$$

$$\boldsymbol{\theta} = \left[\theta_j \right] \quad n \times p$$

$$\mathbf{B}\boldsymbol{\theta} = \begin{bmatrix} b_{11} & b_{12} \\ b_{21} & b_{22} \\ b_{31} & b_{32} \end{bmatrix} \begin{bmatrix} \theta_1 \\ \theta_2 \end{bmatrix}$$

$$\frac{\partial (\mathbf{B}\boldsymbol{\theta})}{\partial \boldsymbol{\theta}'} = \begin{bmatrix} \frac{\partial b_{11}\theta_1}{\partial \theta_1} & \frac{\partial b_{21}\theta_2}{\partial \theta_2} \\ \frac{\partial b_{21}\theta_1}{\partial \theta_1} & \frac{\partial b_{22}\theta_2}{\partial \theta_2} \\ \frac{\partial b_{31}\theta_1}{\partial \theta_1} & \frac{\partial b_{32}\theta_2}{\partial \theta_2} \end{bmatrix} = \mathbf{B}$$

$$\frac{\partial (\mathbf{B}\boldsymbol{\theta})'}{\partial \boldsymbol{\theta}} = \begin{bmatrix} \frac{\partial b_{11}\theta_1}{\partial \theta_1} & \frac{\partial b_{21}\theta_1}{\partial \theta_1} & \frac{\partial b_{31}\theta_1}{\partial \theta_1} \\ \frac{\partial b_{12}\theta_2}{\partial \theta_2} & \frac{\partial b_{22}\theta_2}{\partial \theta_2} & \frac{\partial b_{32}\theta_2}{\partial \theta_2} \end{bmatrix} = \mathbf{B}'$$

A.3 Expectation and Variance

Chapter 4 shows the development of the estimating equations. For fixed effects models, these have the general form

$$\mathbf{X'WX\beta = X'Wy^*}$$

For mixed models, their general form is

$$\begin{bmatrix} \mathbf{X'WX} & \mathbf{X'WZ} \\ \mathbf{Z'WX} & \mathbf{Z'WZ + G^{-1}} \end{bmatrix} \begin{bmatrix} \boldsymbol{\beta} \\ \mathbf{b} \end{bmatrix} = \begin{bmatrix} \mathbf{X'Wy^*} \\ \mathbf{Z'Wy^*} \end{bmatrix}$$

Solving these equations provides estimates of the model effect vectors $\boldsymbol{\beta}$ and **b**. These solutions require matrix inversion (see Appendix A.4). Generically denoting model effect vectors by $\boldsymbol{\theta}$ and noting that $\mathbf{y}^*$ depends on the observation vector **y**, all solutions providing model effect estimates have the form $\tilde{\boldsymbol{\theta}} = \mathbf{Ay}$, where $\tilde{\boldsymbol{\theta}}$ denotes *a* solution.

Chapter 5 begins by stating that inference depends on estimable and predictable functions—linear combinations of $\boldsymbol{\beta}$ and **b**. Generally, we can denote these by $\mathbf{L'}\boldsymbol{\theta}$. For inference purposes, we need the expectation and variance of the estimates of $\mathbf{L'}\boldsymbol{\theta}$ and its distribution, or approximate distribution. Here, we give the general form of the expectation and variance. In Appendix B, the needed distribution results are given.

Let **y** be a random vector with expectation vector $E(\mathbf{y}) = \boldsymbol{\mu}$ and covariance matrix $V(\mathbf{y}) = \boldsymbol{\Sigma} = E[(\mathbf{y}-\boldsymbol{\mu})(\mathbf{y}-\boldsymbol{\mu})'] = E(\mathbf{yy'}) - \boldsymbol{\mu\mu}'$. The following results hold:

- $E(\mathbf{Ay}) = \mathbf{A}\boldsymbol{\mu}$
- $V(\mathbf{Ay}) = \mathbf{A}\boldsymbol{\Sigma}\mathbf{A}'$
- $Cov(\mathbf{Ay}, (\mathbf{By})') = E\{[\mathbf{A}(\mathbf{y}-\boldsymbol{\mu})][\mathbf{B}(\mathbf{y}-\boldsymbol{\mu})]'\} = E(\mathbf{Ayy'B'} - \mathbf{Ay}\boldsymbol{\mu}'\mathbf{B}' - \mathbf{A}\boldsymbol{\mu}\mathbf{y'B'} + \mathbf{A}\boldsymbol{\mu\mu}'\mathbf{B}') = \mathbf{A}\boldsymbol{\Sigma}\mathbf{B}'$

A.4 Inverse and Generalized Inverse

If **X** is full rank and **W** is symmetric and full rank, then **X′WX** is symmetric and full rank.

If **A** is symmetric and full rank, then a unique matrix, denoted $\mathbf{A}^{-1}$, exists such that $\mathbf{A}^{-1}\mathbf{A} = \mathbf{AA}^{-1} = \mathbf{I}$

It follows that if **X** in the linear predictor $\mathbf{X}\boldsymbol{\beta} + \mathbf{Zb}$ of a linear model is of full rank, then a true inverse of $\begin{bmatrix} \mathbf{X'WX} & \mathbf{X'WZ} \\ \mathbf{Z'WX} & \mathbf{Z'WZ + G^{-1}} \end{bmatrix}$ exists and therefore a unique estimate of the effect vector $\begin{bmatrix} \boldsymbol{\beta} \\ \mathbf{b} \end{bmatrix}$ exists.

If **X** is *not* of full rank—which is the case for analysis of variance (ANOVA)-type effects models and analysis of covariance (ANCOVA) models developed from ANOVA-type effects models—then a true inverse is not possible and solving the estimating equations requires *generalized inverse.*

Definition of *generalized inverse*: **G** is a generalized inverse of **A** if **AGA** = **A**

Four properties of a generalized inverse are as follows:

1. $\mathbf{AGA} = \mathbf{A}$ (definition)
2. $\mathbf{GAG} = \mathbf{G}$ (reflexive property)
3. $(\mathbf{GA})' = \mathbf{AG}$
4. $(\mathbf{AG})' = \mathbf{GA}$

These are called the Penrose conditions. Not all generalized inverses have all four properties. A generalized inverse is only *required* to have property 1. For inference to apply as shown in Chapter 5, at least the first two Penrose conditions must hold for the generalized inverse used to solve the estimation equations. That is, we must use a reflexive generalized inverse.

A.4.1 Obtaining a Generalized Inverse Using the SWEEP Operator

One approach to obtaining a generalized inverse is based on Goodnight (1979). All SAS linear model procedures, including GLIMMIX, use this approach. The approach is based on a *sweep operator,* described as follows. Interested readers can implement this operator in SAS PROC IML using the SWEEP function and read more about the operation in the IML online documentation (SAS Institute, 2008). According to the IML documentation, suppose $r \times c$ matrix $\mathbf{A}$ is partitioned as $\mathbf{A} = \begin{bmatrix} \mathbf{R} & \mathbf{S} \\ \mathbf{T} & \mathbf{U} \end{bmatrix}$ where $\mathbf{R}$ is $q \times q$ and $\mathbf{U}$ is $(r - q) \times (c - q)$. The operation $sweep\begin{bmatrix} \mathbf{R} & \mathbf{S} \\ \mathbf{T} & \mathbf{U} \end{bmatrix}$ yields the matrix $\begin{bmatrix} \mathbf{R}^{-1} & \mathbf{R}^{-1}\mathbf{S} \\ -\mathbf{T}\mathbf{R}^{-1} & \mathbf{U} - \mathbf{T}\mathbf{R}^{-1}\mathbf{S} \end{bmatrix}$.

In the context of linear models, for full rank $\mathbf{X}$, $sweep\begin{bmatrix} \mathbf{X}'\mathbf{X} & \mathbf{X}'\mathbf{Y} \\ \mathbf{Y}'\mathbf{X} & \mathbf{Y}'\mathbf{Y} \end{bmatrix}$ yields $\begin{bmatrix} (\mathbf{X}'\mathbf{X})^{-1} & (\mathbf{X}'\mathbf{X})^{-1}\mathbf{X}'\mathbf{Y} \\ -\mathbf{Y}'\mathbf{X}(\mathbf{X}'\mathbf{X})^{-1} & \mathbf{Y}'\left[\mathbf{I} - \mathbf{X}(\mathbf{X}'\mathbf{X})^{-1}\mathbf{X}'\right]\mathbf{Y} \end{bmatrix}$. For the LM, with observation matrix $\mathbf{Y}$ and linear predictor $\mathbf{X}\boldsymbol{\beta}$ and full-rank $\mathbf{X}$, we recognize the upper-right diagonal as the solution to the ordinary least squares estimating equations (a.k.a. normal equations) and the lower-right diagonal as the sum of squares for residual.

For the non-full-rank case, we partition $\mathbf{X} = [\mathbf{X}_1 \mid \mathbf{X}_0]$, such that $\mathbf{X}_1$ is full rank and $\mathbf{X}_0$ consists of the remaining dependent columns. Applying the operator to the independent columns yields $sweep\begin{bmatrix} \mathbf{X}_1'\mathbf{X}_1 & \mathbf{X}_1'\mathbf{X}_0 & \mathbf{X}_1'\mathbf{Y} \\ \mathbf{X}_0'\mathbf{X}_1 & \mathbf{X}_0'\mathbf{X}_0 & \mathbf{X}_0'\mathbf{Y} \\ \mathbf{Y}'\mathbf{X}_1 & \mathbf{Y}'\mathbf{X}_0 & \mathbf{Y}'\mathbf{Y} \end{bmatrix} =$

$$\begin{bmatrix} (\mathbf{X}_1'\mathbf{X}_1)^{-1} & (\mathbf{X}_1'\mathbf{X}_1)^{-1}(\mathbf{X}_1'\mathbf{X}_0) & (\mathbf{X}_1'\mathbf{X}_1)^{-1}\mathbf{X}_1'\mathbf{Y} \\ \mathbf{X}_0'\mathbf{X}_1(\mathbf{X}_1'\mathbf{X}_1)^{-1} & 0 & 0 \\ \mathbf{Y}'\mathbf{X}_1(\mathbf{X}_1'\mathbf{X}_1)^{-1} & 0 & \mathbf{Y}'\left[\mathbf{I} - \mathbf{X}_1(\mathbf{X}_1'\mathbf{X}_1)^{-1}\mathbf{X}_1'\right]\mathbf{Y} \end{bmatrix}$$

Goodnight describes a permutation operator to rearrange $\mathbf{X}$ so that the independent columns all appear first and the dependent columns appear last. Formally, let $\mathbf{M}$ denote the permutation operator. $\mathbf{M}$ must be defined such that $\mathbf{XM} = [\mathbf{X}_1 \mid \mathbf{X}_0]$ and $\mathbf{MM}' = \mathbf{I}$, where $\mathbf{I}$ is the identity matrix. Note that this reexpresses the linear predictor as $\mathbf{XMM'\beta} = [\mathbf{X}_1 \mid \mathbf{X}_0]\begin{bmatrix}\boldsymbol{\beta}_1 \\ \boldsymbol{\beta}_0\end{bmatrix}$, where $\boldsymbol{\beta}_1$ denotes the vector of independent parameters of the linear predictor and $\boldsymbol{\beta}_0$ denotes the vector of dependent parameters. From the sweep operator, we see that the estimates are $\begin{bmatrix}\hat{\boldsymbol{\beta}}_1 \\ \hat{\boldsymbol{\beta}}_0\end{bmatrix} = \begin{bmatrix}(\mathbf{X}_1'\mathbf{X}_1)^{-1}\mathbf{X}_1\mathbf{Y} \\ \mathbf{0}\end{bmatrix}$. The generalized inverse obtained from the sweep operator is thus $\begin{bmatrix}(\mathbf{X}_1'\mathbf{X}_1)^{-1} & \mathbf{0} \\ \mathbf{0} & \mathbf{0}\end{bmatrix}$.

For example, consider a two-factor main effects model with two levels per factor. In matrix form, the linear predictor is $\begin{bmatrix}1 & 1 & 0 & 1 & 0 \\ 1 & 1 & 0 & 0 & 1 \\ 1 & 0 & 1 & 1 & 0 \\ 1 & 0 & 1 & 0 & 1\end{bmatrix}\begin{bmatrix}\eta \\ \alpha_1 \\ \alpha_2 \\ \gamma_1 \\ \gamma_2\end{bmatrix}$. Note that only one column per effect is independent. SAS linear model procedures use the "last effect zero" convention, meaning effects α_2 and γ_2—and hence columns 3 and 5—would be considered dependent. Thus, the permutation matrix $\mathbf{M} = \begin{bmatrix}1 & 0 & 0 & 0 & 0 \\ 0 & 1 & 0 & 0 & 0 \\ 0 & 0 & 0 & 1 & 0 \\ 0 & 0 & 1 & 0 & 0 \\ 0 & 0 & 0 & 0 & 1\end{bmatrix}$ rearranges the linear predictor as $\mathbf{XMM'\beta} = \begin{bmatrix}1 & 1 & 1 & 1 & 0 \\ 1 & 1 & 0 & 1 & 1 \\ 1 & 0 & 1 & 0 & 0 \\ 1 & 0 & 0 & 0 & 1\end{bmatrix}\begin{bmatrix}\eta \\ \alpha_1 \\ \gamma_1 \\ \alpha_2 \\ \gamma_2\end{bmatrix}$. Here, $\mathbf{X}_1 = \begin{bmatrix}1 & 1 & 1 \\ 1 & 1 & 0 \\ 1 & 0 & 1 \\ 1 & 0 & 0\end{bmatrix}$, $\mathbf{X}_0 = \begin{bmatrix}0 & 0 \\ 0 & 1 \\ 1 & 0 \\ 1 & 1\end{bmatrix}$, $\boldsymbol{\beta}_1 = \begin{bmatrix}\eta \\ \alpha_1 \\ \gamma_1\end{bmatrix}$, and $\boldsymbol{\beta}_0 = \begin{bmatrix}\alpha_2 \\ \gamma_2\end{bmatrix}$.

The SAS linear model procedures restore the original order of the parameter vector by applying the permutation matrix to the solution vector, that is, $\hat{\boldsymbol{\beta}} = \mathbf{M}\begin{bmatrix}\hat{\boldsymbol{\beta}}_1 \\ \hat{\boldsymbol{\beta}}_0\end{bmatrix}$ so that the order of the estimates obtained from the sweep operator corresponds to the order in the linear predictor.

Goodnight denotes the generalized inverse obtained in this manner by $\mathbf{G}_2$. The two come from the fact that it satisfies the first two Penrose conditions—that is, it is a reflexive generalized inverse. Note that $\mathbf{G}_2 = \begin{bmatrix}(\mathbf{X}_1'\mathbf{X}_1)^{-1} & \mathbf{0} \\ \mathbf{0} & \mathbf{0}\end{bmatrix} = (\mathbf{M'X'XM})^{-} = \mathbf{M}'(\mathbf{XX})^{-}\mathbf{M}$. From

Chapter 5, recall that the building block for the variance of an estimable function—and hence standard errors and test statistics—is $\mathbf{K}'(\mathbf{X}'\mathbf{X})^{-}\mathbf{K}$. Using the sweep operator, this becomes $\mathbf{K}'\mathbf{M}\mathbf{G}_2\mathbf{M}'\mathbf{K}$. Applying $\mathbf{M}$ to $\mathbf{G}_2$ ensures that the order of the variances along the diagonal corresponds to the order of the parameter vector $\boldsymbol{\beta}$.

For generalized linear models (GLMs), the same partition $\mathbf{XM} = [\mathbf{X}_1 \mid \mathbf{X}_0]$ can be obtained and the sweep operator applied as $sweep\begin{bmatrix} \mathbf{X}_1'\mathbf{W}\mathbf{X}_1 & \mathbf{X}_1'\mathbf{W}\mathbf{X}_0 & \mathbf{X}_1'\mathbf{W}\mathbf{Y} \\ \mathbf{X}_0'\mathbf{W}\mathbf{X}_1 & \mathbf{X}_0'\mathbf{W}\mathbf{X}_0 & \mathbf{X}_0'\mathbf{W}\mathbf{Y} \\ \mathbf{Y}'\mathbf{W}\mathbf{X}_1 & \mathbf{Y}'\mathbf{W}\mathbf{X}_0 & \mathbf{Y}'\mathbf{Y} \end{bmatrix}$ to obtain solution to the GLM estimating equations when $\mathbf{X}$ is singular.

Appendix B: Distribution Theory for Matrices

Inference for GLMMs, meaning inference on estimable functions—$\mathbf{K}'\boldsymbol{\beta}$—or predictable functions—$\mathbf{K}'\boldsymbol{\beta}+\mathbf{M}'\mathbf{b}$—is developed in Chapter 5. These results are based primarily on normality or asymptotic normality and the distribution of quadratic forms based on normally distributed random vectors. This section provides a brief overview of the essential background underlying the development in Chapter 5.

B.1 Variance–Covariance of Estimators

B.1.1 Fixed-Effect-Only Models

We obtain the estimate of the parameter vector from $\hat{\boldsymbol{\beta}} = (\mathbf{X}'\mathbf{W}\mathbf{X})^{-}\mathbf{X}'\mathbf{W}\mathbf{y}^*$. Applying results from Appendix A.3,

- $$E(\hat{\boldsymbol{\beta}}) = E\left[(\mathbf{X}'\mathbf{W}\mathbf{X})^{-}\mathbf{X}'\mathbf{W}\mathbf{y}^*\right] = (\mathbf{X}'\mathbf{W}\mathbf{X})^{-}\mathbf{X}'\mathbf{W}\left[E(\mathbf{y}^*)\right]$$
$$= (\mathbf{X}'\mathbf{W}\mathbf{X})^{-}\mathbf{X}'\mathbf{W}\left\{E\left[\boldsymbol{\eta}+\mathbf{D}(\mathbf{y}-\boldsymbol{\mu})\right]\right\} = (\mathbf{X}'\mathbf{W}\mathbf{X})^{-}\mathbf{X}'\mathbf{W}\mathbf{X}\boldsymbol{\beta} = \boldsymbol{\beta}$$
- $$Var(\hat{\boldsymbol{\beta}}) = Var\left[(\mathbf{X}'\mathbf{W}\mathbf{X})^{-}\mathbf{X}'\mathbf{W}\mathbf{y}^*\right] = (\mathbf{X}'\mathbf{W}\mathbf{X})^{-}\mathbf{X}'\mathbf{W}\left[Var(\mathbf{y}^*)\right]\mathbf{W}\mathbf{X}(\mathbf{X}'\mathbf{W}\mathbf{X})^{-}$$
$$= (\mathbf{X}'\mathbf{W}\mathbf{X})^{-}\mathbf{X}'\mathbf{W}\mathbf{W}^{-1}\mathbf{W}\mathbf{X}(\mathbf{X}'\mathbf{W}\mathbf{X})^{-} = (\mathbf{X}'\mathbf{W}\mathbf{X})^{-}$$

Note that the fixed effect Gaussian linear model is a special case with $\mathbf{y}^*=\mathbf{y}$ and $\mathbf{W}=\mathbf{R}^{-1}$. In the case of independent observations and homoscedastic errors—once known as the "general" linear model—$\mathbf{R}=\mathbf{I}\sigma^2$.

B.1.2 Mixed Models

The covariance derivation is somewhat more involved. Henderson (1975) presents the derivation for the Gaussian linear mixed model. Here, we present Henderson's derivation adapted to the GLMM case. We obtain model effect estimators from
$$\begin{bmatrix}\hat{\boldsymbol{\beta}}\\ \hat{\mathbf{b}}\end{bmatrix} = \begin{bmatrix}\mathbf{X}'\mathbf{W}\mathbf{X} & \mathbf{X}'\mathbf{W}\mathbf{Z}\\ \mathbf{Z}'\mathbf{W}\mathbf{X} & \mathbf{Z}'\mathbf{W}\mathbf{Z}+\mathbf{G}^{-1}\end{bmatrix}^{-}\begin{bmatrix}\mathbf{X}'\mathbf{W}\mathbf{y}^*\\ \mathbf{Z}'\mathbf{W}\mathbf{y}^*\end{bmatrix}$$

Let $\mathbf{C} = \begin{bmatrix}\mathbf{C}^{11} & \mathbf{C}^{12}\\ \mathbf{C}^{21} & \mathbf{C}^{22}\end{bmatrix}$ denote the inverse of $\begin{bmatrix}\mathbf{X}'\mathbf{W}\mathbf{X} & \mathbf{X}'\mathbf{W}\mathbf{Z}\\ \mathbf{Z}'\mathbf{W}\mathbf{X} & \mathbf{Z}'\mathbf{W}\mathbf{Z}+\mathbf{G}^{-1}\end{bmatrix}$. Note that $\mathbf{C}$ is a generalized inverse if $\mathbf{X}$ is not of full rank. Following Henderson, we rewrite the model effect estimates as

$$\begin{bmatrix} \hat{\boldsymbol\beta} \\ \hat{\mathbf{b}} \end{bmatrix} = \mathbf{C}\begin{bmatrix} \mathbf{X'Wy^*} \\ \mathbf{Z'Wy^*} \end{bmatrix} = \begin{bmatrix} \mathbf{Q}_1 \\ \mathbf{Q}_2 \end{bmatrix}\mathbf{y}^*$$

Two results follow:

- $\begin{bmatrix} \mathbf{Q}_1 \\ \mathbf{Q}_2 \end{bmatrix}\begin{bmatrix} \mathbf{X} & \mathbf{Z} \end{bmatrix} = \mathbf{C}\left\{\begin{bmatrix} \mathbf{X'WX} & \mathbf{X'WZ} \\ \mathbf{Z'WX} & \mathbf{Z'WZ}+\mathbf{G}^{-1} \end{bmatrix} - \begin{bmatrix} 0 & 0 \\ 0 & \mathbf{G}^{-1} \end{bmatrix}\right\} = \begin{bmatrix} \mathbf{I} & \mathbf{C}^{12}\mathbf{G}^{-1} \\ 0 & \mathbf{I}-\mathbf{C}^{22}\mathbf{G}^{-1} \end{bmatrix}$
- $\begin{bmatrix} \mathbf{Q}_1 \\ \mathbf{Q}_2 \end{bmatrix}\mathbf{W}^{-1} = \begin{bmatrix} \mathbf{C}^{11}\mathbf{X}'+\mathbf{C}^{12}\mathbf{Z}' \\ \mathbf{C}^{21}\mathbf{X}'+\mathbf{C}^{22}\mathbf{Z}' \end{bmatrix}$

Now, in mixed models, standard errors and test statistics depend on

$$Var\begin{bmatrix} \hat{\boldsymbol\beta} \\ \hat{\mathbf{b}}-\mathbf{b} \end{bmatrix} = \begin{bmatrix} Var(\hat{\boldsymbol\beta}) & Cov\left(\hat{\boldsymbol\beta},(\hat{\mathbf{b}}-\mathbf{b})'\right) \\ & Var(\hat{\mathbf{b}}-\mathbf{b}) \end{bmatrix} = \begin{bmatrix} Var(\hat{\boldsymbol\beta}) & Cov(\hat{\boldsymbol\beta},\hat{\mathbf{b}}')-Cov(\hat{\boldsymbol\beta},\mathbf{b}') \\ & Var(\hat{\mathbf{b}})-2Cov(\hat{\mathbf{b}},\mathbf{b}')+Var(\mathbf{b}) \end{bmatrix}$$

Note that $\hat{\boldsymbol\beta}=\mathbf{Q}_1\mathbf{y}^*$ and $\hat{\mathbf{b}}=\mathbf{Q}_2\mathbf{y}^*$. The following results are obtained:

- $Var(\hat{\boldsymbol\beta}) = \mathbf{Q}_1 Var(\mathbf{y}^*)\mathbf{Q}_1' = \mathbf{Q}_1(\mathbf{ZGZ}'+\mathbf{R}^*)\mathbf{Q}_1'$. Now, noting that $\mathbf{Q}_1\mathbf{Z}=\mathbf{C}^{21}\mathbf{G}^{-1}$, $\mathbf{R}^*=\mathbf{W}^{-1}$ and hence $\mathbf{Q}_1\mathbf{R}^*=\mathbf{C}^{11}\mathbf{X}'+\mathbf{C}^{12}\mathbf{Z}'$, we have $Var(\hat{\boldsymbol\beta})=\mathbf{C}^{12}\mathbf{G}^{-1}\mathbf{G}\mathbf{G}^{-1}\mathbf{C}^{12}+\mathbf{C}^{11}\mathbf{X}'\mathbf{Q}_1'+\mathbf{C}^{12}\mathbf{Z}'\mathbf{Q}_1' = \mathbf{C}^{12}\mathbf{G}^{-1}\mathbf{C}^{12}+\mathbf{C}^{11}\mathbf{I}-\mathbf{C}^{12}\mathbf{G}^{-1}\mathbf{C}^{12}=\mathbf{C}^{11}$
- $Var(\hat{\mathbf{b}}) = \mathbf{Q}_2 Var(\mathbf{y}^*)\mathbf{Q}_2' = \mathbf{Q}_2(\mathbf{ZGZ}'+\mathbf{R}^*)\mathbf{Q}_2'$. Noting $\mathbf{Q}_2\mathbf{Z}-\mathbf{I}-\mathbf{C}^{22}\mathbf{G}^{1}$ and $\mathbf{Q}_1\mathbf{R}^*=\mathbf{C}^{21}\mathbf{X}'+\mathbf{C}^{22}\mathbf{Z}'$, we have

$$
\begin{aligned}
Var(\hat{\mathbf{b}}) &= (\mathbf{I}-\mathbf{C}^{22}\mathbf{G}^{-1})\mathbf{G}(\mathbf{I}-\mathbf{C}^{22}\mathbf{G}^{-1})' + \mathbf{C}^{21}\mathbf{X}'\mathbf{Q}' + \mathbf{C}^{22}\mathbf{Z}'\mathbf{Q}_2' \\
&= \mathbf{G}-\mathbf{G}\mathbf{G}^{-1}\mathbf{C}^{22}-\mathbf{C}^{22}\mathbf{G}^{-1}\mathbf{G}+\mathbf{C}^{22}\mathbf{G}^{-1}\mathbf{G}\mathbf{G}^{-1}\mathbf{C}^{22}+\mathbf{C}^{21}\mathbf{X}\mathbf{Q}_2+\mathbf{C}^{22}\mathbf{Z}\mathbf{Q}_2 \\
&= \mathbf{G}-\mathbf{C}^{22}-\mathbf{C}^{22}-\mathbf{C}^{22}\mathbf{G}^{-1}\mathbf{C}^{22}+0+\mathbf{C}^{22}(\mathbf{I}-\mathbf{C}^{22}\mathbf{G}^{-1})' \\
&= \mathbf{G}-\mathbf{C}^{22}+\mathbf{C}^{22}\mathbf{G}^{-1}\mathbf{C}^{22}-\mathbf{C}^{22}\mathbf{G}^{-1}\mathbf{C}^{22} \\
&= \mathbf{G}-\mathbf{C}^{22}
\end{aligned}
$$

- $Cov(\hat{\boldsymbol\beta},\hat{\mathbf{b}}') = Cov(\mathbf{Q}_1\mathbf{y},\mathbf{y}'\mathbf{Q}_2') = \mathbf{Q}_1[\mathbf{ZGZ}'+\mathbf{R}^*]\mathbf{Q}_2'$

$$
\begin{aligned}
&= \mathbf{Q}_1\mathbf{ZGZ}'\mathbf{Q}_2'+\mathbf{Q}_1\mathbf{R}^*\mathbf{Q}_2' = -\mathbf{C}^{12}\mathbf{G}^{-1}\mathbf{G}(\mathbf{I}-\mathbf{C}^{22}\mathbf{G}^{-1})'+\mathbf{C}^{11}\mathbf{X}'\mathbf{Q}_2'+\mathbf{C}^{21}\mathbf{Z}'\mathbf{Q}_2' \\
&= -\mathbf{C}^{12}(\mathbf{I}-\mathbf{C}^{22}\mathbf{G}^{-1})'+0+\mathbf{C}^{21}(\mathbf{I}-\mathbf{C}^{22}\mathbf{G}^{-1})' = 0
\end{aligned}
$$

In a similar vein, we can show

- $Cov(\hat{\boldsymbol{\beta}}, \mathbf{b}') = -\mathbf{C}^{12}$
- $Cov(\hat{\mathbf{b}}, \mathbf{b}') = \mathbf{G} - \mathbf{C}^{22}$

Assembling these results, we have

$$\begin{bmatrix} Var(\hat{\boldsymbol{\beta}}) & Cov(\hat{\boldsymbol{\beta}}, \hat{\mathbf{b}}') - Cov(\hat{\boldsymbol{\beta}}, \mathbf{b}') \\ & Var(\hat{\mathbf{b}}) - 2Cov(\hat{\mathbf{b}}, \mathbf{b}') + Var(\mathbf{b}) \end{bmatrix} = \begin{bmatrix} \mathbf{C} & \mathbf{0} - (-\mathbf{C}^{12}) \\ & \mathbf{G} - \mathbf{C}^{22} - 2(\mathbf{G} - \mathbf{C}^{22}) + \mathbf{G} \end{bmatrix} = \begin{bmatrix} \mathbf{C}^{11} & \mathbf{C}^{12} \\ & \mathbf{C}^{22} \end{bmatrix}$$

Thus, $Var\begin{bmatrix} \hat{\boldsymbol{\beta}} \\ \hat{\mathbf{b}} - \mathbf{b} \end{bmatrix} = \mathbf{C}$ and hence standard errors and test statistics for estimable and predictable functions in mixed models use **C** as the basic building block.

B.1.3 Quadratic Forms

A quadratic form is defined as $\mathbf{y}'\mathbf{A}\mathbf{y}$. Quadratic forms have many applications in GLMM inference—for example, as Wald statistics, the numerator term in approximate *F*-statistics, variance estimates, etc. The following results are essential for working with these statistics:

1. $E(\mathbf{y}'\mathbf{A}\mathbf{y}) = tr(\mathbf{A}\boldsymbol{\Sigma}) + \boldsymbol{\mu}'\mathbf{A}\boldsymbol{\mu}$.
   a. This result is shown as follows: $E(\mathbf{y}'\mathbf{A}\mathbf{y}) = E[tr(\mathbf{y}'\mathbf{A}\mathbf{y})] = E[tr(\mathbf{A}\mathbf{y}\mathbf{y}')] = tr[E(\mathbf{A}\mathbf{y}\mathbf{y}')] = tr[\mathbf{A}(\boldsymbol{\Sigma} + \boldsymbol{\mu}\boldsymbol{\mu}')] = tr(\mathbf{A}\boldsymbol{\Sigma}) + tr(\mathbf{A}\boldsymbol{\mu}\boldsymbol{\mu}') = tr(\mathbf{A}\boldsymbol{\Sigma}) + \boldsymbol{\mu}'\mathbf{A}\boldsymbol{\mu}$
2. If $\mathbf{y} \sim N(\boldsymbol{\mu}, \boldsymbol{\Sigma})$, then $\mathbf{y}'\mathbf{A}\mathbf{y} \sim \chi^2_{r(\mathbf{A}), \frac{1}{2}\boldsymbol{\mu}'\mathbf{A}\boldsymbol{\mu}}$ if and only if $\mathbf{A}\boldsymbol{\Sigma}$ is idempotent.
3. A matrix is said the be idempotent if it is equal to its square—here, $\mathbf{A}\boldsymbol{\Sigma}$ is idempotent if $(\mathbf{A}\boldsymbol{\Sigma})^2 = \mathbf{A}\boldsymbol{\Sigma}\mathbf{A}\boldsymbol{\Sigma} = \mathbf{A}\boldsymbol{\Sigma}$.
   a. Establishing the distribution of the quadratic form requires properties associated with idempotent matrices and inspection of the moment generating function of the χ^2 distribution. As Searle (1971) notes, we are usually most interested in $\mathbf{A}\boldsymbol{\Sigma}$ idempotent $\Rightarrow \mathbf{y}'\mathbf{A}\mathbf{y} \sim \chi^2$, rather than vice versa. Here, we give the main feature of the proof in this direction. Readers seeking more detail are referred to texts such as Harville (1997) for more extensive treatment of the properties of idempotent matrices and to Searle for a more in-depth presentation of distribution theory.
4. If $\mathbf{y} \sim N(\boldsymbol{\mu}, \boldsymbol{\Sigma})$, then, following Searle, moment generating function $M_{\mathbf{y}'\mathbf{A}\mathbf{y}}(t)$ is

$$|\mathbf{I} - 2t\mathbf{A}\boldsymbol{\Sigma}|^{-\frac{1}{2}} \exp\left\{-\frac{1}{2}\boldsymbol{\mu}'\left[\mathbf{I} - (\mathbf{I} - 2t\mathbf{A}\boldsymbol{\Sigma})^{-1}\right]\boldsymbol{\Sigma}^{-1}\boldsymbol{\mu}\right\}$$

$$= \prod_{i=1}^{n} (1 = 2t\lambda_i)^{-\frac{1}{2}} \exp\left\{-\frac{1}{2}\boldsymbol{\mu}'\left(-\sum_{k=1}^{\infty} (2t)^k (\mathbf{A}\boldsymbol{\Sigma})^k\right)\boldsymbol{\Sigma}^{-1}\boldsymbol{\mu}\right\}$$

where λ_i denotes the ith latent root of $\mathbf{A\Sigma}$. Note that $\mathbf{A\Sigma}$ idempotent implies (1) for $r = rank(\mathbf{A})$ roots, $\lambda_i = 1$ and for the remaining $n-r$ roots, $\lambda_i = 0$ and (2) $(\mathbf{A\Sigma})^r = \mathbf{A\Sigma}$. Therefore,

$$M_{\mathbf{y'Ay}}(t) = \prod_{i=1}^{r}(1-2t)^{-\frac{1}{2}} \exp\left\{-\frac{1}{2}\boldsymbol{\mu}'\left[-\sum_{k=1}^{\infty}(2t)^k\right](\mathbf{A\Sigma})\boldsymbol{\Sigma}^{-1}\boldsymbol{\mu}\right\}$$

$$= (1-2t)^{-\frac{r}{2}} \exp\left\{-\frac{1}{2}\boldsymbol{\mu}'\left[1-(1-2t)^{-1}\right]\mathbf{A}\boldsymbol{\mu}\right\}$$

$$= (1-2t)^{-\frac{r}{2}} \exp\left\{-\frac{1}{2}\boldsymbol{\mu}'\mathbf{A}\boldsymbol{\mu}\left[1-(1-2t)^{-1}\right]\right\}$$

which is the moment-generating function of the $\chi^2_{r,\frac{1}{2}\boldsymbol{\mu}'\mathbf{A}\boldsymbol{\mu}}$ distribution.

B.1.4 Independence Results

- If $\mathbf{y} \sim N(\boldsymbol{\mu}, \boldsymbol{\Sigma})$, then $\mathbf{y'Ay}$ and $\mathbf{By}$ are independent if and only if $\mathbf{B\Sigma A} = \mathbf{0}$.
- If $\mathbf{y} \sim N(\boldsymbol{\mu}, \boldsymbol{\Sigma})$, then $\mathbf{y'Ay}$ and $\mathbf{y'By}$ are independent if and only if $\mathbf{B\Sigma A} = \mathbf{0}$.

We use first independence result in the context of t-statistics for estimable functions, $t = \mathbf{k}'\hat{\boldsymbol{\beta}}/s.e.(\mathbf{k}'\hat{\boldsymbol{\beta}})$. In linear models, the numerator has the form $\mathbf{By}$ and thus has a normal distribution; the denominator has the form $\mathbf{y'Ay}$. Therefore, the statistic $[(\mathbf{k}'\hat{\boldsymbol{\beta}} - \mathbf{k}'\boldsymbol{\beta})/s.e.(\mathbf{k}'\hat{\boldsymbol{\beta}})]$, has the form $[(\mathbf{By} - \mathbf{B}\boldsymbol{\mu})/\sigma_{\mathbf{By}}]/[\mathbf{y'Ay}/\sigma_{\mathbf{By}}]$ and thus has the distributional form $N(0,1)/\chi^2$ which we know has a t-distribution if the numerator and denominator, that is, $\mathbf{y'Ay}$ and $\mathbf{By}$, are independent and $\mathbf{A}$ is idempotent.

We use the second independence result in the context of approximate F-statistics. These, as developed in Chapter 5, are constructed from Wald statistics whose generic matrix form can be written as $\mathbf{y'Ay}/\mathbf{y'By}$, where $\mathbf{A}$ and $\mathbf{B}$ are both idempotent and $\boldsymbol{\mu}'\mathbf{B}\boldsymbol{\mu} = \mathbf{0}$. Assuming numerator and denominator quadratic forms are independent and $rank(\mathbf{B})$ is included in the denominator of $\mathbf{B}$—true for all Wald statistics given in Chapter 5—$Wald/rank(\mathbf{A}) \sim [\chi^2_{r(\mathbf{A})}/r(\mathbf{A})]/[\chi^2_{r(\mathbf{B})}r(\mathbf{B})]$ has an approximate F distribution.

References

Agresti, A. 2002. *Categorical Data Analysis*, 2nd edn. New York: Wiley.

Akaike, H. 1974. A new look at the statistical model identification. *IEEE Transaction on Automatic Control* AC-19, 716–723.

Anderson, R.L. and Nelson, L.A. 1975. A family of models involving intersection straight lines and concomitant experimental designs useful in evaluating response to fertilizer nutrients. *Biometrics* 81, 303–318.

Atkinson, A.C. 1969. The use of residuals as a concomitant variable. *Biometrika* 56, 33–41.

Bartlett, M.S. 1978. Nearest neighbour models in the analysis of field experiments (with discussion). *Journal of the Royal Statistical Society B* 40, 147–174.

Bates, D.G. and Watts, D.M. 1988. *Nonlinear Regression Analysis and Its Applications.* New York: Wiley.

Bello, N.M., Steibel, J.P., and Pursley, J.R. 2006. Optimizing ovulation to first GnRH improved outcomes to each hormonal injection of ovsynch in lactating dairy cows. *Journal of Dairy Science* 89(9), 3413–3424.

Bhattacharya, C.G. 1998. Goodness of Yates-Rao procedure for recovery of inter-block information. *Sankhya A* 61(1), 134–144.

Box, G.E.P. and Behnken, D.W. 1960. Some new three-level designs for the study of quantitative variables. *Technometrics* 2, 455–475.

Box, G.E.P. and Draper, N.R. 1987. *Empirical Model Building and Response Surfaces.* New York: Wiley.

Box, G.E.P., Hunter, J.S., and Hunter, W.G. 2005. *Statistics for Experimenters*, 2nd edn. New York: Wiley.

Box, G.E.P., Jenkins, G.M., and Reinsel, G.C. 2008. *Time Series Analysis: Forecasting and Control*, 3rd edn. New York: Wiley.

Breslow, N.E. and Clayton, D.G. 1993. Approximate inference in generalized linear mixed models. *Journal of the American Statistical Association* 88(421), 9–25.

Burnham, K.P. and Anderson, D.R. 1998. *Model Selection and Inference: A Practical Information-Theoretic Approach.* New York: Springer-Verlag.

Burnham, K.P. and Anderson, D.R. 2002. *Model Selection and Multi-Model Inference.* New York: Springer-Verlag.

Casella, G. and Berger, R.L. 2002. *Statistical Inference*, 2nd edn. Pacific Grove, CA: Duxbury.

Cochran, W.G. and Cox, G.M. 1957. *Experimental Designs*, 2nd edn. New York: Wiley.

Cressie, N.A.C. 1993. *Statistics for Spatial Data.* New York: Wiley.

Cressie, N.A.C. and Wikle, C.K. 2011. *Statistics for Spatio-Temporal Data.* New York: Wiley.

Davidian, M. and Giltinan, D.M. 1995. *Nonlinear Models for Repeated Measurement Data.* New York: Chapman & Hall.

Davidian, M. and Giltinan, D.M. 2003. Nonlinear models for repeated measurement data: An update and overview. *Journal of Agricultural Biological and Environmental Statistics* 8(4), 387–419.

Demidenko, E. 2004. *Mixed Models: Theory and Applications.* New York: Wiley.

Diggle, P.J., Heagerty, P., Liang, K.-Y., and Zeger, S.L. 2002. *Analysis of Longitudinal Data*, 2nd edn. Oxford, U.K.: Oxford Press.

Dobson, A.J. and Barnett, A.G. 2008. *Introduction to Generalized Linear Models*, 3rd edn. Boca Raton, FL: CRC Press.

Dunnett, C.W. 1955. A multiple comparison procedure for comparing several treatments with a control. *Journal of American Statistical Association* 50, 1096–1121.

Eisenhart, C. 1947. The assumptions underlying the analysis of variance. *Biometrics* 3, 1–21.

Federer, W.T. 1955. *Experimental Design—Theory and Applications.* New York: Macmillan Publishing.

Federer, W.T. and King, F. 2007. *Variations on Split Plot and Split Block Experiment Designs.* New York: Wiley.

Ferrari, S.L.P. and Cribari-Neto, F. 2004. Beta regression for modeling rates and proportions. *Journal of Applied Statistics* 31, 799–815.
Fisher, R.A. 1935. Comments in Yates, "Complex experiments". *Journal of the Royal Statistical Society* Supplement 2, 181–223.
Fisher, R.A. and Mackenzie, W.A. 1923. Studies in crop variation II: The manurial response of different potato varieties. *Journal of Agricultural Science* 13, 311–320.
Frenzel, M., Stroup, W.W., and Paparozzi, E.T. 2010. After further review: An update on modeling and design strategies for agricultural dose-response experiments. In *Proceedings of the 22nd Conference on Applied Statistics in Agriculture*. Manhattan, KS: Kansas State University Department of Statistics, pp. 245–266.
Gallant, A.R. 1987. *Nonlinear Statistical Models*. New York: Wiley.
Geisbrecht, F.G. and Burns, J.C. 1985. Two-stage analysis based on a mixed model: Large-sample asymptotic theory and small sample simulation results. *Biometrics* 41(2), 477–486.
Gelman, A. 2005. Analysis of variance—Why it is more important than ever. *Annals of Statistics* 33, 1–53.
Gilmour, A.R. 2007. Mixed model regression mapping for QTL detection in experimental crosses. *Computational Statistics and Data Analysis* 51, 3749–3764.
Goldberger, A. 1962. Best linear unbiased prediction in the generalized linear regression model. *Journal of the American Statistical Association* 57: 369–375.
Goodnight, J. 1979. A tutorial on the SWEEP operator. *American Statistician* 33(3), 149–158.
Gotway, C.A. and Stroup, W.W. 1997. A generalized linear model approach to spatial data and prediction. *Journal of Agricultural Biological and Environmental Statistics* 2, 157–187.
Greenhouse, S.W. and Geisser, S. 1959. On methods in the analysis of profile data. *Psychometrika* 32, 95–112.
Gu, C. 2002. *Smoothing Spline ANOVA Models*. New York: Springer-Verlag.
Guerin, L. and Stroup, W.W. 2000. A simulation study to evaluate PROC MIXED analysis of repeated measures data. In *Proceedings of the 12th Conference on Applied Statistics in Agriculture*. Manhattan, KS: Kansas State University, Department of Statistics, pp. 170–203.
Hardin, J.W. and Hilbe, J.M. 2003. *Generalized Estimating Equations*. Boca Raton, FL: Chapman & Hall.
Harville, D.A. 1976. Extensions of the Gauss-Markov theorem to include the estimation of random effects. *Annals of Statistics* 4(2), 384–395.
Harville, D.A. 1977. Maximum likelihood approaches to variance component estimation and to related problems. *Journal of the American Statistical Association* 72(358), 320–338.
Harville, D.A. 1997. *Matrix Algebra from a Statisticians Perspective*. New York: Springer-Verlag.
Harville, D.A. and Jeske, D.R. 1992. Mean squared error of estimation or prediction under a general linear model. *Journal of the American Statistical Association* 87, 724–731.
Hedecker, D. and Gibbons, R.D. 1994. A random effects ordinal regression model for multilevel analysis. *Biometrics* 50, 933–944.
Henderson, C.R. 1950. Estimation of genetic parameters. *Annals of Mathematical Statistics* 21, 309–310.
Henderson, C.R. 1963. Selection index and expected genetic advance. In W.D. Hanson and H.F. Robinson, Eds. *Statistical Genetics and Plant Breeding*. Washington, DC: National Academy of Science—National Research Council Publication 982, pp. 141–163.
Henderson, C.R. 1975. Best linear unbiased estimation and prediction under a selection model. *Biometrics* 31(2), 423–447.
Henderson, C.R. 1984. *Applications of Linear Models in Animal Breeding*. Guelph, Ontario, Canada: University of Guelph.
Henderson, C.R. 1985. Best linear unbiased prediction using relationship matrices derived from selected base populations. *Journal of Dairy Science* 68, 443–448.
Hilbe, J.M. 2007. *Negative Binomial Regression*. Cambridge, U.K.: Cambridge University Press.
Hsu, J.C. 1992. The factor analytic approach to simultaneous inference in the general linear models. *Journal of Computational and Graphical Statistics* 1, 151–168.

Huynh, H. and Feldt, L.S. 1970. Conditions under which mean square ratios in repeated measurements designs have exact F-distributions. *Journal of the American Statistical Association* 65, 1582–1589.

Isaaks, E.H. and Srivastava, R.M. 1989. *Introduction to Applied Geostatistics*. Oxford, U.K.: Oxford University Press.

Jansen, J. 1990. On the statistical analysis of ordinal data when extravariation is present. *Applied Statistics* 39, 75–84.

Jiang, J. 2007. *Linear and Generalized Linear Mixed Models and Their Applications*. New York: Springer-Verlag.

Joe, H. and Zhu, R. 2005. Generalized Poisson distribution: The property of mixture of Poisson and comparison with negative. *Binomial Distribution Biometrical Journal* 47, 219–229.

Journel, A.G. and Huijbregts, C.J. 1978. *Mining Statistics*. London, U.K.: Academic Press.

Kackar, R.N. and Harville, D.A. 1984. Approximations for standard errors of fixed and random effect in mixed linear models. *Journal of the American Statistical Association* 79(388), 853–862.

Kenward, M.G. and Roger, J.H. 1997. Small sample inference for fixed effects from restricted maximum likelihood. *Biometrics* 53(3), 983–997.

Khuri, A.I. and Cornell, J.A. 1996. *Response Surfaces: Designs and Analyses*. New York: Marcel-Dekker.

Konishi, S. and Kitagawa, G. 2008. *Information Criteria and Statistical Modeling*. New York: Springer-Verlag.

Kullback, S. and Leibler, R.A. 1951. On information and sufficiency. *Annals of Mathematical Statistics* 22, 79–86.

Laird, N.M. and Ware, J.H. 1982. Random effects models for longitudinal data. *Biometrics* 38(4), 963–974.

Lambert, D. 1992. Zero-inflated Poisson regression models with an application to defects in manufacturing. *Technometics* 34, 1–14.

Landes, R.D., Stroup, W.W., Paparozzi, E.T., and Conley, M.E. 1999. Nonlinear models for multifactor plant nutrition experiments. In *Proceedings of the 11th Conference on Applied Statistics in Agriculture*. Manhattan, KS: Kansas State University, Department of Statistics, pp. 105–119.

Lee, Y. and Nelder, J.A. 1996. Hierarchical generalized linear models (with discussion). *Journal of the Royal Statistical Society B* 58, 619–678.

Lee, Y., Nelder, J.A., and Patiwan, Y. 2006. Generalized linear models with random effects: Unified analysis via h-likelihood. Boca Raton, FL: Chapman & Hall.

Liang, K.-Y. and Zeger, S.R. 1986. Longitudinal data analysis using generalized linear models. *Biometrika* 73(1), 13–22.

Linstrom, M.J. and Bates, D.M. 1988. Newton-Raphson and EM algorithms for linear mixed-effects models for repeated-measures data. *Journal of the American Statistical Association* 83, 1014–1022.

Littell, R.C. 1980. Examples of GLM applications SAS Users' Group International. In *Proceedings of the Fifth Annual Conference*. Cary, NC: SAS Institute, Inc., pp. 208–214.

Littell, R.C., Milliken, G.A., Stroup, W.W., Wolfinger, R.D., and Schabenberger, O. 2006. *SAS for Mixed Models*, 2nd edn. Cary, NC: SAS Institute, Inc.

Littell, R.C., Stroup, W.W., and Freund, R.J. 2002. *SAS for Linear Models*, 4th edn. Cary, NC: SAS Institute, Inc.

Lohr, V.I. and O'Brien, R.G. 1984. Power analysis for univariate linear models: SAS makes it easy SAS Users' Group International. In *Proceedings of the Ninth Annual Conference*. Cary, NC: SAS Institute, Inc. pp. 847–852.

McCaffery, D.F. Lockwood, J.R., Koretz, D., Louis, T.A., and Hamilton, L. 2004. Models for value-added modeling of teacher effects. *Journal of Educational and Behavioral Statistics* 29(1), 67–101.

McCullagh, P. and Nelder, J.A. 1989. *Generalized Linear Models*, 2nd edn. New York: Chapman & Hall.

McCulloch, C.E. 1997. Maximum likelihood algorithms for generalized linear mixed models. *Journal of the American Statistical Association* 92, 162–170.

McCulloch, C.E., Searles, S.R., and Neuhaus, J.M. 2008. *Generalized, Linear and Mixed Models*, 2nd edn. New York: Wiley.

McLean, R.A., Sanders, W.L., and Stroup, W.W. 1991. A unified approach to mixed linear models. *American Statistician* 45, 54–64.

Mead, R. 1988. *The Design of Experiments*. Cambridge, U.K.: Cambridge University Press.

Milliken, G.A. and Johnson, D.E. 2008. *Analysis of Messy Data*, Vol. 1, 2nd edn. New York: Chapman & Hall.

Molenberghs, G. and Verbeeke, G. 2006. *Models for Discrete Longitudinal Data*. New York: Springer-Verlag.

Morel, J.G., Bokossa, M.C., and Neerchal, N.K. 2003. Small sample correction for the variance of GEE estimators. *Biometrical Journal* 45(4), 395–409.

Mullahy, J. 1986. Specification and testing of some modified count data models. *Journal of Econometrics* 33, 341–365.

Myers, R.H., Montgomery, D.C., and Anderson-Cook, C.M. 2009. *Response Surface Methodology: Process and Product Optimization Using Designed Experiments*, 3rd edn. New York: Wiley.

Nelder, J.A. 1968. The combination of information in generally balanced designs. *Journal of the Royal Statistical Society B* 30, 303–311.

Nelder, J.A. 1977. A reformulation of linear models. *Journal of the Royal Statistical Society A* 140(1), 48–77.

Nelder, J.A. 1994. The Statistics of linear models: Back to basics. *Statistics and Computing* 4, 221–234.

Nelder, J.A. 1998. The great mixed-model muddle is alive and flourishing, alas! *Food Quality and Preference* 9, 157–159.

Nelder, J.A. and Lee, Y. 1992. Likelihood, quasi-likelihood and pseudo-likelihood: Some comparisons. *Journal of the Royal Statistical Society B* 54(1), 273–284.

Nelder, J.A. and Wedderburn, R.W.M. 1972. Generalized linear models. *Journal of the Royal Statistical Society A* 135(3), 370–384.

Nelson, P.R. 1982. Exact critical points for the analysis of means. *Communications in Statistics* 11, 699–709.

Nelson, P.R. 1991. Numerical evaluation of multivariate normal integrals with correlations. In A. Öztürk, E.C. van der Meulen, E.J. Dudewicz, and P.R. Nelson, Eds. *The Frontiers of Statistical Scientific Theory & Industrial Applications*. Vol. 2. Syracuse, NY: American Sciences Press, pp. 97–114.

Nelson, P.R. 1993. Additional uses for the analysis of means and extended tables of critical values. *Technometrics* 35, 61–71.

O'Brien, R.G. and Lohr, V.I. 1984. Power analysis for linear models: The time has come. In *SAS Users' Group International: Proceedings of the Ninth Annual Conference*. Cary, NC: SAS Institute, Inc., pp. 840–846.

Olson, L.M., Stroup, W.W., Paparozzi, E.T., and Conley, M.E. 2001. Model building in multifactor plant nutrition experiments. In *Proceedings of the18th Conference on Applied Statistics in Agriculture*. Manhattan, KS: Kansas State University Department of Statistics, pp. 183–206.

Papadakis, J.S. 1937. Methode statistique pour des experiences sur champ. *Bulletin de I'lnstitut d'Amelioration des Plantes B Salonique*, 23, 13–28.

Paparozzi, E.T., Stroup, W.W., and Conley, M.E. 2005 How to investigate four-way nutrient interactions in plants: A new look at response surface methods. *Journal of the American Society for Horticultural Science* 130, 459–468.

Patiwan, Y. 2001. *In All Likelihood: Statistical Modelling and Inference Using Likelihood*. Oxford, U.K.: Oxford University Press.

Patterson, H.D. and Thompson, R. 1971. Recovery of inter-block information when block sizes are unequal. *Biometrika* 58, 545–554.

Pinheiro, J.C. and Bates, D.M. 1995. Approximations to the log-likelihood function in the non-linear mixed effects model. *Journal of Computational and Graphical Statistics* 4(1), 12–35.

Prasad, N.G.N. and Rao, J.N.K. 1990. The estimation of the mean squared error of small-area estimators. *Journal of the American Statistical Association* 85(409), 163–171.

Raudenbush, S.M. 2004. What are value-added models estimating and what does this imply for statistical practice? *Journal of Educational and Behavioral Statistics* 29(1), 121–129.

Raudenbush, S.M. 2009. Targets of inference in hierarchical models for longitudinal data. In G. Fitzmaurice, M. Davidian, G. Verbeeke, and G. Molenberghs, Eds. *Longitudinal Data Analysis*. Boca Raton, FL: Chapman & Hall.

Raudenbush, S.M. and Bryk, A.S. 2002. *Hierarchical Linear Models: Applications and Data Analysis Methods*. Newbury Park, CA: Sage Press.

Robinson, G.K. 1991. That BLUP is a good thing. *Statistical Science* 6(1), 15–32.

Ruppert, D., Wand, M.P., and Carroll, R.J. 2003. *Semiparametric Regression*. Cambridge, U.K.: Cambridge University Press.

Sanders, W.L., Saxton, A., and Horn, B. 1997. The Tennessee value-added assessment system: A quantitative outcomes-based approach to educational assessment. In J. Millman, Ed. *Grading Teachers, Grading Schools: Is Student Achievement a Valid Educational Measure*? Thousand Oaks, CA: Corwin Press, Inc., pp. 137–162.

SAS Institute, Inc. 2008. SAS/STAT® 9.2 *User's Guide*. Cary, NC: SAS Institute, Inc.

Satterthwaite, F.E. 1941. Synthesis of variance. *Psychometrika* 6, 309–316.

Satterthwaite, F.E. 1946. An approximate distribution of estimates of variance components. *Biometrics Bulletin* 2, 110–114.

Schabenberger, O. 2008. Aspects of the analysis of split-plot experiments. In *JSM 2008 Section on Physical and Engineering Sciences*. Abstract available: http://www.amstat.org/meetings/jsm/2008/onlineprogram/index.cfm

Schabenberger, O. and Gotway, C.A. 2004. *Statistical Methods for Spatial Data Analysis*. Boca Raton, FL: CRC Press.

Schabenberger, O. and Pierce, F.J. 2002. *Contemporary Statistical Models for the Plant and Soil Sciences*. Boca Raton, FL: CRC Press.

Schall, R. 1991. Estimation in generalized linear models with random effects. *Biometrika* 78(4), 719–727.

Schott, J.R. 2005. *Matrix Analysis for Statisticians*, 2nd edn. New York: Wiley.

Schwarz, G. 1978. Estimating the dimension of a model. *Annals of Statistics* 6, 461–464.

Searle, S.R. 1971. *Linear Models*. New York: Wiley.

Searle, S.R. 1987. *Linear Models for Unbalanced Data*. New York: Wiley.

Searle, S.R., Casella, G., and McCulloch, C.E. 1992. *Variance Component Estimation*. New York: Wiley.

Sidak, Z. 1967. Rectangular confidence regions for the means of multivariate normal distributions. *Journal of the American Statistical Association* 62, 626–633.

Simpson, E.H. 1951. The interpretation of interaction in contingency tables. *Journal of the Royal Statistical Society B* 13, 238–241.

Snedecor, G. and Cochran, W.G. 1989. *Statistical Methods*, 8th edn. Ames, IA: Iowa State University Press.

Speed, T. 2010. And ANOVA thing. *IMS Bulletin* 39(4), 16.

Steel, R.D.G., Torrie, J.H., and Dickey, D.A. 1996. *Principles and Procedures of Statistics: A Biometrical Approach*, 3rd edn. New York: McGraw-Hill.

Stiratelli, R., Laird, N., and Ware, J.H. 1984. Random effects models for serial observations with binary response. *Biometrics* 40(4), 961–971.

Stokes, M.E., Davis, C.S., and Koch, G.C. 2000. *Categorical Data Analysis Using the SAS System*, 2nd edn. Cary, NC: SAS Institute, Inc.

Stroup, W.W. 1999. Mixed model procedures to assess power, precision, and sample size. In *The Design of Experiments 1999 Proceedings of the Biopharmaceutical Section*. Alexandria, VA: American Statistical Association, pp. 15–24.

Stroup, W.W. 2002. Power analysis based on spatial effects mixed models: A tool for comparing design and analysis strategies in the presence of spatial variability. *Journal of Agricultural, Biological, and Environmental Statistics* 7, 491–511.

Stroup, W.W., Baenziger, P.S., and Mulitze, D.K. 1994. Removing spatial variation from wheat yield trials: A comparison of methods. *Crop Science* 86, 62–66.

Stroup, W.W., Guo, S., Paparozzi, E.T., and Conley, M.E. 2006. A comparison of models and designs for experiments with nonlinear dose-response relationships. In *Proceedings of the18th Conference on Applied Statistics in Agriculture*. Manhattan, KS: Kansas State University Department of Statistics, pp. 214–241.

Stroup, W. W. and Littell, R. C. 2002. Impact of variance component estimates on fixed effect inference in unbalanced linear mixed models. In Proceedings of the 14th Annual Kansas State University Conference on Applied Statistics in Agriculture, Manhattan, Kan, USA, pp. 32–48.

Vonesh, E.F. and Chinchilli, V.M. 1997. *Linear and Nonlinear Models for the Analysis of Repeated Measurements*. New York: Marcel-Dekker.

Wedderburn, R.W.M. 1974. Quasi-likelihood functions, generalized linear models and the Gauss-Newton method. *Biometrika* 61(3), 439–447.

Westfall, P.H., Tobias, R.D., Rom, D., Wolfinger, R.D., and Hochberg, Y. 1999. *Multiple Comparisons and Multiple Tests Using the SAS System*. Cary, NC: SAS Institute, Inc.

Williams, R.M. 1952. Experimental designs for serially correlated observations. *Biometrika* 39, 151–167.

Wolfinger, R.D. and O'Connell, M. 1993. Generalized linear mixed models: A pseudo-likelihood approach. *Journal of Statistical Computation and Simulation* 4, 233–243.

Wright, S. 1922. Coefficients of inbreeding and relationship. *The American Naturalist* 56, 330–338.

Wu, L. 2010. Mixed effects models for complex data. Boca Raton, FL: CRC Press.

Yates, F. 1935. Complex experiments. *Journal of the Royal Statistical Society*, Supplement 2, 181–223.

Yates, F. 1940. The recovery of inter-block information in balanced incomplete block designs. *Annals of Eugenics* 10, 317–325.

Young, L.J., Campbell, N.L., and Capuano, G.A. 1998. Analysis of overdispersed count data from single-factor experiments: A comparative study. *Journal of Agricultural, Biological, and Environmental Statistics* 4, 258–275.

Young, L.J. and Young, J.B. 1998. *Statistical Ecology*. New York: Springer-Verlag.

Zeger, S.L. and Liang, K.-Y. 1986. Longitudinal data analysis for discrete and continuous outcomes. *Biometrics* 42, 121–130.

Zeger, S.L., Liang, K.-Y., and Albert, P.S. 1988. Models for longitudinal data: A generalized estimating equation approach. *Biometrics* 44(4), 1049–1060.

Zwillinger, D. (editor). 1996. *CRC Standard Mathematical Tables and Formulae, 30th ed.*, Boca Raton, FL: CRC Press.

Index

I

K

L

M

N

O